STUDENT SOLUTIONS MANUAL AND STUDY GUIDE
to accompany

COLLEGE PHYSICS

FOURTH EDITION

Serway & Faughn

JOHN R. GORDON & RAYMOND A. SERWAY
James Madison University

WITH ASSISTANCE OF

JERRY FAUGHN & CHARLES TEAGUE
Eastern Kentucky University

SAUNDERS GOLDEN SUNBURST SERIES

SAUNDERS COLLEGE PUBLISHING
Harcourt Brace College Publishers

Fort Worth • Philadelphia • San Diego • New York • Orlando
Austin • San Antonio • Toronto • Montreal • London • Sydney • Tokyo

Printed in the United States of America.

Serway: Student Solutions Manual and Study Guide to accompany COLLEGE PHYSICS

ISBN 0-03-003564-3

7 8 9 0 1 2 3 4 5 018 12 11 10 9 8 7 6

Preface

This Student Solution Manual and Study Guide has been written to accompany the textbook **College Physics,** Fourth Edition, by Raymond A. Serway and Jerry S. Faughn. The purpose of this Study Guide is to provide the students with a convenient review of the basic concepts and applications presented in the textbook, together with solutions to selected end-of-chapter problems from the textbook. The Study Guide is not an attempt to rewrite the textbook in a condensed fashion. Rather, emphasis is placed upon clarifying typical troublesome points, and providing further drill for methods of problem solving.

Each chapter of the Study Guide is divided into several parts, and every textbook chapter has a matching chapter in the Study Guide. Very often, reference is made to specific equations or figures in the textbook. Every feature of the Study Guide has been included to insure that it serves as a useful supplement to the textbook. Most chapters contain the following sections:

A. **Notes From Selected Chapter Sections:** This is a summary of important concepts, newly defined physical quantities, and rules governing their behavior.

B. **Equations and Concepts:** This represents a review of the chapter, with emphasis on highlighting important concepts, and describing important equations and formalisms.

C. **Suggestions, Skills, and Strategies:** This offers hints and strategies for solving typical problems that you will often encounter in the course. In some sections, suggestions are made concerning mathematical skills that are necessary in the analysis of problems.

D. **Review Checklist:** This is a list of topics and techniques that you should master after reading the chapter and working the assigned problems.

E. **Solutions to Selected End-of-Chapter Problems:** Solutions are shown for approximately half of the odd-numbered problems from the text. These illustrate the important concepts of the chapter.

F. **Chapter Self-Quiz:** This is a set of twelve multiple-choice questions at the end of each chapter. They can be used to assess your understanding of the concepts presented in the chapter. Answers to the chapter self-quiz questions are given following the Self-Quiz for Chapter 30.

We sincerely hope that this Study Guide will be useful to you in reviewing the material presented in the text, and in improving your ability to solve problems and score well on exams. We welcome any comments or suggestions which could help improve the content of this Study Guide in future editions; and we wish you success in your study.

John R. Gordon
Raymond A. Serway
James Madison University
Harrisonburg, VA 22807

Acknowledgments

It is a pleasure to acknowledge the excellent work of Mrs. Linda Miller who typed and proofread the entire manuscript for this Fourth Edition. The final camera-ready copy is a result of her dedication to the project and careful attention to detail. We thank Michael Rudmin for revising and creating original art work for this edition. His graphics skills and technical expertise have combined to produce illustrations that complement the text and enhance the overall appearance of the Study Guide.

We thank Professors Jerry Faughn and Charles Teague for permission to use their solutions to end-of-chapter problems. The multiple-choice questions in the Chapter Self-Quiz sections were developed by Professor Steve Van Wyk. We also thank the professional staff at Saunders, especially Julie Finegan and Laura Maier for managing all phases of the project. Finally, we express our appreciation to our families for their inspiration, patience, and encouragement.

Suggestions for Study

Very often we are asked "How should I study this subject, and prepare for examinations?" There is no simple answer to this question, however, we would like to offer some suggestions which may be useful to you.

1. It is essential that you understand the basic concepts and principles before attempting to solve assigned problems. This is best accomplished through a careful reading of the textbook before attending your lecture on that material, jotting down certain points which are not clear to you, taking careful notes in class, and asking questions. You should reduce memorization of material to a minimum. Memorizing sections of a text, equations, and derivations does not necessarily mean you understand the material. Perhaps the best test of your understanding of the material will be your ability to solve the problems in the text, or those given on exams.

2. Try to solve as many problems at the end of the chapter as possible. You will be able to check the accuracy of your calculations to the odd-numbered problems, since the answers to these are given at the back of the text. Furthermore, detailed solutions to approximately half of the odd-numbered problems are provided in this Study Guide. Many of the worked examples in the text will serve as a basis for your study.

3. The method of solving problems should be carefully planned. First, read the problem several times until you are confident you understand what is being asked. Look for key words which will help simplify the problem, and perhaps allow you to make certain assumptions. You should also pay special attention to the information provided in the problem. It is a good idea to write down the given information before proceeding with a solution. (For example, v_0 = 5.00 m/s, a = –3.00 m/s^2, and t = 2.00 s are given. Find the velocity, v, and the displacement Δx at t = 2.00 s.) After you have decided on the method you feel is appropriate for the problem, proceed with your solution. If you are having difficulty in working problems, we suggest that you again read the text and your lecture notes. It may take several readings before you are ready to solve certain problems. The solved problems in this Study Guide should be of value to you in this regard.

4. After reading a chapter, you should be able to define any new quantities that were introduced, and discuss the first principles that were used to derive fundamental formulas. A review is provided in each chapter of the Study Guide for this purpose, and the marginal notes in the textbook (or the index) will help you locate these topics. You should be able to correctly associate

with each physical quantity the symbol used to represent that quantity (including vector notation if appropriate) and the SI unit in which the quantity is specified. Furthermore, you should be able to express each important formula or equation in a concise and accurate prose statement.

5. We suggest that you use this Study Guide to review the material covered in the text, and as a guide in preparing for exams. You should also use the Chapter Review, Notes From Selected Chapter Sections, and Equations and Concepts to focus in on any points which require further study. Remember that the main purpose of this Study Guide is to improve upon the efficiency and effectiveness of your study hours and your overall understanding of physical concepts. However, it should not be regarded as a substitute for your textbook or individual study and practice in problem solving.

Table Of Contents

Chapter 1	Introduction	1
Chapter 2	Motion in One Dimension	12
Chapter 3	Vectors and Two-Dimensional Motion	26
Chapter 4	The Laws of Motion	46
Chapter 5	Work and Energy	66
Chapter 6	Momentum and Collisions	85
Chapter 7	Circular Motion and the Law of Gravity	101
Chapter 8	Rotational Equilibrium and Rotational Dynamics	119
Chapter 9	Solids and Fluids	137
Chapter 10	Thermal Physics	157
Chapter 11	Heat	172
Chapter 12	The Laws of Thermodynamics	189
Chapter 13	Vibrations and Waves	210
Chapter 14	Sound	230
Chapter 15	Electric Forces and Electric Fields	251
Chapter 16	Electrical Energy and Capacitance	273
Chapter 17	Current and Resistance	291
Chapter 18	Direct Current Circuits	309
Chapter 19	Magnetism	328
Chapter 20	Induced Voltages and Inductance	348
Chapter 21	Alternating Current Circuits and Electomagnetic Waves	364
Chapter 22	Reflection and Refraction of Light	386
Chapter 23	Mirrors and Lenses	404
Chapter 24	Wave Optics	425
Chapter 25	Optical Instruments	443
Chapter 26	Relativity	461

Chapter 27	Quantum Physics	480
Chapter 28	Atomic Physics	498
Chapter 29	Nuclear Physics	519
Chapter 30	Nuclear Energy and Elementary Particles	539
Answers to Chapter Self-Quiz		556

1
Introduction

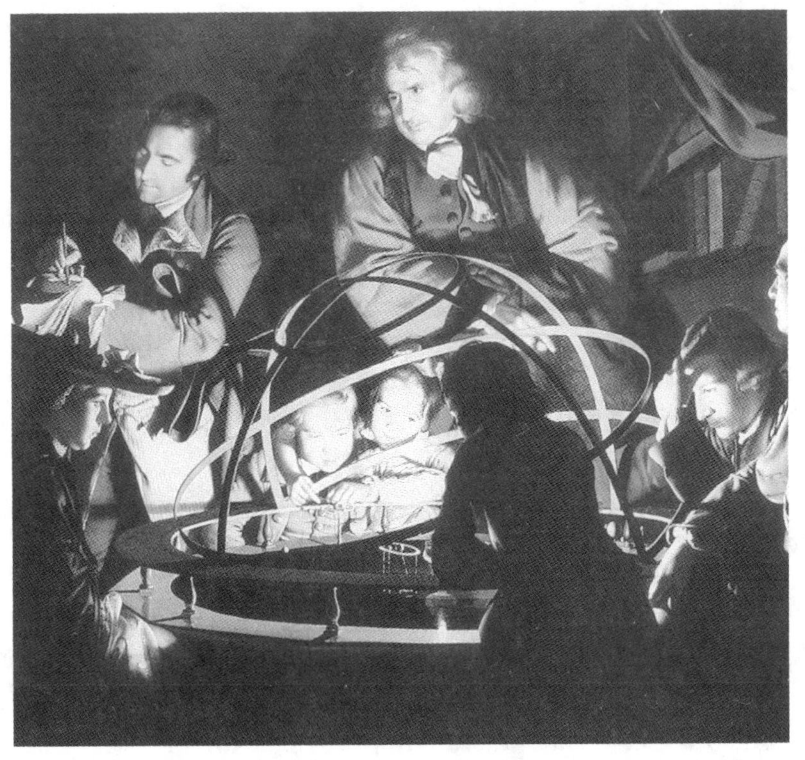

Chapter 1

INTRODUCTION

The goal of physics is to provide an understanding of nature by developing theories based on experiments. The theories are usually expressed in mathematical form. Fortunately, it is possible to explain the behavior of a variety of physical systems with a limited number of fundamental laws.

Since following chapters will be concerned with the laws of physics, we must begin by clearly defining the basic quantities involved in these laws. For example, such physical quantities as force, velocity, volume, and acceleration can be described in terms of more fundamental quantities. In the next several chapters we shall encounter three basic quantities: **length** (L), mass (M), and time (T). In later chapters we will need to add two other standard units to our list, for temperature (the kelvin) and for electric current (the ampere). In our study of mechanics, however, we shall be concerned only with the units of length, mass and time.

NOTES FROM SELECTED CHAPTER SECTIONS

1.1 Standards of Length, Mass, and Time

Until recently, the meter was defined as 1,650,763.73 wavelengths of orange-red light emitted from a krypton-86 lamp. However, in October 1983, the **meter** was redefined to be *the distance traveled by light in a vacuum during a time of 1/299,792,458 second.*

The SI unit of mass, the **kilogram**, is defined as *the mass of a specific platinum-iridium alloy cylinder kept at the International Bureau of Weights and Measures at Sèvres, France.*

The **second** is now defined as 9,192,631,770 *times the period of one oscillation of the cesium atom.*

Systems of units commonly used are the SI *system*, in which the units of mass, length, and time are the kilogram (kg), meter (m), and second (s), respectively; the *cgs* or *gaussian system*, in which the units of mass, length, and time are the gram (g), centimeter (cm), and second, respectively; and the *British engineering system* (sometimes called the conventional system), in which the units of mass, length, and time are the slug, foot (ft), and second, respectively.

1.3 Dimensional Analysis

Dimensional analysis makes use of the fact that *dimensions can be treated as algebraic quantities.* That is, quantities can be added or subtracted only if they have the same dimensions. Furthermore, the quantities on each side of an equation must have the same dimensions.

1.4 Significant Figures

When multiplying several quantities, the number of significant figures in the final answer is the same as the number of significant figures in the *least* accurate of the quantities being multiplied, where "least accurate" means "having the lowest number of significant figures." The same rule applies to division.

When numbers are added (or subtracted), the number of decimal places in the result should equal the smallest number of decimal places of any term in the sum.

1.8 Coordinate Systems and Frames of Reference

A coordinate system used to specify locations in space consists of:

1. A fixed reference point, called the origin
2. A set of specified axes or directions
3. Instructions that tell us how to label a point in space relative to the origin and axes

EQUATIONS AND CONCEPTS

The three most basic trigonometric functions of one of the acute angles of a right triangle are the sine, cosine, and tangent.

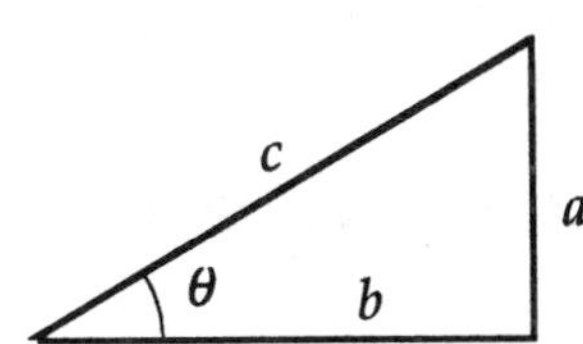

$$\sin\theta = \frac{\text{side opposite to }\theta}{\text{hypotenuse}} = \frac{a}{c}$$

$$\cos\theta = \frac{\text{side adjacent to }\theta}{\text{hypotenuse}} = \frac{b}{c} \qquad (1.1)$$

$$\tan\theta = \frac{\text{side opposite to }\theta}{\text{side adjacent to }\theta} = \frac{a}{b}$$

The Pythagorean theorem states the manner in which the lengths of the sides of a right triangle are related. In this equation, c is the hypotenuse.

$$c^2 = a^2 + b^2 \tag{1.2}$$

SUGGESTIONS, SKILLS, AND STRATEGIES

In developing problem-solving strategies, six basic steps are commonly used:

1. Read the problem carefully at least twice. Be sure you understand the nature of the problem before proceeding further.

2. Draw a suitable diagram with appropriate labels and coordinate axes, if needed.

3. As you examine what is being asked in the problem, identify the basic physical principle (or principles) that are involved, listing the knowns and unknowns.

4. Select a basic relationship or derive an equation that can be used to find the unknown, and symbolically solve the equation for the unknown.

5. Substitute the given values with the appropriate units into the equation.

6. Obtain a numerical value for the unknown. The problem is verified and receives a check mark if the following questions can be properly answered: Do the units match? Is the answer reasonable? Is the plus or minus sign proper or meaningful?

REVIEW CHECKLIST

▷ Discuss the units of length, mass and time and the standards for these quantities in SI units. (Section 1.1). Derive the quantities *force, velocity, volume, acceleration,* etc. from the three basic quantities.

▷ Perform a *dimensional analysis* of an equation containing physical quantities whose individual units are known. (Section 1.3)

▷ *Convert units* from one system to another. (Section 1.5)

▷ Carry out *order-of-magnitude calculations* or guesstimates. (Section 1.6)

▷ Become familiar with the meaning of various mathematical symbols and Greek letters. Identify and properly use mathematical notations such as the following: $\propto$ (is proportional to), $<$ (is less than), $\approx$ (is approximately equal to), Δ (change in value), etc. (Section 1.7)

▷ Describe the coordinates of a point in space using a Cartesian coordinate system. (Section 1.8)

SOLUTIONS TO SELECTED END-OF-CHAPTER PROBLEMS

5. A spherical droplet of blood is drawn into a pipette, as in Figure 1.10 of the text. Write a formula for the height h of the column of blood in the pipette in terms of the radius r of the pipette and the radius R of the original droplet. Check the dimensions of your formula.

Solution We know volume of cylindrical column = volume of spherical drop. Thus,

$$\pi r^2 h = \frac{4}{3}\pi R^3$$

where r = radius of cylinder,
R = radius of drop, and
h = height of cylinder; so

$$h = \frac{4R^3}{3r^2}$$

As a dimension check, we have

$$(\mathrm{L}) = \frac{(\mathrm{L}^3)}{(\mathrm{L}^2)} = (\mathrm{L}) \quad \Diamond$$

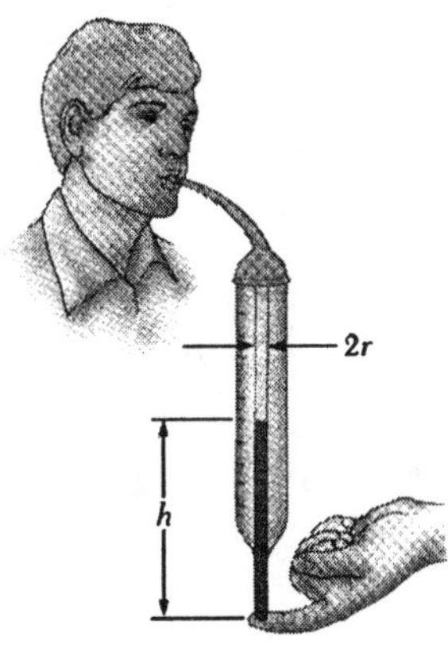

Figure 1.10

11. Calculate (a) the circumference of a circle of radius 3.5 cm and (b) the area of a circle of radius 4.65 cm.

Solution

(a) $c = 2\pi r = 2\pi(3.5\text{ cm})$. The number 2 and π are considered as known to many significant figures, while 3.5 is known only to two. Therefore, $c = 22$ cm. ◊

(b) $A = \pi r^2 = \pi(4.65\text{ cm})^2$ must be rounded to the number of significant figures in 4.65, which gives an answer of $A = 67.9\text{ cm}^2$. ◊

---

19. A painter is to cover the walls in a room that is 8 ft high and 12 ft along a side. What surface area, in square meters, must he cover?

Solution Total area, $A = 4$ (area of one wall) $= 4[(8.00\text{ ft})(12.0\text{ ft})] = 384\text{ ft}^2$
This value for the surface area must be converted to square meters:

$$\text{Area, } A = 384\text{ ft}^2\left(\frac{1.00\text{ m}}{3.28\text{ ft}}\right)^2 = 35.7\text{ m}^2 \quad ◊$$

---

21. One cubic meter (1.00 m^3) of aluminum has a mass of 2.70×10^3 kg, and 1.00 m^3 of iron has a mass of 7.86×10^3 kg. Find the radius of an aluminum sphere whose mass is the same as that of an iron sphere of radius 2.00 cm. (*Note*: Density is defined as the mass of an object divided by its volume.)

Solution We require $(\text{mass})_{\text{al}} = (\text{mass})_{\text{iron}}$. Therefore,

$$(\text{density})_{\text{al}}\left(\frac{4}{3}\pi r^3\right) = (\text{density})_{\text{iron}}\left(\frac{4}{3}\pi(2.00\text{ cm})^3\right)$$

or

$$r^3 = \left(\frac{(\text{density})_{\text{iron}}}{(\text{density})_{\text{al}}}\right)(2.00\text{ cm})^3 = \left(\frac{7.86\text{ kg/m}^3}{2.70\text{ kg/m}^3}\right)(2.00\text{ cm})^3 = 23.3\text{ cm}^3$$

and

$$r = 2.86\text{ cm} \quad ◊$$

---

29. A hamburger chain advertises that it has sold more than 50 billion hamburgers. Estimate how many pounds of hamburger meat must have been used by the restaurant chain and how many head of cattle were required to furnish the meat.

Solution number of pounds = (number of burgers)(weight/burger)

$$= (5.00 \times 10^{10} \text{ burgers})(0.250 \text{ lb/burger}) = 1.25 \times 10^{10} \text{ lb} \quad \Diamond$$

number of head of cattle = (weight needed)/(weight per head)

$$= (1.25 \times 10^{10} \text{ lb})/(300 \text{ lb/head}) = 4.17 \times 10^{7} \text{ head} \quad \Diamond$$

Assumptions are 0.250 lb of meat per burger and 300 lb of meat per head of cattle.

---

35. Two points in a rectangular coordinate system have the coordinates (5, 3) and (–3, 4), where the units are in centimeters. Determine the distance between these points.

Solution The x distance between the two points is 8.00 cm and the y distance between them is 1.00 cm. The distance between them is found from the Pythagorean theorem.

$$\text{distance} = \sqrt{x^2 + y^2} = \sqrt{(8.00 \text{ cm})^2 + (1.00 \text{ cm})^2} = \sqrt{65 \text{ cm}^2} = 8.06 \text{ cm} \quad \Diamond$$

---

39. A right triangle has a hypotenuse of length 3.00 m, and one of its angles is 30.0°. What are the lengths of (a) the side opposite the 30.0° angle and (b) the side adjacent to the 30.0° angle?

Solution

(a) $\sin\theta = \dfrac{\text{side opposite}}{\text{hypotenuse}}$ so side opposite = (sin 30.0°)(3.00 m) = 1.50 m ◊

(b) $\cos\theta = \dfrac{\text{adjacent side}}{\text{hypotenuse}}$ so adjacent side = (cos 30.0°)(3.00 m) = 2.60 m ◊

---

43. The consumption of natural gas by a company satisfies the empirical equation $V = 1.50\,t + 0.008t^2$, where V is the volume in millions of cubic feet and t the time in months. Express this equation in cubic feet and seconds. Put the proper units on the coefficients. Assume a month is 30 days.

Solution The constants in the equation as stated must have units of :

$$\frac{\text{million ft}^3}{\text{month}} \quad \text{and} \quad \frac{\text{million ft}^3}{(\text{month})^2} \quad \text{thus} \quad V = \left(1.5\times10^6\,\frac{\text{ft}^3}{\text{month}}\right)t + \left(0.008\times10^6\,\frac{\text{ft}^3}{(\text{month}^2)}\right)t^2$$

To convert, use 1 month = 2.59×10^6 s, to obtain $V = \left(0.579\frac{\text{ft}^3}{\text{s}}\right)t + \left(1.19\times10^{-9}\,\frac{\text{ft}^3}{(\text{s}^2)}\right)t^2$

$$V = \left(0.579\frac{\text{ft}^3}{\text{s}}\right)t + \left(1.19\times10^{-9}\,\frac{\text{ft}^3}{(\text{s}^2)}\right)t^2 \quad \Diamond$$

---

45. At the time of this book's printing. the U.S. national debt is about \$4 trillion. (a) If this debt were paid at a rate of \$1000 per second, how long (in years) would it take to pay off the debt, assuming no interest were charged? (b) A dollar bill is about 15.5 cm long. If 4 trillion dollars bills were laid end to end around the Earth's equator, how many times would they encircle the Earth? Take the radius of the Earth to be about 6378 km. *Note:* Before doing any of the calculations, try to guess at the answers. You may be very surprised.

Solution (a) Time to repay debt will be calculated by dividing the total debt by the rate at which it is repaid.

$$T = \frac{\$4\text{ trillion}}{\$1000/\text{s}} = \frac{\$4\times10^{12}}{\$1\times10^3/\text{s}} = 4\times10^9\text{ s} \quad \text{and} \quad T = \frac{4\times10^9\text{ s}}{3.16\times10^7\text{ s/y}} = 126.6\text{ y} \quad \Diamond$$

(b) The total length of the end-to-end line of bills would be

$$L = (4\times10^{12}\text{ bills})(0.155\text{ m/bill}) = 6.2\times10^{11}\text{ m}$$

The circumference of the Earth at the equator is

$$C = 2\pi R = 2\pi(6.378\times10^6\text{ m}) = 4.01\times10^7\text{ m}$$

This distance represents a number of "loops" around,

$$N = \frac{L}{C} = \frac{6.2\times10^{11}\text{ m}}{4.01\times10^7\text{ m}} = 1.55\times10^4\text{ times} \quad \Diamond$$

CHAPTER SELF-QUIZ

1. On planet Q, quash is a popular beverage among the natives. The average person on Q consumes 3.6 guppies of quash per month. There are 15 months per year on Q and an estimated population of 200 million. Which of the following represents the best order-of-magnitude estimate value for the total volume of quash consumed per year (in guppy units)?
   a. 10^{12}
   b. 10^{10}
   c. 10^{8}
   d. 10^{6}

2. Note the following mathematical expression: $y = x^2$. Which one of the statements below is most consistent with this expression?
   a. if y doubles, then x quadruples
   b. y is greater than x
   c. if x doubles, then y doubles
   d. if x doubles, then y quadruples

3. On planet N, the standard unit of length is the nose. If a 6.1 foot height astronaut travels to planet N and is measured to have a height of 94 noses, what would be the height in noses of another astronaut who measures 5.5 feet in height?
   a. 74
   b. 104
   c. 81
   d. 85

4. If the displacement of an object is related to velocity, v, according to the relation $x = Av$, the constant, A, has the dimension of which of the following?
   a. acceleration
   b. length
   c. time
   d. area

5. If a submarine on the surface dives at an angle of 15° with respect to the horizontal and follows a straight line path for a distance of 40 m, how far below the surface will it be (measured in m)?
   a. 10
   b. 39
   c. 14
   d. 600

6. The diameter of the Milky Way Galaxy is estimated at about 10^5 light years. A light year is the distance traveled by light in one year; if the speed of light is 3×10^8 m/s, about how far (in meters) is it from one side of the galaxy to the other? (1 year = 3.15×10^7 s)
   a. 10×10^{15}
   b. 1×10^{18}
   c. 9×10^{20}
   d. 300×10^8

7. Suppose it takes you 20 min. to walk a mile. If you trained so that you could walk steadily, day after day, for about 5 hours a day, estimate how long it would take you to walk across America (about 3000 miles).
   a. one month
   b. six months
   c. one year
   d. six years

8. Which point is nearest the x axis?
   a. (3, 4)
   b. (4, 3)
   c. (-5, 2)
   d. (-2, -5)

9. A gallon of paint (volume = 3.78×10^{-3} m^3) covers 25 m^2. What is the thickness of the paint on the wall?
   a. 0.15 mm
   b. 0.30 mm
   c. 0.75 mm
   d. 1.50 cm

10. Assume there are 50 million passenger cars in the United States, and that the average fuel consumption is 20 miles/gallon of gasoline. If the average distance traveled by each car is 10,000 miles/year, how many gallons of gasoline would be saved per year if average fuel consumption could be increased to 25 miles/gallon?
    a. 5×10^6
    b. 5×10^7
    c. 5×10^8
    d. 5×10^9

11. Calculate $(0.41 + 0.021) \times (2.2 \times 10^3)$, keeping only significant figures.
    a. 950
    b. 946
    c. 948
    d. 880

12. $S = \sum_{n=1}^{5} n^3$
    a. S = 45
    b. S = 225
    c. S = 325
    d. S = 3375

2
Motion in One Dimension

Chapter 2

MOTION IN ONE DIMENSION

The branch of physics concerned with the study of the motion of an object and the relationship of this motion to such physical concepts as force and mass is called **dynamics**. *The part of dynamics that describes motion without regard to its causes is called* **kinematics**. In this chapter we shall focus on kinematics and on one-dimensional motion, that is, motion along a straight line. We shall start by discussing displacement, velocity, and acceleration. These concepts will enable us to study the motion of objects undergoing constant acceleration. In Chapter 3 we shall discuss the motions of objects in two dimensions.

NOTES FROM SELECTED CHAPTER SECTIONS

2.2 Average Velocity

The **average velocity** of an object during the time interval t_i to t_f is equal to the slope of the straight line joining the initial and final points on a graph of the position of the object plotted versus time.

2.3 Instantaneous Velocity

The **instantaneous speed** of an object, which is a scalar quantity, is defined as the magnitude of the instantaneous velocity. Hence, by definition, *speed can never be negative.*

The slope of the line tangent to the position-time curve at a point P is defined to be the instantaneous velocity at the corresponding time.

2.4 Acceleration

The **average acceleration** during a given time interval is defined as the change in velocity divided by the time interval during which this change occurs.

The **instantaneous acceleration** of an object at a certain time equals the slope of the tangent to the velocity-time graph at that instant of time.

2.7 Freely Falling Bodies

A freely falling body is an object moving freely under the influence of gravity only, regardless of its initial motion. Objects thrown upward or downward and those released from rest are all falling freely once they are released!

It is important to emphasize that any freely falling object experiences an *acceleration directed downward.* This is true regardless of the initial motion of the object. An object thrown upward (or downward) will experience the same acceleration as an object released from rest.

Once they are in free fall, all objects have an acceleration downward equal to the acceleration due to gravity.

EQUATIONS AND CONCEPTS

The average velocity of an object during a time interval is the ratio of the total displacement to the time interval during which the displacement occurred. Note that the instantaneous velocity of the object during the time interval might have different values from instant to instant.

$$\bar{v} = \frac{x_f - x_i}{t_f - t_i} \quad (2.2)$$

The average acceleration of an object during a time interval is the ratio of the change in velocity to the time interval during which the change in velocity occurs.

$$\bar{a} = \frac{v_f - v_i}{t_f - t_i} \quad (2.4)$$

These equations are called the equations of kinematics and can be used to describe one-dimensional motion along the x axis with constant acceleration. Note that each equation shows a different relationship among physical quantities: initial velocity, final velocity, acceleration, time, and position. Also in the form that they are shown, it is assumed that the object whose motion is described is located at the origin ($x = 0$) when $t = 0$.

$$v = v_0 + at \quad (2.6)$$

$$\bar{v} = \frac{v_0 + v}{2} \quad (2.7)$$

$$x = \frac{1}{2}(v + v_0)t \quad (2.8)$$

$$x = v_0 t + \frac{1}{2}at^2 \quad (2.9)$$

$$v^2 = v_0^2 + 2ax \quad (2.10)$$

The equations of kinematics for an object in free fall ($a = -g$) along the y axis.

$$v = v_0 - gt$$

$$y = \frac{1}{2}(v + v_0)t$$

$$y = v_0 t - \frac{1}{2}gt^2$$

$$v^2 = v_0^2 - 2gy$$

SUGGESTIONS, SKILLS, AND STRATEGIES

The following procedure is recommended for solving problems involving acceleration motion:

1. Make sure all the units in the problem are consistent. That is, if distances are measured in meters, be sure that velocities have units of m/s and accelerations have units of m/s^2.

2. Choose a coordinate system.

3. Make a list of all the quantities given in the problem and a separate list of those to be determined.

4. Select from the list of kinematic equations the one or ones that will enable you to determine the unknowns.

5. Construct an appropriate motion diagram and check to see if your answers are consistent with the diagram.

REVIEW CHECKLIST

▷ Define the displacement and average velocity of a particle in motion. (Sections 2.1 and 2.2)

▷ Define the instantaneous velocity and understand how this quantity differs from average velocity. (Section 2.3)

▷ Define average acceleration and instantaneous acceleration. (Section 2.4)

▷ Construct a graph of position versus time (given a function such as $x = 5 + 3t - 2t^2$) for a particle in motion along a straight line. From this graph, you should be able to determine both average and instantaneous values of velocity by calculating the slope of the tangent to the graph. (Section 2.6)

▷ Describe what is meant by a body in *free fall* (one moving under the influence of gravity--where air resistance is neglected). Recognize that the equations of kinematics apply directly to a freely falling object and that the acceleration is then given by $a = -g$ (where $g = 9.8\ \mathrm{m/s^2}$). (Section 2.7)

▷ Apply the equations of kinematics to any situation where the motion occurs under constant acceleration.

SOLUTIONS TO SELECTED END-OF-CHAPTER PROBLEMS

5. An athlete swims the length of a 50.0-m pool in 20.0 s and makes the return trip to the starting position in 22.0 s. Determine her average velocities in (a) the first half of the swim, (b) the second half of the swim, and (c) the round trip.

Solution

(a) In the first half of the trip, the average velocity is

$$\bar{v} = \frac{x_2 - x_1}{20.0\ \mathrm{s}} = \frac{+50.0\ \mathrm{m}}{20.0\ \mathrm{s}} = +2.50\ \mathrm{m/s} \quad \lozenge$$

(b) On the return leg, we have

$$\bar{v} = \frac{x_3 - x_2}{22.0\ \mathrm{s}} = \frac{0 - 50.0\ \mathrm{m}}{22.0\ \mathrm{s}} = -2.27\ \mathrm{m/s} \quad \lozenge$$

(c) For the entire trip,

$$\bar{v} = \frac{x_3 - x_1}{42.0\ \mathrm{s}} = \frac{0}{42.0\ \mathrm{s}} = 0 \quad \lozenge$$

9. The position-time graph for a bug crawling along the x axis is shown in Figure 2.16. Determine whether the velocity is positive, negative, or zero for the times (a) t_1, (b) t_2, (c) t_3, and (d) t_4.

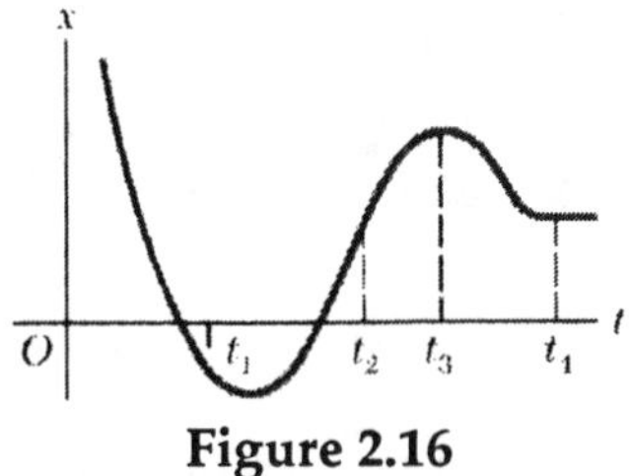

Figure 2.16

Solution Consider the slope of the line tangent to the graph at the point of interest.

(a) $v < 0$ (slope is negative) ◊

(b) $v > 0$ (slope positive) ◊

(c) $v = 0$ (zero slope) ◊

(d) $v = 0$ (zero slope) ◊

---

11. As a race car moves such that its position fits the relation

$$x = (5.00 \text{ m/s})t + (0.750 \text{ m/s}^3)t^3$$

x is measured in meters and t in seconds. (a) Plot a graph of position versus time. (b) Determine the instantaneous velocity at t = 4.00 s, using time intervals of 0.400 s, 0.200 s, and 0.100 s. (c) Compare the average velocity during the first 4.00 s with the results of (b).

Solution

(a) A few typical values calculated from the equation for x vs t are

t(s)	x(m)
1.0	5.75
2.0	16.0
3.0	35.3
4.0	68.0
5.0	119
6.0	192

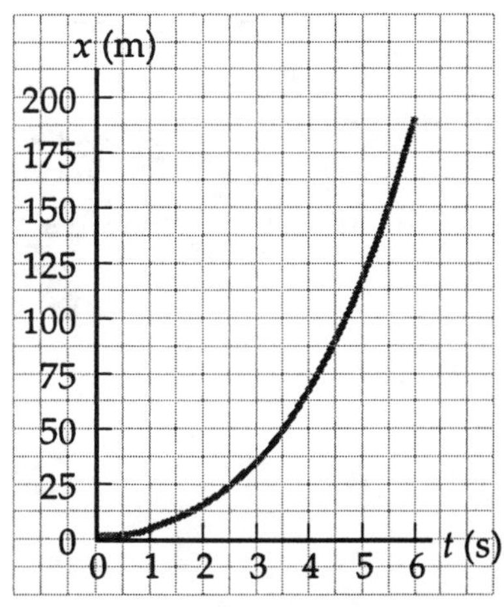

(b) We will use a 0.400-s interval centered at $t = 4.00$ s. We find
at $t = 3.80$ s, $x = 60.154$ m
at $t = 4.20$ s, $x = 76.566$ m

so $$v = \frac{\Delta x}{\Delta t} = \frac{16.412 \text{ m}}{0.400 \text{ s}} = 41.03 \text{ m/s} \quad \Diamond$$

Using a time interval of 0.200 s, we find the corresponding values to be
at $t = 3.90$ s, $x = 63.989$ m
at $t = 4.10$ s, $x = 72.191$ m

and $$v = \frac{\Delta x}{\Delta t} = \frac{8.201 \text{ m}}{0.200 \text{ s}} = 41.008 \text{ m/s} \quad \Diamond$$

For a time interval of 0.100 s, the values are
at $t = 3.95$ s, $x = 65.972$ m
at $t = 4.05$ s, $x = 70.073$ m

and $$v = \frac{\Delta x}{\Delta t} = \frac{4.1002 \text{ m}}{0.100 \text{ s}} = 41.002 \text{ m/s} \quad \Diamond$$

(c) at $t = 4.00$ s, $x = 68$ m. Thus, for the first 4 s,

$$\bar{v} = \frac{\Delta x}{\Delta t} = \frac{68 \text{ m}}{4.00 \text{ s}} = 17 \text{ m/s}$$

This value is considerably less than the instantaneous velocity at $t = 4.00$ s. ◊

---

13. A tennis player moves in a straight-line path as shown in Figure 2.17. Find her average velocity in the time interval (a) 0 to 1 s, (b) 0 to 4 s, (c) 1 s to 5 s, (d) 0 to 5 s.

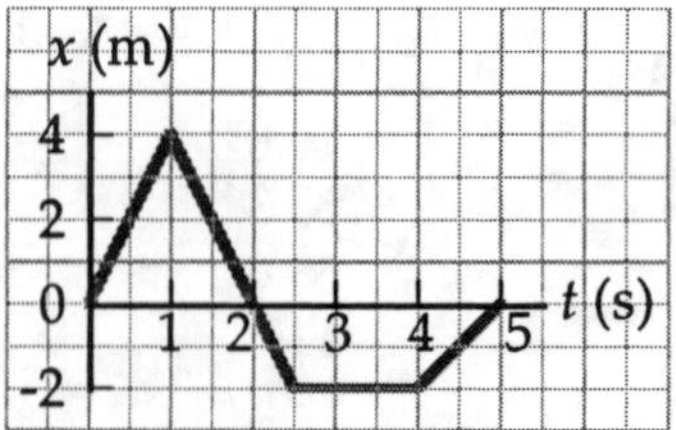

Figure 2.17

Solution

(a) $$\bar{v}_{0,1} = \frac{x_1 - x_0}{\Delta t} = \frac{4 \text{ m} - 0}{1 \text{ s}} = 4 \text{ m/s} \quad \Diamond$$

(b) $\bar{v}_{0,4} = \dfrac{x_4 - x_0}{\Delta t} = \dfrac{-2\text{ m} - 0}{4\text{ s}} = -0.5\text{ m/s}$ ◊

(c) $\bar{v}_{1,5} = \dfrac{x_5 - x_1}{\Delta t} = \dfrac{0 - 4\text{ m}}{4\text{ s}} = -1\text{ m/s}$ ◊

(d) $\bar{v}_{0,5} = \dfrac{x_5 - x_0}{\Delta t} = \dfrac{0 - 0}{\Delta t} = 0\text{ m/s}$ ◊

---

19. A car traveling initially at +7.00 m/s accelerates at the rate of +0.800 m/s^2 for an interval of 2.00 s. What is its velocity at the end of the acceleration?

Solution From the definition of acceleration, $\Delta v = a(\Delta t) = (0.800\text{ m/s}^2)(2.00\text{ s}) = 1.60\text{ m/s}$. From this, the final velocity is

$$v_f = 7.00\text{ m/s} + 1.60\text{ m/s} = 8.60\text{ m/s}$$ ◊

---

31. A jet plane lands with a velocity of +100 m/s and can accelerate at a maximum rate of -5.00 m/s^2 as it comes to rest. (a) From the instant it touches the runway, what is the minimum time needed before it can come to rest? (b) Can this plane land on a small island airport where the runway is 0.800 km long?

Solution

(a) $t = \dfrac{v - v_0}{a} = \dfrac{0 - 100\text{ m/s}}{-5.00\text{ m/s}^2} = 20.0\text{ s}$ ◊

(b) $x = \bar{v}t = \left(\dfrac{v + v_0}{2}\right)t = \dfrac{100\text{ m/s} + 0}{2}(20.0\text{ s}) = 1000\text{ m} = 1.00\text{ km}$

Therefore, the minimum distance to stop exceeds the length of the runway, so *it cannot land safely.* ◊

---

33. A car starts from rest and travels for 5.00 s with a uniform acceleration of $+1.50\text{ m/s}^2$. The driver then applies the brakes, causing a uniform negative acceleration of -2.00 m/s^2. If the brakes are applied for 3.00 s, how fast is the car going at the end of the braking period, and how far has it gone?

Solution The velocity when the brakes are applied is

$$v_i = v_0 + at = 0 + (1.50\ \text{m/s}^2)(5.00\ \text{s}) = 7.50\ \text{m/s}$$

(a) After braking

$$v_f = v_i + at = 7.50\ \text{m/s} + (-2.00\ \text{m/s}^2)(3.00\ \text{s}) = 1.50\ \text{m/s} \quad \Diamond$$

(b) $x = x_1 + x_2 = v_1 t_1 + v_2 t_2$

$$\bar{v}_1 = \frac{0 + 7.50\ \text{m/s}}{2.00} = 3.75\ \text{m/s}$$

$$\bar{v}_2 = \frac{7.50\ \text{m/s} + 1.50\ \text{m/s}}{2.00} = 4.50\ \text{m/s}$$

Thus, $x = (3.75\ \text{m/s})(5.00\ \text{s}) + (4.50\ \text{m/s})(3.00\ \text{s}) = 32.3\ \text{m}$ ◊

---

41. A peregrine falcon dives at a pigeon. The falcon starts downward from rest and falls with the free-fall acceleration. If the pigeon is 76.0 m below the initial position of the falcon, how long does it take the falcon to reach the pigeon? Assume that the pigeon remains at rest.

Solution Use $y = v_{y0}t + \frac{1}{2}at^2$ where $a = -g$

Choose the origin at the initial position of the falcon so that $y = -76.0$ m and $v_{y0} = 0$.

$$t = \sqrt{\frac{2y}{-g}} = \sqrt{\frac{2(-76.0\ \text{m})}{-9.80\ \text{m/s}^2}} \quad \text{or} \quad t = 3.94\ \text{s} \ \Diamond$$

---

43. A rocket moves upward, starting from rest with an acceleration of +29.4 m/s^2 for 4.00 s. It runs out of fuel at the end of this 4.00 s and continues to move upward. How high does it rise?

Solution We shall first find y_1, the height of the rocket, and its velocity, v_1, at the instant it runs out of fuel. The height of the rocket is found from $y = v_0 t + \frac{1}{2}at^2$.

$$y_1 = 0 + \frac{1}{2}(29.4 \text{ m/s}^2)(4.00 \text{ s})^2 = 235 \text{ m}$$

The velocity is found from $v = v_0 + at$. We have

$$v_1 = 0 + (29.4 \text{ m/s}^2)(4.00 \text{ s}) = 117.6 \text{ m/s}$$

At this point, the rocket begins to behave as a *freely falling body*. We shall now find h, the distance the rocket rises after its fuel is exhausted. To do this, we use the equation $v^2 = v_0^2 + 2ay$.

$$0 = (117.6 \text{ m/s})^2 + 2(-9.8 \text{ m/s}^2)h \qquad \text{and} \qquad h = 706 \text{ m}$$

Thus, the total height reached is $y_{max} = y_1 + h = 235 \text{ m} + 706 \text{ m} = 941 \text{ m}$ ◊

---

49. One swimmer in a relay race has a 0.500-s lead and is swimming at a constant speed of 4.00 m/s. He has 50.0 m to swim before reaching the end of the pool. A second swimmer moves in the same direction as the leader. What constant speed must the second swimmer have in order to catch up to the leader at the end of the pool?

Solution The initial distance the leader is ahead of the second swimmer equals the distance the leader swims in 0.500 s, which is (4.00 m/s)(0.500 s) = 2.00 m. Thus, the leader has 50.0 m to go and the second swimmer has 52.0 m to the end of the pool. The time for the leader to reach the end of the pool is

$$t = \frac{x}{v} = \frac{50.0 \text{ m}}{4.00 \text{ m/s}} = 12.5 \text{ s}$$

In this same time, the second swimmer must swim 52.0 m. The minimum velocity is

$$v = \frac{x}{t} = \frac{52.0 \text{ m}}{12.5 \text{ s}} = 4.16 \text{ m/s} \quad ◊$$

---

57. An ice sled powered by a rocket engine starts from rest on a large frozen lake and accelerates at +40 ft/s^2. After some time t_1, the rocket engine is shut down and the sled moves with constant velocity **v** for a time of t_2. If the total distance traveled by the sled is 17 500 ft and the total time is 90 s, find (a) the times t_1 and t_2 and (b) the velocity **v**. At the 17 500-ft mark, the sled begins to accelerate at –20 ft/s^2. (c) What is the final position of the sled when it come to rest? (d) How long does it take to come to rest?

Solution Consider the motion of the sled during three time intervals:

1) Constant positive acceleration for t_1
2) Constant velocity for t_2
3) Constant negative acceleration for t_3

(a) For time interval t_1, $v_{01} = 0$

$$x_1 = v_{01}t_1 + \frac{1}{2}a_1t_1^2 = \frac{1}{2}a_1t_1^2 \quad \text{and} \quad v_1 = v_{01} + a_1t_1 = a_1t_1$$

For time interval t_2, $a_2 = 0$; $v_{02} = v_2 = v_1$

$$x_2 = v_{02}t_2 + \frac{1}{2}a_2t_2^2 = v_1t_2 = a_1t_1t_2$$

But $x_1 + x_2 = 17\,500$ ft and $t_2 = (90\text{ s} - t_1)$ so $17\,500\text{ ft} = \frac{1}{2}a_1t_1^2 + a_1t_1t_2$

The resulting quadric equation is $t_1^2 + (180\text{ s})t_1 - 875\text{ s}^2 = 0$

Solve to find $t_1 = 4.74$ s and $t_2 = 85.26$ s ◊

(b) The constant velocity, v, during the second interval equals the final velocity at the end of the first interval,

$$v = v_1 = a_1t_1 = (40\text{ ft/s}^2)(4.74\text{ s}); \quad v = 190\text{ ft/s} \quad ◊$$

(c) For the third time interval, $v_3^2 = v_{03}^2 + 2a_3x_3$ and $x_3 = \frac{v_3^2 - v_{03}^2}{2a_3}$

but $v_3 = 0$ and $v_{03} = v_2 = 190\text{ ft/s}$ so $x_3 = \frac{-(190\text{ ft/s})^2}{(2)(-20.0\text{ ft/s}^2)}$; $x_3 = 903$ ft

Find distance from starting point, d = 17 500 ft + x_3; d = 18 400 ft ◊

(d) Find the third time interval, $t_3 = \frac{v_3 - v_{03}}{a_3} = \frac{0 - 190\text{ ft/s}}{-20\text{ ft/s}^2} = 9.50\text{ s}$

Total time to come to rest, T = 99.5 s ◊

CHAPTER SELF-QUIZ

1. A European sports car dealer claims that his product will accelerate at a constant rate from rest to a speed of 90 km/hr in 8 s. If so, what is the acceleration (in m/s^2)? (First, convert the speed to m/s.)
   a. 3.1
   b. 6.2
   c. 11.3
   d. 16.9

2. A rock, released at rest from the top of a tower, hits the ground after falling for 2.00 s. What is the height of the tower? ($g = 9.80\ m/s^2$ and air resistance is negligible)
   a. 15 m
   b. 20 m
   c. 31 m
   d. 39 m

3. A rock is thrown downward from the top of a tower with an initial speed of 12 m/s. If the rock hits the ground after 2.00 s, what is the speed of the rock as it hits the ground? ($g = 9.80\ m/s^2$ and air resistance is negligible)
   a. 32 m/s
   b. 39 m/s
   c. 64 m/s
   d. 78 m/s

4. A rock is released at rest from the top of a 40 m tower. If $g = 9.8\ m/s^2$ and air resistance is negligible, how long does it take the rock to reach the ground?
   a. 2.00 s
   b. 2.90 s
   c. 4.10 s
   d. 5.10 s

5. Human reaction time is usually greater than 0.10 s. If your lab partner holds a ruler between your finger and thumb and releases it without warning, how far can you expect the ruler to fall before you catch it? At least
   a. 2 cm
   b. 3 cm
   c. 4 cm
   d. 4.9 cm

6. Two stones are thrown from the top edge of a building with a speed of 10 m/s, one straight down and the other straight up. The first stone hits the street below in 4 s. How much later is it before the second stone hits the street?
   a. 5 s
   b. 4 s
   c 3 s
   d. 2 s

7. A ball is released and it rolls down a hill with constant acceleration. At the end of one second, it has traveled 3 m. How far has it traveled after four seconds?
   a. 12 m
   b. 16 m
   c. 36 m
   d. 48 m

8. A car travels forward along a straight road at 40 m/s for 1000 m and then travels at 50 m/s for the next 1000 m. What is the average velocity?
   a. 45 m/s
   b. 44.4 m/s
   c. 46 m/s
   d. 0

9. A locomotive slows from 28 m/s to zero in 12 s. What distance does it travel?
   a. 168 m
   b. 196 m
   c. 336 m
   d. 392 m

10. A drag racer starts her car from rest and accelerates at 10 m/s^2 for the entire distance of 400 m (1/4 mile). How long did it take the race car to run the quarter?
   a. 8.94 s
   b. 7.74 s
   c. 6.93 s
   d. 5.99 s

11. A motorist traveling at 110 km/hr is being chased by a police car at 130 km/hr. If the police car starts from 1 kilometer back, how long does it take the police to intercept the motorist?
   a. 3 minutes
   b. 5 minutes
   c. 7.5 minutes
   d. 10 minutes

12. A toy rocket, launched from the ground, rises vertically with an acceleration of 20 m/s^2 for 2 s when it runs out of fuel. Disregarding air resistance, how high will the rocket rise?
   a. 61 m
   b. 81 m
   c. 121 m
   d. 141 m

3
Vectors and Two-Dimensional Motion

VECTORS AND TWO-DIMENSIONAL MOTION

In our discussion of one-dimensional motion in Chapter 2, we used the concept of vectors only to a very limited extent. As we progress in our study of motion, the ability to manipulate vector quantities will become increasingly important. As a result, much of this chapter will be devoted to techniques for adding vectors, subtracting them, and so forth. We will then apply our newfound skills to a special case of two-dimensional motion--projectiles. We shall also see that an understanding of vector manipulation is necessary in order to work with and understand relative motion.

NOTES FROM SELECTED CHAPTER SECTIONS

3.1 Vectors and Scalars Revisited

A ***scalar*** has only magnitude and no direction. On the other hand, a ***vector*** is a physical quantity that requires the specification of both direction and magnitude.

3.2 Some Properties of Vectors

Two vectors are *equal* if they have both the same magnitude and direction. When two or more vectors are *added*, they must have the same units. Two vectors which are the *negative* of each other have the same magnitude but opposite directions. When a vector is *multiplied* or *divided* by a positive (negative) scalar, the result is a vector in the same (opposite) direction. The magnitude of the resulting vector is equal to the product of the scalar and the magnitude of the original vector.

3.3 Components of a Vector

Any vector can be completely described by its ***components***. See the recommended procedure for adding vectors algebraically in the Skills section.

3.5 Projectile Motion

In the case of projectile motion, if it is assumed that air resistance is negligible and that the rotation of the Earth does not affect the motion, then . . .

(1) the horizontal component of velocity, v_x, remains constant because there is no horizontal component of acceleration;

(2) the vertical component of acceleration is equal to the acceleration due to gravity, g;

(3) the vertical component of velocity, v_y, and the displacement in the y direction are identical to those of a freely falling body;

(4) projectile motion can be described as a superposition of the two motions in the x and y directions.

Review the recommended procedure in the Skills section solving projectile motion problems.

3.6 Relative Velocity

Observations made by observers in different frames of reference can be related to one another through the techniques of the transformation of relative velocities.

EQUATIONS AND CONCEPTS

Vector quantities obey the commutative law of addition. In order to add vector **A** to vector **B** using the graphical method, first construct **A**, and then draw **B** such that the tail of **B** starts at the head of **A**. The sum of **A** + **B** is the vector that completes the triangle by connecting the tail of **A** to the head of **B**.

$$\mathbf{A} + \mathbf{B} = \mathbf{B} + \mathbf{A}$$

When more than two vectors are to be added, they are all connected head-to-tail in any order and the resultant or sum is the vector which joins the tail of the first vector to the head of the last vector.

$$\mathbf{R} = \mathbf{A} + \mathbf{B} + \mathbf{C} + \mathbf{D}$$

When two or more vectors are to be added, all of them must represent the same physical quantity--that is, have the same units. In the graphical or geometrical method of vector addition, the length of each vector corresponds to the magnitude of the vector according to a chosen scale. Also, the direction of each vector must be along a direction which makes the proper angle relative to the others.

Comment on vector addition

The operation of vector subtraction utilizes the definition of the negative of a vector. The negative of vector **A** is the vector which has a magnitude equal to the magnitude of **A**, but acts or points along a direction opposite the direction of **A**.

$$\mathbf{A} - \mathbf{B} = \mathbf{A} + (-\mathbf{B}) \tag{3.1}$$

A vector **A** in a two-dimensional coordinate system can be resolved into its components along the x and y directions. The projection of **A** onto the x axis is the x component of **A**; and the projection of **A** onto the y axis is the y component of **A**.

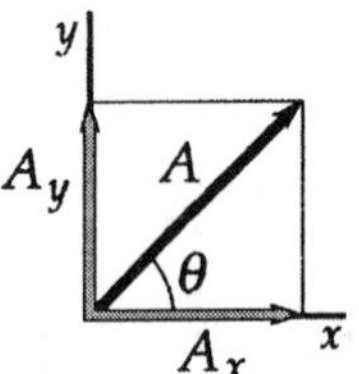

The magnitude of **A** and the angle, θ, which the vector makes with the positive x axis can be determined from the values of the x and y components of **A**.

$$A_x = A\cos\theta \tag{3.2}$$

$$A_y = A\sin\theta$$

$$A = \sqrt{A_x^2 + A_y^2} \tag{3.3}$$

$$\tan\theta = \frac{A_y}{A_x} \tag{3.4}$$

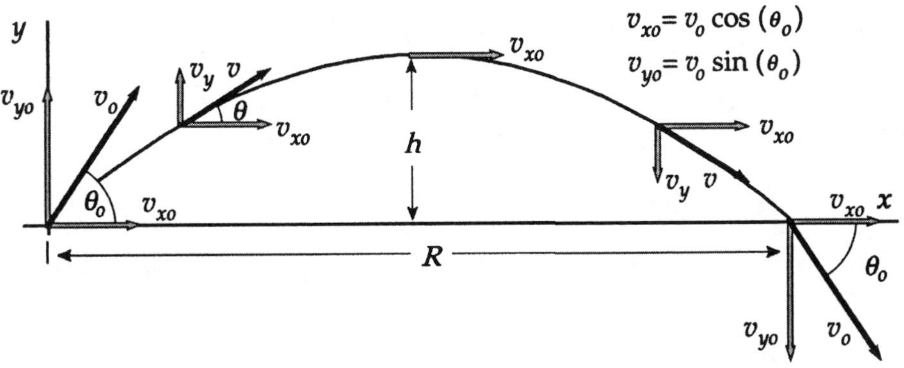

The initial horizontal and vertical components of velocity of a projectile depend on the value of the initial velocity and the initial angle of launch.

$$v_{x0} = x_0 \cos\theta_0$$

$$v_{y0} = v_0 \sin\theta_0$$

The horizontal component of velocity for a projectile remains constant ($a_x = 0$); while the vertical component decreases uniformly with time ($a_y = -g$).

$$v_x = v_0 \cos\theta_0 = \text{constant} \qquad (3.8)$$

$$v_y = v_{y0} - gt \qquad (3.10)$$

$$v_y = v_0 \sin\theta_0 - gt$$

The x and y coordinates of the position of a projectile as functions of the elapsed time.

$$x = (v_0 \cos\theta_0)t \qquad (3.9)$$

The positive direction for the vertical motion is assumed to be upward.

$$y = v_{y0}t - \frac{1}{2}gt^2 \qquad (3.11)$$

$$y = (v_0 \sin\theta_0)t - \frac{1}{2}gt^2$$

The initial and final velocities in the y direction as a function of vertical position.

$$v_y{}^2 = v_{y0}{}^2 - 2gy \qquad (3.12)$$

SUGGESTIONS, SKILLS, AND STRATEGIES

When two or more vectors are to be added, the following step-by-step procedure is recommended:

1. Select a coordinate system.

2. Draw a sketch of the vectors to be added (or subtracted), with a label on each vector.

3. Find the x and y components of all vectors.

4. Find the resultant components (the algebraic sum of the components) in both the x and y directions.

5. Use the Pythagorean theorem to find the magnitude of the resultant vector.

6. Use a suitable trigonometric function to find the angle the resultant vector makes with the x axis.

To add vector **A** to vector **B** graphically, first construct **A**, and then draw **B** such that the tail of **B** starts at the head of **A**. The sum **A** + **B** is the vector that completes the triangle as shown in the figure below. The procedure for adding more than two vectors (the polygon rule) is also illustrated.

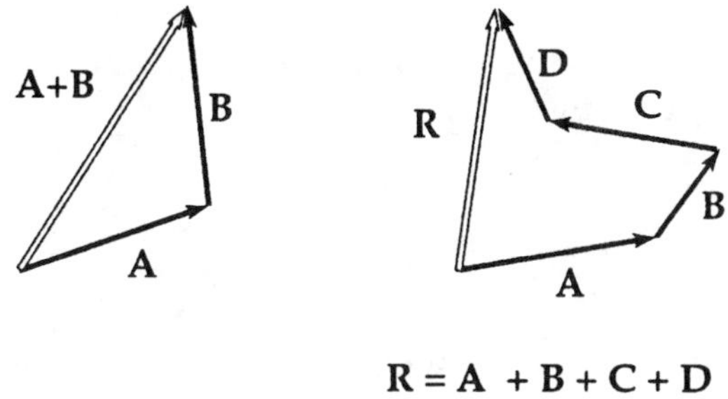

Figure 3.1 Adding vectors by (a) the triangle rule and (b) the polygon rule.

In order to subtract two vectors *graphically*, recognize that **A** – **B** is equivalent to the operation **A** + (–**B**). Since the vector –**B** is a vector whose magnitude is B and is opposite in direction to **B**, the construction shown in Figure 3.2 is obtained:

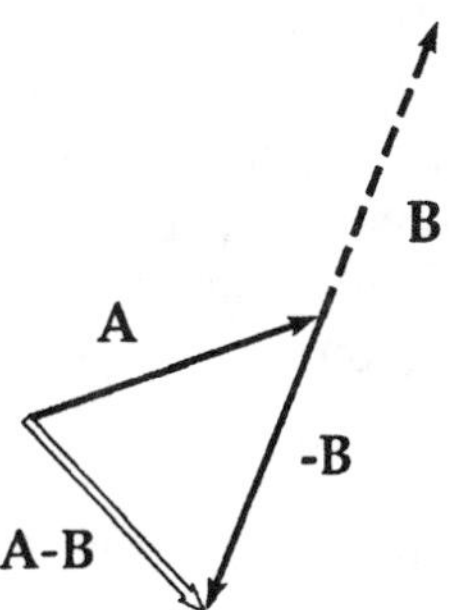

Figure 3.2 Subtracting two vectors graphically

The following procedure is recommended for solving projectile motion problems:

1. Select a coordinate system.

2. Resolve the initial velocity vector into x and y components.

3. Treat the horizontal motion and the vertical motion independently.

4. Follow the techniques for solving problems with constant velocity to analyze the horizontal motion of the projectile.

5. Follow the techniques for solving problems with constant acceleration to analyze the vertical motion of the projectile.

REVIEW CHECKLIST

▷ Recognize that two-dimensional motion in the xy plane with constant acceleration is equivalent to two independent motions along the x and y directions with constant acceleration components a_x and a_y.

▷ Sketch a typical trajectory of a particle moving in the xy plane and draw vectors to illustrate the manner in which the displacement, velocity, and acceleration of the particle changes with time.

▷ Recognize the fact that if the initial speed v_0 and initial angle θ_0 of a projectile are known at a given point at $t = 0$, the velocity components and coordinates can be found at any later time t. Furthermore, one can also calculate the horizontal range R and maximum height h if v_0 and $\theta_0$are known.

▷ Understand and describe the basic properties of vectors, resolve a vector into its rectangular components, use the rules of vector addition (including graphical solutions for addition of two or more vectors), and determine the magnitude and direction of a vector from its rectangular components.

▷ Practice the technique demonstrated in the text to solve relative velocity problems.

SOLUTIONS TO SELECTED END-OF-CHAPTER PROBLEMS

5. A roller coaster moves 200 ft horizontally, then rises 135 ft at an angle of 30° above the horizontal. It then travels 135 ft at an angle of 40° downward. What is its displacement from its starting point at the end of this movement? Use graphical techniques.

Solution Your sketch when drawn to scale should look somewhat like the one at the right. The distance R and the angle θ can be measured to give, upon use of your scale factor, the values of:

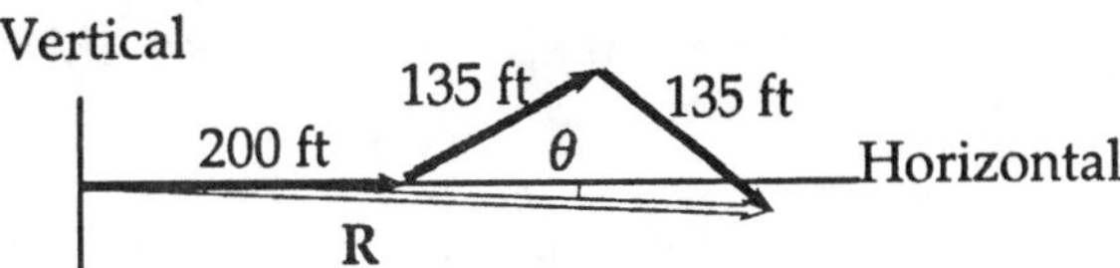

$R = 420$ ft at about 3° below the horizontal. ◊

---

7. A jogger runs 100 m due west, then changes direction for the second leg of the run. At the end of the run, she is 175 m away from the starting point at an angle of 15° north of west. What were the direction and length of her second displacement? Use graphical techniques.

Solution The vector diagram sketched for this problem should look like the one shown at the right. The initial displacement **A** = 100 m and the resultant **R** = 175 m are both known. In order to reach the end point of the run following the initial displacement. the jogger must follow the path shown as **B**. The distance can be found from the scale used for your sketch and the angle θ measured. The results should be about 83 m at 33° north of west. ◊

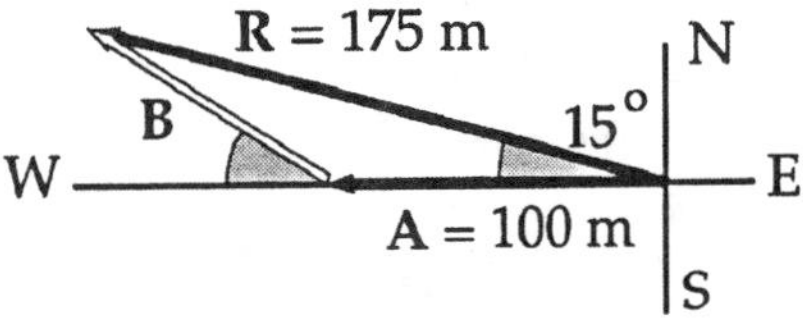

---

9. A roller coaster travels 135 ft at an angle of 40° above the horizontal. How far does it move horizontally and vertically?

Solution Moving for 135 ft at an angle of 40° above the horizontal moves the roller coaster along the horizontal a distance x given by

$$x = (135 \text{ ft})(\cos 40°) = 103 \text{ ft} \quad \Diamond$$

and the coaster rises vertically a distance y given by

$$y = (135 \text{ ft})(\sin 40°) = 86.8 \text{ ft} \quad \Diamond$$

---

13. A shopper pushing a cart through a store moves 40.0 m down one aisle, then makes a 90° turn and moves 15.0 m. He then makes another 90° turn and moves 20.0 m. Where is the shopper relative to his original position, in magnitude and direction? Note that you are not given the direction moved after any of the 90° turns. *As a result, there could be more than one answer.*

Solution The two basic situations are shown in the figures below.

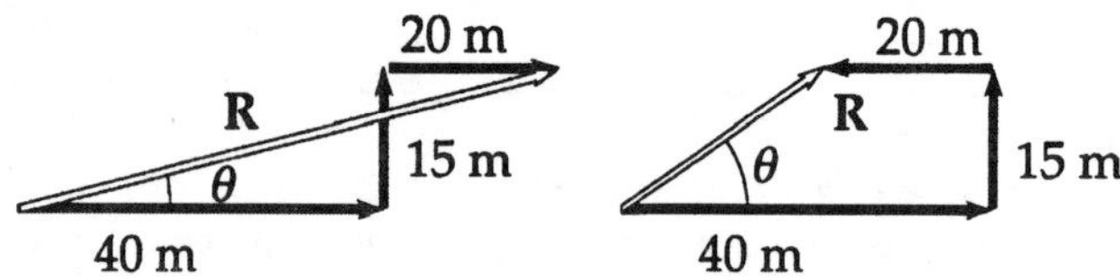

In Case (a) the net displacement in the x direction is

$$R_x = 40.0 \text{ m} + 20.0 \text{ m} = 80.0 \text{ m}$$

and the net displacement in the y direction is 15.0 m. The resultant is found by the Pythagorean theorem as

$$R = \sqrt{(R_x)^2 + (R_y)^2} = \sqrt{(60.0 \text{ m})^2 + (15.0 \text{ m})^2} = 61.8 \text{ m} \quad \Diamond$$

and the angle θ is found as

$$\tan\theta = \frac{15.0 \text{ m}}{60.0 \text{ m}} = 0.250 \qquad \text{from which} \qquad \theta = 14.0° \quad \Diamond$$

(Case b) For the situation depicted in Case (b), the resultant x displacement is 20.0 m and the resultant y displacement is 15.0 m. Using the Pythagorean theorem and trig functions as above,

$$R = 25.0 \text{ m} \quad \text{and} \quad \theta = 36.9° \quad \Diamond$$

---

19. A place kicker must kick a football from a point 36.0 m (about 40 yd) from the goal, and the ball must clear the crossbar, which is 3.05 m high. When kicked, the ball leaves the ground with a speed of 20.0 m/s at an angle of 53° to the horizontal. (a) By how much does the ball clear or fall short of clearing the crossbar? (b) Does the ball approach the crossbar while still rising or while falling?

Solution

(a) We use the equation

$$y = x\tan\theta_0 - \frac{gx^2}{2v_0^2\cos^2\theta_0} \quad \text{where} \quad x = 36.0 \text{ m},\ v_0 = 20.0 \text{ m/s, and } \theta = 53°$$

So, $$y = (36.0 \text{ m})(\tan 53°) - \frac{(9.80 \text{ m/s}^2)(36.0 \text{ m})^2}{(2)(20.0 \text{ m/s})^2\cos^2 53°} = 47.77 - 43.83 = 3.94 \text{ m}$$

The ball clears the bar by (3.94 - 3.05) m = 0.890 m. ◊

(b) The time the ball takes to reach the maximum height is

$$t_1 = \frac{v_0\sin\theta_0}{g} = \frac{(20.0 \text{ m/s})(\sin 53°)}{9.80 \text{ m/s}^2} = 1.63 \text{ s}$$

The time to travel 36.0 m horizontally is $t_2 = \dfrac{x}{v_{0x}}$

$$t_2 = \frac{36.0 \text{ m}}{(20.0 \text{ m/s})(\cos 53°)} = 2.99 \text{ s}$$

Since $t_2 > t_1$, the ball clears the goal on its way down. ◊

---

21. A ball is thrown straight upward and returns to the thrower's hand after 3.00 s in the air. A second ball is thrown at an angle of 30° with the horizontal. At what speed must the second ball be thrown so that it reaches the same height as the one thrown vertically?

Solution We shall first find the initial velocity of the ball thrown vertically upward. At its maximum height, $v = 0$ and $t = 1.50$ s.

$$v_1 = (v_1)_0 + at \quad \text{and} \quad a = -g \quad \text{so} \quad (v_1)_0 = -gt$$

or $$(v_1)_0 = -(-9.80\ \text{m/s}^2)(1.50\ \text{s}) = 14.7\ \text{m/s}$$

In order for the second ball to reach the same vertical height as the first, the second must have the same initial vertical velocity. The vertical component of the velocity of the second ball is

$$v_{2y} = (v_2)_0 \sin\theta_0 \quad \text{where} \quad \text{v}_{2y} = (v_1)_0 = 14.7\ \text{m/s}$$

Therefore $$(v_2)_0 = \frac{v_{2y}}{\sin\theta} = \frac{14.7\ \text{m/s}}{\sin 30°} = 29.4\ \text{m/s} \quad \lozenge$$

---

27. An escalator is 20.0 m long. If a person stands on the "up" side, it takes 50.0 s to ride to the top. (a) If a person walks up the moving escalator with a speed of 0.500 m/s relative to the escalator, how long does it take to get to the top? (b) If a person walks down the "up" escalator with the same relative speed as in (a), how long does it take to reach the bottom?

Solution We have $v_{eg} = \dfrac{20.0\ \text{m}}{50.0\ \text{s}} = 0.400\ \text{m/s}$ (speed of escalator relative to the ground)

and $v_{pe} = \pm 0.500\ \text{m/s}$ (speed of person relative to the escalator)

(The plus sign is used when the person is going up; the negative sign when the person is going down.) Also, v_{pg} is the speed of the person relative to ground.

We have $$v_{pg} = v_{pe} + v_{eg}$$

(a) When he walks up the escalator, $$v_{pg} = 0.500\ \text{m/s} + 0.400\ \text{m/s} = 0.900\ \text{m/s}$$

and the time is $$t = \frac{20.0\ \text{m}}{0.900\ \text{m/s}} = 22.2\ \text{s} \quad \lozenge$$

(b) When he walks down the escalator, $$v_{pg} = -0.500\ \text{m/s} + 0.400\ \text{m/s} = -0.100\ \text{m/s}$$

and the time is $$t = \frac{-20.0\ \text{m}}{-0.100\ \text{m/s}} = 200\ \text{s} \quad \lozenge$$

---

31. A river flows due east at 1.50 m/s. A boat crosses the river from the south shore to the north shore by maintaining a constant velocity of 10.0 m/s due north relative to the water. (a) What is the velocity of the boat relative to shore? (b) If the river is 300 m wide, how far downstream has the boat moved by the time it reaches the north shore?

Solution

(a) We have $\mathbf{v}_{bw}$ = 10.0 m/s is the speed of the boat relative to the water, and is directed northward.

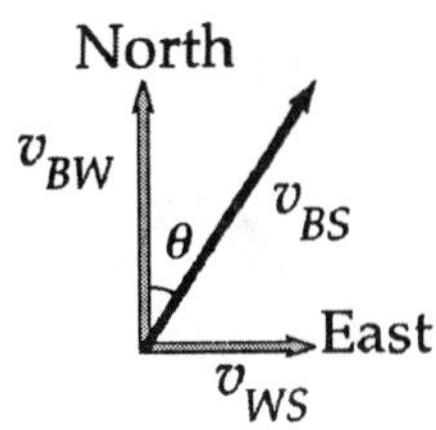

$\mathbf{v}_{ws}$ = 1.50 m/s, the speed of the water relative to the shore, and is directed east.

$\mathbf{v}_{bs}$ = the speed of the boat relative to the shore, and is directed at an angle of θ, relative to the northward direction.

$$\mathbf{v}_{bs} = \mathbf{v}_{bw} + \mathbf{v}_{ws}$$

Northward component of $\mathbf{v}_{bs}$ is $\quad v_{bs}\cos\theta = 10.0\text{ m/s} \quad (1)$

Eastward component is $\quad v_{bs}\sin\theta = 1.50\text{ m/s} \quad (2)$

Dividing (2) by (1) gives $\quad \tan\theta = 0.150$ and $\theta = 8.53°$

Then, from (1) $\quad v_{bs} = \dfrac{10.0\text{ m/s}}{\cos 8.53} = 10.1\text{ m/s}$ ◊

(b) The time to cross the river is $\quad t = \dfrac{300\text{ m}}{v_{bs}\cos\theta} = \dfrac{300\text{ m}}{10.0\text{ m/s}}$ 30.0 s

The eastward drift = $(v_{bs}\sin\theta)t = (1.50\text{ m/s})(30.0\text{ s}) = 45.0\text{ m}$ ◊

---

35. A science student is riding on a flatcar of a train traveling along a straight horizontal track at a constant speed of 10.0 m/s. The student throws a ball along a path that she judges to make an initial angle of 60° with the horizontal and to be in line with the track. The student's professor, who is standing on the ground nearby, observes the ball to rise vertically. How high does the ball rise?

Solution $\mathbf{v}_{bc}$ = the velocity of the ball relative to the car
$\mathbf{v}_{be}$ = the velocity of the ball relative to the Earth
$\mathbf{v}_{ce}$ = velocity of the car relative to the Earth = 10.0 m/s

and $\mathbf{v}_{be} = \mathbf{v}_{bc} + \mathbf{v}_{ce}$

Since $\mathbf{v}_{be}$ has zero horizontal component, we have $0 = -v_{bc}\cos 60° + v_{ce}$

So, $v_{bc} = \dfrac{10.0 \text{ m/s}}{0.500} = 20.0 \text{ m/s}$ and the vertical components give

$$v_{be} = v_{bc}\sin 60° + 0 = (20.0 \text{ m/s})(0.866) = 17.3 \text{ m/s}$$

This is the initial velocity of the ball relative to the Earth.

Now from $v_y^2 = v_{0y}^2 + 2ay$, we have $0 = (17.3 \text{ m/s})^2 + 2(-9.8 \text{ m/s}^2)h$

to give h = 15.3 m, the maximum height the ball rises. ◊

41. If a person can jump a horizontal distance of 3.00 m on the Earth, how far could he jump on the Moon, where the acceleration due to gravity is $g/6.00$ and g = 9.80 m/s^2? Repeat for Mars, where the acceleration due to gravity is 0.380 g.

Solution An expression for the horizontal range can be found from $x = v_{0x}t$. The time of flight is found from $y = v_{0y}t + \frac{1}{2}at^2$ with y = 0, when $t = \frac{2v_{0y}}{g}$. This gives general expression for the range as

$$x = v_{0x}\left(\frac{2v_{0y}}{g}\right)$$

On Earth this becomes $x_e = v_{0x}\left(\frac{2v_{0y}}{g_e}\right)$ and on the Moon, $x_m = v_{0x}\left(\frac{2v_{0y}}{g_m}\right)$

Dividing x_m by x_e, we have $x_m = \left(\frac{g_e}{g_m}\right)(x_e)$; with $g_m = \frac{1}{6}g_e$, we find $x_m = 18$ m ◊

For Mars, $g_{\text{mars}} = 0.380\ g_e$ and we find $x_{\text{mars}} = 7.89$ m ◊

43. Towns A and B in Figure 3.22 are 80.0 km apart. A couple arranges to drive from town A and meet a couple driving from town B at the Lake, L. The two couples leave simultaneously and drive for 2.50 h in the directions shown. Car 1 has a speed of 90.0 km/h. If the cars arrive simultaneously at the lake, what is the speed of car 2?

Solution The displacement vectors are shown in the vector diagram to the right.

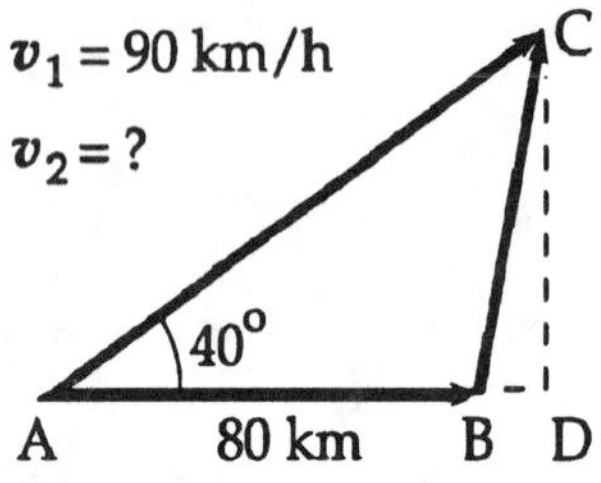

$$\mathbf{AC} = v_1 t = (90.0\text{ km/h})(2.50\text{ h}) = 225\text{ km}$$

$$\mathbf{AD} = \mathbf{AC}\cos 40^\circ = 172.4\text{ km}, \quad \text{and}$$

$$\mathbf{CD} = \mathbf{AD}\sin 40^\circ = 144.6\text{ km}$$

$$\mathbf{BD} = \mathbf{AD} - \mathbf{AB} = 172.4\text{ km} - 80.0\text{ km} = 92.4\text{ km}$$

$v_1 = 90$ km/h

$v_2 = ?$

40°

A 80 km B D

(Not to scale)

From the triangle BCD, $$BC = \sqrt{(BD)^2 + (CD)^2}$$

or $$BC = \sqrt{(92.4\text{ km})^2 + (144.6\text{ km})^2} = 171.6\text{ km}$$

Since car 2 travels this distance in 2.50 h, its constant speed is

$$v_2 = \frac{171.6\text{ km}}{2.5\text{ h}} = 68.6\text{ km/h} \quad \lozenge$$

---

49. A projectile is fired with an initial speed of v_0 at an angle of θ_0 to the horizontal, as in Figure 3.12. (See sketch of projectile trajectory in Equations and Concepts section.) When it reaches its peak, it has (x, y) coordinates given by $(R/2, h)$, and when it strikes the ground, its coordinates are $(R, 0)$, where R is called the horizontal range.

(a) Show that it reaches a maximum height, h, given by $$h = \frac{v_0^2 \sin^2\theta_0}{2g}$$

(b) Show that its horizontal range is given by $$R = \frac{v_0^2 \sin 2\theta_0}{g}$$

Solution

(a) At the top of the arc, $v_y = 0$; and from $v_y = v_{0y} - gt$, we find the time to reach the top of the arc to be

$$t = \frac{v_0 \sin\theta}{g} \qquad (1)$$

The vertical height, h, reached in this time is found from $y = v_{0y}t - \frac{1}{2}gt^2$ as

$$h = (v_0 \sin\theta)t - \frac{1}{2}gt^2 \qquad (2)$$

Substitute into (2) for t from(1), and we find $h = \frac{v_0^2 \sin^2\theta}{2g}$ ◊

(b) The total time of flight, T, is twice the time given in (1) above. The horizontal range, R, is found from $x = v_{0x}t$ as

$$R = (v_0 \cos\theta)\frac{2v_0 \sin\theta}{g} \quad \text{or} \quad R = \frac{v_0^2 \sin(2\theta)}{g}$$ ◊

---

51. An enemy ship is on the east side of a mountainous island as shown in Figure 3.25. The enemy ship can maneuver to within 2500 m of the 1800-m-high mountain peak and can shoot projectiles with an initial speed of 250 m/s. If the western shoreline is horizontally 300 m from the peak, what are the distances from the western shore at which a ship can be safe from the bombardment of the enemy ship?

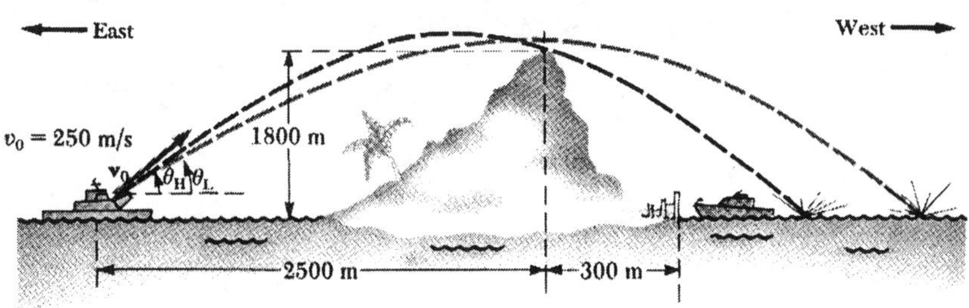

Figure 3.25

Solution Find highest elevation, θ_H , that will clear the mountain peak; this will yield the range of the closest point of bombardment. Next find lowest elevation, θ_L, that will clear the mountain peak; this will yield the maximum range under these conditions if both θ_H and θ_L, are > 45°: $x = 2500$ m, $y = 1800$ m, $v_0 = 250$ m/s

$$y = v_{y0}t - \frac{1}{2}gt^2 = v_0(\sin\theta)t - \frac{1}{2}gt^2 \quad \text{and} \quad x = v_{x0}t = v_0(\cos\theta)t \quad \text{thus} \quad t = \frac{x}{v_0(\cos\theta)}$$

Substitute this expression for t into the expression for y.

$$y = \frac{v_0(\sin\theta)x}{v_0(\cos\theta)} - \frac{1}{2}g\left(\frac{x}{v_0(\cos\theta)}\right)^2 = x\tan\theta - \left[\frac{gx^2}{2v_0^2}\right][\tan^2\theta + 1] \quad \text{and}$$

$$0 = \frac{gx^2}{2v_0^2\tan^2\theta} - x\tan\theta + \frac{gx^2}{2v_0^2} + y$$

Substitute values, use the quadratic formula, and find

$$\tan\theta = 3.905 \quad \text{or} \quad 1.197 \quad \text{which gives} \quad \theta_H = 75.6° \quad \text{and} \quad \theta_L = 50.1°$$

Range (at θ_H) $= v_0^2 \sin 2\theta_{H/g} = 3065$ m from enemy ship

3065 – 2500 – 300 = 265 m from shore.

Range (at θ_L) $= v_0^2 \sin 2\theta_{L/g} = 6276$ m from enemy ship.

6276 – 2500 – 200 = 3576 from shore.

Therefore, safe distance is < 265 m or > 3480 m from shore. ◊

CHAPTER SELF-QUIZ

1. Vector **A** points north and vector **B** points east. If we subtract **A** from **B**, the direction of the vector (**B** – **A**) points
   a. north of east
   b. south of east
   c. north of west
   d. south of west

2. A baseball is thrown by the center fielder (from shoulder level) to home plate where it is caught (on the fly at shoulder level) by the catcher. At what point is the ball's speed at a minimum? (Air resistance is negligible.)
   a. just after leaving the center fielder's hand
   b. just before arriving at the catcher's mitt
   c. at the top of the trajectory
   d. speed is constant during entire trajectory

3. A baseball is thrown by the center fielder (from shoulder level) to home plate where it is caught (on the fly at shoulder level) by the catcher. At what point does the magnitude of the horizontal component of velocity have its minimum value? (Air resistance is negligible.)
   a. just after leaving the center fielder's hand
   b. just before arriving at the catcher's mitt
   c. at the top of the trajectory
   d. magnitude of the horizontal component of velocity is constant

4. A city jogger runs four blocks due east, eight blocks due south, and another two blocks due east. Assume all blocks are of equal size. What is the magnitude of the jogger's displacement, start to finish?
   a. 14.0 blocks
   b. 8.4 blocks
   c. 2.0 blocks
   d. 10.0 blocks

5. A student adds two vectors with magnitudes of 100 and 40. Of the following, which one is the only possible choice for the magnitude of the resultant?
   a. 150
   b. 120
   c. 50
   d. 25

6. A string attached to an airborne kite is maintained at an angle of 35° with the horizontal. If a total of 80 m of string is reeled in while bringing the kite back to the ground, what is the horizontal displacement of the kite in the process (assume the kite string doesn't sag)?
   a. 114 m
   b. 56 m
   c. 46 m
   d. 66 m

7. Find the resultant of the following two vectors: i) 100 units due east and ii) 200 units 30° north of west.
   a. 150 units 30° north of west
   b. 308 units 15° north of west
   c. 273 units 60° north of west
   d. 124 units 54° north of west

8. A bullet is fired from a gun at 300 m/s. It hits the ground 3 s later. At what angle above the horizon was the bullet fired? (Ignore air friction.)
   a. 3°
   b. 10°
   c. 30°
   d. 80°

9. A jet airliner moving at 600 mph due east moves into a region where the wind is blowing at 100 mph in a direction 30° north of east. What is the new velocity and direction of the aircraft?
   a. 606 mph, 7.1° N of E
   b. 629 mph, 5.6° N of E
   c. 650 mph, 4.7° N of E
   d. 688 mph, 3.3° N of E

10. A river flows due east at 2 m/s. A boat crosses the 300-m wide river by maintaining a constant velocity of 10 m/s due north relative to the water If no correction is made for the current, how far downstream does the boat move by the time it reaches the far shore?
   a. 6 m
   b. 30 m
   c. 60 m
   d. 90 m

11. A golfer wants to drive a golf ball a distance of 310 yards (283 m). If the 4-wood launches the ball at 15 ° above the horizontal, what must be the initial speed of the ball to achieve the required distance? (Ignore air friction and use g = 9.8 m/s^2).
   a. 74.5 m/s
   b. 57.7 m/s
   c. 44.1 m/s
   d. 37.2 m/s

12. A baseball leaves the bat with a speed of 40 m/s and an angle of 37° above the horizontal. A very high fence is located at a horizontal distance of 128 m from the point where the ball is struck. Assuming the ball leaves the bat near ground level, how far above ground level does the ball strike the fence?
   a. 4.4 m
   b. 8.8 m
   c. 13.2 m
   d. 17.6 m

4
The Laws of Motion

Chapter 4

THE LAWS OF MOTION

In the foregoing chapters on kinematics, we used the definitions of displacement, velocity, and acceleration to describe motion, without concerning ourselves with the causes of that motion. In this chapter we shall investigate motion in terms of the forces that cause it. We shall then discuss the three fundamental laws of motion, which are based on experimental observations and were formulated by Sir Isaac Newton three centuries ago.

NOTES FROM SELECTED CHAPTER SECTIONS

4.2 The Concept of Force

Equilibrium is the condition under which the *net force* (vector mass of all forces) acting on an object is zero. An object in equilibrium has a zero acceleration (velocity is constant or equals zero).

Fundamental forces in nature are: (1) gravitational (attractive forces between objects due to their masses), (2) electromagnetic forces (between electric charges at rest or in motion), (3) strong nuclear forces (between subatomic particles), and (4) weak nuclear forces (accompanying the process of radioactive decay).

Classical physics is concerned with contact forces (which are the result of physical contact between two or more objects) and action-at-a-distance forces (which act through empty space and do not involve physical contact).

4.3 Newton's First Law

Newton's first law is called the *law of inertia* and states that an object at rest will remain at rest and an object in motion will remain in motion with a constant velocity unless acted on by a *net external force.*

Mass and *weight* are two different physical quantities. The weight of a body is equal to the *force of gravity* acting on the body and varies with location in the Earth's gravitational field. Mass is an *inherent property* of a body and is a measure of the body's inertia (resistance to change in its state of motion). The SI unit of mass is the *hologram* and the unit of weight is the *newton.*

4.4 Newton's Second Law

Newton's second law, the *law of acceleration,* states that the acceleration of an object is directly proportional to the resultant force acting on it and inversely proportional to its mass. The direction of the acceleration is in the direction of the net force.

4.5 Newton's Third Law

Newton's third law, the *law of action-reaction,* states that when two bodies interact, the force which body "A" exerts on body "B" (the *action force*) is equal in magnitude and opposite in direction to the force which body "B" exerts on body "A" (the *reaction force).* A consequence of the third law is that forces occur in *pairs.* Remember that the action force and the reaction force act on *different objects.*

4.6 Some Applications of Newton's Laws

Construction of a *free-body diagram* is an important step in the application of Newton's laws of motion to the solution of problems involving bodies in equilibrium or accelerating under the action of external forces. The diagram should include an arrow labeled to identify each of the external forces acting on the body whose motion (or condition of equilibrium) is to be studied. Forces which are the *reactions* to these external forces must *not* be included. When a system consists of more than one body or mass, you must construct a free-body diagram for each mass.

EQUATIONS AND CONCEPTS

A quantitative measurement of mass (the term used to measure inertia) can be made by comparing the accelerations that a given force will produce on different bodies. If a given force acting on a body of mass m_1 produces an acceleration a_1 and the same force acting on a body of mass m_2 produces an acceleration a_2, the ratio of the two masses equals the inverse of the ratio of the two accelerations.

$$\frac{m_1}{m_2} = \frac{a_2}{a_1}$$

The acceleration of an object is proportional to the resultant force acting on it and inversely proportional to its mass. This is a statement of Newton's second law.

$$\Sigma \mathbf{F} = m\mathbf{a} \quad \text{or} \quad \mathbf{a} = \frac{\Sigma \mathbf{F}}{m} \tag{4.1}$$

When several forces act on an object, it is often convenient to write the equation expressing Newton's second law as component equations. The orientation of the coordinate system can often be chosen so that the object has a nonzero acceleration along only one direction.

$$\Sigma F_x = ma_x \tag{4.2}$$

$$\Sigma F_y = ma_y$$

$$\Sigma F_z = ma_z$$

Calculations with Equation 4.3 must be made using a consistent set of units for the quantities' force, mass, and acceleration. The SI unit of force is the newton (N), defined as the force that, when acting on a 1-kg mass, produces an acceleration of 1 m/s^2.

$$1\text{ N} \equiv 1\text{ kg}\cdot\text{m/s}^2 \tag{4.3}$$

$$1\text{ dyne} \equiv 1\text{ g}\cdot\text{cm/s}^2 \tag{4.4}$$

$$1\text{ lb} \equiv 4.448\text{ N} \tag{4.5}$$

Weight is not an inherent property of a body, but depends on the local value of g and varies with location.

$$w = mg \tag{4.6}$$

This is a statement of Newton's third law, which states that forces always occur in pairs; and the force exerted on body 1 by body 2 is equal in magnitude and opposite in direction to the force exerted on body 2 by body 1.

$$\mathbf{F}_{12} = \mathbf{F}_{21}$$

The force of static friction between two surfaces in contact but not in motion, relative to each other, cannot be greater than $\mu_s n$, where n is the normal (perpendicular) force between the two surfaces and μ_s (coefficient of static friction) is a dimensionless constant which depends on the nature of the pair of surfaces.

$$f_s \le \mu_s n \tag{4.9}$$

When two surfaces are in relative motion, the force of kinetic friction on each body is directed opposite to the direction of motion of the body.

$$f_k = \mu_k n \tag{4.10}$$

SUGGESTIONS, SKILLS, AND STRATEGIES

The following procedure is recommended for problems involving objects in equilibrium.

1. Make a sketch of the object under consideration.

2. Draw a free-body diagram and label all external forces acting on the object. Try to guess the correct direction for each force. If you select a direction that leads to a negative sign in your solution for a force, do not be alarmed; this merely means that the direction of the force is the opposite of what you assumed.

3. Resolve all forces into x and y components, choosing a convenient coordinate system.

4. Use the equations $\Sigma F_x = 0$ and $\Sigma F_y = 0$. Remember to keep track of the signs of the various force components.

5. Application of Step 4 above leads to a set of equations with several unknowns. All that is left is to solve the simultaneous equations for the unknowns in terms of the known quantities.

The following procedure is recommended when dealing with problems involving the application of Newton's second law:

1. Draw a simple, neat diagram of the system.

2. Isolate the object of interest whose motion is being analyzed. Draw a free-body diagram for this object; that is, a diagram showing *all external forces acting on the object*. For systems containing more than one object, draw *separate* diagrams for each object. Do not include forces that the object exerts on its surroundings.

3. Establish convenient coordinate axes for each body and find the components of the forces along these axes.

4. Apply Newton's second law, $\Sigma \mathbf{F} = m\mathbf{a}$, in the x and y directions for each object under consideration.

5. Solve the component equations for the unknowns. Remember that you must have as many independent equations as you have unknowns in order to obtain a complete solution.

6. Often in solving such problems, one must also use the equations of kinematics (motion with constant acceleration) to find all the unknowns.

REVIEW CHECKLIST

▷ State in your own words a description of Newton's laws of motion, recall physical examples of each law, and identify the action-reaction force pairs in a multiple-body interaction problem as specified by Newton's third law.

▷ Express the normal force in terms of other forces acting on an object and write out the equation which relates the coefficient of friction, force of friction and normal force between an object and surface on which it rests or moves.

▷ Apply Newton's laws of motion to various mechanical systems using the recommended procedure discussed in Section 4.6. Most important, you should identify all external forces acting on the system, draw the *correct* free-body diagrams which apply to each body of the system, and apply Newton's second law, $\mathbf{F} = m\mathbf{a}$, in *component* form.

▷ Apply the equations of kinematics (which involve the quantities' displacement, velocity, and acceleration) as described in Chapter 2 along with those methods and equations of Chapter 4 (involving mass, force, and acceleration) to the solutions of problems where *both* the kinematical and dynamic aspects are present.

▷ Be familiar with solving several linear equations simultaneously for the unknown quantities. Recall that you must have as many *independent* equations as you have unknowns.

SOLUTIONS TO SELECTED END-OF-CHAPTER PROBLEMS

3. After falling from rest at a height of 30.0 m, a 0.500-kg ball rebounds upward, reaching a height of 20.0 m. If the contact between ball and ground lasted 2.00 ms, what average force was exerted on the ball?

Solution The impulse imparted to the ball is equal to its change in momentum while in contact with the ground.

$$\overline{\mathbf{F}}(\Delta t) = \Delta \mathbf{p} = \Delta(m\mathbf{v}) \qquad \text{or} \qquad \overline{\mathbf{F}} = \frac{m}{\Delta t}(\mathbf{v_f} - \mathbf{v_i})$$

The velocity before impact ($\mathbf{v_i}$) and the velocity following impact ($\mathbf{v_f}$) can be calculated from the distances of fall and rebound using the equation $v^2 = v_0{}^2 - 2gy$. We find (taking the positive direction upward)

$$v_f = \pm\sqrt{0 - (2)(9.80 \text{ m/s}^2)(-30.0 \text{ m})} = -24.25 \text{ m/s}$$

$$v_i = \sqrt{2gy} = \sqrt{(2)(9.80 \text{ m/s}^2)(20.0 \text{ m})} = +19.8 \text{ m/s}$$

$$\overline{\mathbf{F}} = \frac{0.500 \text{ kg}}{2.00 \times 10^{-3} \text{ s}}[19.8 \text{ m/s} - (-24.25 \text{ m/s})] = 1.10 \times 10^4 \text{ N} \quad \Diamond$$

---

7. A 6.00-kg object undergoes an acceleration of 2.00 m/s². (a) What is the magnitude of the resultant force acting on it? (b) If this same force is applied to a 4.00-kg object, what acceleration is produced?

Solution

(a) $F_R = ma = (6.00 \text{ kg})(2.00 \text{ m/s}^2) = 12.0 \text{ N} \quad \Diamond$

(b) $a = \dfrac{F_R}{m} = \dfrac{12.0 \text{ N}}{4.00 \text{ kg}} = 3.00 \text{ m/s}^2 \quad \Diamond$

---

13. Two forces are applied to a car in an effort to move it, as shown in Figure 4.18. (a) What is the resultant of these two forces? (b) If the car has a mass of 3000 kg, what acceleration does it have?

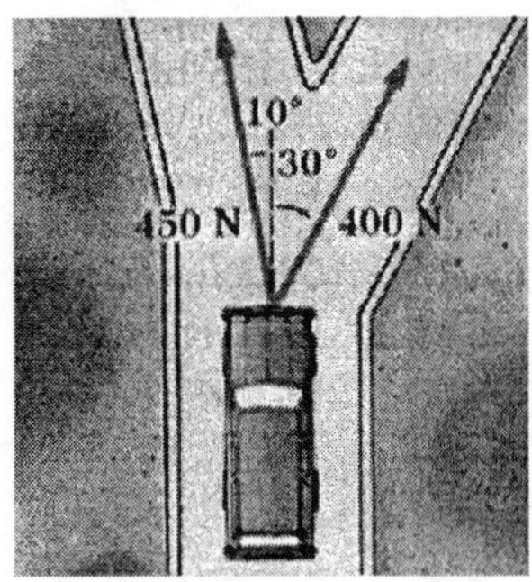

Figure 4.18

Solution

(a) We resolve the forces shown into their components as

	x comp	y comp
400 N:	200. N	346. N
450 N:	-78.1 N	443. N
F_R:	122 N	789. N

The magnitude of the resultant force is found from the Pythagorean theorem as

$$F_R = \sqrt{\left(\Sigma F_x\right)^2 + \left(\Sigma F_y\right)^2} = \sqrt{(122\text{ N})^2 + (789\text{ N})^2} \qquad \text{or} \qquad F_R = 798\text{ N} \quad \lozenge$$

and $$\tan\theta = \frac{\Sigma F_y}{\Sigma F_x} = \frac{789}{122} = 6.47 \qquad \text{from which} \qquad \theta = 81.2° \quad \lozenge$$

Thus, the resultant force is at an angle of 8.8° to the right of the forward direction.

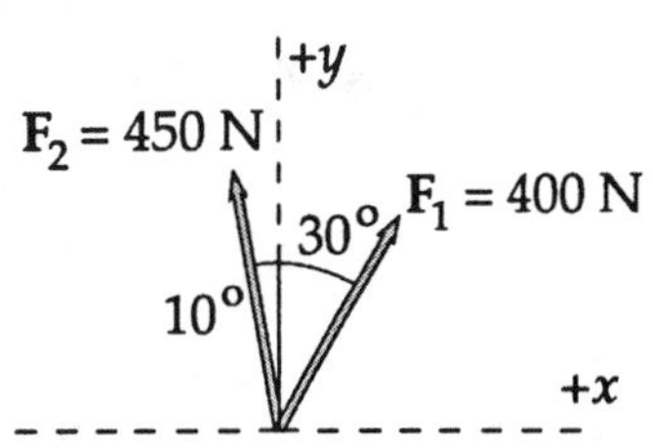

(b) The acceleration is in the same direction as F_R and is given by

$$a = \frac{F_R}{m} = \frac{798\text{ N}}{3000\text{ kg}} = 0.266\text{ m/s}^2 \quad \lozenge$$

17. Find the tension in the two wires that support the 100-N light fixture in Figure 4.21.

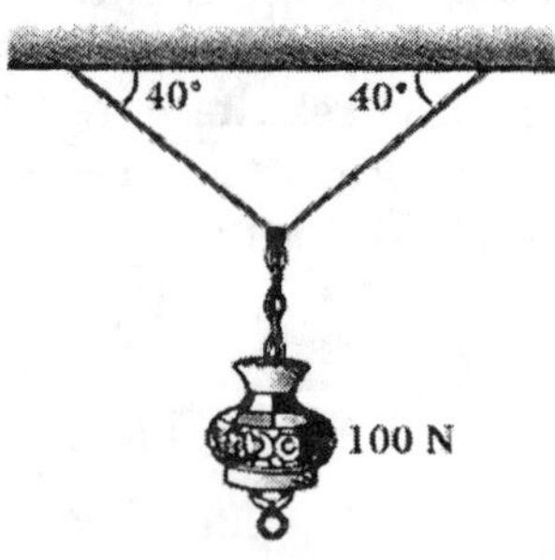

Figure 4.21

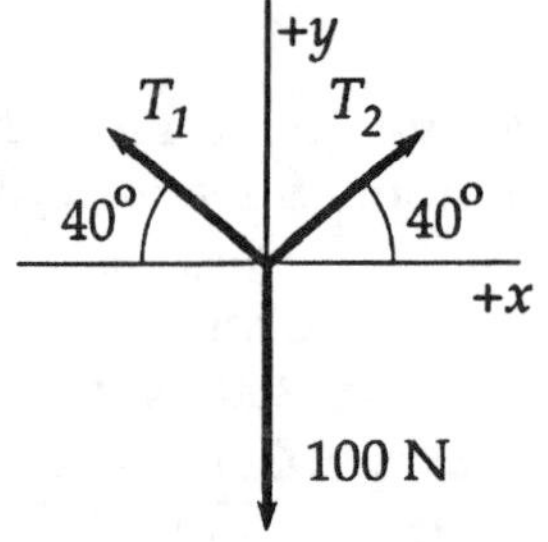

Solution Use $\Sigma F_x = 0$ as

$$T_1 \cos 40° - T_2 \cos 40° = 0$$

or $\quad T_1 = T_2$

From $\Sigma F_y = 0$, we have

$$T_1 \sin 40° + T_2 \sin 40° - 100 \text{ N} = 0$$

Thus, $\quad T_1 = T_2 = 77.8 \text{ N}$ ◊

---

27. A 40.0-kg wagon is towed up a hill inclined at 18.5° with respect to the horizontal. The tow rope is parallel to the incline and has a tension of 140 N in it. Assume that the wagon starts from rest at the bottom of the hill, and neglect friction. How fast is the wagon going after moving 80.0 m up the hill?

Solution Choosing our x axis along the incline with the positive direction up the incline, we have

$$\Sigma F_x = ma_x = T - w \sin 18.5° = (40.0 \text{ kg})a_x$$

which gives $\quad a_x = \dfrac{140 \text{ N} - 124.4 \text{ N}}{40.0 \text{ kg}} = 0.390 \text{ m/s}^2$

Since we have constant acceleration, we find

$$v^2 = v_0{}^2 + 2ax \quad \text{becomes} \quad v^2 = 0 + 2(0.390\ \text{m/s}^2)(80.0\ \text{m}) \quad \text{and} \quad v = 7.90\ \text{m/s}\ \Diamond$$

---

31. Two people pull as hard as they can on ropes attached to a 200-kg boat. If they pull in the same direction, the boat has an acceleration of 1.52 m/s^2 to the right. If they pull in opposite directions, the boat has an acceleration of 0.518 m/s^2 to the left. What is the force exerted by each person on the boat? (Disregard any other forces on the boat.)

Solution Let us call the forces exerted by each of the men $\mathbf{F_1}$ and $\mathbf{F_2}$. Thus, when pulling in the same direction, Newton's second law becomes

$$F_1 + F_2 = (200\ \text{kg})(1.52\ \text{m/s}^2) \quad \text{or} \quad F_1 + F_2 = 304\ \text{N} \qquad (1)$$

When pulling in opposite directions,

$$F_1 - F_2 = (200\ \text{kg})(-\ 0.518\ \text{m/s}^2) \quad \text{or} \quad F_1 - F_2 = -\ 103.6\ \text{N} \qquad (2)$$

Solving simultaneously, we find

$$F_1 = 100\ \text{N}\ \Diamond \quad \text{and} \quad F_2 = 204\ \text{N}\ \Diamond$$

---

35. Two blocks are fastened to the ceiling of an elevator as in Figure 4.28. The elevator accelerates upward at 2.00 m/s^2. Find the tension in each rope.

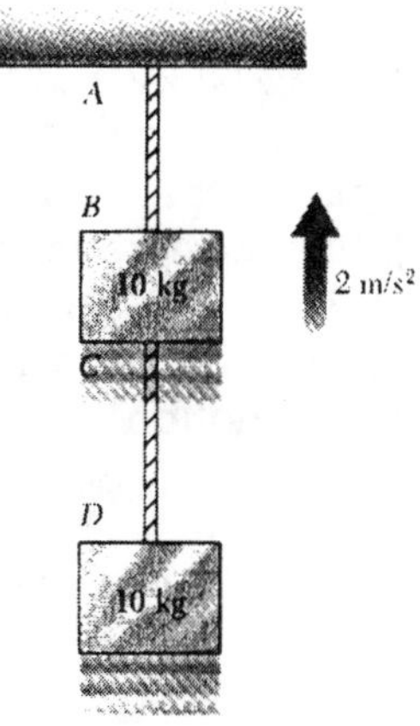

Figure 4.28

Solution First, consider the forces on the upper block. They are $\mathbf{T_1}$, the tension in the cable connecting the block to the ceiling of the elevator; $\mathbf{T_2}$, the tension connecting the blocks; and the weight of the block, 98.0 N.

T_1
Upper Block
$m_1 = 10$ kg
T_2 $w_1 = 98$ N

T_2
Lower Block
$m_2 = 10$ kg
$w_2 = 98$ N

$$\Sigma F_y = ma_y$$

$$T_1 - T_2 - 98.0 \text{ N} = (10.0 \text{ kg})(2.00 \text{ m/s}^2) \quad (1)$$

The forces on the lower block are $\mathbf{T_2}$ and its weight, 98.0 N.

$$\Sigma F_y = ma_y$$

$$T_2 - 98.0 \text{ N} = (10.0 \text{ kg})(2.00 \text{ m/s}^2) \quad (2)$$

From (2), we find $T_2 = 118$ N ◊

and using this in (1), we find $T_1 = 236$ N ◊

---

37. Two masses of 3.00 kg and 5.00 kg are connected by a light string that passes over a frictionless pulley as in Figure 4.29. Determine (a) the tension in the string, (b) the acceleration of each mass, and (c) the distance each mass will move in the first second of motion if both masses start from rest.

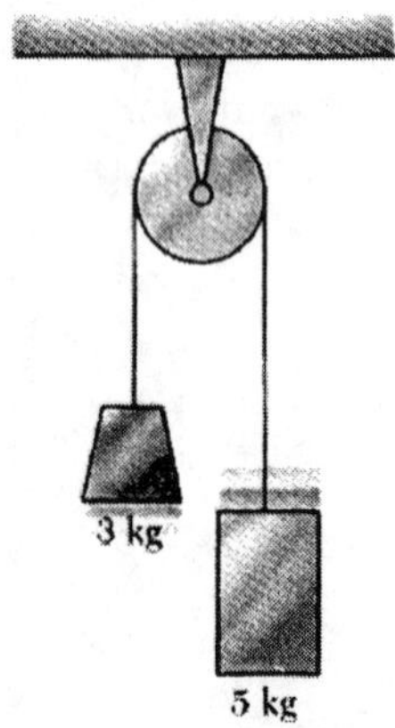

Figure 4.29

Solution First, consider the 3.00-kg rising mass. The forces on it are the tension, **T**, and its weight, 29.4 N. With the upward direction as positive, the second law becomes

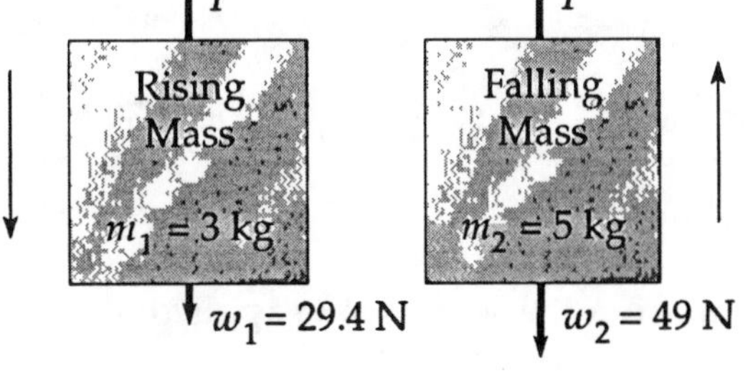

$$\Sigma F_y = ma_y$$

$$T - 29.4 \text{ N} = (3.00 \text{ kg})a \qquad (1)$$

The forces on the falling 5.00-kg mass are its weight and **T**, and its acceleration is the same as that of the rising mass. Calling the positive direction down for this mass, we have

$$F = ma$$

$$49.0 \text{ N} - T = (5.00 \text{ kg})a \qquad (2)$$

Equations (1) and (2) can be solved simultaneously to give

(a) the tension as $T = 36.8 \text{ N}$ ◊

(b) and the acceleration as $a = 2.45 \text{ m/s}^2$ ◊

(c) Consider the 3.00-kg mass. We have

$$y = v_0 t + \frac{1}{2}at^2 = 0 + \frac{1}{2}(2.45 \text{ m/s}^2)(1 \text{ s})^2 = 1.23 \text{ m} \quad ◊$$

---

41. A box of books weighing 300 N is shoved across the floor of an apartment by a force of 400 N exerted downward at an angle of 35.2° below the horizontal. If the coefficient of kinetic friction between box and floor is 0.57, how long does it take to move the box 4.0 m, starting from rest?

Solution In the vertical direction, we have

$$n - 300 \text{ N} - (400 \text{ N})\sin 35.2° = 0$$

from which $n = 530 \text{ N}$

Therefore, $f = \mu_k n = (0.57)(530 \text{ N}) = 302 \text{ N}$

Then, from the second law, applied along the horizontal direction, we have

$$(400 \text{ N})\cos 35.2° - 302 \text{ N} = (30.6 \text{ kg})a_x$$

from which $a_x = 0.80 \text{ m/s}^2$

Then, from $x = v_{0x}t + \frac{1}{2}a_x t^2$

we have $4.0 \text{ m} = 0 + \frac{1}{2}(0.80 \text{ m/s}^2)t^2$

which gives $t = 3.2 \text{ s}$ ◊

Note: The number of significant figures in the result is limited by the value for coefficient of friction.

---

45. The board sandwiched between two other boards in Figure 4.32 weighs 95.5 N. If the coefficient of friction between the boards is 0.663, what must be the magnitude of the compression forces (assume horizontal) acting on both sides of the center board to keep it from slipping?

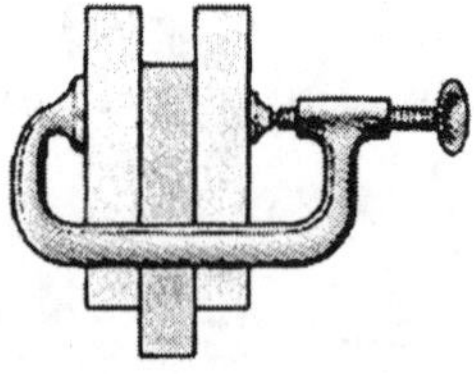

Figure 4.32

Solution The forces on the center board are shown at the right. In the horizontal direction, we see that

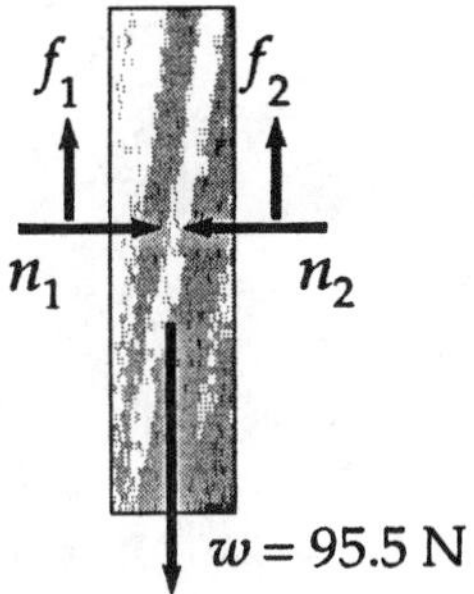

$$n_1 = n_2 = n$$

Therefore, $f_1 = \mu_s n_1$ and $f_2 = \mu_s n_2$

so $f_1 = f_2 = f$

In the vertical direction, we have $\Sigma F_y = 0$, which gives $2f - 95.5 \text{ N} = 0$

Thus, $$f = 47.75 \text{ N}$$

$$n = \frac{f}{\mu_s} = \frac{47.75 \text{ N}}{0.663} = 72.0 \text{ N} \quad \lozenge$$

---

49. A car is traveling at 50 km/h on a flat highway. (a) If the coefficient of friction between road and tires on a rainy day is 0.10, what is the minimum distance in which the car will stop? (b) What is the stopping distance when the surface is dry and the coefficient of friction is 0.60?

Solution

(a) The force of friction is found as $f = \mu_k n = \mu_k m g$

Now, choose the positive direction of the x axis in the direction of f and apply the second law. We have

$$f = ma_x \quad \text{or} \quad a_x = \frac{f}{m} = \mu_k g$$

Now choose the positive direction of the x axis to be in the direction of motion of the car and use $v^2 = v_0^2 + 2ax$, with $v = 0$, $v_0 = 50$ km/(h = 13.9 m/s). We have

$$0 = (13.9 \text{ m/s})^2 + 2(-\mu_k g)x \qquad (1)$$

With $\mu_k = 0.10$, this gives a value for x of $x = 99 \text{ m}$ $\lozenge$

(b) With $\mu_k = 0.60$, (1) above gives $x = 16\text{ m}$ ◊

---

57. A hockey puck is hit on a frozen lake and starts moving with a speed of 12.0 m/s. Five seconds later, its speed is 6.00 m/s. (a) What is its average acceleration? (b) What is the average value of the coefficient of kinetic friction between puck and ice? (c) How far does the puck travel during this 5-s interval?

Solution

(a) $a = \dfrac{v_f - v_0}{t} = \dfrac{6.00\text{ m/s} - 12.0\text{ m/s}}{\text{s}} = -1.20\text{ m/s}^2$ ◊

(b) We also know that $n = mg$. Now apply the second law:

$$-f = m(-1.20\text{ m/s}^2) \quad \text{or} \quad f = m(1.20\text{ m/s}^2) \qquad (1)$$

But, also, $f = \mu_k n = \mu_k mg$. So, from (1), we have

$$\mu_k mg = m(1.20\text{ m/s}^2)$$

so

$$\mu_k = \frac{1.20\text{ m/s}^2}{9.80\text{ m/s}^2} = 0.122 \quad ◊$$

(c) and

$$x = \bar{v}t = \left(\frac{v_0 + v_f}{2}\right)t = \left(\frac{12.0\text{ m/s} + 6.00\text{ m/s}}{2}\right)(5.00\text{ s}) = 45\text{ m} \quad ◊$$

---

65. As a protest against the umpire's calls, a baseball pitcher throws a ball straight up into the air at a speed of 20.0 m/s. In the process, he moves his hand through a distance of 1.50 m. If the ball has a mass of 0.150 kg, find the force he exerts on the ball to give it this upward speed. Note that the force of gravity is acting against the motion of the ball.

Solution The acceleration of the ball is found as

$$v^2 = v_0^2 + 2ax$$

$$(20.0\text{ m/s})^2 = 0 + 2a(1.50\text{ m})$$

From which, $a = 133.3\text{ m/s}^2$

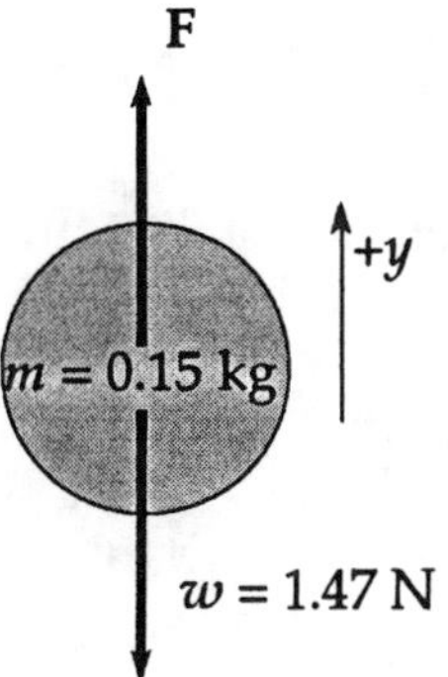

The resultant force on the ball is the upward force, **F**, exerted by the thrower, less the weight of the ball, 1.47 N, downward. The second law becomes

$$F - 1.47\text{ N} = (0.150\text{ kg})(133.3\text{ m/s}^2)$$

$$F = 21.5\text{ N} \quad \lozenge$$

---

75. A 3-kg block starts from rest at the top of a 30° incline and slides 2.00 m down the incline in 1.50 s. Find (a) the acceleration of the block. (b) the coefficient of kinetic friction between it and the incline, (c) the frictional force acting on the block, and (d) the speed of the block after it has slid 2.00 m.

Solution

(a) $x = v_0 t + \frac{1}{2}at^2$ or $2.00\text{ m} = 0 + \frac{1}{2}a_x(1.50\text{ s})^2$ gives $a_x = 1.78\text{ m/s}^2 \quad \lozenge$

(b) Perpendicular to the plane, we have equilibrium, so

$$n = (29.4\text{ N})\cos 30° = 25.5\text{ N}$$

Then $f = \mu_k n$ or $\mu_k = \frac{f}{n} = \frac{9.37\text{ N}}{25.5\text{ N}} = 0.368 \quad \lozenge$

(c) Newton's second law along the incline gives

$$(29.4\text{ N})\sin 30° - f = (3.00\text{ kg})(1.78\text{ m/s}^2)$$

We find $f = 9.37\text{ N} \quad \lozenge$

(d) and finally $v^2 = v_0^2 + 2ax$ becomes $v^2 = 0 + 2(1.78 \text{ m/s}^2)(2.00 \text{ m})$

from which $v = 2.67 \text{ m/s}$ ◊

---

81. In Figure 4.53, the coefficient of kinetic friction between the two blocks is 0.30. The table surface and the pulleys are frictionless. (a) Draw a free-body diagram for each block. (b) Determine the acceleration of each block. (c) Find the tension in the strings.

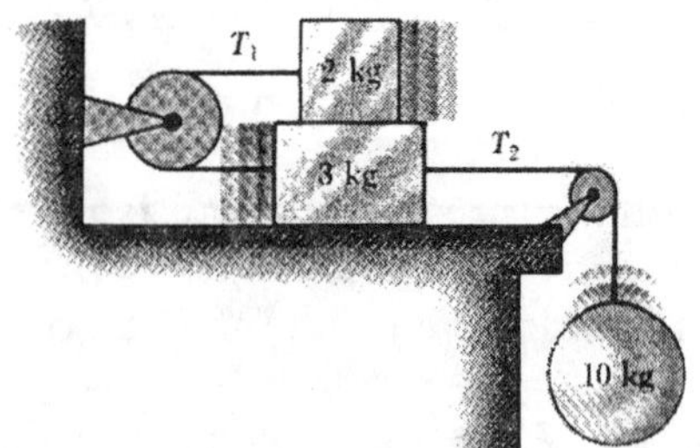

Figure 4.53

Solution

(a)

(b) and (c) $\Sigma F = ma,$ therefore

(1) $T_1 - \mu m_1 g = m_1 a$ or $T_1 - (0.30)(2)(9.80) = 2a$

(2) $T_2 - T_1 - \mu m_1 g = m_2 a$ or $T_2 - T_1 - 2(9.80)(0.30) = 3a$

(3) $T_2 - m_3 g = -m_3 a$ or $T_2 - 10(9.80) = -10a$

Solving the above gives: $T_1 = 17 \text{ N},$ $T_2 = 41 \text{ N},$ and $a = 5.8 \text{ m/s}^2$ ◊

CHAPTER SELF-QUIZ

1. There are six books in a stack, each with a weight of 3 N. The coefficient of friction between each pair of books is 0.1. With what horizontal force must I push to start sliding the top five books off the bottom one?
   a. 0.3 N
   b. 1.5 N
   c. 1.8 N
   d. 2.7 N

2. As a 3.0-kg bucket is being lowered into a 10-m deep well, starting from the top, the tension in the rope is 9.8 N. The time to reach the bottom is
   a. 1.7 s
   b. 2.5 s
   c. 1.4 s
   d. 1.1 s

3. A woman at an airport is pulling a 15-kg suitcase with wheels at constant speed by pulling on a strap at an angle θ above the horizontal. She pulls on the strap with a 30-N force, and the frictional force is 24 N. What is θ?
   a. 30°
   b. 37°
   c. 45°
   d. 53°

4. It is 50 m between telephone poles. When a 1-kg bird lands on the telephone wire midway between the poles, the wire sags 0.2 m. How much tension in the wire does the bird produce? Ignore the weight of the wire.
   a. 1200 N
   b. 600 N
   c. 120 N
   d. 9.8 N

5. A box is dropped onto a conveyor belt moving at 2 m/s. If the coefficient of friction between the box and the belt is 0.3, how long before the box moves without slipping?
   a. 0.7 s
   b. 1.4 s
   c. 2.1 s
   d. 3.0 s

6. Two unequal masses are falling through the air with the heavier one below the lighter one. They are connected by a string. (Ignore air friction.) The tension in the string will be equal to
   a. the weight of the heavier mass
   b. the weight of the lighter mass
   c. the difference in weight of the two masses
   d. zero

7. A 15-kg block rests on a level frictionless surface and is attached by a light string to a 5.0-kg hanging mass where the string passes over a massless frictionless pulley. If $g = 9.8\ m/s^2$, what is the acceleration of the system when released?
   a. $2.45\ m/s^2$
   b. $7.35\ m/s^2$
   c. $3.27\ m/s^2$
   d. $9.8\ m/s^2$

8. A puck, upon being struck by a hockey stick, is given an initial speed of 8 m/s and continues to move in a straight path for a distance of 16 m before coming to rest. What is the coefficient of kinetic friction between puck and ice?
   a. 0.05
   b. 0.1
   c. 0.2
   d. 0.08

9. Find the tension in an elevator cable if the 1000-kg elevator is descending with an acceleration of 1.6 m/s^2, downward.
   a. 5,700 N
   b. 8,200 N
   c. 9,800 N
   d. 11,400 N

10. Two perpendicular forces, one of 30 N directed due north; and the second, 40 N directed due east; act simultaneously on an object with mass of 35 kg. What is the magnitude of the resultant acceleration of the object?
    a 1.4 m/s^2
    b. 155 m/s^2
    c. 3.5 m/s^2
    d. 0.7 m/s^2

11. Two blocks of masses 4 and 6 kg, respectively, rest on a frictionless horizontal surface and are connected by a string. A second string attached only to the 6-kg block, has a horizontal force of 20 N applied to it, causing both blocks to accelerate. What is the tension in the string between the two blocks?
    a. 12 N
    b. 28 N
    c. 8 N
    d. 10 N

12. A 150-N sled is pulled up a 28-degree slope at a constant speed by a force of 100 N parallel to the hill What force directed up the hill will allow the sled to move downhill at a constant speed?
    a. 81 N
    b. 70 N
    c. 30 N
    d. 41 N

5
Work and Energy

Chapter 5

WORK AND ENERGY

Energy is present in the Universe in a variety of forms including mechanical energy, chemical energy, electromagnetic energy, heat energy, and nuclear energy. Although energy can be transformed from one form to another, the total amount of energy in the Universe remains the same. If an isolated system loses energy in some form, then, by the principle of conservation of energy, the system must gain an equal amount of energy in other forms.

In this chapter we are concerned only with mechanical energy. We introduce the concept of *kinetic energy,* which is defined as the energy associated with motion, and the concept of *potential energy,* the energy associated with position. We shall see that the ideas of work and energy can be used in place of Newton's laws to solve certain problems.

We begin by defining *work,* a concept that provides a link between force and energy. With this as a foundation, we can then discuss the principle of conservation of energy and apply it to problems.

NOTES FROM SELECTED CHAPTER SECTIONS

5.1 Work

In order for work to be accomplished, an object must undergo a displacement; the force associated with the work must not be perpendicular to the direction of the displacement. Work is a scalar quantity; the SI unit of work is the newton-meter (N·m) or joule (J).

5.2 Kinetic Energy and the Work-Energy Theorem

Any object which has mass m and speed v has kinetic energy. Kinetic energy is a scalar quantity and has the same units as work and speed; therefore kinetic energy of an object will change only if work is done on the object by some external force. The relationship between work and change in kinetic energy is stated in the work-energy theorem.

5.3 Potential Energy

The work done on an object by the force of gravity is equal to the object's initial potential energy minus its final potential energy. The gravitational potential energy associated with an object depends only on the object's weight and its vertical height above the surface of the Earth. If the height above the surface increases, the potential energy will also increase; but the work done by the gravitational force will be negative. In working problems involving gravitational potential energy, it is necessary to choose an arbitrary reference level (or location) at which the potential energy is zero.

5.4 Conservative and Nonconservative Forces

A force is *conservative* if the work it does on an object moving between two points is independent of the path the object takes between the points. The work done on an object by a conservative force depends only on the initial and final positions of the object.

A force is *nonconservative* if the work it does on an object moving between two points depends on the path taken.

5.5 Conservation of Mechanical Energy

The sum of the kinetic energy plus the potential energy is called the total mechanical energy.

The law of conservation of mechanical energy states that the mechanical energy of a system remains constant if the only forces that do work on the system are *conservative forces*.

5.9 Work Done by a Varying Force

The total work done by a varying force is equal to the area under the force-displacement curve.

EQUATIONS AND CONCEPTS

The work done on a body by a force **F**, which is constant in both magnitude and direction, is defined to be the product of the component of the force in the direction of the displacement and the magnitude of the displacement. Note that the work done by a force can be positive, negative, or zero, depending on the value of θ, the angle between the direction of the force and the direction of the displacement. If

$$W \equiv (F\cos\theta)s \qquad (5.1)$$

$0 \le \theta < 90°$, W is positive; if $90° < \theta < 180°$, W is negative; and if $\theta = 90°$ (**F** perpendicular to **s**), then $W = 0$. In the special case where **F** and **s** are parallel, $\theta = 0$, $\cos\theta = 4$, and $W = Fs$.

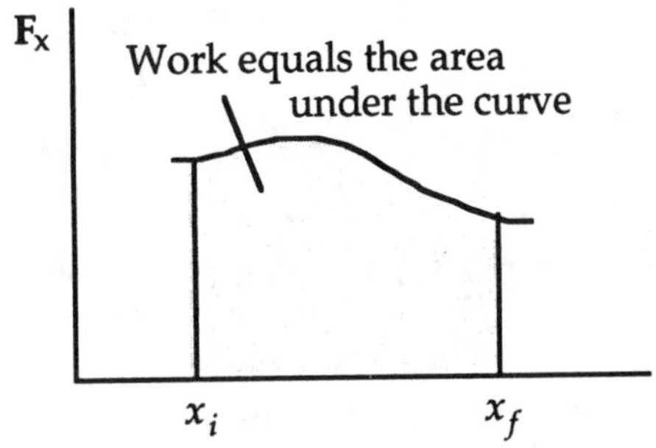

Work is a scalar quantity and the SI unit of work is the newton-meter or joule. See the summary of units of work in Table 5.1 of the textbook.

$$1 \text{ N·m} = 1 \text{ J}$$

When a body of mass m experiences an acceleration due to a *net* force, the work done by the net force can be expressed in terms of the acceleration, mass, and the distance over which the acceleration is achieved.

$$W_{\text{net}} = (ma)s \qquad (5.3)$$

The work done on a body by the net or resultant force can be expressed in terms of the charge in the kinetic energy of the body. Kinetic energy is a scalar quantity and is the energy associated with an object's motion. Equation 5.6 is a statement of the work-energy theorem, where it must be remembered that the work calculated by the equation is the work done by the net or resultant force acting on the body.

$$W_{\text{net}} = \frac{1}{2}mv^2 - \frac{1}{2}mv_0^2 \qquad (5.4)$$

$$KE \equiv \frac{1}{2}mv^2 \qquad (5.5)$$

$$W_{\text{net}} = KE_f - KE_i \qquad (5.6)$$

A conservative force is one for which the work done by the force in moving between any two points is independent of the path followed from the initial to final point. The force of gravity, mg, is an example of a conservative force. The work done on a

$$W_g = mgy_i - mgy_f$$

$$PE \equiv mgy \qquad (5.7)$$

$$W_g = PE_i - PE_f = -\Delta PE \qquad (5.8)$$

body by the force of gravity can be expressed in terms of initial and final values of the body's y-coordinates.

Work done by the gravitational force can also be expressed in terms of the gravitational potential energy function: the work done by the force of gravity equals the negative of the change in the gravitational potential energy function.

The units of energy (kinetic and potential) are the same as the units of work.

Comment on units.

In calculating the work done by the gravitational force, remember that the difference in potential energy between two points is independent of the location of the origin. Choose an origin which is convenient to calculate PE_i and PE_f for a particular situation.

Comment on reference level for potential energy.

When only conservative forces act on a system, the total mechanical energy *(KE + PE)* of the system remains constant; this is a statement of the law of conservation of mechanical energy.

$$KE_i + PE_i = KE_f + PE_f \qquad (5.10)$$

$$KE_i + \sum PE_i = KE_f + \sum PE_f$$

If the only conservative force is the gravitational force, the equation for conservation of mechanical energy takes a special form.

$$\frac{1}{2}mv_i^2 + mgy_i = \frac{1}{2}mv_f^2 + mgy_f \qquad (5.11)$$

The work done by a force in stretching or compressing a spring is stored in the spring as elastic potential energy. For a given displacement from the equilibrium position, the potential energy in the spring depends on the spring constant, k.

$$PE_s \equiv \frac{1}{2}kx^2 \qquad (5.12)$$

If both conservative forces and nonconservative forces act on a system, the total mechanical energy will not remain constant. In this case, the work done by all nonconservative forces equals the change in the total mechanical energy of the system.

$$W_{nc} + W_c = \Delta KE$$

$$W_{nc} = (KE_f - KE_i) + (PE_f - PE_i) \quad (5.14)$$

The average power supplied by a force is the ratio of the work done by the force to the time interval over which the force acts. The average power can also be expressed in terms of the force and the average speed of the object on which the force acts.

$$\overline{P} = \frac{\Delta W}{\Delta t} \quad (5.15)$$

$$\overline{P} = F\overline{v} \quad (5.16)$$

The SI unit of power is the watt; in the British engineering system, the unit of power is the horsepower.

$$1\ \mathrm{W} = 1\ \mathrm{J/s} = 1\ \mathrm{kg \cdot m^2/s^3} \quad (5.17)$$

$$1\ \mathrm{hp} = 550\frac{\mathrm{ft \cdot lb}}{\mathrm{s}} = 746\ \mathrm{W} \quad (5.18)$$

When the force acting on an object does not have a constant magnitude, the work done by the force during a given displacement can be calculated by finding the area under the force versus displacement curve.

$$W \approx \sum F_x \Delta x \quad (5.19)$$

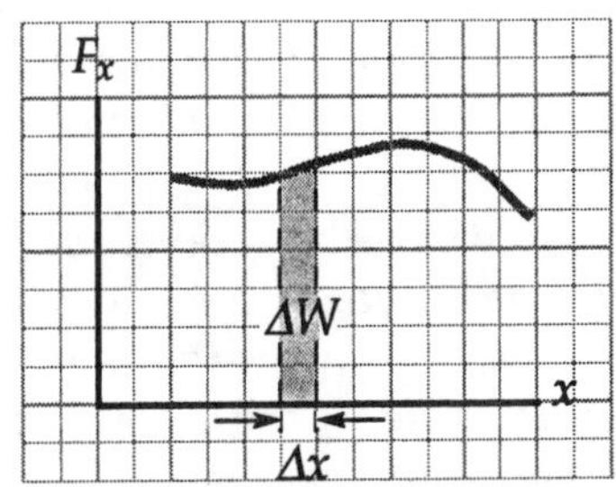

SUGGESTIONS, SKILLS, AND STRATEGIES

Choosing a Zero Level

In working problems involving gravitational potential energy, it is always necessary to choose a location at which the gravitational potential energy is zero. This choice is completely arbitrary because the important quantity is the *difference* in potential energy, and that difference is independent of the location of zero. It is often convenient, but not essential, to choose the surface of the Earth as the reference position for zero potential energy. In most cases, the statement of the problem suggests a convenient level to use.

Conservation of Energy

Take the following steps in applying the principle of conservation of energy:

1. Define your system, which may consist of more than one object.
2. Select a reference position for the zero point of gravitational potential energy.
3. Determine whether or not nonconservative forces are present.
4. If mechanical energy is conserved (that is, if only conservative forces are present), you can write the total initial energy at some point as the sum of the kinetic and potential energies at that point. Then, write an expression for the total final energy, $KE_f + PE_f$, at the final point of interest. Since mechanical energy is conserved, you can equate the two total energies and solve for the unknown.
5. If nonconservative forces such as friction are present (and thus mechanical energy is not conserved), first write expressions for the total initial and total final energies. In this case, the difference between the two total energies is equal to the work done by the nonconservative force(s). That is, you should apply Equation 5.14.

REVIEW CHECKLIST

▷ Define the work done by a constant force and work done by a force which varies with position. (Recognize that the work done by a force can be positive, negative, or zero; describe at least one example of each case.)

▷ Recognize that the gravitational potential energy function, $PE_g = mgy$, can be positive, negative, or zero, depending on the location of the reference level used to measure y. Be aware of the fact that although *PE* depends on the origin of the coordinate system, the *change* in potential energy, $(PE)_f - (PE)_i$, is *independent* of the coordinate system used to define *PE*.

▷ Understand that a force is said to be *conservative* if the work done by that force on a body moving between any two points is independent of the path taken. *Nonconservative* forces are those for which the work done on a particle moving between two points depends on the path. Account for nonconservative forces acting on a system using the work-energy theorem. In this case, the work done by all nonconservative forces equals the change in total mechanical energy of the system.

Understand the distinction between kinetic energy (energy associated with motion), potential energy (energy associated with the position or coordinates of a system), and the total mechanical energy of a system. State the law of conservation of mechanical energy, noting that mechanical energy is conserved only when conservative forces act on a system. This extremely powerful concept is most important in all area of physics.

▷ Relate the work done by the net force on an object to the *change* in kinetic energy. The relation $W = \Delta KE = KE_f - KE_i$ is called the work-energy theorem, and is valid whether or not the (resultant) force is constant. That is, if we know the net work done on a particle as it undergoes a displacement, we also know the *change* in its kinetic energy. This is the most important concept in this chapter, so you must understand it thoroughly.

SOLUTIONS TO SELECTED END-OF-CHAPTER PROBLEMS

5. A shopper in a supermarket pushes a cart with a force of 35.0 N directed at an angle of 25° downward from the horizontal. Find the work done by the shopper as she moves down a 50-m length of aisle.

Solution The component of force along the direction of motion is

$$F\cos\theta = (35.0\text{ N})\cos 25° = 31.7\text{ N}$$

The work done by this force is $W = (F\cos\theta)s = (31.7\text{ N})(50.0\text{ m}) = 1.59\times10^3\text{ J}$ ◊

---

7. Batman, whose mass is 80.0 kg, holds onto the free end of a 12-m rope, the other end of which is fixed to a tree limb above. He puts the rope in motion as only Batman knows how, eventually getting it to swing enough that he can reach a ledge when the rope makes a 60° angle with the downward vertical. How much work is done against the force of gravity in this maneuver?

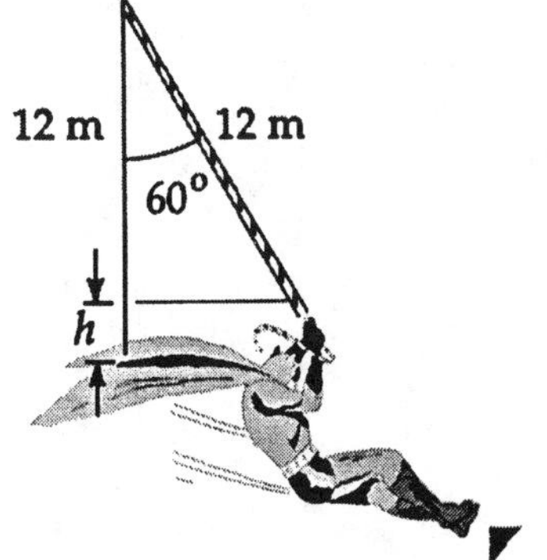

Solution Batman's weight $= mg$
$= (80.0 \text{ kg})(9.80 \text{ m/s}^2)$
$= 784 \text{ N}.$

The work done is equal to the weight of Batman times the vertical height his weight is raised. This height, h, is

$$h = 12.0 \text{ m} - (12.0 \text{ m})\cos 60° = 6.00 \text{ m}$$

Thus, $W = (784 \text{ N})(6.00 \text{ m}) = 4700 \text{ J}$ ◊

---

13. A 0.6-kg particle has a speed of 2.00 m/s at point A and kinetic energy of 7.50 J at point B. What is (a) its kinetic energy at A? (b) its speed at point B? (c) the total work done on the particle as it moves from A to B?

Solution

(a) $KE_A = \frac{1}{2}mv_A{}^2 = \frac{1}{2}(0.600 \text{ kg})(2.00 \text{ m/s})^2 = 1.20 \text{ J}$ ◊

(b) $KE_B = \frac{1}{2}mv_B{}^2 = 7.50 \text{ J} = \frac{1}{2}(0.600 \text{ kg})v_B{}^2$ Thus, $v_B = 5.00 \text{ m/s}$ ◊

(c) $\text{Work} = \Delta KE = KE_B - KE_A = 7.50 \text{ J} - 1.20 \text{ J} = 6.30 \text{ J}$ ◊

---

19. A 10-kg crate is pulled up a rough incline with an initial speed of 1.50 m/s. The pulling force is 100 N parallel to the incline, which makes an angle of 20° with the horizontal. If the coefficient of kinetic friction is 0.40 and the crate is pulled a distance of 5.00 m, (a) how much work is done by the gravitational force? (b) How much work is done by the 100-N force? (c) What is the change in kinetic energy of the crate? (d) What is the speed of the crate after it is pulled 5.00 m?

Solution

(a) The work done by the gravitational force is equal to the product of the weight of the object times the vertical height through which it has been raised. The work is also negative because the force and displacement are in opposite directions $(\cos\theta = \cos 180°)$.

$$W = -ws\sin 20° = -(98.0 \text{ N})(5.00 \text{ m})\sin 20° = -168 \text{ J} \quad ◊$$

(b) W(applied force) = Fs = (100 N)(5.00 m) = 500 J ◊

(c) Use $\Delta KE = W_{net} = W_{grav} + W_{friction} + W_{applied}$

We know W_{grav} and $W_{applied}$ from parts (a) and (b).

$W_f = fs\cos 180^\circ$ where $f = \mu n = \mu mg\cos\theta = 36.8\text{ N}$

so $W_f = (36.8\text{ N})(5.00\text{ m})(-1) = -180\text{ J}$

and therefore $\Delta KE = -168\text{ J} - 180\text{ J} + 500\text{ J} = 150\text{ J}$ ◊

(d) $\Delta KE = \frac{1}{2}mv_f^2 - \frac{1}{2}mv_i^2$ or $v_f = \left[\frac{2}{m}\left(\Delta KE - \frac{1}{2}mv_i^2\right)\right]^{\frac{1}{2}}$

$$v_f = \left[\left(\frac{2}{10.0\text{ kg}}\right)\left(150\text{ J} - \left(\frac{1}{2}\right)(10.0\text{ kg})(1.50\text{ m/s})^2\right)\right]^{\frac{1}{2}} = 5.30\text{ m/s} \quad ◊$$

---

23. A 2-kg ball is attached to a ceiling by a 1-m-long string. The height of the room is 3.00 m. What is the gravitational potential energy of the ball relative to (a) the ceiling? (b) the floor? (c) a point at the same elevation as the ball?

Solution

(a) Relative to the ceiling, $h = -1.00$ m. Thus,

$$PE = mgy = (2.00\text{ kg})(9.80\text{ m/s}^2)(-1.00\text{ m}) = -19.6\text{ J} \quad ◊$$

(b) Relative to the floor, $h = 2.00$ m

$$PE = mgy = (2.00\text{ kg})(9.80\text{ m/s}^2)(2.00\text{ m}) = 39.2\text{ J} \quad ◊$$

(c) Relative to the height of the ball, $h = 0$, so $PE = 0$ ◊

---

27. A 3-kg particle moves from the origin to the position $x = 5.00$ m, $y = 5.00$ m under the influence of gravity acting in the negative y direction (Fig. 5.20). Using Equation 5.1, calculate the work done by gravity in movements from O to C along the paths (a) OAC, (B) OBC, and (c) OC. Your results should all be identical. Why?

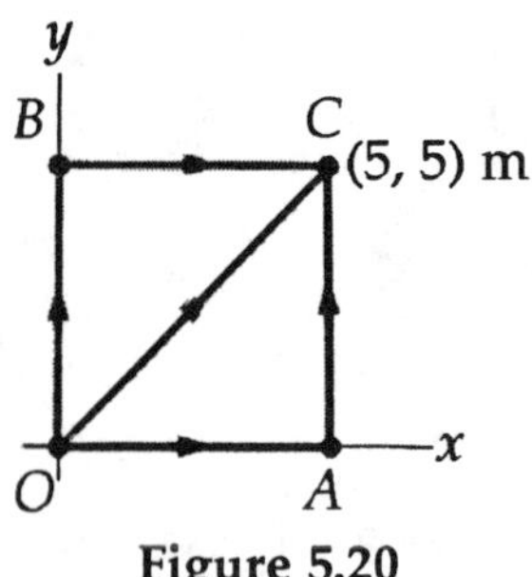

Figure 5.20

Solution $w = mg = (3.00\text{ kg})(9.80\text{ m/s}^2) = 29.4\text{ N}$

(a) $\text{Work along} OAC = \text{work along} OA + \text{work along} AC$

$$= w(OA)\cos 270° + w(AC)\cos 180°$$

$$= (29.4\text{ N})(5.00\text{ m})(0) + (29.4\text{ N})(5.00\text{ m})(-1) = -147\text{ J} \quad \Diamond$$

(b) $\text{Work along} OBC = w \text{ along} OB + w \text{ along} BC$

$$= (29.4\text{ N})(5.00\text{ m})\cos 180° + (29.4\text{ N})(5.00\text{ m})\cos 90° = -147\text{ J} \quad \Diamond$$

(c) $\text{Work along} OC = w(OC)\cos 135° = (29.4\text{ N})\left(5\sqrt{2}\text{ m}\right)(-0.707) = -147\text{ J} \quad \Diamond$

The results should all be the same since gravitational forces are conservative.

---

33. Tarzan swings on a 30-m-long vine initially inclined at an angle of 37° with the vertical. What is his speed at the bottom of the swing (a) if he starts from rest? (b) if he pushes off with a speed of 4.00 m/s?

Solution (a) We choose the zero level for potential energy at the bottom of the arc. The initial height of Tarzan above this level is shown in the sketch to be

$$(30.0\text{ m})(1 - \cos 37°) = 6.04\text{ m}$$

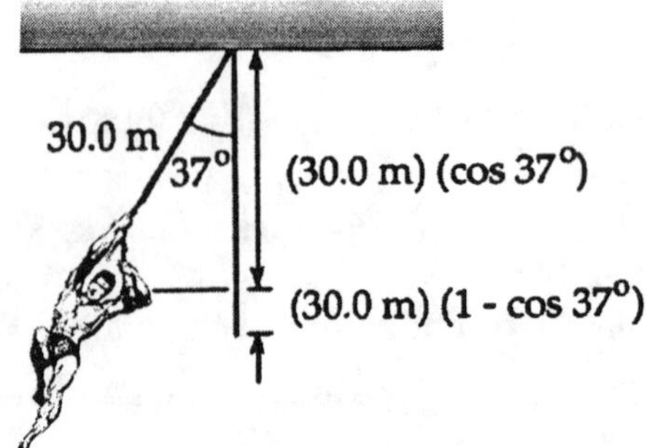

We use conservation of mechanical energy.

$$\frac{1}{2}mv_i^2 + mgy_i = \frac{1}{2}mv_f^2 + mgy_f$$

$$0 + mg(6.04\text{ m}) = \frac{1}{2}mv^2 + 0 \qquad \text{from which} \qquad v = 10.9\text{ m/s} \quad \Diamond$$

(b) In this case, conservation of mechanical energy becomes

$$\frac{1}{2}mv_i^2 + mgy_i = \frac{1}{2}mv_f^2 + mgy_f$$

$$\frac{1}{2}m(4.00\text{ m/s})^2 + mg(6.04\text{ m}) = \frac{1}{2}mv^2 + 0 \qquad \text{and} \qquad v = 11.6\text{ m/s} \quad \Diamond$$

---

39. A 25-kg child on a 2-m-long swing is released from rest when the swing supports make an angle of 30° with the vertical. (a) Neglecting friction, find the child's speed at the lowest position. (b) If the speed of the child at the lowest position is 2.00 m/s, what is the mechanical energy lost due to friction?

Solution

(a) The initial vertical height of the child above the zero level for gravitational energy at the bottom of the arc is $(2.00\text{ m})(1-\cos 30°) = 0.268\text{ m}$ (See problem 33.). In the absence of friction, we use conservation of mechanical energy as

$$\frac{1}{2}mv_i^2 + mgy_i = \frac{1}{2}mv_f^2 + mgy_f$$

$$v_i = 0 \qquad \text{and} \qquad y_f = 0, \qquad \text{therefore} \qquad v_f = \sqrt{2gy_i}$$

or

$$v_f = \left[(2)(9.80\text{ m/s}^2)(0.268\text{ m})\right]^{\frac{1}{2}} = 2.29\text{ m/s} \quad \Diamond$$

(b) In the presence of friction, we use $W_{nc} = \Delta KE + \Delta PE$

$$W_{nc} = \frac{1}{2}mv_f^2 - \frac{1}{2}mv_i^2 + mgy_f - mgy_i$$

$$W_{nc} = \frac{1}{2}(25.0\text{ kg})(2.00\text{ m/s})^2 - 0 + 0 - (25.0\text{ kg})g(0.268\text{ m}) = -15.6\text{ J} \quad \Diamond$$

---

45. A 1500-kg car accelerates uniformly from rest to 10.0 m/s in 3.00 s. Find (a) the work done on the car in this time interval, (b) the average power delivered by the engine in this time interval, and (c) the instantaneous power delivered by the engine at t = 2.00 s.

Solution (a) We use the work-kinetic energy theorem as

$$W = \frac{1}{2}mv^2 - \frac{1}{2}mv_0^2$$

so

$$W = \frac{1}{2}mv^2 - 0 = \frac{1}{2}(1500\text{ kg})(10.0\text{ m/s})^2 = 7.50 \times 10^4\text{ J} \quad \Diamond$$

(b) The average power is given by

$$\overline{P} = \frac{W}{\Delta t} = 7.50 \times 10^4\text{ J/3 s} = 2.50 \times 10^4\text{ W} = 33.5\text{ hp} \quad \Diamond$$

(c) $a = \frac{\Delta v}{\Delta t} = \frac{10.0\text{ m/s} - 0}{3.00\text{ s}} = 3.33\text{ m/s}^2$

$F = ma$ becomes $F = (1500\text{ kg})(3.33\text{ m/s}^2) = 5000\text{ N}$

at $t = 2.00$ s, $v = v_0 + at$ gives $v = 0 + (3.33\text{ m/s}^2)(2.00\text{ s}) = 6.66\text{ m/s}$

so

$$P_{\text{(instantaneous)}} = Fv = (5000\text{ N})(6.66\text{ m/s}) = 3.33 \times 10^4\text{ W} = 44.7\text{ hp} \quad \Diamond$$

---

53. A 650-kg elevator starts from rest. It moves upward for 3.00 s with constant acceleration until it reaches its cruising speed, 1.75 m/s. (a) What is the average power of the elevator motor during this period? (b) How does this compare with its power during an upward cruise with constant speed?

Solution

(a) First, calculate the acceleration of the elevator,

$$a = \frac{v_f - v_0}{t} = \frac{1.75\text{ m/s} - 0}{3\text{ s}} = 0.583\text{ m/s}^2$$

The vertical forces on the elevator (assuming no frictional forces are present) are the upward tension, T, in the cable (force exerted by the motor) and the downward force of gravity. From the second law $\Sigma F = ma$ and $T - mg = ma$ so $T = m(a+g)$ or

$$T = (650\text{ kg})(0.583 + 9.80)\text{ m/s}^2 = 6750\text{ N}$$

$$\overline{P} = T\overline{v} = (6750\text{ N})\left(\frac{1.75\text{ m/s} + 0}{2}\right) = 5.91\times10^3\text{ W} \quad \text{or} \quad \overline{P} = 7.92\text{ hp} \quad \Diamond$$

(b) When moving upward at constant speed (v = 1.75 m/s), the applied force equals the weight = (650 kg)(9.80 m/s^2) = 6370 N. Therefore,

$$P = Fv = (6370\text{ N})(1.75\text{ m/s}) = 11{,}100\text{ W} = 14.9\text{ hp} \quad \Diamond$$

---

55. A 1-kg object initially at rest is acted upon by a nonconstant force that causes it to move 3.00 m. The force varies with position as shown in Figure 5.23. (a) How much work is done on the object by this force? (b) What is the speed of the object at x = 3.00 m?

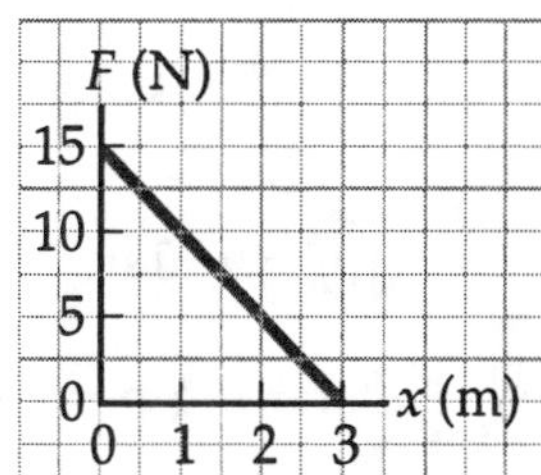

Figure 5.23

Solution

(a) The work done by a variable force equals the area under the force versus displacement graph. This area is in the shape of a triangle. Therefore,

$$W = \frac{1}{2}(\text{base})(\text{height}) = \frac{1}{2}(15.0\text{ N})(3.00\text{ m}) = 22.5\text{ J} \quad \Diamond$$

(b) But we know from the work-kinetic energy theorem that

$$W = \frac{1}{2}mv^2 - \frac{1}{2}mv_0^2 \qquad \text{or} \qquad 22.5\text{ J} = \frac{1}{2}(1\text{ kg})v^2 - 0$$

Thus, $v = 6.71\text{ m/s}$ ◊

---

65. If the wire in Problem 61 (Fig. 5.28) is frictionless between points A and B and rough between B and C, and if the bead starts from rest at A, (a) find its speed at B. (b) If the bead comes to rest at C, find the loss in mechanical energy as it goes from B to C.

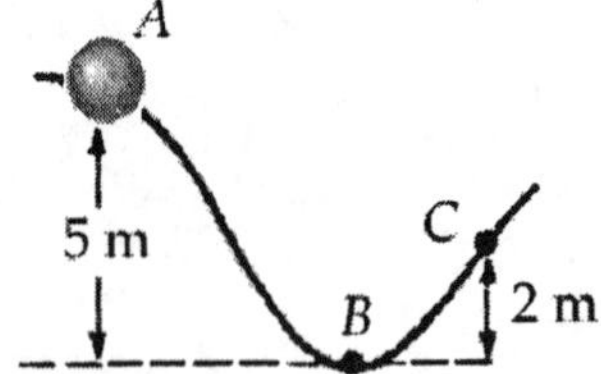

Figure 5.28

Solution

(a) Choose the zero level for potential energy at the level of B. Between A and B we can use

$$\frac{1}{2}mv_i^2 + mgy_i = \frac{1}{2}mv_f^2 + mgy_f$$

$$0 + (0.4\text{ kg})g(5.00\text{ m}) = \frac{1}{2}(0.400\text{ kg})v_B^2 + 0$$

$$v_B = 9.90\text{ m/s} \quad ◊$$

(b) We choose the starting point at B, the zero level at B, and the end point at C.

$$W_{nc} = \frac{1}{2}mv_f^2 - \frac{1}{2}mv_i^2 + mgy_f - mgy_i$$

$$W_{nc} = 0 - \frac{1}{2}(0.400\text{ kg})(9.90\text{ m/s})^2 = (0.400\text{ kg})g(2.00\text{ m}) - 0$$

$$W_{nc} = -11.8\text{ J} \quad ◊$$

---

69. (a) A 75-kg man jumps from a window 1.00 m above a sidewalk. What is his speed just before his feet strike the pavement? (b) If the man jumps with his knees and ankles locked, the only cushion for his fall is an approximately 0.500-cm give in the pads of his feet. Calculate the average force exerted on him by the ground in this situation. This average force is sufficient to cause cartilage damage in the joints or to break bones.

Solution

(a) $\frac{1}{2}mv_i^2 + mgy_i = \frac{1}{2}mv_f^2 + mgy_f$; $\qquad 0 + (75.0 \text{ kg})g(1.00 \text{ m}) = \frac{1}{2}(75.0 \text{ kg})v_f^2 + 0$

Thus, the velocity with which he hits is $v_f = 4.43 \text{ m/s}$ ◊

(b) $W_{nc} = -Fs = -F(5\times10^{-3} \text{ m})$ where **F** is the force exerted on the man by the floor.

$$W_{nc} = \frac{1}{2}mv_f^2 - \frac{1}{2}mv_i^2 + mgy_f - mgy_i$$

$$-F(5\times10^{-3} \text{ m}) = 0 - \frac{1}{2}(75.0 \text{ kg})(4.43 \text{ m/s})^2 + 0 - (75.0 \text{ kg})g(5\times10^{-3} \text{ m})$$

Thus, $\qquad F = 1.47\times10^5 \text{ N}$ ◊

---

75. A projectile of mass m is shot horizontally with an initial speed of v_0 from a height of h above a flat desert surface. At the instant before the projectile hits the surface, find (a) the work done on the projectile by gravity, (b) the change in kinetic energy since the projectile was fired, and (c) the final kinetic energy of the projectile.

Solution

(a) The work done on the projectile by gravity is the product of the weight and the vertical distance through which the projectile falls. Thus, $W = mgh$ ◊

(b) $W = \Delta KE = mgh$ ◊

(c) $\Delta KE = KE_f - KE_i = W$ so $KE_f = W + KE_i = mgh + \frac{1}{2}mv_0^2$ ◊

CHAPTER SELF-QUIZ

1. A ball hits a wall and bounces back at half the original speed. What part of the original kinetic energy of the ball did it lose in the collision?
   a. 1/4
   b. 1/2
   c. 3/4
   d. It did not lose kinetic energy

2. I decide to make a pile out of the 100 boards that are lying out in the yard, so I stack them on top of each other. Each board is 0.05 m thick and weights 40 N. If the boards originally had zero potential energy, what is the potential energy of the new pile?
   a. 200 J
   b. 100 J
   c. 20000 J
   d. 10000 J

3. A 3-kg object starting at rest falls from a height of 10 m to the ground. In this instance, the force of the air is not negligible so that the magnitude of work done by this frictional force is 20 J. What is the object's kinetic energy prior to hitting the ground?
   a. 78 J
   b. 118 J
   c. 49 J
   d. 274 J

4. A 40-N crate is pulled 5 m up an inclined plane at a constant velocity. If the plane is inclined at an angle of 37° to the horizontal, what is the magnitude of the work done on the crate by the force of gravity?
   a. 120 J
   b. 6 J
   c. 1180 J
   d. 200 J

5. A worker pushes a 250-N weight wheelbarrow up a ramp 6.00 m in length and inclined at 20° with the horizontal. What potential energy change does the wheelbarrow experience?
   a. 513 J
   b. 1500 J
   c. 3550 J
   d. 4500 J

6. A 20-N crate starting at rest slides down a rough 3-m long ramp, inclined at 30° with the horizontal. The force of friction between crate and ramp is 6 N. What is the kinetic energy of the crate at the bottom of the ramp?
   a. zero
   b. 8 J
   c. 12 J
   d. 32 J

7. A baseball catcher puts on an exhibition by catching a 0.15-kg ball dropped from a helicopter at a height of 61 m. If the catcher "gives" with the ball for a distance of 0.75 m while catching it, what average force is exerted on the mitt by the ball? ($g = 9.80\ m/s^2$)
   a. 47 N
   b. 94 N
   c. 119 N
   d. 376 N

8. A simple pendulum, 2.00 m in length, is released from rest when the support string is at an angle of 25° from the vertical. What is the speed of the suspended mass at the bottom of the swing? (Ignore air resistance, $g = 9.80\ m/s^2$.)
   a. 0.60 m/s
   b. 0.80 m/s
   c. 1.30 m/s
   d. 1.90 m/s

9. A 1500-kg car travels along a highway at a speed of 20 m/s. What is its kinetic energy?
   a. 15.0×10^5 J
   b. 2.5×10^5 J
   c. 3.0×10^5 J
   d. 6.0×10^5 J

10. A pulley-cable system on a crate hoists a bucket of cement with a total weight of 20 000 N to a height of 40 m. If this is accomplished in 2 min., what is the power output by the pulley-cable system?
    a. 6.7 kW
    b. 3.3 kW
    c. 13.3 kW
    d. 400 kW

11. A horizontal force of 200 N is applied to move a 55-kg cart across a 10-m level surface. If the cart accelerates at 2.00 m/s^2, starting from rest, then what kinetic energy does it gain while moving the 10-m distance?
    a. 2200 J
    b. 1100 J
    c. 550 J
    d. 880 J

12. A 15.0-kg crate, initially at rest, slides down a ramp 2.00 m long and inclined at an angle of 20° with the horizontal. If there is a constant force of kinetic friction of 25 N between the crate and ramp, what kinetic energy would the crate have at the bottom of the ramp? (g = 9.80 m/s^2)
    a. 50 J
    b. 638 J
    c. 171 J
    d. 5.2 J

6
Momentum and Collisions

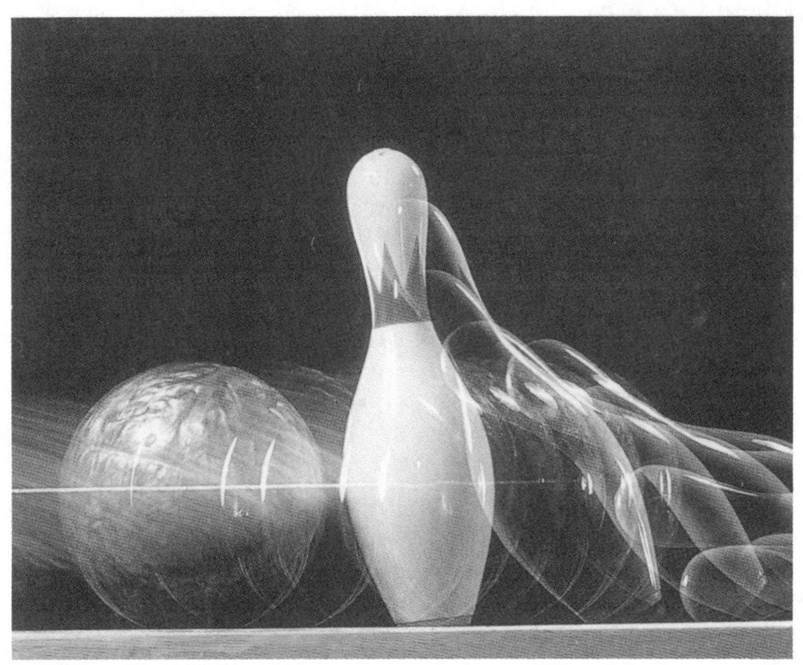

Chapter 6

MOMENTUM AND COLLISIONS

One of the main objectives of this chapter is to help you understand how to analyze collisions and other events which involve objects which experience forces and accelerations over very short time intervals. As a first step, we shall introduce the term *momentum*. We often use the concept of momentum which is defined as the product of mass and velocity in describing objects in motion.

The concept of momentum leads us to a second conservation law: conservation of momentum. This law is especially useful for treating problems that involve collisions between objects.

NOTES FROM SELECTED CHAPTER SECTIONS

6.1 Momentum and Impulse

The *time rate of change of the momentum* of a particle is equal to the resultant force on the particle. The *impulse* of a force equals the change in momentum of the particle on which the force acts. Under the *impulse approximation,* it is assumed that one of the forces acting on a particle is of short time duration but of much greater magnitude than any of the other forces.

6.2 Conservation of Momentum

If two particles of masses m_1 and m_2 form an *isolated system,* then the total momentum of the system remains constant.

6.3 Collisions

For *any type of collision,* the total momentum before the collision equals the total momentum just after the collision.

In an *inelastic collision,* the total momentum is conserved; however, the total kinetic energy is not conserved.

In a *perfectly inelastic* collision, the two colliding objects stick together following the collision.

In an *elastic collision,* both momentum and kinetic energy are conserved.

6.4 Glancing Collisions

The law of conservation of momentum is not restricted to one-dimensional collisions. If two masses undergo a *two-dimensional* (glancing) *collision* and there are no external forces acting, the total momentum in each of the x, y, and z directions is conserved.

EQUATIONS AND CONCEPTS

An object of mass, m, and velocity, **v**, is characterized by a vector quantity called linear momentum. The SI units of linear momentum are kg·m/s. Since momentum is a vector quantity, the defining equation can be written in component form.

$$\mathbf{p} \equiv m\mathbf{v} \tag{6.1}$$

Since momentum is a vector quantity, the components of the vector along the x direction and the y direction can be calculated.

$$p_x = mv_x$$

$$p_y = mv_y$$

The resultant force acting on an object equals the time rate of change of the object's momentum. This equation is a mathematical expression of Newton's second law.

$$\Sigma\mathbf{F} = \frac{\Delta\mathbf{p}}{\Delta t} \tag{6.2}$$

Note that as a special case in Equation 6.2, if the resultant force $\Sigma\mathbf{F} = 0$, then the momentum does not change.

Comment.

This equation is a mathematical statement of the important impulse-momentum theorem. The product of the force acting on an object and the time interval during which it acts is called the impulse imparted to the object by the force.

$$\mathbf{F}\Delta t = m\mathbf{v}_f - m\mathbf{v}_i \tag{6.4}$$

The impulse imparted by a force during a time interval Δt is equal to the area under the force-time graph from the beginning to the end of the time interval. When the force varies in time, as illustrated in the figure, it is often convenient to define an average force, $\overline{\mathbf{F}}$, which is a constant force that imparts the same impulse in a time interval Δt as the actual time varying force.

Comment on impulse of force.

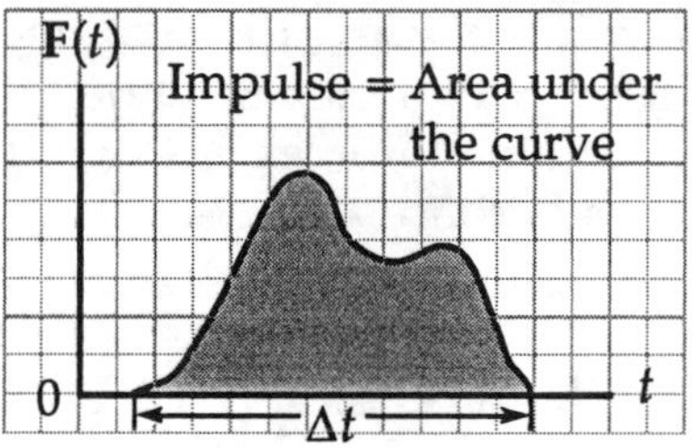

When two objects interact in a collision (and exert forces on each other) and no external forces act on the two-object system, the total momentum of the system before the collision equals the total momentum after the collision.

$$m_1\mathbf{v}_{1i} + m_2\mathbf{v}_{2i} = m_1\mathbf{v}_{1f} + m_2\mathbf{v}_{2f} \qquad (6.5)$$

In general, when the external force acting on any system of objects is zero, the linear momentum of the system is conserved.

Comment on law of conservation of momentum.

An *elastic collision* is one in which both momentum and kinetic energy are conserved.

Comment on types of collisions.

An *inelastic collision* is one in which momentum is conserved but kinetic energy is not.

A *perfectly inelastic collision* is a collision in which the colliding objects stick together so that their final velocities are equal.

Conservation of momentum for a one-dimensional perfectly inelastic collision.

$$m_1 v_{1i} + m_2 v_{2i} = (m_1 + m_2) v_f \qquad (6.6)$$

Note that v_{1i}, v_{2i}, and v_f are actually components of velocity vectors and hence they may have positive or negative values --determined by the direction of motion relative to the chosen coordinate origin.

Comment on signs.

In an elastic head-on collision, both momentum and kinetic energy are conserved.

$$m_1v_{1i} + m_2v_{2i} = m_1v_{1f} + m_2v_{2f} \qquad (6.7)$$

$$\tfrac{1}{2}m_1v_{1i}^{\ 2} + \tfrac{1}{2}m_2v_{2i}^{\ 2} = \tfrac{1}{2}m_1v_{1f}^{\ 2} + \tfrac{1}{2}m_2v_{2f}^{\ 2} \qquad (6.8)$$

This equation states another characteristic of a perfectly elastic collision: the relative velocity of the two objects before the collision equals the negative of the relative velocity of the two objects after the collision.

$$v_{1i} - v_{2i} = -(v_{1f} - v_{2f}) \qquad (6.11)$$

In a glancing collision, the colliding masses rebound along a direction which is not parallel to the direction of motion before the collision. This is an example of a two-dimensional collision. If no net external force acts on the system of colliding masses, momentum is conserved along each coordinate direction.

$$\sum p_{ix} = p_{fx} \qquad (6.12)$$

$$\sum p_{iy} = p_{fy} \qquad (6.13)$$

SUGGESTIONS, SKILLS, AND STRATEGIES

The following procedure is recommended when dealing with problems involving collisions between two objects:

1. Set up a coordinate system and define your velocities with respect to that system. That is, objects moving in the direction selected as the positive direction of the x axis are considered as having a positive velocity and negative if moving in the negative x direction. It is convenient to have the x axis coincide with one of the initial velocities.

2. In your sketch of the coordinate system, draw all velocity vectors with labels and include all the given information.

3. Write expressions for the momentum of each object before and after the collision. (In two-dimensional collision problems, write expressions for the x and y components of momentum before and after the collision.) Remember to include the appropriate signs for the velocity vectors.

4. Now write expressions for the *total* momentum *before* and *after* the collision and equate the two. (For two-dimensional collisions, this expression should be written for the momentum in both the x and y directions.) It is important to emphasize that it is the momentum of the *system* (the two colliding objects) that is conserved, not the momentum of the individual objects.

5. If the collision is *inelastic*, you should then proceed to solve the momentum equations for the unknown quantities.

6. If the collision is *elastic*, kinetic energy is also conserved, so you can equate the total kinetic energy before the collision to the total kinetic energy after the collision. This gives an additional relationship between the various velocities. The conservation of kinetic energy for elastic collisions leads to the expression $v_{l1} - v_{2i} = -(v_{1f} - v_{2f})$, which is often easier to use in solving elastic collision problems than is an expression for conservation of kinetic energy.

REVIEW CHECKLIST

- ▷ The impulse of a force acting on a particle during some time interval equals the *change* in momentum of the particle, and the impulse equals the area under the Force-Time graph.

- ▷ The momentum of any isolated system (one for which the net external force is zero) is conserved, regardless of the nature of the forces between the masses which comprise the system.

- ▷ There are two types of collisions that can occur between two particles, namely elastic and inelastic collisions. Recognize that a *perfectly* inelastic collision is an inelastic collision in which the colliding particles stick together after the collision, and hence move as a composite particle.

▷ The conservation of linear momentum applies not only to head-on collisions (one-dimensional), but also to glancing collisions (two- or three-dimensional). For example, in a two-dimensional collision, the total momentum in the x direction is conserved and the total momentum in the y direction is conserved.

▷ The equations for momentum and kinetic energy can be used to calculate the final velocities in a two-body head-on elastic collision; and to calculate the final velocity and the change of kinetic energy in a two-body system for a completely inelastic collision.

SOLUTIONS TO SELECTED END-OF-CHAPTER PROBLEMS

1. Calculate the magnitude of the linear momentum for the following cases:

(a) a proton with mass 1.67×10^{-27} kg, moving with a speed of 5.00×10^{6} m/s;
(b) a 15.0-g bullet moving with a speed of 300 m/s;
(c) a 75.0-kg sprinter running with a speed of 10.0 m/s;
(d) the Earth (mass = 5.98×10^{24} kg) moving with an orbital speed equal to 2.98×10^{4} m/s.

Solution Use $p = mv$:

(a) $p=(1.67\times10^{-27}\text{ kg})(5.00\times10^{6}\text{ m/s})=8.35\times10^{-21}\text{ kg}\cdot\text{m/s}$ ◊

(b) $p=(1.50\times10^{-2}\text{ kg})(3.00\times10^{2}\text{ m/s})=4.50\text{ kg}\cdot\text{m/s}$ ◊

(c) $p=(75.0\text{ kg})(10.0\text{ m/s})=750\text{ kg}\cdot\text{m/s}$ ◊

(d) $p=(5.98\times10^{24}\text{ kg})(2.98\times10^{4}\text{ m/s})=1.78\times10^{29}\text{ kg}\cdot\text{m/s}$ ◊

---

11. Find the average force exerted on the particle graphed in Figure 6.20 (on the following page) for the time interval $t_i = 0$ to $t_f = 5.0$ s.

Solution Impulse = area under curve = (two triangular areas each of altitude 4.0 N and base 2.0 s) + (one rectangular area of width 1.0 s and height of 4.0 N). Thus,

$$\text{Impulse} = 2\left[\frac{1}{2}(2.0\text{ s})(4.0\text{ N})\right]+(1.0\text{ s})(4.0\text{ N})=12\text{ N}\cdot\text{s}$$

The average force, $\overline{\mathbf{F}}$, is that constant force which, if acting during the entire 5.0-s interval, would result in the same impulse (area under the curve) as the actual force. Therefore,

$$\overline{F} = \frac{\text{impulse}}{\text{time}} = \frac{12\ \text{N}\cdot\text{s}}{5.0\ \text{s}} = 2.4\ \text{N} \quad \Diamond$$

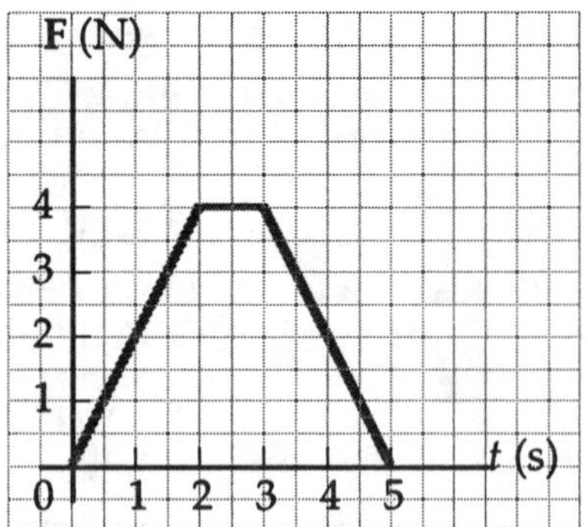

Figure 6.20

---

13. A 0.50-kg object is at rest at the origin of a coordinate system. A 3.0-N force in the $+x$ direction acts on the object for 1.5 s. (a) What is the velocity at the end of this time? (b) At the end of this time, a constant force of 4.0 N is applied in the $-x$ direction for 3.0 s. What is the velocity at the end of the 3.0 s?

Solution (a) From $F\Delta t = \Delta p = mv_f - mv_i$, we have $(3.0\ \text{N})(1.5\ \text{s}) = (0.50\ \text{kg})v - 0$, or $v = 9.0\ \text{m/s}$ ◊

(b) Again use $F\Delta t = \Delta p = mv_f - mv_i$.

$$(-4.0\ \text{N})(3.0\ \text{s}) = (0.50\ \text{kg})v - (0.50\ \text{kg})(9.0\ \text{m/s}) \quad \text{or} \quad v = -15.0\ \text{m/s} \quad \Diamond$$

---

25. A 1.20-kg skateboard is coasting along the pavement as a speed of 5.00 m/s when a 0.800-kg cat drops from a tree vertically downward onto the skateboard. What is the speed of the skateboard-cat combination?

Solution The momentum of the cat is initially along the y-direction. Therefore, if we require conservation of momentum along the horizontal direction,

$$P_x(\text{before}) = P_x(\text{after})$$

$(1.20\ \text{kg})(5.00\ \text{m/s}) = (1.20\ \text{kg} + 0.800\ \text{kg})v$ and for the combination, $v = 3\ \text{m/s}$ ◊

---

27. A 3.00-kg sphere makes a perfectly inelastic collision with a second sphere that is initially at rest. The composite system moves with a speed equal to one third the original speed of the 3.00-kg sphere. What is the mass of the second sphere?

Solution No *external* forces act on the system of two spheres. Therefore,

$$P(\text{after}) = P(\text{before})$$

$$(3.00 \text{ kg} + \text{M})\frac{v}{3} = (3.00 \text{ kg})v$$

Cancel v and solve for M: $M = 6.00 \text{ kg}$ ◊

---

29. A 10.0-g object moving to the right at 20.0 cm/s makes an elastic head-on collision with a 15.0-g object moving in the opposite direction at 30.0 cm/s. Find the velocity of each object after the collision.

Solution First, require conservation of momentum:

$$m_1v_{1i} + m_2v_{2i} = m_1v_{1f} + m_2v_{2f}$$

$$(10.0 \text{ g})(20.0 \text{ cm/s}) + (15.0 \text{ g})(-30.0 \text{ cm/s}) = (10.0 \text{ g})v_{1f} + (15.0 \text{ g})v_{2f} \qquad (1)$$

Now use Equation 6.11 which states that, in a perfectly elastic head-on collision, the *relative* velocity of the two objects before the collision equals the relative velocity after the collision.

$$v_{1i} - v_{2i} = -(v_{1f} - v_{2f})$$

$$20.0 \text{ cm/s} - (-30.0 \text{ cm/s}) = -(v_{1f} - v_{2f}) \qquad (2)$$

Solve (1) and (2) simultaneously:

$$v_{1f} = -40.0 \text{ cm/s} \quad \text{and} \quad v_{2f} = 10.0 \text{ cm/s}$$ ◊

---

41. A 0.40-kg soccer ball approaches a player horizontally with a speed of 15.0 m/s. The player illegally strikes the ball with her hand and causes it to move in the opposite direction with a speed of 22.0 m/s. What impulse was delivered to the ball by the player?

Solution Impulse = $F\Delta t = \Delta p = mv_f - mv_i$

Impulse = $0.400\text{ kg}(-22.0\text{ m/s} - 15.0\text{ m/s}) = -14.8\text{ kg}\cdot\text{m/s}$

Thus, an impulse of $14.8\text{ kg}\cdot\text{m/s}$ in the direction opposite the initial velocity is delivered to the ball. ◊

---

45. An 80-kg man drops from a 3.0-m diving board. Two (2.0) seconds after reaching the water, the man comes to rest. What average force did the water exert on him?

Solution We will first find the speed as he enters the water. We select the origin of the coordinate system at the initial position of the diver and calculate his speed as he enters the water.

$$v^2 = v_0^2 - 2gy$$

$$v^2 = 0 - (2)(9.8\text{ m/s}^2)(-3.0\text{ m}) \quad \text{and} \quad v = -7.67\text{ m/s}$$

Now consider the deceleration period. From the impulse-momentum theorem, we have

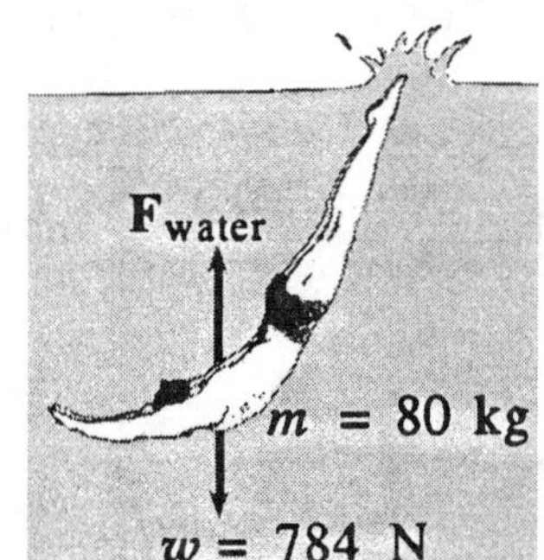

$$\overline{F}\Delta t = \Delta p = mv_f - mv_i$$

$$\overline{F}(2.0\text{ s}) = 0 - (80\text{ kg})(-7.67\text{ m/s})$$

$$\overline{F} = 3.07 \times 10^2\text{ N}$$

This is the net force acting on the man. But

$$\overline{F} = \overline{F}_{\text{water}} - mg = \overline{F}_{\text{water}} - 784\text{ N} = 307\text{ N}$$

Thus, $\overline{F}_{\text{water}} = 1.10 \times 10^3\text{ N}$ ◊

---

49. An 80.0-kg astronaut is working on the engines of his ship, which is drifting through space with a constant velocity. The astronaut, wishing to get a better view of the Universe, pushes against the ship and later finds himself 30.0 m behind the ship and moving so slowly that he can be considered to be at rest Without a thruster, the only way to return to the ship is to throw his 0.500-kg wrench directly away from the ship. If he throws the wrench with a speed of 20.0 m/s, how long does it take him to reach the ship?

Solution No external force acts on the system (astronaut plus wrench), so the total momentum is constant. Since the final momentum (wrench plus astronaut) must be zero, we have final momentum = initial momentum = 0, or

$$m_{\text{wrench}}v_{\text{wrench}} + m_{\text{astronaut}}v_{\text{astronaut}} = 0$$

Thus $$v_{\text{astronaut}} = -\frac{m_{\text{wrench}}v_{\text{wrench}}}{m_{\text{astronaut}}} = -\frac{(0.500\text{ kg})(20.0\text{ m/s})}{80.0\text{ kg}} = -0.125\text{ m/s}$$

or, the astronaut drifts toward the ship at 0.125 m/s. At this speed, the time to travel to the ship is

$$t = \frac{30.0\text{ m}}{0.125\text{ m/s}} = 240\text{ s} = 4.00\text{ minutes} \quad \Diamond$$

---

51. A 2000-kg car moving east at 10.0 m/s collides with a 3000-kg car moving north. The cars stick together and move as a unit after the collision, at an angle of 40° north of east and at a speed of 5.22 m/s. Find the velocity of the 3000-kg car before the collision.

Solution We use a coordinate system with the positive direction of the x axis toward the east and positive y toward the north. We set the initial velocity of the 3000-kg car as v_0, and use conservation of momentum. First, we consider the momentum along the y direction:

$$(3000\text{ kg})v_0 = (5000\text{ kg})(5.22\text{ m/s})\sin 40° \qquad \text{giving} \qquad v_0 = 5.59\text{ m/s} \quad \Diamond$$

A quick check will reveal that with this value of v_0, momentum is also conserved in the x direction.

---

53. Tarzan, whose mass is 80.0 kg, swings from a 3.00-m vine that is horizontal when he starts. At the bottom of his arc, he picks up 60.0-kg Jane in an inelastic collision. What is the maximum height, relative to the bottom of the arc, where they can land on a tree limb after their upward swing?

Solution Treat the overall process in three steps:

(1) Conservation of energy from the top of the arc to the bottom (take the zero level of gravitational potential energy at the bottom of the arc) and let v_0 equal Tarzan's velocity at the instant of the inelastic collision. This gives

$$\frac{1}{2}mv_i^2 + mgy_i = \frac{1}{2}mv_f^2 + mgy_f$$

or $$0 + \frac{1}{2}(80.0\text{ kg})(9.80\text{ m/s}^2)(3.00\text{ m}) = \frac{1}{2}(80.0\text{ kg})v_0^2 + 0$$

From this equation, we find $v_0 = 7.67\text{ m/s}$

(2) Now, use conservation of momentum to find the velocity, **v**, of Tarzan + Jane just after he picks her up.

$$m_T v_0 + 0 = (m_T + m_J)v \quad \text{or} \quad (80.0\text{ kg})(7.67\text{ m/s}) = (140\text{ kg})v$$

and $$v = 4.38\text{ m/s}$$

(3) Finally, we use conservation of mechanical energy from just after he picks her up to the end of their upward swing to determine the maximum height, H, reached.

$$\frac{1}{2}mv_i^2 + mgy_i = \frac{1}{2}mv_f^2 + mgy_f$$

$$\frac{1}{2}(140\text{ kg})(4.38\text{ m/s})^2 + 0 = 0 + (140\text{ kg})(9.80\text{ m/s}^2)H$$

and $$H = 0.979\text{ m} \quad \Diamond$$

---

55. The force shown in the force-time diagram in Figure 6.27 acts on a 1.5-kg mass. Find (a) the impulse of the force, (b) the final velocity of the mass if it is initially at rest, and

(c) the final velocity of the mass if it is initially moving along the x axis with a velocity of –2.0 m/s.

Solution (a) The impulse equals the area under the F versus t graph. This area is the sum of the area of the rectangle of base = 3.0 s and height = 2.0 N plus the area of the triangle of base = 2.0 s and height = 2.0 N. Thus,

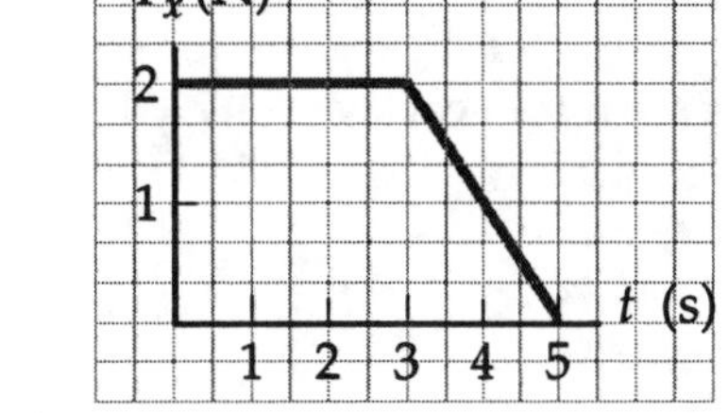

Figure 6.27

$$\text{Impulse} = (3.0\text{ s})(2.0\text{ N}) + \frac{1}{2}(2.0\text{ s})(2.0\text{ N}) = 8.0\text{ N}\cdot\text{s} \quad \lozenge$$

(b) $F\Delta t = \Delta p = mv_f - mv_i$

$8.0\text{ N}\cdot\text{s} = (1.5\text{ kg})v_f - 0$

$v_f = 5.3\text{ m/s} \quad \lozenge$

(c) $F\Delta t = \Delta p = mv_f - mv_i = (1.5\text{ kg})v_f - (1.5\text{ kg})(-2.0\text{ m/s})$ and $v_f = 3.3\text{ m/s} \quad \lozenge$

---

59. A 0.11-kg tin can is resting on top of a 1.7-m-high fence post. A 0.002-kg bullet is fired horizontally at the can. It strikes the can with a speed of 900 m/s, passes through it, and emerges with a speed of 720 m/s. When the can hits the ground, how far is it from the fence post? Disregard friction while the can is in contact with the post.

Solution Use conservation of momentum to find the velocity v_0 of the can immediately after the bullet passes through.

$(0.11\text{ kg})(v_0) + (0.002\text{ kg})(720\text{ m/s}) = (0.002\text{ kg})(900\text{ m/s})$ or $v_0 = 3.27\text{ m/s}$

Now, treat the motion of the can as a projectile with an initial velocity v_0 along the horizontal direction. We have:

$$y = v_{0y}t - \frac{1}{2}gt^2 \quad \text{or} \quad -1.7\text{ m} = 0 + \frac{1}{2}(-9.8\text{ m/s}^2)t^2$$

which gives $t = 0.589$ s as the time to reach the ground. From its horizontal motion,

$$x = v_{0x}t = (3.27\text{ m/s})(0.589\text{ s}) = 1.9\text{ m} \quad \lozenge$$

CHAPTER SELF-QUIZ

1. A 0.6-kg tennis ball, initially moving at a speed of 12 m/s, is struck by a racket causing it to rebound in the opposite direction at a speed of 18 m/s. A high speed movie film determines that the racket and ball are in contact for 0.05 s. What is the average net force exerted on the ball by the racket?
   a. 32.4 N
   b. 36.0 N
   c. 1.44 N
   d. 0.72 N

2. During a snowball fight, two balls, with masses of 0.4 and 0.6 kg respectively, are thrown in such a manner that they meet head-on and combine to form a single mass of 1 kg. The magnitude of initial velocity for each is 15 m/s. The final kinetic energy of the system just after the collision is what percentage of kinetic energy just before the collision?
   a. 100%
   b. 50 %
   c. 33%
   d. 4%

3. A railroad freight car, mass 15,000 kg, is allowed to coast along a level track at a speed of 2 m/s. It collides and couples with a 50,000 kg second car, initially at rest and with brakes released. What is the speed of the two cars after coupling?
   a. 1.2 m/s
   b. 86 m/s
   c. 0.60 m/s
   d. 0.46 m/s

4. Three masses are arrayed in a two-dimensional coordinate system as follows: 4 kg at (5, –5), 6 kg at (5, 5) and 5 kg at (–4, –5). What are the coordinates of the center of mass of the system?
   a. (3, –2)
   b (4, –1)
   c. (2, 6)
   d. (2, –1)

5. A miniature, spring-loaded, radio-controlled gun is mounted on an air puck. The gun's bullet has a mass of 0.005 kg and the gun and puck have a combined mass of 0.120 kg. With the system initially at rest, the radio-controlled trigger releases the bullet causing the puck and empty gun to move with a speed of 0.5 m/s. After release, what is the speed of the center of mass of the gun-puck-bullet system?
   a. Zero
   b. 36 m/s
   c. 12 m/s
   d. 4.8 m/s

6. A 0.02-kg bullet is fired into, and becomes embedded in, a 1.7-kg wooden ballistic pendulum. If the pendulum rises a vertical distance of 0.05 m, what is the initial speed of the bullet?
   a. 167 m/s
   b. 85 m/s
   c. 67 m/s
   d. 35 m/s

7. A 0.10-kg object moving initially with a velocity of +0.20 m/s makes an elastic head-on collision with a 0.15-kg object initially at rest. What is the velocity of the 0.15-kg object after the collision?
   a. +0.16 m/s
   b. –0.16 m/s
   c. +0.04 m/s
   d. –0.045 m/s

8. A 10-gram bullet is fired into a 100-gram block of wood at rest on a horizontal surface. After impact, the block slides 8 m before coming to rest. If the coefficient of friction, $\mu = 0.6$, find the speed of the bullet before impact.
   a. 106 m/s
   b. 212 m/s
   c. 318 m/s
   d. 424 m/s

9. A firehose directs a steady stream of 15 kg/sec of water with velocity 28 m/s against a flat plate. What force is required to hold the plate in place?
   a. 110 N
   b. 420 N
   c. 1100 N
   d. 4116 N

10. Two skaters, both of mass 50 kg, are at rest on skates on a frictionless ice pond. One skater throws a 0.2-kg Frisbee at 5 m/s to her friend, who catches it and throws it back at 5 m/s. When the first skater has caught the returned Frisbee, what is the velocity of each of the two skaters?
    a. 0.02 m/s, moving apart
    b. 0.04 m/s, moving apart
    c. 0.02 m/s, moving towards each other
    d. 0.04 m/s, moving towards each other

11. A 1.00-kg duck is flying overhead at 1.50 m/s when a hunter fires straight up. The 0.010-kg bullet is moving 100 m/s when it hits the duck and stays lodged in the duck's body. What is the momentum of the duck and bullet immediately after the hit?
    a. 1.50 $kg \cdot m/s$
    b. 2.50 $kg \cdot m/s$
    c. 0.05 $kg \cdot m/s$
    d. 1.80 $kg \cdot m/s$

12. A railroad car of 1000 kg is rolling with a speed of 2.0 m/s when 2000 kg of coal is dropped from a height of 1.0 m into the car. What is the final speed of the cart after the coal is in the cart?
    a. 0.25 m/s
    b. 0.67 m/s
    c. 1.0 m/s
    d. 3.6 m/s

7
Circular Motion and the Law of Gravity

Chapter 7

CIRCULAR MOTION AND THE LAW OF GRAVITY

In this chapter we shall investigate circular motion, a specific type of two-dimensional motion. We shall encounter such terms as *angular velocity, angular acceleration, centripetal acceleration,* and *centripetal force.* The results derived here will help you understand the motions of a diverse range of objects in our environment, from a car moving around a circular race track to clusters of galaxies orbiting a common center.

We shall also introduce Newton's universal law of gravitation, one of the fundamental laws in nature, and show how this law, together with Newton's laws of motion, enables us to understand a variety of familiar phenomena.

NOTES FROM SELECTED CHAPTER SECTIONS

7.1 Angular Velocity and Angular Acceleration

Pure rotational motion refers to the motion of a rigid body about a fixed axis.

One *radian* (rad) is the angle subtended by an arc length equal to the radius of the arc.

In the case of *rotation about a fixed axis,* every particle on the rigid body has the same angular velocity and the same angular acceleration.

7.2 Rotational Motion Under Constant Angular Acceleration

The *kinematic expressions* for rotational motion under constant angular acceleration are of the *same form* as those for linear motion under constant linear acceleration with the substitutions $x \rightarrow \theta$, $v \rightarrow \omega$, and $a \rightarrow \alpha$.

7.3 Relations Between Angular and Linear Quantities

When a rigid body rotates about a fixed axis, every point in the object moves along a circular path which has its center at the axis of rotation. The instantaneous velocity of each point is directed along a tangent to the circle. Every point on the object experiences the same *angular* speed; however, points that are different distances from the axis of rotation have different tangential speeds. The value of each of the linear quantities [displacement (s), velocity (v), and acceleration (a_t)] is equal to the radial distance from the axis multiplied by the corresponding angular quantity, θ, ω, and x.

7.4 Centripetal Acceleration

In circular motion, the centripetal acceleration is directed inward toward the center of the circle and has a magnitude given either by $\frac{v^2}{r}$ or $r\omega^2$.

7.5 Centripetal Force

All centripetal forces act toward the center of the circular path along which the object moves.

7.6 Describing Motion of a Rotating System

One must be very careful to distinguish real forces from fictitious ones in describing motion in an accelerating frame. An observer in a car rounding a curve is in an accelerating frame and invents a fictitious outward force to explain why he or she is thrown outward. A stationary observer outside the car, however, considers only real forces on the passenger. To this observer, the mysterious outward force *does not exist!* The only real external force on the passenger is the centripetal (inward) force due to friction or the normal force of the door.

7.7 Newton's Universal Law of Gravitation

There are several features of the universal law of gravity that deserve some attention:

1. The gravitational force is an action-at-a distance force that always exists between two particles regardless of the medium that separates them.
2. The force varies as the inverse square of the distance between the particles and therefore decreases rapidly with increasing separation.
3. The force is proportional to the product of their masses.

7.8 Kepler's Laws

Kepler's laws applied to the solar system are:

1. All planets move in elliptical orbits with the Sun at one of the focal points.
2. A line drawn from the Sun to any planet sweeps out equal areas in equal time intervals.
3. The square of the orbital period of any planet is proportional to the cube of the average distance from the planet to the Sun.

EQUATIONS AND CONCEPTS

When a particle moves along a circular path of radius r, the distance traveled by the particle is called the arc length, s. The radial line from the center of the path to the particle sweeps out an angle, θ.

$$\theta \equiv \frac{s}{r} \tag{7.1}$$

The angle θ in Equation 7.1 is the ratio of two lengths (arc length to radius) and hence is a dimensionless quantity. It is common practice to refer to the angle as being in units of radians. In calculations using Equation 7.1, the value of θ must be in radians as given by

Comment on radian measure.

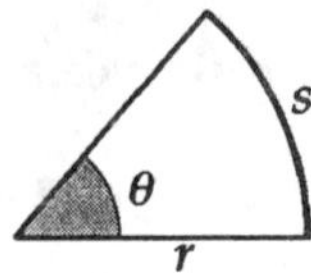

$$\theta(\text{rad}) = \left(\frac{\pi}{180^\circ}\right)\theta(\text{deg})$$

The average angular velocity of a rotating object is the ratio of the angular displacement to the time interval during which the angular displacement occurs. In this equation θ must be measured in radians.

$$\overline{\omega} = \frac{\Delta\theta}{\Delta t} \tag{7.2}$$

The average angular acceleration of a rotating object is the ratio of change in angular velocity to the time interval during which the change in velocity occurs.

$$\overline{\alpha} = \frac{\Delta\omega}{\Delta t} \tag{7.4}$$

The tangential velocity and tangential acceleration of a given point on a rotating object are related to the corresponding angular quantities via the radius of the circular path along which the point moves.

$$v_t = r\omega \tag{7.8}$$

$$a_t = r\alpha \tag{7.9}$$

Comment on rotational velocity and acceleration.

The tangential velocity and tangential acceleration are along directions which are tangent to the circular path (and therefore perpendicular to the radius corresponding to a given point in a rotating object). Also note that every point on a rotating object has the same value of ω and the same value of α; however, points which are at different distances from the axis of rotation have different values of v_t and a_t.

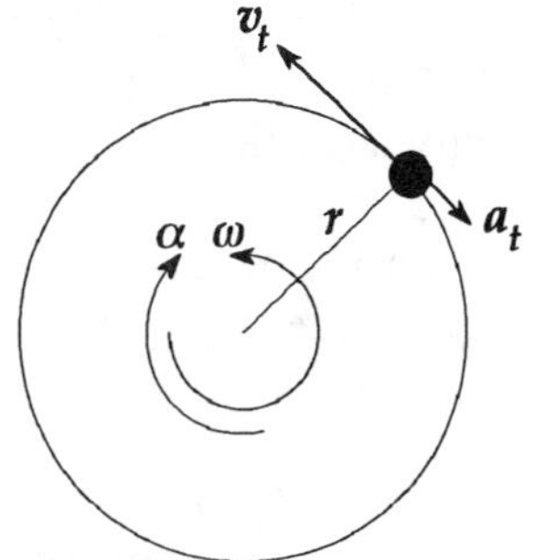

The equations of rotational kinematics apply in the case of a particle or body which rotates about a fixed axis with constant angular acceleration. In solving problems in rotational kinematics, it will be helpful to keep in mind the analogous structure of the equations for rotational kinematics and those for linear motion.

$$\omega = \omega_0 + \alpha t \tag{7.5}$$

$$v = v_0 + at$$

$$\theta = \omega_0 t + \frac{1}{2}\alpha t^2 \tag{7.6}$$

$$x = v_0 t + \frac{1}{2}at^2$$

$$\omega^2 = \omega_0^2 + 2\alpha\theta \tag{7.7}$$

$$v^2 = v_0^2 + 2ax$$

An object in circular motion has a centripetal acceleration which is directed toward the center of the circle and has a magnitude which depends on the values of the tangential velocity and the radius of the path.

$$a_c = \frac{v^2}{r} \tag{7.11}$$

or

$$a_c = r\omega^2 \tag{7.14}$$

An object which is moving along a circular path with increasing or decreasing speed has both a tangential component of acceleration (tangent to the instantaneous direction of travel) and a centripetal component of acceleration (perpendicular to the instantaneous direction of travel). The magnitude and direction of the total acceleration can be found by the usual methods of vector addition.

$$a = \sqrt{a_t^2 + a_c^2} \quad (7.15)$$

$$\theta = \tan^{-1}\left(\frac{a_t}{a_c}\right)$$

Forces which maintain motion along a circular path are directed toward the center of the path and are called centripetal forces.

$$F_c = ma_c = m\frac{v_t^2}{r} \quad (7.16)$$

The universal law of gravity states that every particle in the Universe attracts every other particle with a force that is directly proportional to the product of their masses and inversely proportional to the square of the distance between them. The constant G is called the universal gravitational constant.

$$F = G\frac{m_1 m_2}{r^2} \quad (7.17)$$

$$G = 6.673 \times 10^{-11} \frac{\mathrm{N \cdot m^2}}{\mathrm{kg^2}} \quad (7.18)$$

The gravitational force between two masses m_1 and m_2 is one of attraction; each mass exerts a force of attraction on the other. These two forces form an action-reaction pair. The magnitudes of the forces of gravitational attraction on the two masses are equal regardless of the relative values of m_1 and m_2.

Comments on the gravitational force.

$\mathbf{F}_1$ $\mathbf{F}_2$ m_1 m_2

Kepler's third law states that the square of the orbital period of a planet is proportional to the cube of the mean distance from the planet to the Sun.

$$T^2 = \left(\frac{4\pi^2}{GM_s}\right)r^3 = K_s r^3 \quad (7.19)$$

K_s is independent of the mass of the planet.

$$K_s = \frac{4\pi^2}{GM_s} = 2.97 \times 10^{-19}\ \mathrm{s^2/m^3}$$

For an Earth satellite, M_s in Equation 7.19 must be replaced by M_e, the mass of the Earth.

Comment on Kepler's third law.

SUGGESTIONS, SKILLS, AND STRATEGIES

The following features involving centripetal forces and centripetal accelerations should be kept in mind:

1. Draw a free-body diagram of the object(s) under consideration, showing all forces that act on it.

2. Choose a coordinate system with one axis perpendicular to the circular path followed by the object and with one axis tangent to the circular path.

3. Find the net force toward the center of the circular path. This is the centripetal force.

4. From this point onward, the steps are virtually identical to those encountered when solving Newton's second law problems with $\Sigma \mathbf{F} = m\mathbf{a}$. Also, you should note that the magnitude of the centripetal acceleration can always be written as $a_c = v^2/r$.

REVIEW CHECKLIST

▷ Quantitatively, the angular displacement, velocity, and acceleration for a rigid body system in rotational motion is related to the distance traveled, tangential velocity, and tangential acceleration. The linear quantity is calculated by multiplying the angular quantity by the radius arm for an object or point in that system.

▷ If a body rotates about a fixed axis, every particle on the body has the same angular velocity and angular acceleration. For this reason, rotational motion can be simply described using these quantities. The formulas which describe angular motion are analogous to the corresponding set of formulas pertaining to linear motion.

▷ The nature of the acceleration of a particle moving in a circle with constant speed is such that, although $\mathbf{v}$ = constant, the *direction* of $\mathbf{v}$ varies in time, which is the origin of the radial, or centripetal acceleration.

▷ There are two components of acceleration for a particle moving on a curved path, when both the magnitude and direction of $\mathbf{v}$ are changing with time. In this case, the particle has a tangential component of acceleration and a radial component of acceleration.

▷ The forces acting on a vehicle on a banked curve can be resolved into components which are either parallel or perpendicular to the radial direction. Newton's second law can then be used to find an expression for the proper banking angle corresponding to a given speed and radius of path.

▷ Newton's universal law of gravitation is an example of an inverse-square law, and it describes an *attractive* force between two *particles* separated by a distance *r*.

SOLUTIONS TO SELECTED END-OF-CHAPTER PROBLEMS

5. Find the angular speed of the Earth about the Sun in radians per second and degrees per day.

Solution The Earth moves through one revolution in its orbit or 2π rad in one year $(3.156\times10^7\text{ s})$. Thus,

$$\omega = \frac{2\pi\text{ rad}}{3.156\times10^7\text{ s}} = 1.99\times10^{-7}\text{ rad/s} \quad \Diamond$$

Alternatively, the Earth moves through 360° in one year (365.242 days). Thus,

$$\omega = \frac{360°}{365.2\text{ days}} = 0.986\text{ deg/day} \quad \Diamond$$

---

9. A dentist's drill starts from rest. After 3.20 s of constant angular acceleration, it turns at a rate of 2.51×10^4 rev/min. (a) Find the drill's angular acceleration. (b) Determine the angle (in radians) through which the drill rotates during this period.

Solution $\omega_f = 2.51\times10^4\text{ rev/min} = 2.63\times10^3\text{ rad/s}$

(a) $\alpha = \dfrac{\omega_f - \omega_0}{t} = \dfrac{2.63\times10^3\text{ rad/s} - 0}{3.20\text{ s}} = 8.22\times10^2\text{ rad/sec}^2 \quad \Diamond$

(b) $\theta = \omega_0 t + \dfrac{1}{2}\alpha t^2 = 0 + \dfrac{1}{2}\left(8.22\times10^2\text{ rad/s}^2\right)(3.20\text{ s})^2 = 4.21\times10^3\text{ rad} \quad \Diamond$

---

13. The tub of a washer goes into its spin-dry cycle, starting from rest and reaching an angular speed of 5.0 rev/s in 8.0 s. At this point the person doing the laundry opens the lid, and a safety switch turns off the washer. The tub slows to rest in 12.0 s. Through how many revolutions does the tub turn? Assume constant angular acceleration while it is starting and stopping.

Solution $\omega_f = 5.00 \text{ rev/s} = 10\pi \text{ rad/s}$

We will break the motion into two stages: (1) an acceleration period and (2) a deceleration period. While speeding up,

$$\theta_1 = \bar{\omega}t = \frac{0 + 10\pi \text{ rad/s}}{2}(8.00 \text{ s}) = 125.7 \text{ rad}$$

While slowing down,

$$\theta_2 = \bar{\omega}t = \frac{10\pi \text{ rad/s} + 0}{2}(12.0 \text{ s}) = 188.5 \text{ rad}$$

So, $\theta_{\text{total}} = \theta_1 + \theta_2 = 314 \text{ rad} = 50.0 \text{ rev}$ ◊

---

21. A car is traveling at 17.0 m/s on a straight horizontal highway, which we assume extends from our left to our right. The wheels of the car have radii of 48.0 cm. If the car speeds up with an acceleration of 2.00 m/s^2 for 5.00 s, find the number of revolutions of 2.00 m/s^2 for 5.00 s, find the number of revolutions of the wheels during this period.

Solution $s = v_0 t + \frac{1}{2}at^2 = (17.0 \text{ m/s})(5.00 \text{ s}) + \frac{1}{2}(2.00 \text{ m/s}^2)(5.00 \text{ s})^2 = 110 \text{ m}$

$s = r\theta$ and $r = 48.0 \text{ cm} = 0.480 \text{ m}$

Thus, $\theta = \frac{s}{r} = \frac{110 \text{ m}}{0.480 \text{ m}} = 229.2 \text{ rad} = 36.5 \text{ rev}$ ◊

---

27. (a) What is the tangential acceleration of a bug on the rim of a 78-rpm, 10-in. diameter record if the record moves from rest to its final angular speed in 3.0 s?
(b) When the record is at its final speed, what is the tangential velocity of the bug?

(c) One second after the bug starts from rest, what are its tangential acceleration, radial acceleration, and total acceleration?

Solution

(a) The final angular velocity is 78 rev/min = 8.17 rad/s, and the radius of the record is equal to 5 in = 0.127 m.

$$\alpha = \frac{\Delta\omega}{\Delta t} = \frac{8.17 \text{ rad/s}}{3.0 \text{ s}} = 2.72 \text{ rad/s}^2$$

and $$a_t = r\alpha = 3.5 \times 10^{-1} \text{ m/s}^2 \quad \Diamond$$

(b) $v_t = r\omega$

$$v_t = (1.27 \times 10^{-1} \text{ m})(8.17 \text{ rad/s}) = 1.0 \text{ m/s} \quad \Diamond$$

(c) After $t = 1$ s from rest, the tangential acceleration will be constant; so from part (a) above, $a_t = 3.5 \times 10^{-1} \text{ m/s}^2$. $\Diamond$

The radial acceleration, $a_r = \frac{v_t^2}{r}$ but $v_t = r\omega$ so $a_r = r\omega^2$ when

$$\omega = \omega_0 t = \alpha t = 2.72 \text{ rad/s} \text{ so } a_r = (0.127 \text{ m})(2.72 \text{ rad/s})^2 = 0.94 \text{ m/s}^2 \quad \Diamond$$

Total acceleration $a_{\text{total}} = \sqrt{a_t^2 + a_r^2}$

$$a_{\text{total}} = \left[\left(0.345 \text{ m/s}^2\right)^2 + \left(0.939 \text{ m/s}^2\right)^2\right] = 1.0 \text{ m/s}^2 \quad \Diamond$$

The direction of the total acceleration (relative to the radial direction) is θ where

$$\theta = \tan^{-1}\left(\frac{a_t}{a_r}\right) = \tan^{-1}(0.367) = 20° \quad \Diamond$$

31. A 2000-kg car rounds a circular turn of radius 20 m. If the road is flat and the coefficient of friction between tires and road is 0.70, how fast can the car go without skidding?

Solution The friction force must supply the required centripetal force. Thus, we must have

$$f = F_c, \text{ where } f = \mu mg \text{ and } F_c = \frac{mv^2}{r}$$

Therefore, $$\mu mg = \frac{mv^2}{r}$$

The mass cancels, and we have $$v^2 = \mu rg = 0.70(20 \text{ m})(9.80 \text{ m/s}^2)$$

from which, $$v = 12 \text{ m/s} \quad \lozenge$$

---

35. An airplane is flying in a horizontal circle at a speed of 100 m/s. The 80.0-kg pilot does not want his radial acceleration to exceed 7.00 g. (a) What is the minimum radius of the circular path? (b) At this radius, what are the *net* centripetal forces exerted on the pilot by the seat belt, the friction between him and the seat, and so forth?

Solution

(a) The radial acceleration must not exceed 7.00 g = 7.00(9.80 m/s^2) = 66.6 m/s^2.

Thus, from $a_r = \frac{v_t^2}{r}$, we have $r = \frac{(100 \text{ m/s})^2}{68.6 \text{ m/s}^2} = 1.46 \times 10^2 \text{ m} \quad \lozenge$

(b) $F = ma_r = m(7.00 \text{ g}) = 7.00(mg) = 7.00(\text{pilot's weight})$

$$= 7.00(80.0 \text{ kg})(9.80 \text{ m/s}^2) = 5490 \text{ N} \quad \lozenge$$

---

39. A pail of water is rotated in a vertical circle of radius 1.00 m (the approximate length of a person's arm). What must be the minimum speed of the pail at the top of the circle if no water is to spill out?

Solution At the top of the vertical circle, the speed must be great enough so the necessary centripetal force is equal to or greater than the weight, mg, of the water. That is,

$$F_c \geq mg \quad \text{so} \quad \frac{mv^2}{r} \geq mg \quad \text{or} \quad v^2 \geq rg$$

At the minimum speed, we have $v^2_{\text{min}} = rg$ or

$$v_{\text{min}} = \sqrt{rg} = \sqrt{(1.00 \text{ m})(9.80 \text{ m/s}^2)} = 3.13 \text{ m/s} \quad \Diamond$$

---

47. A satellite is in a circular orbit just above the surface of the Moon. (The radius of the Moon is 1738 kg.) What are the satellite's (a) acceleration and (b) speed? (c) What is the period of the satellite orbit?

Solution (a) $F_c = ma_c = F_G$. Therefore, $ma_c = \frac{GmM_{\text{moon}}}{r^2}$ and

$$a_c = \frac{GM_{\text{moon}}}{r^2} = \frac{(6.67 \times 10^{-11} \text{ N} \cdot \text{m}^2/kg^2)(7.36 \times 10^{22} \text{ kg})}{(1.738 \times 10^6 \text{ m})^2} = 1.63 \text{ m/s}^2 \quad \Diamond$$

(b) $a_c = \frac{v^2}{r}$ so $v^2 = a_c r = (1.63 \text{ m/s}^2)(1.738 \times 10^6 \text{ m})$ which yields

$$v = 1.68 \times 10^3 \text{ m/s} \quad \Diamond$$

(c) The time to orbit or the period $T = \frac{\text{distance around}}{\text{speed}} = \frac{2\pi r}{v}$ so

$$T = \frac{2\pi(1.738 \times 10^6 \text{ m})}{1.68 \times 10^3 \text{ m/s}} = 6.50 \times 10^3 \text{ s} \quad \text{or} \quad T = 1.81 \text{ h} \quad \Diamond$$

---

51. Use the data of Table 7.2 to find the point between the Earth and the Sun at which an object can be placed so that the net gravitational force exerted on it by these two objects is zero.

Solution At the equilibrium position, the magnitude of the force exerted by the Earth on the object is equal to the magnitude of the force exerted by the Sun on the object. Thus, if we let r equal the distance from the object to the Sun, the equation becomes

$$\frac{GM_E m}{(1.50 \times 10^{11} - r)^2} = \frac{GM_S m}{r^2}$$

giving
$$\frac{r^2}{(1.50 \times 10^{11} - r)^2} = \frac{M_S}{M_E} = \frac{1.991 \times 10^{30}}{5.98 \times 10^{24}} = 3.33 \times 10^5 \text{ m/s}$$

Taking the square root of both sides gives $\frac{r}{(1.50 \times 10^{11} - r)} = 577$ which gives

$$r = 1.50 \times 10^{11} \text{ m} \quad \Diamond$$

---

53. An athlete swings a 5.00-kg ball horizontally on the end of a rope. The ball moves in a circle of radius 0.800 m at an angular speed of 0.500 rev/s. What are (a) the tangential velocity of the ball and (b) its centripetal acceleration? (c) If the maximum tension the rope can withstand before breaking is 100 N, what is the maximum tangential velocity the ball can have?

Solution $\omega = 0.500 \text{ rev/s} = \pi \text{ rad/s}$

(a) $v_t = r\omega = (0.800 \text{ m})(\pi \text{ rad/s}) = 2.51 \text{ m/s} \quad \Diamond$

(b) $a_r = r\omega^2 = \frac{v_t^2}{r} = \frac{(2.51 \text{ m/s})^2}{0.800 \text{ m}} = 7.90 \text{ m/s}^2 \quad \Diamond$

(c) $F = \frac{mv^2}{r}$

and if the maximum value of the centripetal force, T, is 100 N, then we have

$$100 \text{ N} = \frac{(5.00 \text{ kg})v^2}{0.800 \text{ m}} \qquad \text{from which} \qquad v = 4.00 \text{ m/s} \quad \Diamond$$

---

55. A rotating bicycle wheel has an angular velocity of 3.00 rad/s at some instant of time. It is then given an angular acceleration of 1.50 rad/s^2. A chalk line drawn on the wheel is horizontal at t = 0. (a) What angle does this line make with its original direction at t = 2.00 s? (b) What is the angular velocity of the wheel at t = 2.00 s?

Solution

(a) Use $\theta = \omega_0 t + \frac{1}{2}\alpha t^2$

$$\theta = (3.00 \text{ rad/s})(2.00 \text{ s}) + \frac{1}{2}(1.50 \text{ rad/s}^2)(2.00 \text{ s})^2$$

$\theta = 9.00 \text{ rad} = 516°$ or $\theta = 516° - 360° = 156°$ ◊

(b) The angular velocity at 2.00 s is $\omega = \omega_0 + \alpha t$

$$\omega = 3.00 \text{ rad/s} + (1.50 \text{ rad/s}^2)(2.00 \text{ s}) = 6.00 \text{ rad/s} \quad ◊$$

---

65. Because of the Earth's rotation about its axis, a point on the equator experiences a centripetal acceleration of 0.034 m/s^2 while a point at the poles experiences no centripetal acceleration. (a) Show that at the equator the gravitational force on an object (the true weight) must exceed the object's apparent weight. (b) What are the apparent weights at the equator and at the poles of a 75.0-kg person? (Assume the Earth is a uniform sphere, and take g = 9.800 m/s^2.)

Solution (a) Consider a person at the equator standing on a pair of scales. There are two forces on him: w, his true weight; and the upward force exerted on him by the scale, w', which we shall call his apparent weight since, by Newton's third law, it is the apparent weight that the scale will read.

We have $w - w' = \frac{mv^2}{r}$ or $w' = w - \frac{mv^2}{r}$ so $w > w'$ ◊

(b) At the poles, $v = 0$ and $w' = w = mg = (75.0 \text{ kg})(9.80 \text{ m/s}^2) = 735 \text{ N}$ ◊

At the equator $w' = 735 \text{ N} - (75.0 \text{ kg})(0.034 \text{ m/s}^2) = 732.5 \text{ N}$ ◊

---

67. A frictionless roller coaster is given an initial velocity of v_0 at height h, as in Figure 7.29. The radius of curvature of the track at point A is R. (a) Find the maximum value of v_0 so that the roller coaster stays on the track at A solely because of gravity. (b) Using the value of v_0 calculated in (a), determine the value of h' that is necessary if the roller coaster is to just make it to point B.

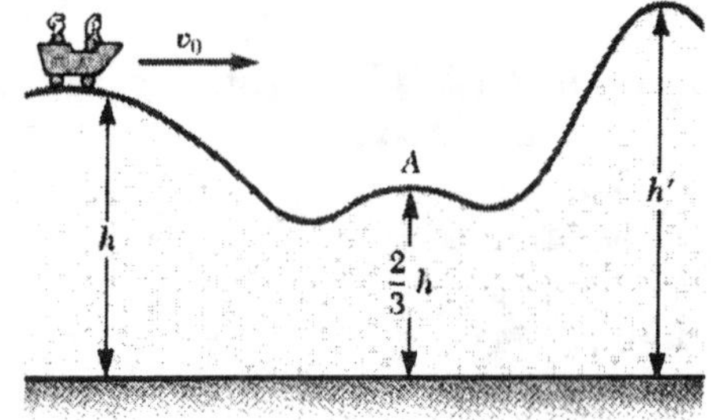

Figure 7.29

Solution

(a) In order for the coaster to stay on the track at point A, the weight must be equal to centripetal force or

$$mg = \frac{mv_A^2}{R} \qquad \text{or} \qquad v_A = \sqrt{Rg}$$

Conservation of energy between the initial point and point A yields the equation

$$\frac{1}{2}mv_0^2 + mgh = \frac{1}{2}mv_A^2 + mg\left(\frac{2h}{3}\right) \qquad \text{or} \qquad v_0 = \left[v_A^2 - \frac{2gh}{3}\right]^{\frac{1}{2}}$$

Substituting for v_A, we find for the maximum initial velocity

$$v_0 = \sqrt{g\left(R - \frac{2h}{3}\right)} \quad \lozenge$$

(b) By conservation of energy, we have

$$\frac{1}{2}mv_0^2 + mg\left(\frac{2h}{3}\right) = 0 + mgh'$$

Substituting for v_0 and solving for h', we find

$$h' = R + \frac{2h}{3} \quad \lozenge$$

CHAPTER SELF-QUIZ

1. A point on a wheel rotating at 5 rev/s and located 0.2 m from the axis has what tangential velocity?
   a. 3.8 m/s
   b. 6.3 m/s
   c. 1.2 m/s
   d. 0.104 m/s

2. A point on a wheel rotating at 5 rev/s and located 0.2 m from the axis experiences what centripetal acceleration?
   a. 192 m/s^2
   b. 48 m/s^2
   c. 0.05 m/s^2
   d. 1.35 m/s^2

3. A Ferris wheel, rotating initially at an angular velocity of 0.50 rad/s, accelerates over a 5-s interval at a rate of 0.04 rad/s^2. What angular displacement does the Ferris wheel undergo in this 5-s interval?
   a. 4.1 rad
   b. 2.5 rad
   c. 3.0 rad
   d. 0.50 rad

4. A 0.3-kg mass, attached to the end of a 0.75-m string, is whirled around in a circular horizontal path. If the maximum tension that the string can withstand is 250 N, then what maximum velocity can the mass have if the string is not to break?
   a. 375 m/s
   b. 22.4 m/s
   c. 19.4 m/s
   d. 25 m/s

5. A point on the rim of a 0.20-m-radius rotating wheel has a centripetal acceleration of 4.0 m/s^2. What is the angular velocity of the wheel?
   a. 0.89 rad/s
   b. 1.6 rad/s
   c. 3.2 rad/s
   d. 4.5 rad/s

6. When a point on the rim of a 0.25-m-radius wheel experiences a centripetal acceleration of 4.0 m/s^2, what tangential acceleration does that point experience?
   a. 1.0 m/s^2
   b. 2.0 m/s^2
   c. 4.0 m/s^2
   d. Cannot determine with information given

7. A Ferris wheel, starting at rest, builds up to a final angular velocity of 0.71 rad/s while rotating through an angular displacement of 2.5 rad. What is its average angular acceleration?
   a. 0.20 rad/s^2
   b. 0.10 rad/s^2
   c. 1.8 rad/s^2
   d. 3.6 rad/s^2

8. A roller coaster, loaded with passengers, has a mass of 500 kg; the radius of curvature of the track at the bottom point of the dip is 12 m. If the vehicle has a speed of 18 m/s at this point, what force is exerted on the vehicle by the track? (g = 9.8 m/s^2)
   a. 1.6×10^4 N
   b. 1.8×10^4 N
   c. 0.5×10^4 N
   d. 1.0×10^4 N

9. Consider a point on a bicycle tire that is momentarily in contact with the ground as the bicycle rolls across the ground with constant speed. The direction for the acceleration for this point at that moment is
   a. upward
   b. down toward the ground
   c. forward
   d. at that moment the acceleration is zero

10. What angular velocity (in revolutions/second) is needed for a centrifuge to produce an acceleration of 1000g at a radius arm of 10 cm?
    a. 50 rev/s
    b. 75 rev/s
    c. 100 rev/s
    d. 150 rev/s

11. Geosynchronous satellites orbit the Earth at a distance of 42,000 km from the Earth's center. Their angular velocity at this height is the same as the rotation of the Earth, so they appear stationary at certain locations in the sky. What is the force acting on a 1000-kg satellite at this height?
    a. 57 N
    b. 222 N
    c. 304 N
    d. 431 N

12. What is the angular velocity about the center of the Earth for a person standing on the equator?
    a. 7.3×10^{-5} rad/s
    b. 3.6×10^{-5} rad/s
    c. 6.28×10^{-5} rad/s
    d. 3.14×10^{-5} rad/s

8
Rotational Equilibrium and Rotational Dynamics

Chapter 8

ROTATIONAL EQUILIBRIUM AND ROTATIONAL DYNAMICS

We shall begin this chapter by studying objects that are at rest or moving with constant velocity; such objects are said to be in equilibrium. Knowledge of the conditions that prevail when an object is in equilibrium is important in a variety of fields. Newton's first law, as discussed in Chapter 4, is the basis for much of our work in this chapter. In addition, in order to fully understand objects in equilibrium, we must consider torque. This concept will also play a key role in our discussion of rotational motion.

In this chapter we will complete our study of rotational motion. We shall build on the definitions of angular speed and angular acceleration encountered in Chapter 7 by examining the relationship between these concepts and the forces that produce rotational motion. Specifically, we shall find the rotational analog of Newton's second law and define a term that needs to be added to our equation for conservation of mechanical energy: rotational kinetic energy. One of the central purposes of this chapter is to develop the concept of angular momentum, a quantity that plays a key role in rotational motion. Finally, just as we found that linear momentum is conserved, we shall also find that the angular momentum of any isolated system is always conserved.

NOTES FROM SELECTED CHAPTER SECTIONS

8.1 Torque

Torque is the physical quantity which is a measure of the tendency of a force to cause rotation of a body about a specified axis. It is important to remember that torque must be defined with respect to a *specific axis* of rotation. Torque which has the SI *units* of N·m must not be confused with force.

8.2 Torque and the Second Condition of Equilibrium

A body in static equilibrium must satisfy two conditions:

1. The resultant external force must be zero.
2. The resultant external torque must be zero.

You should note that it does not matter where you pick the axis of rotation for calculating the net torque if the object is in equilibrium; since the object is not rotating, the location of the axis is completely arbitrary.

8.3 The Center of Gravity

In order to calculate the torque due to the weight (gravitational force) on a rigid body, the entire weight of the object can be considered to be concentrated at a single point called the center of gravity. The center of gravity of a homogeneous, symmetric body must lie along an axis of symmetry.

8.5 Relationship Between Torque and Angular Acceleration

The angular acceleration of an object is proportional to the net torque acting on it. The moment of inertia of the object is the proportionality constant between the net torque and the angular acceleration. The force and mass in linear motion correspond to torque and moment of inertia in rotational motion. Moment of inertia of an object depends on the location of the axis of rotation and upon the manner in which the mass is distributed relative to that axis (e.g., a ring has a greater moment of inertia than a disk of the same mass and radius).

8.6 Rotational Kinetic Energy

In linear motion, the energy concept is useful in describing the motion of a system. The energy concept can be equally useful in simplifying the analysis of rotational motion. We now have expressions for three types of energy: *gravitational potential energy,* PE_g*;* *translational kinetic energy,* KE_t; and *rotational kinetic energy,* KE_r. We must include all these forms of energy in the equation for conservation of mechanical energy.

EQUATIONS AND CONCEPTS

This is the first condition for equilibrium. An object will remain at rest or move with uniform motion along a straight line when no net external force acts on it. This condition corresponds to translational equilibrium. The vector equation for translational equilibrium can be written in component form (in this case for a two-dimensional situation).

$$\Sigma \mathbf{F} = 0$$

$$\Sigma F_x = 0$$

$$\Sigma F_y = 0$$

A set of two or more forces are concurrent if their lines of action when extended intersect at a single point. When an object is being acted on by concurrent forces, the first condition ($\Sigma\mathbf{F} = 0$) is sufficient to ensure equilibrium.
REVIEW THE RECOMMENDED SOLUTION PROCEDURE AND EXAMPLE PROBLEMS IN THE TEXTBOOK.

Comment on concurrent forces.

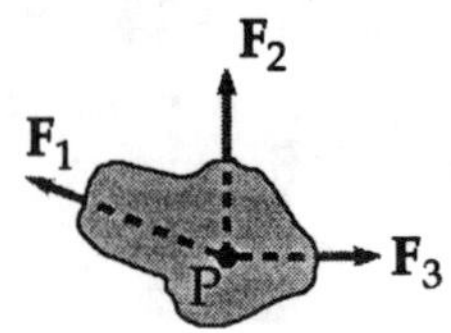

When nonconcurrent forces act on a body, the first condition is not sufficient to ensure complete equilibrium. In this case it is necessary to consider the net torque acting on the body relative to some axis. For a given force, the magnitude of the corresponding torque is the product of the magnitude of the force and the lever arm. For the object shown in the figure, $\tau_1 = +F_1 d_1$ and $\tau_2 = -F_2 d_2$.

$$\tau = Fd \qquad (8.1)$$

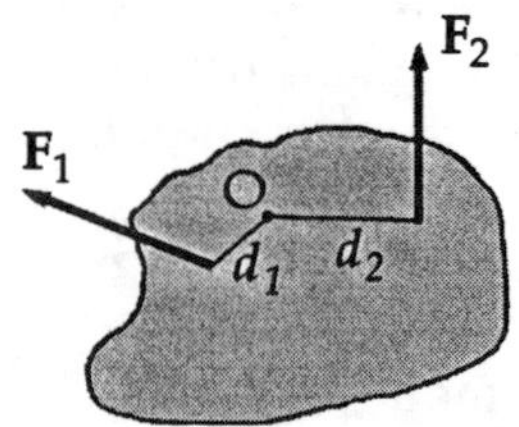

The sign of a torque due to a force is considered positive if the force has a tendency to rotate the body counterclockwise about the chosen axis, and negative if the tendency for rotation is clockwise.

Comment on sign conventions.

The second condition for equilibrium requires that the net torque acting on a body equals zero. This condition will ensure that the body is in rotational equilibrium.

$$\Sigma\tau = 0 \qquad (8.2)$$

In order to compute the torque due to the force of gravity (an object's weight), the total weight can be considered as being concentrated at a single point called the center of gravity and having coordinates x_{cg}, y_{cg}.

$$x_{cg} = \frac{\Sigma m_i x_i}{\Sigma m_i} \qquad (8.3)$$

$$y_{cg} = \frac{\Sigma m_i y_i}{\Sigma m_i} \qquad (8.4)$$

The angular acceleration of a point mass, moving in a path of radius r, is proportional to the net torque acting on the mass.

$$\tau = mr^2\alpha \qquad (8.6)$$

The moment of inertia of a rigid body depends on the distribution of mass relative to the axis of rotation.

$$I \equiv \Sigma mr^2 \qquad (8.8)$$

The angular acceleration of an extended object is proportional to the net torque acting on the object. The proportionality constant, I, is called the moment of inertia of the object.

$$\Sigma\tau = I\alpha \qquad (8.9)$$

The moment of inertia of a system depends on the mass and the manner in which the mass is distributed relative to the axis of rotation.

Comment on moment of inertia.

For a point mass, m, moving in a path of radius, r, $I = mr^2$.

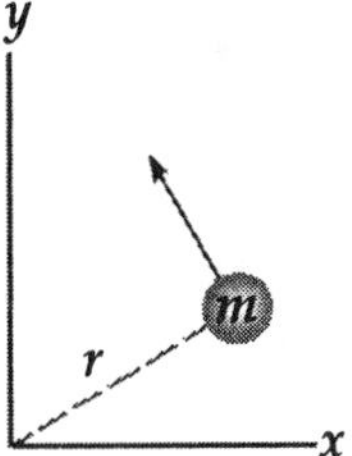

For a collection of discrete point masses, each with its own corresponding value of r, $I = \Sigma mr^2$.

The value of I can be calculated for an extended object with good symmetry. Expressions for the moments of inertia for a number of objects of common shape are given in Table 8.1 of your text.

The SI units of moment of inertia are $kg{\cdot}m^2$.

A rigid body or mass in rotational motion has kinetic energy due to its motion. Note that this is not a new form of energy but is a convenient form for representing kinetic energy associated with rotational motion.

$$KE_r \equiv \frac{1}{2}I\omega^2 \qquad (8.11)$$

When only conservative forces act on a system, the total mechanical energy of the system is conserved.

$$(KE_t + KE_r + PE_g)_i = (KE_t + KE_r + PE_g)_f \quad (8.12)$$

An object in rotational motion is characterized by a quantity, L, which is called angular momentum. Angular momentum is a vector quantity and has the direction of the angular velocity.

$$L \equiv I\omega \quad (8.13)$$

The net external torque acting on an object equals the time rate of change of its angular momentum. Note that this is the rotational analog of Newton's second law for transitional motion.

$$\tau = \frac{\Delta L}{\Delta t} \quad (8.14)$$

When the net external torque acting on a system is zero, the angular momentum of the system is conserved.

If $\Sigma\tau = 0$, then

$$I_i\omega_i = I_f\omega_f \quad (8.15)$$

SUGGESTIONS, SKILLS, AND STRATEGIES

Problem-Solving Strategy for Objects in Equilibrium:

1. Draw a simple, neat diagram of the system.

2. Isolate the object of interest being analyzed. Draw a free-body diagram for this object showing all external forces acting on the object. For systems containing more than one object, draw *separate* diagrams for each object. Do not include forces that the object exerts on its surroundings.

3. Establish convenient coordinate axes for each body and find the components of the forces along these axes. Now apply the first condition of equilibrium for each object under consideration; namely, that the net force on the object in the x and y directions must be zero.

4. Choose a convenient origin for calculating the net torque on the object. Now apply the second condition of equilibrium that says that the net torque on the object about any origin must be zero. Remember that the choice of the origin for the torque equation is arbitrary; therefore, choose an origin that will simplify your calculation

as much as possible. Note that a force that acts along a line passing through the point chosen as the axis of rotation gives zero contribution to the torque.

5. The first and second conditions for equilibrium will give a set of simultaneous equations with several unknowns. To complete your solution, all that is left is to solve for the unknowns in terms of the known quantities.

Problem-Solving Strategy for Rotational Motion:

The following facts and procedures should be kept in mind when solving rotational motion problems.

1. There are actually very few new techniques that need to be learned in order to solve rotational motion problems. For example, problems involving the equation $\Sigma\tau = I\alpha$ are very similar to those encountered in Newton's second law problems, $\Sigma\mathbf{F} = m\mathbf{a}$. Note the correspondences between linear and rotational quantities in that **F** is replaced by τ, m by I, and a by α.

2. Other analogs between rotational quantities and linear quantities include the replacement of x by θ and v by ω. These are helpful as memory devices for such rotational motion quantities as rotational kinetic energy, $KE_r = \frac{1}{2}I\omega^2$, and angular momentum, $L = I\omega$.

3. With the analogs mentioned in Step 2, conservation of energy techniques remain the same as those examined in Chapter 5, except for the fact that a new kind of energy, rotational kinetic energy, must be included in the expression for the conservation of energy.

4. Likewise, the techniques for solving conservation of angular momentum problems are essentially the same as those used in solving conservation of linear momentum problems, except you are equating total angular momentum before to total angular momentum after as $I_i\omega_i = I_f\omega_f$.

REVIEW CHECKLIST

▷ There are two necessary conditions for equilibrium of a rigid body: $\Sigma F = 0$ and $\Sigma\tau = 0$. Torques which cause counterclockwise rotations are positive and those causing clockwise rotations are negative.

(See Errata Sheet on page 558)

9. A window washer is standing on a scaffold supported by a vertical rope at each end. The scaffold weighs 200 N and is 3.00 m long. What is the tension in each rope when the 700-N worker stands 1.00 m from one end?

Solution Taking torques about the left end of the scaffold, we have

$$T_1(0) - (700\text{ N})(1.00\text{ m}) - (200\text{ N})(1.50\text{ m}) + T_2(3.00\text{ m}) = 0$$

from which $T_2 = 333\text{ N}$ ◊

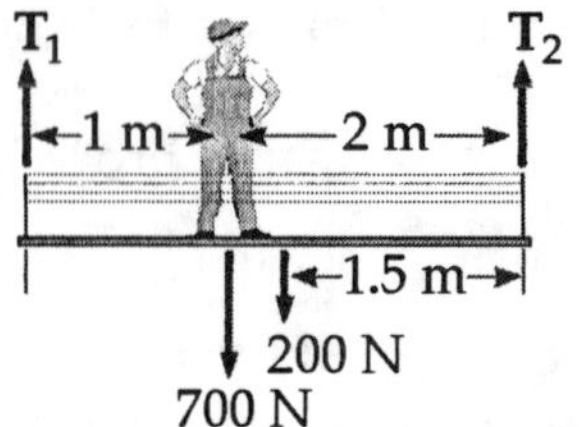

Then from $\Sigma F_y = 0,$ we have

$$T_1 + T_2 - 700\text{ N} - 200\text{ N} = 0$$

Since $T_2 = 333\text{ N},$ we find $T_1 = 567\text{ N}$ ◊

---

15. The chewing muscle, the masseter, is one of the strongest in the human body. It is attached to the mandible (lower jawbone) as shown in Figure 8.38a. The jawbone is pivoted about a socket just in front of the auditory canal. The forces acting on the jawbone are equivalent to those acting on the curved bar in Figure 8.38b. **C** is the force exerted against the jawbone by the food being chewed, **T** is the tension in the masseter, and **R** is the force exerted on the mandible by the socket. Find **T** and **R** if you bite down on a piece of steak with a force of 50 N.

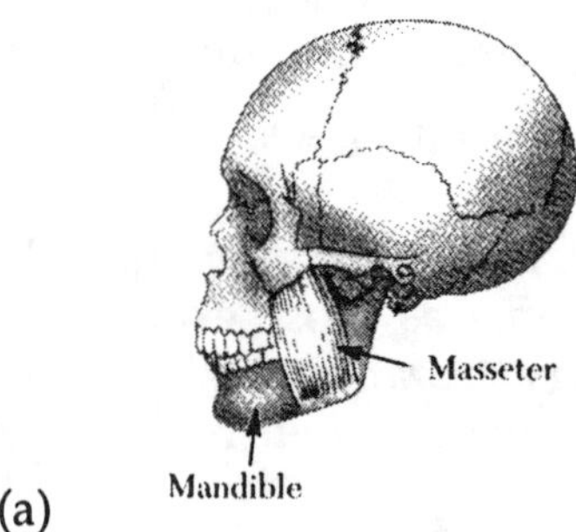

(a)

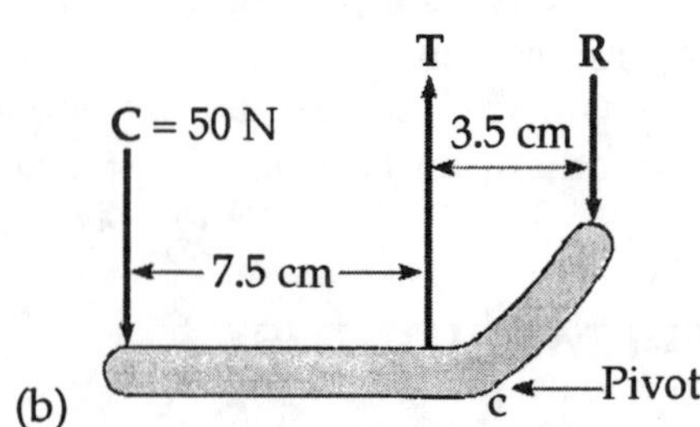

(b)

Figure 8.38

Solution Choosing the pivot point at the point shown, $\Sigma\tau = 0$ becomes

$$(50\ \text{N})(7.5\ \text{cm}) + T(0) - R(3.5\ \text{cm}) = 0$$

Thus, $R = 110\ \text{N}$ ◊

Now, apply $\Sigma F_y = 0$ to obtain

$$-50\ \text{N} + T - 107\ \text{N} = 0 \quad \text{and} \quad T = 160\ \text{N}\ \Diamond$$

---

27. If the system shown in Figure 8.47 is set in rotation about each of the axes mentioned in Problem 26, find the torque that will produce an angular acceleration of 1.50 rad/s² in each case.

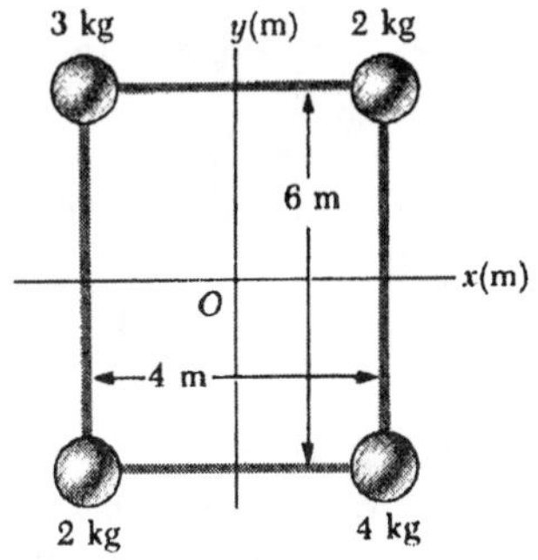

Figure 8.47

Solution From Problem 26, the values of moment of inertia are calculated to be

$$I_x = 99.0\ \text{kg}\cdot\text{m}^2 \qquad I_y = 44.0\ \text{kg}\cdot\text{m}^2$$

and $$I_0 = 143\ \text{kg}\cdot\text{m}^2$$

Using these values and the equation $\tau = I\alpha$, we find

$$\tau_x = I_x\alpha = (99.0\ \text{kg}\cdot\text{m}^2)(1.50\ \text{rad/s}^2) = 149\ \text{N}\cdot\text{m}\ \Diamond$$

$$\tau_y = I_y\alpha = (44.0\ \text{kg}\cdot\text{m}^2)(1.50\ \text{rad/s}^2) = 66.0\ \text{N}\cdot\text{m}\ \Diamond$$

$$\tau_0 = I_0\alpha = (143\ \text{kg}\cdot\text{m}^2)(1.50\ \text{rad/s}^2) = 215\ \text{N}\cdot\text{m}\ \Diamond$$

---

31. A cylindrical fishing reel has a mass of 0.85 kg and a radius of 4.0 cm. A friction clutch in the reel exerts a restraining torque of 1.3 N·m if a fish pulls on the line. The fisherman gets a bite, and the reel begins to spin with an angular acceleration of 66.0 rad/s². (a) What is the force of the fish on the line? (b) How much line unwinds in 0.50 s?

Solution (a) $I = \frac{1}{2}mr^2 = \frac{1}{2}(0.85\text{ kg})(4.0\times10^{-2}\text{ m})^2 = 6.80\times10^{-4}\text{ kg}\cdot\text{m}^2$

$$\tau_{net} = I\alpha = (6.80\times10^{-4}\text{ kg}\cdot\text{m}^2)(66\text{ rad/s}^2) = 4.49\times10^{-2}\text{ kg}\cdot\text{m}^2$$

The torque exerted by the fish = Fr, so the net torque is

$$\tau_{net} = Fr - 1.3\text{ N}\cdot\text{m} = 4.49\times10^{-2}\text{ kg}\cdot\text{m}^2$$

From which, $F = \dfrac{1.345\text{ N}\cdot\text{m}}{4.0\times10^{-2}\text{ m}} = 34\text{ N}$ ◊

(b) $\theta = \omega_0 t + \frac{1}{2}\alpha t^2 = 0 + \frac{1}{2}(66\text{ rad/s}^2)(0.50\text{ s})^2 = 8.25\text{ rad}$ and

$s = r\theta = (4.0\times10^{-2}\text{ m})(8.25\text{ rad}) = 0.33\text{ m} = 33\text{ cm}$ ◊

---

35. A cable passes over a pulley. Because of friction, the tension in the cable is not the same on opposite sides of the pulley. The force on one side is 120 N, and the force on the other side is 100 N. Assuming that the pulley is a uniform disk of mass 2.1 kg and radius 0.811 m, determine its angular acceleration.

Solution The resultant torque is given by

$$(120\text{ N})(0.811\text{ m}) - (100\text{ N})(0.811\text{ m}) = 16.2\text{ N}\cdot\text{m}$$

The moment of inertia is:

$$I = \frac{1}{2}mr^2 = \frac{1}{2}(2.1\text{ kg})(0.811\text{ m})^2 = 0.691\text{ kg}\cdot\text{m}^2$$

Then, $\tau = I\alpha$ gives $\alpha = \dfrac{\tau}{I} = \dfrac{16.2\text{ N}\cdot\text{m}}{0.691\text{ kg}\cdot\text{m}^2} = 23.5\text{ rad/s}^2$ ◊

---

39. An airliner lands with a speed of 50.0 m/s. Each wheel of the plane has a radius of 1.25 m and a moment of inertia of 110 kg·m². At touchdown the wheels begin to spin under the action of friction. Each wheel supports a weight of 1.40×10^4 N, and the wheels attain the angular speed of rolling without slipping in 0.480 s. What is the coefficient of kinetic friction between the wheels and the runway? Assume that the speed of the plane is constant.

Solution The initial angular velocity of the wheels is zero, and the final angular velocity is

$$\omega_f = \frac{v}{r} = \frac{50.0 \text{ m/s}}{1.25 \text{ m}} = 40.0 \text{ rad/s}$$

Thus,
$$\alpha = \frac{\omega_f - \omega_0}{t} = \frac{40.0 \text{ rad/s} - 0}{0.480 \text{ s}} = 83.3 \text{ rad/s}^2$$

$$\tau_{\text{center of a wheel}} = fr \quad \text{so} \quad \tau = I\alpha \quad \text{gives}$$

$$f = \frac{I\alpha}{r} = \frac{(110 \text{ kg}\cdot\text{m}^2)(83.3 \text{ rad/s}^2)}{1.25 \text{ m}} = 7.33 \times 10^3 \text{ N}$$

and
$$\mu_k = \frac{f}{n} = \frac{7.33 \times 10^3 \text{ N}}{1.40 \times 10^4 \text{ N}} = 0.524 \quad \Diamond$$

---

49. A car is designed to get its energy from a rotating flywheel with a radius of 2.00 m and a mass of 500 kg. Before a trip, the flywheel is attached to an electric motor, which brings the flywheel's rotational speed up to 5000 rev/min. (a) Find the kinetic energy stored in the flywheel. (b) If the flywheel is to supply energy to the car as would a 10-hp motor, find the length of time the car could run before the flywheel would have to be brought back up to speed.

Solution

(a) The angular velocity of the flywheel is 5000 rev/min = 524 rad/s, and its moment of inertia is

$$I = \frac{1}{2}mr^2 = \frac{1}{2}(500 \text{ kg})(2.00 \text{ m})^2 = 10^3 \text{ kg}\cdot\text{m}^2$$

Therefore, the stored energy is

$$KE = \frac{1}{2}I\omega^2 = \frac{1}{2}(10^3 \text{ kg}\cdot\text{m}^2)(524 \text{ rad/s})^2 = 1.37\times10^8 \text{ J} \quad \lozenge$$

(b) A 10-hp motor is equivalent to 7460 W. Therefore, if energy is used at the rate of 7460 J/s, the stored energy will last for

$$t = \frac{E}{P} = \frac{1.37\times10^8 \text{ J}}{7.46\times10^3 \text{ J/s}} = 1.84\times10^4 \text{ s} = 5.10 \text{ h} \quad \lozenge$$

---

61. The puck in Figure 8.52 has a mass of 0.120 kg. Its original distance from the center of rotation is 40.0 cm. The string is pulled downward 15.0 cm through the hole in the frictionless table at a speed of 80.0 cm/s. Determine the work done on the puck. (*Hint:* Consider the change of kinetic energy.)

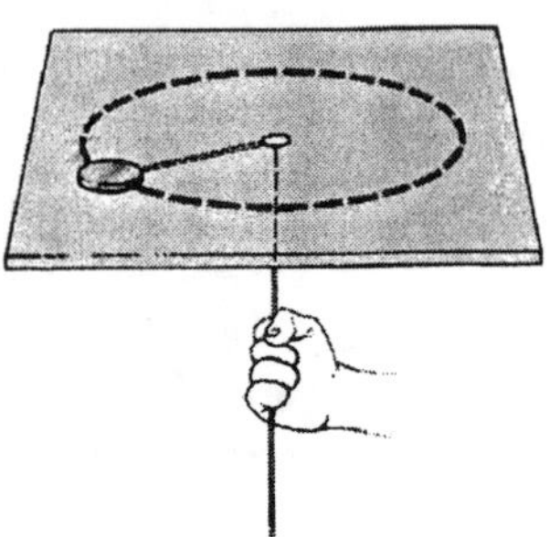

Figure 8.52

Solution

$$I_0 = mr_0^2 = (0.120 \text{ kg})(0.400 \text{ m})^2 = 1.92\times10^{-2} \text{ kg}\cdot\text{m}^2$$

$$I_f = mr_f^2 = (0.120 \text{ kg})(0.250 \text{ m})^2 = 7.50\times10^{-3} \text{ kg}\cdot\text{m}^2$$

and $$\omega_0 = \frac{v_0}{r_0} = \frac{0.800 \text{ m/s}}{0.400 \text{ m}} = 2.00 \text{ rad/s}$$

Now, use conservation of angular momentum:

$$\omega_f = \frac{I_0}{I_f}\omega_0 = \frac{1.92\times10^{-2} \text{ kg}\cdot\text{m}^2}{7.50\times10^{-3} \text{ kg}\cdot\text{m}^2}(2.00 \text{ rad/s}) = 5.12 \text{ rad/s}$$

The work done = $\Delta KE = \frac{1}{2}I_f\omega_f^2 - \frac{1}{2}I_0\omega_0^2$

Substituting the appropriate values found earlier, we have work done,

$$W = 5.99\times10^{-2} \text{ J} \quad \lozenge$$

---

65. A 240-N sphere 0.20 m in radius rolls 6.0 m down a ramp that is inclined at 37° with the horizontal. What is the angular velocity of the sphere at the bottom of the hill if it starts from rest?

Solution

We will use conservation of energy in the form

$$\frac{1}{2}mv_i^2 + \frac{1}{2}I\omega_i^2 + mgy_i = \frac{1}{2}mv_f^2 + \frac{1}{2}I\omega_f^2 + mgy_f$$

$$0 + 0 + mg(6.0 \text{ m})\sin 37° = \frac{1}{2}m(0.20 \text{ m})^2\omega^2 + \left(\frac{1}{2}\right)\frac{2}{5}m(0.20)^2\omega^2 + 0$$

where we have used $v_f = r\omega$ and $I = \frac{2}{5}mR^2$

The mass cancels from the equation, and we find the angular velocity to be

$$\omega = 36 \text{ rad/s} \quad \Diamond$$

---

67. A uniform 10.0-N picture frame is supported as shown in Figure 8.56. Find the tension in the cords and the magnitude of the horizontal force at P that are required to hold the frame in the position shown.

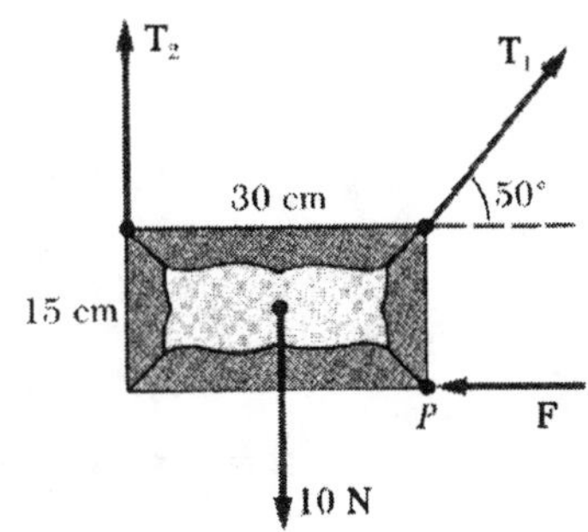

Figure 8.56

Solution Use the lower left-hand corner as the pivot point, and apply $\Sigma\tau = 0$ or

$$\Sigma\tau = -(10.0 \text{ N})(0.150 \text{ m}) - (T_1\cos 50°)(0.150 \text{ m}) + (T_1\sin 50°)(0.300 \text{ m}) = 0$$

which gives, $T_1 = 11.2 \text{ N}$ ◊

Now use $\Sigma F_x = 0$:

$$\Sigma F_x - F + (11.2)\cos 50° = 0 \quad \text{and} \quad F = 7.23 \text{ N} \;◊$$

Finally, $\Sigma F_y = 0$ yields $T_2 - 10.0 \text{ N} + (11.2 \text{ N})\sin 50° = 0$

so $T_2 = 1.39 \text{ N}$ ◊

---

71. A 12.0-kg mass is attached to a cord that is wrapped around a wheel of radius r = 10.0 cm (Fig. 8.59). The acceleration of the mass down the frictionless incline is measured to be 2.00 m/s². Assuming the axle of the wheel to be frictionless, determine (a) the tension in the rope, (b) the moment of inertia of the wheel, and (c) the angular speed of the wheel 2.00 s after it begins rotating, starting from rest.

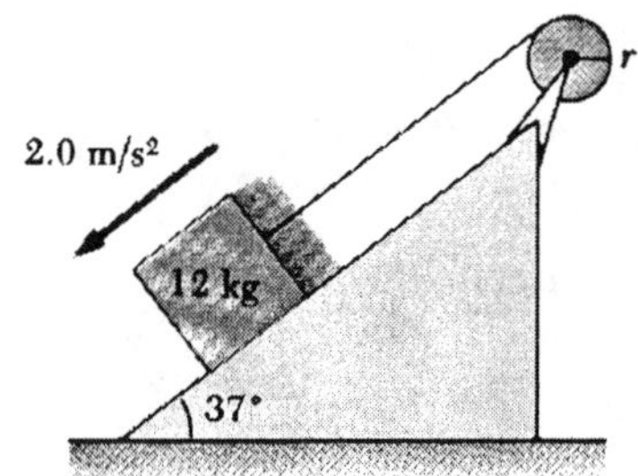

Figure 8.59

Solution From Newton's second law, we have

$mg\sin 37° - T = ma$ which gives

$$(12.0 \text{ kg})(9.80 \text{ m/s}^2)\sin 37° - T = (12.0 \text{ kg})a$$

or $T = 70.56 \text{ N} - (12.0 \text{ kg})a$ (1)

Also $\alpha = \dfrac{a}{r} = \dfrac{2.00 \text{ m/s}^2}{0.100 \text{ m}} = 20.0 \text{ rad/s}^2$ (2)

(a) Thus, from (1), $T = 70.56 \text{ N} - (12.0 \text{ kg})(2.00 \text{ m/s}^2) = 46.6 \text{ N}$ ◊

(b) From $\tau = I\alpha$, we have: $I = \dfrac{\tau}{\alpha} = \dfrac{Tr}{\alpha} = \dfrac{(46.6 \text{ N})(0.100 \text{ m})}{20.0 \text{ rad/s}} = 0.233 \text{ kg} \cdot \text{m}^2$ ◊

(c) $\omega = \omega_0 + \alpha t$ becomes $\omega = 0 + (20.0 \text{ rad/s}^2)(2.00 \text{ s}) = 40.0 \text{ rad/s}$ ◊

---

75. A 4.00-kg mass is connected by a light cord to a 3.00-kg mass on a smooth surface (Fig. 8.62). The pulley rotates about a frictionless axle and has a moment of inertia of 0.500 kg·m^2 and a radius of 0.30 m. Assuming that the cord does not slip on the pulley, find (a) the acceleration of the two masses and (b) the tensions T_1 and T_2.

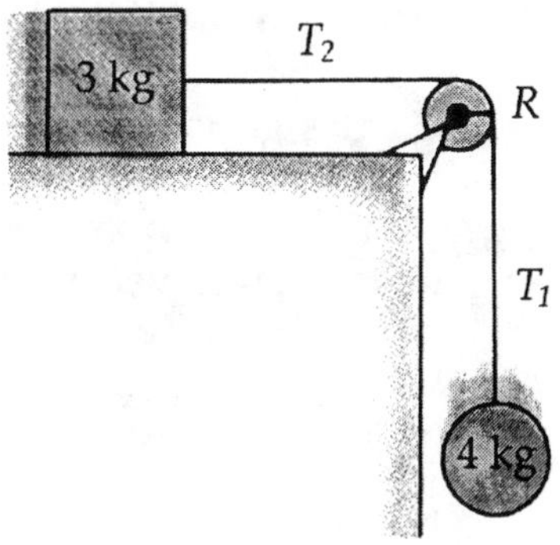

Figure 8.62

Solution

(a) and (b) First, we write down Newton's second law for the 3.00-kg mass.

$$T_2 = (3.00 \text{ kg})a \qquad (1)$$

The second law for the 4.00-kg mass is

$$39.2 \text{ N} - T_1 = (4.00 \text{ kg})a \qquad (2)$$

Now apply $\tau = I\alpha$ to the pulley (axis of rotation at its center).

$$(T_1 - T_2)r = (0.500 \text{ kg} \cdot \text{m}^2)\left(\frac{a}{r}\right)$$

or

$$T_1 - T_2 = (5.56 \text{ kg})a \qquad (3)$$

Equations (1), (2), and (3) can be solved simultaneously to give

$$a = 3.12 \text{ m/s}^2 \quad \lozenge$$

$$T_1 = 26.7 \text{ N} \quad \lozenge$$

and

$$T_2 = 9.37 \text{ N} \quad \lozenge$$

CHAPTER SELF-QUIZ

1. The Earth's gravity exerts no torque on a satellite orbiting the Earth in an elliptical orbit. Compare the motion at the point nearest the Earth (perigee) to the motion at the point farthest from the Earth (apogee). At these two points
   a. the tangential velocities are the same
   b. the angular velocities are the same
   c. the angular momenta are the same
   d. the kinetic energies are the same

2. A majorette takes two batons and fastens them together in the middle at right angles to make an "x" shape. Each baton was 0.8 m long and each ball on the end is 0.30 kg. (Ignore the mass of the rods.) What is the moment of inertia if the arrangement is spun around an axis formed by one of the batons?
   a. $0.048\ \text{kg}\cdot\text{m}^2$
   b. $0.096\ \text{kg}\cdot\text{m}^2$
   c. $0.19\ \text{kg}\cdot\text{m}^2$
   d. $0.38\ \text{kg}\cdot\text{m}^2$

3. A solid cylinder $\left(I = \dfrac{MR^2}{2}\right)$, a hoop $(I = MR^2)$, and a sphere $\left(I = \dfrac{2MR^2}{5}\right)$ are released and roll down a slope. The mass of all objects is the same, but the cylinder has twice the radius of the other two. Which reaches the bottom first?
   a. the cylinder
   b. the hoop
   c. the sphere
   d. at least two reach the bottom at the same time

4. An astronaut is on a 200-m life-line outside a spaceship, circling the ship with an angular velocity of 0.10 rad/s. How far can she be pulled in before the centripetal acceleration reaches $5g = 49\ \text{m/s}^2$? The final length of the rope will be
   a. 40 m
   b. 69 m
   c. 571 m
   d. 10.1 m

5. A bus is designed to draw its power from a rotating flywheel that is brought up to its maximum speed (3000 rpm) by an electric motor. The flywheel is a solid cylinder of mass 1000 kg and diameter 1 m $\left(I_{\text{cylinder}} = \frac{1}{2}MR^2\right)$. If the bus requires an average power of 10 kilowatts, how long will the flywheel rotate?
   a. 330 s
   b. 625 s
   c. 975 s
   d. 1250 s

6. A solid cylinder with mass, M, and radius, R, rolls along a level surface without slipping with a linear velocity, v. What is the ratio of its rotational kinetic energy to its linear kinetic energy? (For a solid cylinder, $I = 0.5MR^2$.)
   a. 1/4
   b. 1/2
   c. 1/1
   d. 2/1

7. A phonograph turntable has a moment of inertia of $2.50 \times 10^{-2}\ \text{kg} \cdot \text{m}^2$ and spins freely on a frictionless bearing at 33.3 rev/min. A 0.25-kg ball of putty is dropped vertically on the turntable and sticks at a point 0.20 m from the center. What is the new rate of rotation of the 0.20 system?
   a. 40.8 rev/min
   b. 23.8 rev/min
   c. 33.3 rev/min
   d. 27.2 rev/min

8. A solid sphere with a mass of 3.0 kg and radius of 0.2 m, starts from rest at the top of a ramp, inclined at 15°, and rolls to the bottom. The upper end of the ramp is 1.2 m higher than the lower end. What is the linear velocity of the sphere when it reaches the bottom of the ramp? (*Note:* $I = 0.4MR^2$ for a solid sphere and $g = 9.8\ \text{m/s}^2$.)
   a. 4.7 m/s
   b. 4.1 m/s
   c. 3.4 m/s
   d. 2.4 m/s

9. A bowling ball has a mass of 4.0 kg, a moment of inertia of 1.60×10^{-2} $kg \cdot m^2$ and a radius of 0.20 m. If it rolls down the lane without slipping at a linear speed of 4.0 m/s, what is its angular velocity?
   a. 0.8 rad/s
   b. 10.0 rad/s
   c. 0.05 rad/s
   d. 20.0 rad/s

10. A bucket filled with water has a mass of 23 kg and is attached to a rope which in turn is wound around a 0.05 m radius cylinder at the top of a well. The bucket and water are first raised to the top of the well and then released. The bucket and water are moving with a speed of 7.9 m/s upon hitting the water surface in the well below. What is the angular velocity of the cylinder at this instant?
    a. 39 rad/s
    b. 79 rad/s
    c. 118 rad/s
    d. 158 rad/s

11. A phonograph record and turntable combination have a moment of inertia of 250×10^{-6} (SI units) and rotate with an angular velocity of 6 rad/sec. What net torque must be applied to bring the system to rest within 3 s?
    a. 4500×10^{-6} $N \cdot m$
    b. 750×10^{-6} $N \cdot m$
    c. 500×10^{-6} $N \cdot m$
    d. 250×10^{-6} $N \cdot m$

12. If a long rod of length L is hinged at one end, the moment of inertia as the rod rotates around that hinge is $\frac{ML^2}{3}$. A 4-m rod with a mass of 3 kg, hinged at one end, is held in a horizontal position, and then released as the free end is allowed to fall. What is the angular acceleration as it is released?
    a. 3.7 rad/s^2
    b. 7.35 rad/s^2
    c. 2.45 rad/s^2
    d. 4.9 rad/s^2

9
Solids and Fluids

Chapter 9

SOLIDS AND FLUIDS

In this chapter we consider some properties of solids and fluids (both liquids and gases). We spend some time looking at properties that are peculiar to solids, but much of our emphasis is on the properties of fluids. We open the fluids part of the chapter, with a study of fluids at rest, and finish with a discussion of the properties of fluids in motion.

NOTES FROM SELECTED CHAPTER SECTIONS

9.1 States of Matter

Matter is generally classified as being in one of three states: solid, liquid, or gas.

In a *crystalline solid,* the atoms are arranged in an ordered periodic structure; while in an *amorphous solid* (i.e. glass), the atoms are present in a disordered fashion.

In the *liquid state,* thermal agitation is greater than in the solid state, the molecular forces are weaker, and molecules wander throughout the liquid in a random fashion.

The molecules of a *gas* are in constant random motion and exert weak forces on each other. The distances separating molecules are large compared to the dimensions of the molecules.

9.2 The Deformation of Solids

The elastic properties of solids are described in terms of stress and strain. Stress is a quantity that is related to the force causing a deformation; strain is a measure of the degree of deformation. It is found that, for sufficiently small stresses, stress is proportional to strain, and the constant of proportionality depends on the material being deformed and on the nature of the deformation. We call this proportionality constant the elastic modulus.

We shall consider three types of deformation and define an elastic modulus for each:

1. *Young's modulus,* which measures the resistance of a solid to a change in its length.
2. *Shear modulus,* which measures the resistance to displacement of the planes of a solid sliding past each other.
3. *Bulk modulus,* which measures the resistance that solids or liquids offer to changes in their volume.

9.3 Density and Pressure

The *density*, ρ, of a substance of uniform composition is defined as its *mass per unit volume* and has units of kilograms per cubic meter (kg/m^3) in the SI system.

The *specific gravity* of a substance is a dimensionless quantity which is the ratio of the density of the substance to the density of water.

The *pressure*, P, in a fluid is the force per unit area that the fluid exerts on an object immersed in the fluid.

9.4 Variation of Pressure with Depth

In a *fluid at rest*, all points at the same depth are at the same pressure. *Pascal's law* states that a change in pressure applied to an *enclosed* fluid is transmitted undiminished to every point in the fluid and the walls of the containing vessel.

9.5 Pressure Measurements

The *absolute pressure* of a fluid is the sum of the *gauge pressure* and atmospheric pressure. The SI unit of pressure is the Pascal (Pa). Note that $1\ \text{Pa} \equiv 1\ \text{N/m}^2$.

9.6 Buoyant Forces and Archimedes' Principle

Any object partially or completely submerged in a fluid experiences a buoyant force equal in magnitude to the weight of the fluid displaced by the object and acting vertically upward through the point which was the center of gravity of the displaced fluid.

9.7 Fluids in Motion

Many features of fluid motion can be understood by considering the behavior of an ideal fluid, which satisfies the following conditions:

1. *The fluid is nonviscous;* that is, there is no internal friction force between adjacent fluid layers.
2. *The fluid is incompressible,* which means that its density is constant.
3. *The fluid motion is steady,* meaning that the velocity, density, and pressure at each point in the fluid do not change in time.
4. *The fluid moves without turbulence.* This implies that each element of the fluid has zero angular velocity about its center; that is, there can be no eddy currents present in the moving fluid.

Fluids which have the "ideal" properties stated above obey two important equations:

> The *equation of continuity* states that the flow rate through a pipe is constant (i.e. the product of the cross-sectional area of the pipe and the speed of the fluid is constant).
>
> *Bernoulli's equation* states that the sum of the pressure, kinetic energy per unit volume, and the potential energy per unit volume has a constant value at all points along a streamline.

EQUATIONS AND CONCEPTS

A body of matter can be deformed (experience change in size or shape) by the application of external forces. Stress is a quantity which is proportional to the force which causes the deformation; strain is a measure of the degree of deformation. The elastic modulus is a general characterization of the deformation. Particular deformations are characterized by specific moduli.

$$\text{Elastic modulus} \equiv \frac{\text{stress}}{\text{strain}} \qquad (9.1)$$

The SI units of pressure are newtons per square meter, or Pascal (Pa).

$$1\text{ Pa} \equiv 1\text{ N/m}^2 \qquad (9.2)$$

Young's modulus is a measure of the resistance of a body to elongation.

$$Y = \frac{F/A}{\Delta L/L_0} \qquad (9.3)$$

Within a limited range of values, the graph of stress vs. strain for a given substance will be a straight line. When the stress exceeds the elastic limit (at the yield point), the stress-strain curve will no longer be linear.

Comment on experimental observations.

The **Shear modulus** is a measure of the deformation which occurs when a force is applied along a direction parallel to one surface of a body.

$$S = \frac{F/A}{\Delta x/h} \tag{9.4}$$

The **Bulk modulus** characterizes the response of a body to uniform pressure (or squeezing) on all sides. Note that when ΔP is positive (increase in pressure), the ratio $\Delta V/V$ will be negative (decrease in volume) and vice versa. Therefore, the negative sign in the equation ensures that B will always be positive.

$$B = -\frac{\Delta P}{\Delta V/V} \tag{9.5}$$

The **density** of a homogeneous substance is defined as its ratio of mass per unit volume. The value of density is characteristic of a particular type of material and independent of the total quantity of material in the sample.

$$\rho \equiv \frac{M}{V} \tag{9.6}$$

The SI units of density are kg per cubic meter.

$1\ \text{g/cm}^3 = 1000\ \text{kg/m}^3$ — Conversion of Units.

The (average) pressure of a fluid is defined as the normal force per unit area acting on a surface immersed in the fluid.

$$P \equiv F/A \tag{9.7}$$

Atmospheric pressure is often expressed in other units: — Conversion of Units.

atmosphere: $1\ \text{atm} = 1.01 \times 10^5\ \text{Pa}$

mm of mercury (Torr): $1\ \text{Torr} = 133.3\ \text{Pa}$

pounds per sq. inch: $1\ \text{lb/in}^2 = 6.9 \times 10^3\ \text{Pa}$

The absolute pressure, P, at a depth, h, below the surface of a liquid which is open to the atmosphere is greater than atmospheric pressure, P_a, by an amount which depends on the depth below the surface.

$$P = P_a + \rho g h \qquad (9.10)$$

Comments on fluid pressure.

The quantity $\rho g h$ is called the gauge pressure and P is the absolute pressure. Therefore,

absolute pressure = atmospheric pressure + gauge pressure

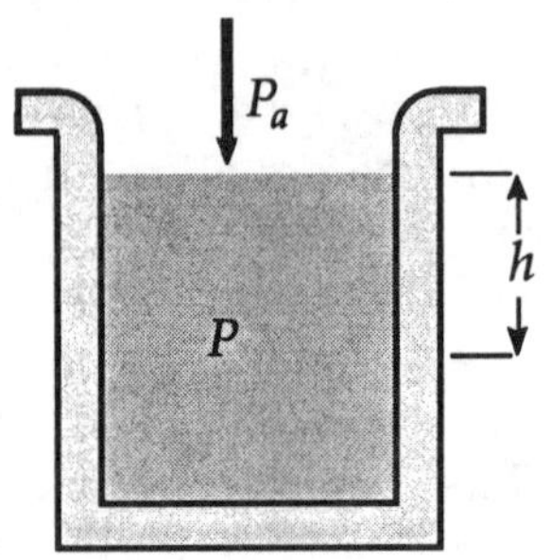

Pascal's law states that pressure applied to an enclosed fluid (liquid or gas) is transmitted undiminished to every point within the fluid and over the walls of the vessel which contain the fluid.

Archimedes' principle states that when an object is partially or fully submersed in a fluid, the fluid exerts an upward buoyant force on the object. The magnitude of the buoyant force depends on the density of the fluid and the volume of displaced fluid, V. In particular, note that B equals the weight of the displaced fluid.

$$B = \rho_f V g \qquad (9.11)$$

Archimedes' principle

Fluid dynamics, the treatment of fluids in motion, is greatly simplified under the assumption that the fluid is ideal with the following characteristics:

Comments on ideal fluids.

(a) nonviscous - internal friction between adjacent fluid layers is negligible

(b) incompressible - the density of the fluid is constant throughout

(c) steady - the velocity, density, and pressure at each point in the fluid are constant in time

(d) irrotational (without turbulence) - there are no eddy currents within the fluid (each element of the fluid has zero angular velocity about its center)

Equation 9.13 is the equation of continuity. For an *incompressible fluid* (ρ = constant), the equation of continuity can be written as Equation 9.14. The product Av is called the flow rate. Therefore, the flow rate at any point along a pipe carrying an incompressible fluid is constant.

$$\rho A_1 v_1 = \rho A_2 v_2 \qquad (9.13)$$

$$A_1 v_1 = A_2 v_2 \qquad (9.14)$$

Bernoulli's equation is the most fundamental law in fluid mechanics. The equation is a statement of the law of conservation of mechanical energy as applied to a fluid. Bernoulli's equation states that the sum of pressure, kinetic energy per unit volume, and potential energy per unit volume remains constant along a streamline of an ideal fluid.

$$P + \frac{1}{2}\rho v^2 + \rho g y = \text{constant} \qquad (9.16)$$

REVIEW CHECKLIST

▷ Describe the three types of deformations that can occur in a solid, and define the elastic modulus that is used to characterize each: (Young's modulus, Shear modulus, and Bulk modulus).

▷ Understand the concept of pressure at a point in a fluid, and the variation of pressure with depth. Understand the relationships among absolute, gauge, and atmospheric pressure values; and know the several different units commonly used to express pressure.

▷ Understand the origin of buoyant forces; and state and explain Archimedes' principle.

▷ State and understand the physical significance of the *equation of continuity* (constant flow rate) and *Bernoulli's equation* for fluid flow (relating *flow velocity, pressure,* and *pipe elevation).*

SOLUTIONS TO SELECTED END-OF-CHAPTER PROBLEMS

1. The heels on a pair of women's shoes have radii of 0.50 cm at the bottom. If 30% of the weight of a woman weighing 480 N is supported by each heel, find the stress on each heel.

Solution

$$\text{Stress} = \frac{F}{A}, \quad \text{where} \quad F = 0.3(\text{weight}) = 0.3(480\text{ N}) = 144\text{ N}, \quad \text{and}$$

$$A = \pi r^2 = \pi(0.50\times10^{-2}\text{ m})^2 = 7.85\times10^{-5}\text{ m}^2$$

Thus,

$$\text{Stress} = \frac{144\text{ N}}{7.85\times10^{-5}\text{ m}^2} = 1.8\times10^6\text{ Pa} \quad \Diamond$$

---

5. A child slides across a floor in a pair of rubber-soled shoes. The friction force acting on each foot is 20 N, the cross-sectional area of each foot is 14 cm^2, and the height of the soles is 5.0 mm. Find the horizontal distance traveled by the sheared face of the sole. The shear modulus of the rubber is 3.0×10^6 Pa.

Solution From the defining equation for the shear modulus, $S = \dfrac{F/A}{\Delta x/h}$, we find Δx:

$$\Delta x = \frac{Fh}{SA} = \frac{(20\text{ N})(5.0\times10^{-3}\text{ m})}{(3.0\times10^6\text{ Pa})(14\times10^{-4}\text{ m}^2)} = 2.4\times10^{-5}\text{ m}$$

or

$$\Delta x = 2.4\times10^{-2}\text{ mm} \quad \Diamond$$

---

13. Determine the elongation of the rod in Figure 9.31 if it is under a tension of 5.8×10^3 N.

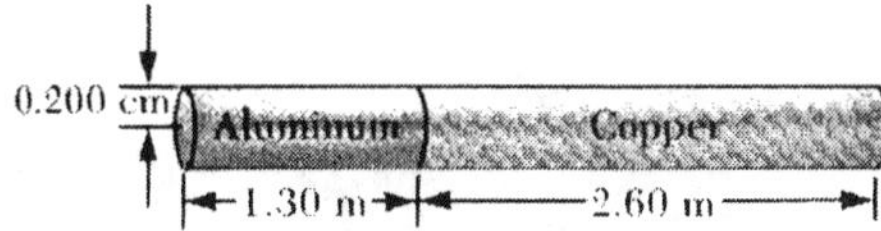

Figure 9.31

Solution

The cross-sectional area of the wire

$$A = \pi r^2 = \pi(2.0 \times 10^{-3}\ \text{m})^2 = 1.256 \times 10^{-5}\ \text{m}^2, \qquad \text{and the tension} = 5.8 \times 10^3\ \text{N}$$

throughout both pieces. Let us compute the elongation of each part separately:

For aluminum

$$\Delta L_{al} = \frac{L_0 F}{YA} = \frac{(1.3\ \text{m})(5.8 \times 10^3\ \text{N})}{(7.0 \times 10^{10}\ \text{Pa})(1.256 \times 10^{-5}\ \text{m}^2)} = 8.57 \times 10^{-3}\ \text{m}$$

Similarly, for the copper part

$$\Delta L_{cu} = \frac{L_0 F}{YA} = \frac{(2.6\ \text{m})(5.8 \times 10^3\ \text{N})}{(11 \times 10^{10}\ \text{Pa})(1.256 \times 10^{-5}\ \text{m}^2)} = 10.9 \times 10^{-3}\ \text{m}$$

So, the total elongation = 19.5×10^{-3} m = 2.0 cm ◊

---

17. A pipe contains water at 5.00×10^5 Pa above atmospheric pressure. If the only material you have available to patch a 4-mm-diameter hole in the pipe is a piece of bubble gum, how much force must the gum be able to withstand?

Solution From the definition of pressure, we have $F = PA$. The area of the hole is

$$A = \frac{\pi d^2}{4} = \frac{\pi(4.00 \times 10^{-3}\ \text{m})^2}{4} = 1.26 \times 10^{-5}\ \text{m}^2$$

Therefore, $$F = (5.00 \times 10^5\text{Pa})(1.26 \times 10^{-5}\ \text{m}^2) = 6.28\ \text{N} \quad ◊$$

---

21. A collapsible plastic bag (Fig. 9.33) contains a glucose solution. If the average gauge pressure in the artery is 1.33×10^4 Pa, what must be the minimum height, h, of the bag in order to infuse glucose into the artery? Assume that the specific gravity of the solution is 1.02.

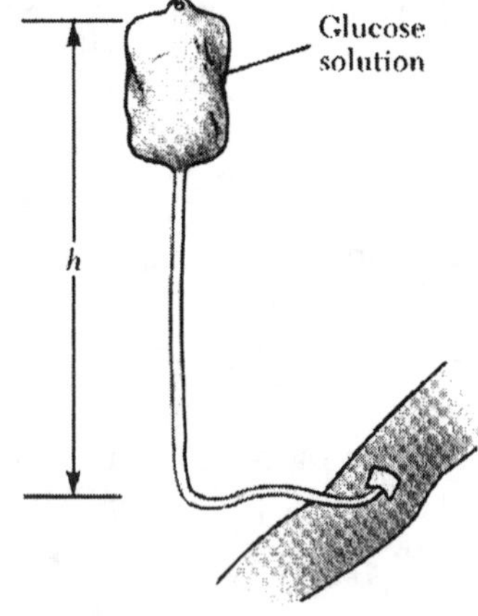

Figure 9.33

Solution The gauge pressure of the fluid at the level of the needle must equal the gauge pressure in the artery.

$$P_{\text{gauge}} = \rho g h = 1.333 \times 10^4 \text{Pa}$$

So,

$$h = \frac{1.333 \times 10^4 \text{ Pa}}{(1.02 \times 10^3 \text{ kg/m}^3)(9.8 \text{ m/s}^2)} = 1.33 \text{ m} \quad \Diamond$$

---

25. Piston 1 in Figure 9.37 has a diameter of 0.250 in.; piston 2 has a diameter of 1.50 in. In the absence of friction, determine the force, **F**, necessary to support the 500-lb weight.

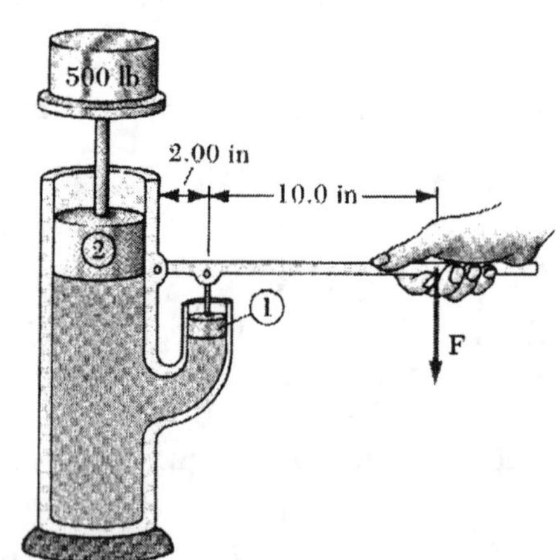

Figure 9.37

Solution Pascal's principle, $\dfrac{F_2}{A_2} = \dfrac{F_1}{A_1}$

becomes

$$F_1 = \frac{A_1}{A_2} F_2 = \frac{\frac{\pi}{4}(0.250 \text{ in})^2}{\frac{\pi}{4}(1.50 \text{ in})^2}(500 \text{ lbs}) = 13.9 \text{ lb}$$

Now, consider the jack handle with the pivot point at the left end.

$$\Sigma\tau = 0 = (13.9 \text{ lb})(2.00 \text{ in}) - F(12.0 \text{ in}) = 0 \quad \text{and} \quad F = 2.31 \text{ lb} \quad \Diamond$$

---

37. A hollow brass tube (diam. = 4.00 cm) is sealed at one end and loaded with lead shot to give a total mass of 0.200 kg. When the tube is floated in pure water, what is the depth, z, of its bottom end?

Solution When the system floats, the buoyant force, $\mathbf{F_B}$ (the weight of the displaced water), equals the weight of the object. Therefore, $F_B = mg$, or

$$\rho(\pi r^2 z)g = 1.96 \text{ N} \qquad \text{where} \qquad \pi r^2 z = \text{volume of displaced water}$$

$$z = \frac{1.96 \text{ N}}{(10^3 \text{ kg/m}^3)\pi(2.00\times10^{-2} \text{ m})^2(9.80 \text{ m/s}^2)} = 0.159 \text{ m} = 15.9 \text{ cm} \quad \Diamond$$

---

41. A cowboy at a dude ranch fills a horse trough that is 1.5 m long, 60 cm wide, and 40 cm deep. He uses a 2-cm-diameter hose from which water emerges at 1.5 m/s. How long does it take him to fill the trough?

Solution The cross-sectional area of the 2-cm diameter hose is $A = \pi r^2 = \pi \times 10^{-4} \text{ m}^2$, and the flow rate is:

$$Av = (\pi \times 10^{-4} \text{ m}^2)(1.5 \text{ m/s}) = 1.5\pi \times 10^{-4} \text{ m}^3/\text{s}$$

The volume to be filled is (1.5 m)(0.60 m)(0.40 m) = 3.6×10^{-1} m^3. The time required to fill the trough is:

$$\text{time} = \frac{\text{volume}}{\text{flow rate}} = \frac{3.6\times10^{-1} \text{ m}^3}{1.5\pi\times10^{-4} \text{ m}^3/\text{s}} = 760 \text{ s} = 13 \text{ min} \quad \Diamond$$

---

45. A liquid ($\rho = 1.65 \text{ g/cm}^3$) flows through two horizontal sections of tubing joined end to end. In the first section, the cross-sectional area is 10.0 cm^2, the flow speed is 275 cm/s, and the pressure is 1.20×10^5 Pa. In the second section, the cross-sectional area is 2.50 cm^2. Calculate the smaller section's (a) flow speed and (b) pressure.

Solution (a) We find the flow velocity in the second section from the continuity equation:

$$v_2 = \frac{A_1 v_1}{A_2} = \frac{10.0}{2.50} v_1 = 4(2.75 \text{ m/s}) = 11.0 \text{ m/s} \quad \lozenge$$

(b) Choosing the zero level for y along the common center line of the pipes and using Bernoulli's equation, we have

$$P_1 + \frac{1}{2}\rho v_1^2 = P_2 + \frac{1}{2}\rho v_2^2 \quad \text{or} \quad P_2 = P_1 + \frac{1}{2}\rho(v_1^2 - v_2^2) \quad \text{giving}$$

$$P_2 = (1.20 \times 10^5 \text{ Pa}) + \frac{1}{2}(1650 \text{ kg/m}^3)[(2.75 \text{ m/s})^2 - (11.0 \text{ m/s})^2]$$

$$\text{and} \quad P_2 = 2.64 \times 10^4 \text{ Pa} \quad \lozenge$$

---

49. What is the net upward force on an airplane wing of area 20 m^2 if the speed of flow is 300 m/s across the top of the wing and 280 m/s across the bottom?

Solution We select point 1 just above the wing and point 2 just below it. As a result, the difference in vertical heights between these two points is negligible, and Bernoulli's equation reduces to

$$P_2 - P_1 = \frac{1}{2}\rho(v_1^2 - v_2^2) = \frac{1}{2}(1.3 \text{ kg/m}^3)[(300 \text{ m/s})^2 - (280 \text{ m/s})^2]$$

or

$$P_2 - P_1 = 7540 \text{ Pa}$$

The net upward force is therefore

$$\mathbf{F} = (P_2 - P_1)A = (7540 \text{ Pa})(20 \text{ m}^2) = 1.50 \times 10^5 \text{ N upward} \quad \lozenge$$

---

53. A jet of water squirts out horizontally from a hole in the bottom of the tank in Figure 9.46 on the following page. If the hole has a diameter of 3.50 mm, what is the height, h, of the water level in the tank?

Solution First, consider the path from the standpoint of projectile motion to find the speed at which the water emerges from the tank. The time to drop one meter with an initial vertical velocity of zero is:

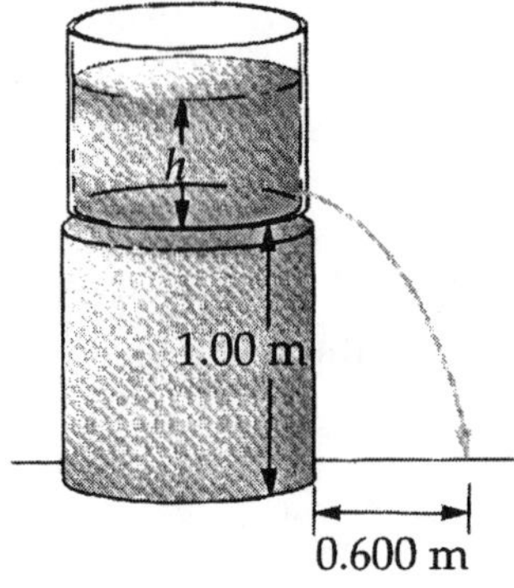

Figure 9.46

$$y = v_{0y}t + \frac{1}{2}at^2 \qquad \text{or} \qquad 1\text{ m} = 0 + \frac{1}{2}(9.80\text{ m/s}^2)t^2$$

or $t = 0.452$ s, and from the horizontal motion

$$v_x = v_0 = \frac{\Delta x}{t} = \frac{0.600\text{ m}}{0.452\text{ s}} = 1.33\text{ m/s}$$

We now use Bernoulli's equation, with point 1 at the top of the tank and point 2 at the level of the hole. With $P_1 = P_2 = P_{atm}$ and v_1 approximately equal to zero, we have

$$\frac{1}{2}\rho v_2^2 = \rho g(y_1 - y_2) = \rho g h, \qquad \text{giving}$$

$$h = \frac{v_0^2}{2g} = \frac{(1.33\text{ m/s})^2}{2(9.80\text{ m/s}^2)} = 9.00 \times 10^{-2}\text{ m} = 9.00\text{ cm} \quad \Diamond$$

---

57. A water tank open to the atmosphere at the top has two holes punched in its side, one above the other. The holes are 5.00 cm and 12.00 cm above the floor. How high does water stand in the tank if the two streams of water hit the floor at the same place?

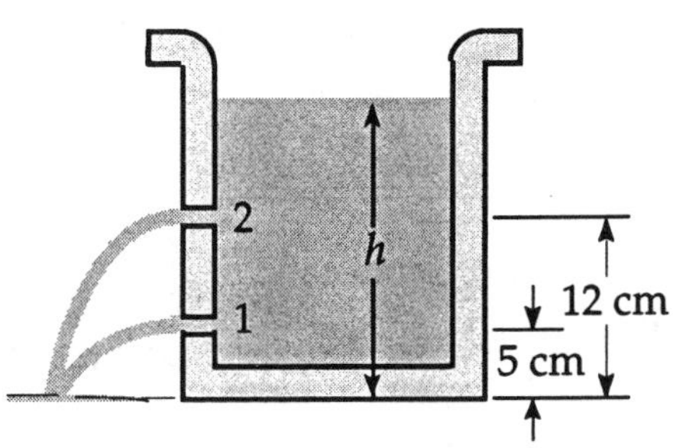

Solution We call the position of the lower hole point 1 and the position of the higher hole point 2. *Both of these points are at atmospheric pressure.* From our earlier study of projectile motion, the range is given by

$$R = v_1 t_1 = v_2 t_2 \qquad (1)$$

An expression for the time for the water to reach the floor can also be found through the projectile motion equations. We have

$$y = v_{0y}t + \frac{1}{2}at^2 = 0 + \frac{1}{2}gt^2, \qquad \text{or} \qquad t = \sqrt{\frac{2}{g}h}$$

where h is the height through which the projectile falls before it reaches ground level. Thus, we have the times, t_1 and t_2, for the water from the two holes to reach the floor to be

$$t_2 = \sqrt{\frac{2}{g}(0.120\text{ m})} \qquad \text{and} \qquad t_1 = \sqrt{\frac{2}{g}(0.050\text{ m})}$$

Substitute these two values for the times into Equation (1) above to give

$$v_1^2 = 2.40 v_2^2 \qquad (2)$$

Since the pressure cancels from Bernoulli's equation, it reduces to

$$(v_1^2 - v_2^2) = 2g(y_2 - y_1) \qquad (3)$$

Given: $y_2 = 0.120\text{ m}, \quad y_1 = 0.050\text{ m}$

Then, with the use of Equations (2) and (3), we find $v_2 = 0.990\text{ m/s}$

Let us now consider Bernoulli's Equation once again, with a point 3 *at the top of the tank* and point 2 still at the position of the top hole. Both these points are at atmospheric pressure, and the velocity of fall of water at the top of the tank is negligibly small. We find

$$\rho g y_3 = \frac{1}{2}\rho v_2^2 + \rho g y_2 \qquad \text{or} \qquad \rho g(y_3 - y_2) = \rho g h = \frac{1}{2}\rho v_2^2$$

from which $h = \dfrac{v_2^2}{2g} = \dfrac{(0.990\text{ m/s})^2}{2(9.80\text{ m/s}^2)} = 5.00 \times 10^{-2}\text{ m} = 5.00\text{ cm}$ ◊

Thus, the water surface is 5.00 cm above the top hole or 17.0 cm above the bottom of the tank.

---

61. The density of ice is 920 kg/m^3, and that of sea water is 1030 kg/m^3. What fraction of the total volume of an iceberg is exposed?

Solution The buoyant force, **B**, on the iceberg must be equal to its weight, w, in order for it to float. Thus, $B = w$. But, the buoyant force is equal to the weight of the water displaced.

Therefore, $\rho_w V_{(uw)} g = \rho_{\text{ice}} V_{(\text{total})} g$ where $V_{(uw)}$ is the volume of the iceberg under water and $V_{(total)}$ is the total volume of the berg.

We have
$$\frac{V_{(uw)}}{V_{(\text{total})}} = \frac{\rho_{\text{ice}}}{\rho_w} = \frac{920}{1000} = 0.920 \quad \lozenge$$

Therefore, 92.0% of the volume is submerged and 8.00% of the volume is exposed.

---

67. A 600-kg weather balloon is designed to lift a 400-kg package. What volume should the balloon have after being inflated with helium at standard temperature and pressure, in order that the total load can be lifted?

Solution The forces on the balloon while in flight are B_a, the buoyant force of the air; w_{He}, the weight of the helium; w_B, the weight of the balloon; and w_L, the weight of the load. These quantities are found as follows:

$$B_a = \rho_a V_{(\text{balloon})} g$$

$$w_{He} = \rho_{He} V_{(\text{balloon})} g$$

$$w_B = (600 \text{ kg})g$$

and
$$w_L = (4000 \text{ kg})g$$

When floating in equilibrium, we have $B_a = w_{He} + w_B + w_L$ or

$$\rho_a V_{(\text{balloon})} g = \rho_{He} V_{(\text{balloon})} g + (600 \text{ kg})g + (4000 \text{ kg})g$$

We are given the density of Helium as 0.178 kg/m^3 and the density of air as 1.29 kg/m^3. Thus, we can solve for the volume of the balloon to find

$$V_{(\text{balloon})} = 4.14 \times 10^3 \text{ m}^3 \quad \lozenge$$

---

73. A small sphere 0.60 times as dense as water is dropped from a height of 10 m above the surface of a smooth lake. Determine the maximum depth to which the sphere will sink. Neglect any energy transferred to the water during impact and sinking.

Solution First, find the speed of the sphere just before impact by use of the free-fall equations:

$$v^2 = v_0^2 + 2ay = 0 + 2(-9.80 \text{ m/s}^2)(-10 \text{ m})$$

or $$v = -14 \text{ m/s}$$

This is the initial velocity of the sphere as it enters the water. Once under water, the forces acting on the sphere are the buoyant force of the water, **B**, and its weight, w. Thus, the net force on the object is

$$F_{\text{net}} = B - w = \rho_w Vg - \rho Vg \qquad (1)$$

But, we are given that the density of the sphere is $\rho = 0.60\rho w$. Thus, from (1) we see that the net force on the sphere is

$$F_{\text{net}} = \rho_w Vg - 0.60\rho_w Vg \qquad \text{or} \qquad F_{\text{net}} = 0.4\rho_w Vg$$

From Newton's second law, the acceleration of the sphere is

$$a = \frac{F_{\text{net}}}{m} = \frac{F_{\text{net}}}{\rho V} = \frac{F_{\text{net}}}{0.6\rho_w V} = \frac{0.4\rho_w Vg}{0.6\rho_w V} = \left(\frac{2}{3}\right)g$$

We can now find the stopping distance, h, using

$$v^2 = v_0^2 + 2ay = (-14 \text{ m/s})^2 + 2\left(\frac{2}{3}\right)gh$$

giving $$h = -15 \text{ m}$$

The object sinks 15 m into the water. ◊

---

79. Water at a pressure of 3.00×10^5 Pa flows through a horizontal pipe at a speed of 1.00 m/s. If the pipe narrows to one fourth its original diameter, find (a) the flow speed in the narrow section and (b) the pressure in the narrow section.

Solution (a) The speed at the narrow section is found from the equation of continuity:

$$v_2 = \frac{A_1 v_1}{A_2} = \frac{d_1^2}{d_2^2} v_1 = 16.0 v_1 = 16.0 \text{ m/s} \quad \Diamond$$

(b) Choose the y = 0 level at the common center line of the horizontal pipes; and use Bernoulli's equation to calculate the pressure in the narrow section.

$$P_2 = P_1 + \frac{1}{2}\rho(v_1^2 - v_2^2)$$

$$P_2 = 3.00\times10^5\ \text{Pa} + \frac{1}{2}(1000\ \text{kg/m}^3)\left[(1.00\ \text{m/s})^2 - (16.0\ \text{m/s})^2\right]$$

$$P_2 = 1.73\times10^5\ \text{Pa} \quad \Diamond$$

---

81. A siphon is a device that allows a fluid to seemingly defy gravity (Fig. 9.52). The flow must be initiated by a partial vacuum in the tube, as in a drinking straw. (a) Show that the speed at which the water emerges from the siphon is given by $v = \sqrt{2gh}$. (b) For what values of y will the siphon work?

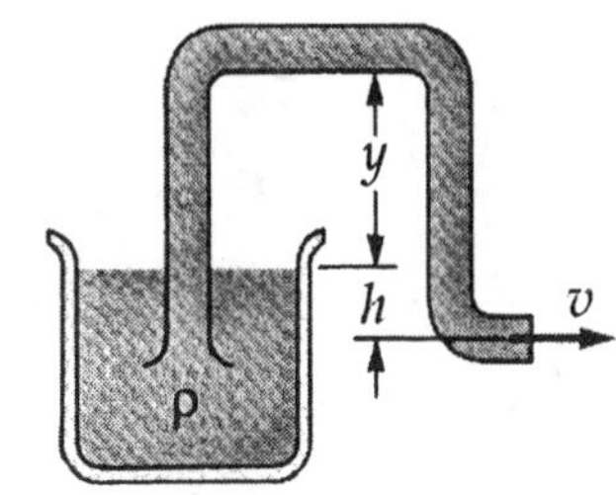

Figure 9.52

Solution (a) Using Bernoulli's equation, we have

$$P_1 + \frac{1}{2}\rho v_1^2 - \rho g y_1 = P_2 + \frac{1}{2}\rho v_2^2 - \rho g y_2 \qquad \text{where} \qquad P_1 = P_2 = P_a$$

For a large tank, the velocity at the top, v_1, is approximately equal to zero , and $y_1 - y_2 = h$.

This gives $\qquad v_2 = \sqrt{2gh} \quad \Diamond$

(b) For $y = y_{\text{max}}$, the fluid will be at rest at the top of the system; v_2 becomes zero, and

$y_2 - y_2 = y_{\text{max}}$, so that $P_a = P + \rho g y$. The minimum value of P is 0. Therefore,

$$y_{\text{max}} = \frac{P_a}{\rho g} \quad \Diamond$$

CHAPTER SELF-QUIZ

1. A liquid-filled tank has a hole on its vertical surface just above the bottom edge. If the surface of the liquid is 0.4 m above the hole, at what speed will the stream of liquid emerge from the hole?
a. 2.8 m/s
b. 7.84 m/s
c. 2.0 m/s
d. 1.7 m/s

2. An aluminum cube, 0.10 m on an edge, is subjected to a shear force of 750 N while the opposite side is clamped. What is the resultant shear strain?
(For aluminum, $S = 2.5 \times 10^{10}$ N/m^2.)
a. 3.0×10^{-6}
b. 0.40×10^{-6}
c. 16.0×10^{-6}
d. 0.07×10^{-6}

3. A 4.0-kg cylinder made of solid iron is supported by a string while submerged in water. What is the tension in the string? (Densities of iron and water in x 10^3 kg/m^3 units, respectively, are 7.86 and 1.0; $g = 9.8$ m/s^2.)
a. 2.5 N
b. 19.6 N
c. 23.7 N
d. 34.2 N

4. A solid rock, suspended in air by a spring scale, has a measured mass of 13.5 kg. When the rock is submerged in water, the scale reads 2.2 kg. What is the density of the rock? (Water density = 1.0×10^3 kg/m^3)
a. 4.55×10^3 kg/m^3
b. 3.5×10^3 kg/m^3
c. 1.2×10^3 kg/m^3
d. 2.4×10^3 kg/m^3

5. A hydraulic lift raises a 2000-kg automobile when a 500 N force is applied to the smaller piston. If the smaller piston has an area of 10 cm^2, what is the cross-sectional area of the larger piston?
   a. 40 cm^2
   b. 80 cm^2
   c. 196 cm^2
   d. 392 cm^2

6. A wire of length 4 meters, cross-sectional area $0.2 \times 10^{-4}\ m^2$ and Young's modulus $8 \times 10^{10}\ N/m^2$ has a 200-kg load hung on it. What is its increase in length? ($g = 9.8\ m/s^2$)
   a. 0.10×10^{-3} m
   b. 0.49×10^{-3} m
   c. 4.9×10^{-3} m
   d. 9.8×10^{-3} m

7. What is the total pressure at the bottom of a 5-m deep swimming pool? (Note the pressure contribution from the atmosphere is $1.01 \times 10^5\ N/m^2$, the density of water is $10^3\ kg/m^3$, and $g = 9.8\ m/s^2$.)
   a. $1.5 \times 10^5\ N/m^2$
   b. $0.72 \times 10^5\ N/m^2$
   c. $0.49 \times 10^5\ N/m^2$
   d. $1.01 \times 10^5\ N/m^2$

8. Water (density = $1 \times 10^3\ kg/m^3$) is flowing through a pipe whose radius is 0.04 m with a speed of 15 m/s. This same pipe goes up to the second floor of the building, 3 m higher, and the pressure remains unchanged. What is the radius of the pipe on the second floor?
   a. 0.046 m
   b. 0.043 m
   c. 0.037 m
   d. 0.034 m

9. A uniform pressure of $3.5\times10^5\ \text{N/m}^2$ is applied to all six sides of an aluminum cube. What is the percentage change in volume of the cube?
(For aluminum, $B=7\times10^{10}\ \text{N/m}^2$)
a. $2.4\times10^{-2}\%$
b. $0.4\times10^{-2}\%$
c. $8.4\times10^{-2}\%$
d. $0.5\times10^{-3}\%$

10. When a force **F** is applied to a certain rod, there is a 1 cm change in length. If that same force is applied to a rod that is three times bigger in each dimension, what will be the change in length?
a. 9 cm
b. 3 cm
c. 1 cm
d. 0.33 cm

11. A glass container is half filled with mercury and a steel ball is floating on the mercury. If water is then poured on top of the steel ball and mercury, filling the glass, what will happen to the steel ball?
a. It will float higher in the mercury.
b. It will float at the same height.
c. It will float lower in the mercury.
d. It will sink to the bottom of the mercury.

12. If the flow rate of a liquid going through a 2.00-cm radius pipe is measured at $0.8\times10^{-3}\ \text{m}^3/\text{s}$, the average fluid velocity in the pipe is which of the following?
a. 0.64 m/s
b. 2.0 m/s
c. 0.04 m/s
d. 6.3 m/s

10
Thermal Physics

Chapter 10

THERMAL PHYSICS

Our study thus far has focused exclusively on mechanics. Such concepts as mass, force, and kinetic energy have been carefully defined in order to make the subject quantitative. We now move to a new branch of physics, thermal physics. Here we shall find that quantitative descriptions of thermal phenomena require careful definitions of the concepts of temperature, heat and internal energy. We shall examine the process of linear expansion in which a temperature increase or decrease for an object results in a change in size.

This chapter concludes with a study of ideal gases. We approach this study on two levels. The first examines ideal gases on the macroscopic scale. Here we are concerned with the relationships among such quantities as pressure, volume, and temperature. On the second level we examine gases on a microscopic scale, using a model that pictures the components of a gas as small particles. This latter approach, called the kinetic theory of gases, helps us understand what happens on the atomic level to affect such macroscopic properties as pressure and temperature.

NOTES FROM SELECTED CHAPTER SECTIONS

10.1 Temperature and the Zeroth Law of Thermodynamics

The zeroth law of thermodynamics (or the equilibrium law) can be stated as follows:

> If bodies A and B are separately in thermal equilibrium with a third body, C, then A and B will be in thermal equilibrium with each other if placed in thermal contact.

Two objects in thermal equilibrium with each other are at the same temperature.

10.2 Thermometers and Temperature Scales

The physical property used in a constant volume gas thermometer is the pressure variation with temperature of a fixed volume of gas. The temperature readings are nearly independent of the substance used in the thermometer.

The triple point of water, which is *the single temperature and pressure at which water, water vapor, and ice can coexist in equilibrium,* was chosen as a convenient and reproducible reference temperature for the Kelvin scale. It occurs at a temperature of 0.01°C and a pressure of 4.58 mm of mercury. The temperature at the triple point of water on the Kelvin scale has been assigned a value of

273.16 kelvins (K). Thus, the SI unit of temperature, the kelvin, is defined as *1/273.16 of the temperature of the triple point of water.*

The temperature 0 K is often referred to as absolute zero although this temperature has never been achieved.

10.5 Avogadro's Number and the Ideal Gas Law

Equal volumes of gas at the same temperature and pressure contain the same numbers of molecules.

Single moles of any gases at standard temperature and pressure contain the same numbers of molecules.

10.6 The Kinetic Theory of Gases

A microscopic *model of an ideal gas* is based on the following assumptions:

1. *The number of molecules is large, and the average separation between them is large* compared with their dimensions. Therefore, the molecules occupy a negligible volume compared with the volume of the container.

2. *The molecules obey Newton's laws of motion, but the individual molecules move in a random fashion.* By random fashion, we mean that the molecules move in all directions with equal probability and with various speeds. This distribution of velocities does not change in time, despite the collisions between molecules.

3. *The molecules undergo elastic collisions with each other.* Thus, the molecules are considered to be structureless (that is, point masses), and in the collisions both kinetic energy and momentum are conserved.

4. *The forces between molecules are negligible, except during a collision.* The forces between molecules are short-range, so that the only time the molecules interact with each other is during a collision.

5. *The gas under consideration is a pure gas.* That is, all molecules are identical.

6. *The gas is in thermal equilibrium with the walls of the container.* Hence, the wall will eject as many molecules as it absorbs, and the ejected molecules will have the same average kinetic energy as the absorbed molecules.

The pressure exerted by each component of a mixture of gases is the same as the pressure that component would exert if it were alone in the volume occupied by the mixture. This means that each component acts virtually independently of the rest of the mixture.

Vapor pressure is *the pressure exerted by a gas when it is in equilibrium with its liquid form.*

The dew point is the temperature at which the air becomes saturated with water vapor.

Boiling begins when the vapor pressure is equal to atmospheric pressure.

EQUATIONS AND CONCEPTS

T_C is the Celsius temperature and T is the Kelvin temperature (sometimes called the absolute temperature). The size of a degree on the Kelvin scale is identical to the size of a degree on the Celsius scale.

$$T_C = T - 273.15 \qquad (10.1)$$

These equations are useful in converting temperature values between Fahrenheit and Celsius scales.

$$T_F = \frac{9}{5}T_C + 32 \qquad (10.2)$$

$$T_C = \frac{5}{9}(T_C - 32)$$

The SI unit of temperature is the kelvin, K, which is defined as $\frac{1}{273.16}$ of the temperature of the triple point of water. Note that the notations C° and F° refer to temperature differences in Celsius degrees and Fahrenheit degrees; and the notations °C and °F refer to actual temperature values in degrees Celsius and degrees Fahrenheit.

Comment on units and notation.

These are two forms of the basic equation for the thermal expansion of a solid. Note that the change in length is proportional to the original length and to the change in temperature. The constant α is characteristic of a particular type of material and is called the average temperature coefficient of linear expansion.

$$\Delta L = \alpha L_0 \Delta T \qquad (10.4a)$$

$$L - L_0 = \alpha L_0 (T - T_0) \qquad (10.4b)$$

If the temperature of a body changes, both the surface area and volume of the body will change by amounts which are proportional to the changes in temperature. The quantity γ (gamma) is the average temperature coefficient of area expansion and β (beta) is the average temperature coefficient of volume expansion.

$$\Delta A = \gamma A_0 \Delta T \qquad (10.5)$$

$$\Delta V = \beta V_0 \Delta T \qquad (10.6)$$

$$\gamma = 2\alpha$$

$$\beta = 3\alpha$$

The number of moles in a sample of element or compound is the ratio of the mass, m, of the sample to the atomic or molecular mass, M, of the material. Also, one mole of a substance contains Avogadro's number of molecules.

$$n = \frac{m}{M} \qquad (10.7)$$

$$N_A = 6.022 \times 10^{23} \ \frac{\text{molecules}}{\text{mole}}$$

Boyle's law states that when a gas is maintained at constant temperature, its pressure is inversely proportional to its volume; the law of Charles-Gay-Lussac states that when the pressure is maintained constant, the volume of a gas is directly proportional to its absolute temperature.

$$P \,\alpha \left(\frac{1}{V}\right); \quad T = \text{constant}$$

$$V \,\alpha\, T; \quad P = \text{constant}$$

This is the equation of state of an ideal gas. In this equation, T must be the temperature in kelvins. R is the universal gas constant and must be expressed in units which are consistent with those used for pressure P and volume V.

$$PV = nRT \qquad (10.8)$$

$$R = 8.31 \text{ J/mol}\cdot\text{K} \qquad (10.9)$$

$$R = 0.0821 \text{ L}\cdot\text{atm/mol}\cdot\text{K}$$

The ideal gas law can also be expressed in this alternate form where N is the total number of molecules in the sample and k is Boltzmann's constant.

$$PV = Nk_BT \tag{10.12}$$

$$k_B = \frac{R}{N_A} = 1.38 \times 10^{-23} \text{ J/K} \tag{10.13}$$

In an ideal gas, the pressure of the gas is proportional to the number of molecules per unit volume and proportional to the average kinetic energy of the molecules.

$$P = \frac{2}{3}\left(\frac{N}{V}\right)\left(\frac{1}{2}m\overline{v^2}\right) \tag{10.14}$$

Equations 10.12 and 10.14 can be combined to present an important result: the average kinetic energy of gas molecules is directly proportional to the absolute temperature of the gas.

$$\frac{1}{2}m\overline{v^2} = \frac{3}{2}k_BT \tag{10.16}$$

This expression for the root mean square (rms) speed shows that, at a given temperature, lighter molecules move faster on the average than heavier ones.

$$v_{rms} = \sqrt{\overline{v^2}} = \sqrt{\frac{3RT}{M}} \tag{10.19}$$

REVIEW CHECKLIST

▷ Describe the operation of the constant-volume gas thermometer and how it is used to determine the Kelvin temperature scale. Convert between the various temperature scales, especially the conversion from degrees Celsius into kelvins, degrees Fahrenheit into kelvins, and degrees Celsius into degrees Fahrenheit.

▷ Define the linear expansion coefficient and volume expansion coefficient for an isotropic solid, and understand how to use these coefficients in practical situations involving expansion or contraction.

▷ Understand the assumptions made in developing the molecular model of an ideal gas; and apply the equation of state for an ideal gas to calculate pressure, volume, temperature, or number of moles.

▷ Define each of the following terms: *molecular weight, mole, Avogadro's number, universal gravitational constant, and Boltzmann's constant.*

SOLUTIONS TO SELECTED END-OF-CHAPTER PROBLEMS

1. The pressure in a constant-volume gas thermometer is 0.700 atm at 100°C and 0.512 atm at 0°C. (a) What is the temperature when the pressure is 0.0400 atm? (b) What is the pressure at 450°C?

Solution When volume is constant, we have a linear relation between pressure and temperature. That is, at temperature T, the pressure is given by $P = A + BT$, where A and B are constants. To find A and B, we use the given data:

$$0.700 \text{ atm} = A + (100°\text{C})B \qquad (1)$$

$$0.512 \text{ atm} = A + 0 \qquad (2)$$

Thus, from (2): $A = 0.512$ atm

Then from (1): $B = 1.88 \times 10^{-3}$ atm/°C

Therefore, $P = 0.512 \text{ atm} + (1.88 \times 10^{-3} \text{ atm/°C})T$

(a) When $P = 0.0400$ atm, we have $0.040 \text{ atm} = 0.512 \text{ atm} + (1.88 \times 10^{-3} \text{ atm/°C})T$, yielding $T = -251°\text{C}$. ◊

(b) When $T = 450°\text{C}$, $P = 0.512 \text{ atm} + (1.88 \times 10^{-3} \text{ atm/°C})(450°\text{C}) = 1.358 \text{ atm}$. ◊

---

9. Show that the temperature −40° is unique in that it has the same numerical value on the Celsius and Fahrenheit scales.

Solution Let us use $T_C = \frac{5}{9}(T_F - 32°\text{C})$ with $T_F = -40°\text{C}$.

$$T_C = \frac{5}{9}(-40 - 32)°\text{C} = -40°\text{C} \quad ◊$$

---

11. The New River Gorge bridge in West Virginia is a 518-m-long steel arch. How much will its length change between temperature extremes of −20°C and 35°C?

Solution Use $L = L_0[1 + \alpha(T - T_0)]$

$$L_{(-20)} = L_0 + \alpha L_0(-20°C) - L_0T_0 \qquad \text{and} \qquad L_{(35)} = L_0 + \alpha L_0(35°C) - L_0T_0$$

$$\Delta L = L_{(35)} - L_{(-20)} = \alpha L_0(55°C) \qquad \text{or}$$

$$\Delta L = (11 \times 10^{-6}/°C)(518\text{ m})(55°C) = 0.313\text{ m} = 31\text{ cm} \quad \Diamond$$

---

17. A cylindrical brass sleeve is to be shrink-fitted over a brass shaft whose diameter is 3.212 cm at 0°C. The diameter of the sleeve is 3.196 cm at 0°C. (a) To what temperature must the sleeve be heated before it will slip over the shaft? (b) Alternatively, to what temperature must the shaft be cooled before it will slip into the sleeve?

Solution (a) We must raise the diameter (a linear dimension) of the sleeve to match the diameter of the shaft. It is necessary to increase the diameter of the sleeve by $\Delta L = 6 \times 10^{-2}$ cm.

$$\Delta L = \alpha L_0 \Delta T \qquad \text{or} \qquad \Delta T = \frac{\Delta L}{\alpha L_0} = \frac{6 \times 10^{-2}\text{ cm}}{[19.0 \times 10^{-6}(°C)^{-1}](3.196\text{ cm})}$$

$$\Delta T = 263.5\text{ C°} \quad \Diamond$$

(b) We must cool the shaft to reduce its diameter from 3.212 cm to 3.196 cm, a change of -1.6×10^{-2} cm.

$$\Delta L = \alpha L_0 \Delta T; \qquad \Delta T = \frac{\Delta L}{\alpha L_0} = \frac{1.6 \times 10^{-2}\text{ cm}}{[19 \times 10^{-6}(°C)^{-1}](3.212\text{ cm})} = -262.2\text{ C°} \quad \Diamond$$

---

23. An automobile fuel tank is filled to the brim with 45.0 L (11.9 gal) of gasoline at 10.0°C. Immediately afterward, the vehicle is parked in the Sun, where the temperature is 35.0°C. How much gasoline overflows from the tank as a result of the expansion? (Neglect the expansion of the tank.)

Solution $V_0 = 45.0\text{ liters} = 45.0 \times 10^{-3}\text{ m}^3$

When the gasoline warms to 35.0°C, its new volume can be found from $\Delta V = \beta V_0 \Delta T$. This is the volume which must overflow.

$$\Delta V = (9.60 \times 10^{-4}\ \text{C}^{-1})(45.0 \times 10^{-3}\ \text{m}^3)(25.0°\text{C})$$

or $$\Delta V = 1.08 \times 10^{-3}\ \text{m}^3 = 1.08\ \text{liters (about 0.29 gallons)} \quad \Diamond$$

---

33. An air bubble has a volume of 1.50 cm³ when it is released by a submarine 100 m below the surface of a lake. What is the volume of the bubble when it reaches the surface? Assume that the temperature of the air in the bubble remains constant during ascent.

Solution

First find the pressure of air in the bubble at a depth of 100 m using $P_i = P_a + \rho g h$.

$$P_i = 1.013 \times 10^5\ \text{Pa} + (1000\ \text{kg/m}^3)(9.80\ \text{m/s}^2)(100\ \text{m}) = 1.08 \times 10^6\ \text{Pa}$$

Now use $\frac{P_f V_f}{T_f} = \frac{P_i V_i}{T_i}$ or at constant temperature $P_f V_f = P_i V_i$ so

$$V_f = \left(\frac{P_i}{P_f}\right) V_i = \left(\frac{1.08 \times 10^6}{1.013 \times 10^5}\right)(1.50\ \text{cm}^3) = 16.0\ \text{cm}^3 \quad \Diamond$$

---

37. A cylindrical diving bell, 3.00 m in diameter and 4.00 m tall with an open bottom, is submerged to a depth of 220 m in the ocean. The surface temperature is 25.0°C and the temperature 220 m down is 5.00°C. The density of sea water is 1025 kg/m³. How high does the sea water rise in the bell when it is submerged?

Solution Original volume of air (at depth of 220 m), $V_i = \pi r^2 h_i$, where h_i = "height" of air in cylindrical bell.

$$V_i = \pi(1.50\ \text{m})^2(4.00\ \text{m}) = 28.3\ \text{m}^3$$

The pressure P_i at depth of 220 m is

$$P_i = P_a + \rho g h = 1.013 \times 10^5 \text{ Pa} + (1025 \text{ kg/m}^3)(9.80 \text{ m/s}^2)(220 \text{ m})$$

$$P_i = 2.311 \times 10^6 \text{ Pa}$$

From $\dfrac{P_i V_i}{T_i} = \dfrac{P_f V_f}{T_f}$ and $V_f = \left(\dfrac{P_i}{P_f}\right)\left(\dfrac{T_f}{T_i}\right)V_i$

$$V_f = \left(\frac{1.013 \times 10^5 \text{ Pa}}{2.311 \times 10^6 \text{ Pa}}\right)\left(\frac{278 \text{ K}}{298 \text{ K}}\right)(28.3 \text{ m}^3) = 1.16 \text{ m}^3$$

but $V_f = \pi r^2 h'$ where h' = "height" of air column at depth of 220 m.

$$h' = \frac{V_f}{\pi r^2} = \frac{1.16 \text{ m}^3}{\pi(1.50 \text{ m})^2} = 0.164 \text{ m}$$

Water rises a distance $d = h - h'$.

$$d = 4.00 \text{ m} - 0.164 \text{ m} = 3.84 \text{ m} \quad \Diamond$$

---

43. A sealed cubical container 20.0 cm on a side contains three times Avogadro's number of molecules at a temperature of 20.0°C. Find the force exerted by the gas on one of the walls of the container.

Solution We first find the pressure exerted by the gas on the wall of the container.

$$P = \frac{NkT}{V} = \frac{3N_a kT}{V} = \frac{3RT}{V} = \frac{3(8.31 \text{ N}\cdot\text{m/mol}\cdot\text{K})(293 \text{ K})}{8.00 \times 10^{-3} \text{ m}^3} = 9.13 \times 10^5 \text{ Pa}$$

Thus, the force on one of the walls of the cubical container is

$$F = PA = (9.13 \times 10^5 \text{ Pa})(4.00 \times 10^{-2} \text{ m}^2) = 3.65 \times 10^4 \text{ N} \quad \Diamond$$

---

49. A cylinder contains a mixture of helium and argon gas in equilibrium at a temperature of 150°C. (a) What is the average kinetic energy of each type of molecule? (b) What is the rms speed of each type of molecule?

Solution (a) $<KE> = \frac{3}{2}kT = \frac{3}{2}(1.38\times10^{-23}\text{ J/K})(423\text{ K}) = 8.76\times10^{-21}\text{ J/molecule}$ ◊

(b) $<KE> = \frac{1}{2}mv^2_{rms} = 8.76\times10^{-21}\text{ J}$ so $v_{rms} = \sqrt{\frac{1.75\times10^{-20}\text{ J/molecule}}{m}}$ (1)

For helium:

$$m = \frac{4.00\text{ gm/mol}}{6.02\times10^{23}\text{molecules/mol}} = 6.64\times10^{-24}\text{ gm/molecule}$$

or $m = 6.64\times10^{-27}\text{ kg/molecule}$

Similarly for argon:

$$m = \frac{39.9\text{ gm/mol}}{6.02\times10^{23}\text{ molecules/mol}} = 6.63\times10^{-23}\text{ gm/molecule}$$

and $m = 6.63\times10^{-26}\text{ kg/molecule}$

Substituting in (1) above, we find:

$v_{rms} = 1620\text{ m/s}$ (for helium) and $v_{rms} = 514\text{ m/s}$ (for argon) ◊

---

53. The active element of a certain laser is an ordinary glass rod 20.0 cm long and 1.00 cm in diameter. If the temperature of the rod increases by 75.0°C, find its increase in (a) length, (b) diameter, and (c) volume.

Solution (a) The change in length of the rod is

$$\Delta L = \alpha L_0 \Delta T = [9.00\times10^{-6}(^\circ\text{C})^{-1}](20.0\text{ cm})(75.0^\circ\text{C}) = 1.35\times10^{-2}\text{ cm} \quad ◊$$

(b) The change in diameter is

$$\Delta L = \alpha L_0 \Delta T = [9.00\times10^{-6}(^\circ\text{C})^{-1}](1.00\text{ cm})(75.0^\circ\text{C}) = 6.75\times10^{-4}\text{ cm} \quad ◊$$

(c) The change in volume is

$$\Delta V = 3\alpha V_0 \Delta T = 3[9.00\times10^{-6}(°C)^{-1}](15.7\text{ cm}^3)(75.0°C) = 3.18\times10^{-2}\text{ cm}^3 \quad \Diamond$$

---

57. A liquid with coefficient of volume expansion β just fills a spherical shell of volume V at temperature T (Fig. 10.16). The shell is made of a material that has a coefficient of linear expansion of α. The liquid is free to expand into a capillary of cross-sectional area A at the top. (a) If the temperature increases by ΔT, show that the liquid rises in the capillary by the amount $\Delta h = (V/A)(\beta - 3\alpha)\Delta T$. (b) For a typical system, such as a mercury thermometer, why is it a good approximation to neglect the expansion of the shell?

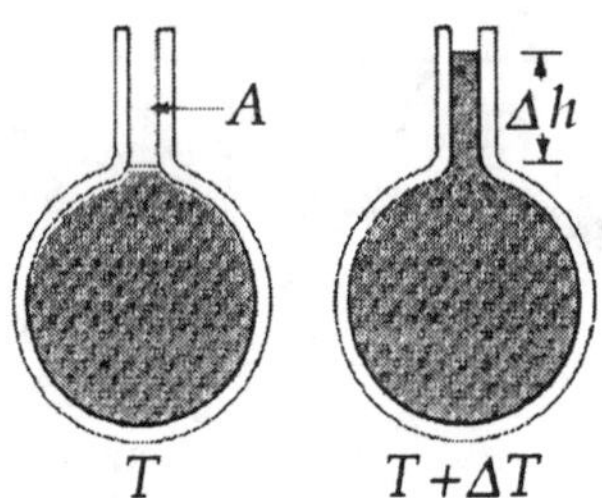

Figure 10.16

Solution (a) The volume of the liquid increases as $\Delta V_{\text{liq}} = V\beta\Delta T$. The volume of the flask increases as $\Delta V_g = V3\alpha\Delta T$. Therefore, the overflow into the capillary is

$$V_c = \Delta V_{\text{liq}} - \Delta V_g = V\Delta T(\beta - 3\alpha) \quad \text{but} \quad V_c = A\Delta h$$

Therefore, $$\Delta h = \left(\frac{V}{A}\right)(\beta - 3\alpha)\Delta T \quad \Diamond$$

(b) For a mercury thermometer $\beta(Hg) = 1.82\times10^{-4}(°C)^{-1}$

and for glass, $3\alpha = 3\times9\times10^{-6}(°C)^{-1} = 2.7\times10^{-5}(°C)^{-1}$ so $3\alpha \ll \beta$ or $(\beta - 3\alpha) \approx \beta$ $\Diamond$

---

61. A hollow aluminum cylinder is to be fitted over a steel piston. At 20.0°C, the inside diameter of the cylinder is 99% of the outside diameter of the piston. To what common temperature should the two pieces be heated in order that the cylinder just fits over the piston?

Solution We have $$d_{0c} = 0.99 d_{0p} \qquad (1)$$

The diameter is a linear dimension. Thus, we use $L=(1+\alpha\Delta T)L_0$, with the ultimate goal of finding the temperature at which the final diameter of the piston and cylinder are equal $(d_{fc}=d_{fp})$.

Thus, $$(1+\alpha_c\Delta T)d_{0c}=(1+\alpha_p\Delta T)d_{0p} \qquad (2)$$

Using (1) in (2), we have $(1+\alpha_c\Delta T)(0.99)=(1+\alpha_p\Delta T)$ or

$$\Delta T=\frac{0.010}{0.990\alpha_c-\alpha_p}=\frac{0.010°\text{C}}{(0.990)(24.0\times10^{-6})-11.0\times10^{-6}}=784°\text{C}$$

Therefore, $$T_f=804°\text{C} \quad \lozenge$$

---

65. A vertical cylinder of cross-sectional area 0.050 m^2 is fitted with a tight-fitting, frictionless piston of mass 5.00 kg (Fig. 10.17). If there are 3.00 mol of an idea gas in the cylinder at 500 K, determine the height, h, at which the piston will be in equilibrium under its own weight.

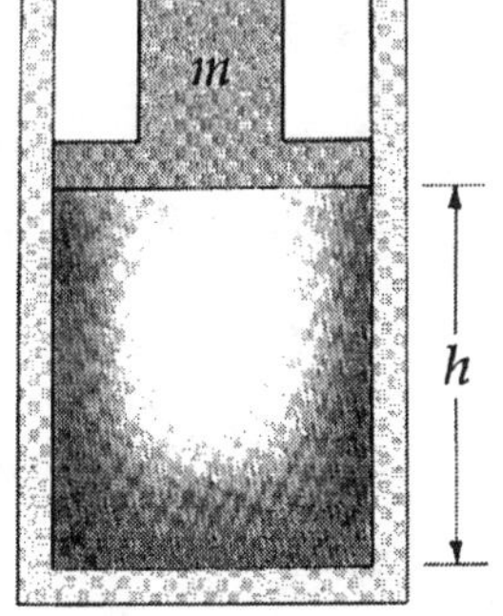

Figure 10.17

Solution The pressure of the gas in the cylinder is due to the weight of the piston plus atmospheric pressure. Therefore, in equilibrium,

$$P_{\text{gas}}=\frac{mg}{A}+P_a \quad \text{and} \quad PV=nRT \quad \text{where} \quad V=hA$$

Therefore,

$$\frac{nRT}{hA}=\frac{mg}{A}+P_a \quad \text{or} \quad h=\frac{nRT}{(mg+P_aA)} \quad \text{so}$$

$$h=\frac{(3.00\text{ mol})(8.314)(500\text{ K})}{(5.00\text{ kg})(9.80)+(1.01\times10^5\text{ Pa})(0.050\text{ m}^2)}=2.45\text{ m} \quad \lozenge$$

CHAPTER SELF-QUIZ

1. Consider two containers that have the same volume and temperature. The first contains a certain number of oxygen molecules while the second contains the same number of hydrogen molecules.
   a. The heavier oxygen molecules will exert greater pressure on the walls of the first bottle.
   b. The faster hydrogen molecules will exert greater pressure on the walls of the second bottle.
   c. The pressure on the walls of both bottles will be equal.
   d. Not enough information is given to allow an accurate calculation of relative pressure.

2. Suppose for a brief moment that the gas molecules hitting a wall stuck to the wall instead of bouncing off the wall. How would the pressure on the wall be affected during that brief time?
   a. The pressure would be zero.
   b. The pressure would be half as big.
   c. The pressure would remain unchanged.
   d. The pressure would be twice as big.

3. Three moles of an ideal gas are confined to a 30-liter container at a pressure of 2 atmospheres. What is the gas temperature? (R = 0.0821 L·atm/mole·K)
   a. 975 K
   b. 487 K
   c. 365 K
   d. 244 K

4. Suppose one sphere is made of a metal that has twice the coefficient of linear expansion of a second sphere. The coefficient of volume expansion for the first sphere will be bigger than the second by a factor of
   a. 2
   b. 3
   c. 4
   d. 8

5. Suppose the ends of a 30-m-long steel rail are rigidly clamped at 0°C to prevent expansion. The rail has a cross-sectional area of 30 cm^2. What force does the rail exert when it is heated to 40°C? ($\alpha_{steel} = 1.1 \times 10^{-5}$ /°C, $Y_{steel} = 2 \times 10^{11}$ N/m^2)
   a. 2.6×10^5 N
   b. 5.6×10^4 N
   c. 1.3×10^3 N
   d. 650 N

6. Oxygen condenses into a liquid at approximately 90° Kelvin. What temperature, in degrees Fahrenheit does this correspond to?
   a. −118°F
   b. −193°F
   c. −265°F
   d. −297°F

7. An auditorium has dimensions 10 m × 20 m × 30 m. How many molecules of air are needed to fill the auditorium at standard temperature and pressure?
   a. 1.6×10^{29}
   b. 1.6×10^{27}
   c. 1.6×10^{25}
   d. 1.6×10^{23}

8. A helium-filled balloon has a volume of 1 m^3. As it rises in the Earth's atmosphere, its volume expands. What will be its new volume (in cubic meters) if its original temperature and pressure are 20°C and 1 atm, and it final temperature and pressure are −40°C and 0.1 atm?
   a. 4 m^3
   b. 6 m^3
   c. 8 m^3
   d. 10 m^3

9. A spherical air bubble originating from a scuba diver has a radius of 5 mm at some depth, *h*. When the bubble reaches the surface, it has a radius of 7 mm. Assuming constant temperature, what is the depth of the diver?
   a. 8 m
   b. 13 m
   c. 18 m
   d. 23 m

10. If the rms velocity of a helium atom at room temperature is 1350 m/s, what is the rms velocity of an oxygen (O_2) molecule?
    a. 675 m/s
    b. 477.3 m/s
    c. 337.5 m/s
    d. 168.7 m/s

11. The density of gasoline is 730 kg/m³ at 0°C. Its volume expansion coefficient is 9.6×10^{-4} /°C. If 1 gallon occupies 0.0038 m³, how many extra kilograms of gasoline do you get when you buy 10 gallons of gasoline at 0°C rather than at 20°C?
    a. 0.78 kg
    b. 0.52 kg
    c. 0.26 kg
    d. Zero

12. An automobile tire is filled with air to a gauge pressure of 30 PSI at 10°C. After driving into desert country, the temperature of the tire is 29°C. What is the pressure (in PSI) at that temperature? (Assume tire volume doesn't change.)
    a. 31 PSI
    b. 32 PSI
    c. 33 PSI
    d. 34 PSI

11
Heat

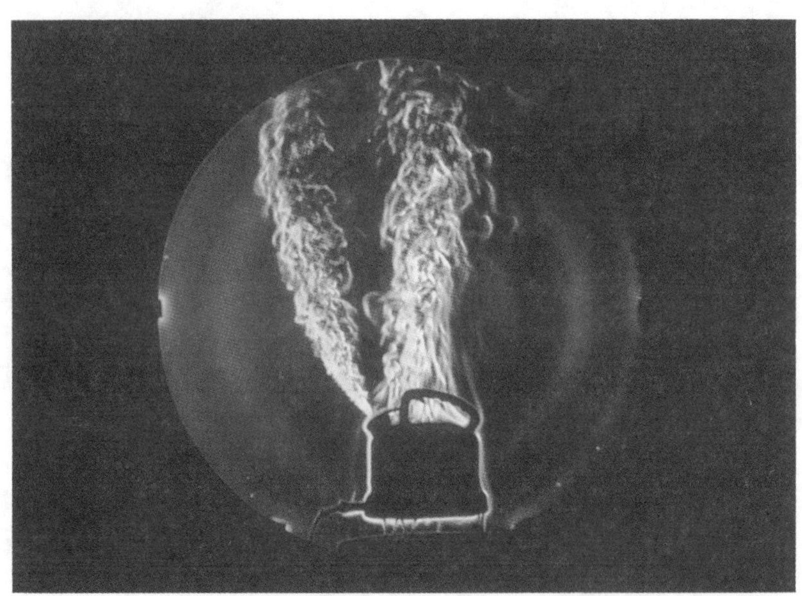

Chapter 11

HEAT

It is an experimentally established fact that when two objects at different temperatures are placed in thermal contact with each other, the temperature of the warmer object decreases and the temperature of the cooler object increases. If the two are left in contact for some time, they eventually reach a common equilibrium temperature that is intermediate between the two initial temperatures. When such a process occurs, we say that heat is transferred from the object at the higher temperature to the one at the lower temperature.

The principle of conservation of energy is a universal principle of nature in which heat is treated as another form of energy that can be transformed into mechanical energy. It has been demonstrated that whenever heat is gained or lost by a system during some process, the gain or loss can be accounted for by an equivalent quantity of mechanical work done on the system.

The focus of this chapter is to introduce the concept of heat and some of the processes that enable heat to be transferred between a system and its surroundings.

NOTES FROM SELECTED CHAPTER SECTIONS

11.1 The Mechanical Equivalent of Heat

When two systems at different temperatures are in contact with each other, energy will transfer between them until they reach the same temperature (that is, when they are in thermal equilibrium with each other). This energy is called heat, or thermal energy, and the term "heat flow" refers to an energy transfer as a consequence of a temperature difference.

The unit of heat is the *calorie* (cal), defined as the amount of heat necessary to increase the temperature of 1 g of water from 14.5°C to 15.5°C. The *mechanical equivalent of heat,* first measured by Joule, is given by 1 cal = 4.186 J.

11.2 Specific Heat

The specific heat, *c,* of any substance is defined as the amount of heat required to increase the temperature of that substance by one Celsius degree. Its units are cal/°C.

11.4 Latent Heat and Phase Changes

The *latent heat of fusion* is a parameter used to characterize a solid-to-liquid phase change; the *latent heat of vaporization* characterizes the liquid-to-gas phase change.

11.5 Heat Transfer by Conduction
11.6 Convection
11.7 Radiation

There are three basic processes of heat transfer. These are (1) conduction, (2) convection, and (3) radiation.

Conduction is a heat transfer process which occurs when there is a *temperature gradient* across the body. That is, conduction of heat occurs only when the body's temperature is *not* uniform. For example, if you heat a metal rod at one end with a flame, heat will flow from the hot end to the colder end. The rate of flow of heat along the rod, sometimes called the *heat current*, is proportional to the cross-sectional area of the rod, the temperature gradient, and k, the thermal conductivity of the material of which the rod is made.

When heat transfer occurs as the result of the motion of material, such as the mixing of hot and cold fluids, the process is referred to as *convection*. Convection heating is used in conventional hot-air and hot-water heating systems. Convection currents produce changes in weather conditions when warm and cold air masses mix in the atmosphere.

Heat transfer by *radiation* is the result of the continuous emission of electromagnetic radiation by all bodies.

EQUATIONS AND CONCEPTS

When two systems initially at different temperatures are placed in contact with each other, energy will be transferred from the system at higher temperature to the system at lower temperature until the two systems reach a common temperature (thermal equilibrium). The energy transferred from one system to the other is called heat or thermal energy; and the

Comment on heat energy.

term "heat flow" refers to the process of energy transfer due to a temperature difference.

The unit of heat energy is the calorie (cal), defined as the quantity of heat energy required to increase the temperature of 1 g of water from 14.5°C to 15.5°C.

Comment on units.

The **mechanical equivalent of heat** was first measured by Joule.

$$1 \text{ cal} = 4.186 \text{ J} \qquad (11.1)$$

The quantity of heat energy required to increase the temperature of a given mass by a specified amount varies from one substance to another. Every substance is characterized by a unique value of specific heat, c.

$$Q = mc\,\Delta T \qquad (11.3)$$

The heat energy required to cause a quantity of substance of mass, m, to undergo a phase change (solid to liquid or liquid to gas) depends on the value of the latent heat of the substance. The latent heat of fusion, L_f, is used when the phase change is from solid to liquid and the latent heat of vaporization, L_v, when the phase change is from liquid to gas.

$$Q = mL \qquad (11.5)$$

There are three basic processes of heat transfer: (1) conduction, (2) convection, and (3) radiation.

Comment on heat transfer processes.

This is the basic **law of heat conduction.** The constant k is called the thermal conductivity and is characteristic of a particular material. The rate of flow of heat energy along the conducting material is the heat current, H. Conduction of heat occurs only when there is a temperature

$$H = \frac{Q}{\Delta t} = kA\left(\frac{T_2 - T_1}{L}\right) \qquad (11.7)$$

Conduction process

gradient (or temperature difference between two points in the conducting material).

The heat current, H, can be expressed in watts (where 1 W = 1 J/s), cal/s, or Btu/h.

Comment on units.

The rate at which heat is transferred by a fluid to a surface of area A depends on the temperature difference and the convection coefficient h.

$$\frac{\Delta Q}{\Delta t} = hA\,\Delta T$$

Convection process

When heat transfer occurs as the result of the movement of a heated substance (usually a fluid) between points at different temperatures, the process is called convection.

Comment on convection.

The rate of emission of heat energy by radiation (the power radiated) is given by Stefan's law. The radiated power is proportional to the fourth power of the absolute temperature.

$$P = \sigma A e T^4 \qquad (11.12)$$

Radiation process

In Equation 11.12, σ is a universal constant, $\sigma = 5.6696 \times 10^{-8}$ W/m^2, T is the absolute temperature in K, and e (the emissivity of the radiating body) can have a value between 0 and 1 depending on the nature of the surface.

Comment on Stefan's law.

An object at temperature T in surroundings at temperature T_0 will experience a net radiated ($T > T_0$) or absorbed ($T < T_0$) power. At thermal equilibrium ($T = T_0$), an object radiates and absorbs energy at the same rate and the temperature of the object remains constant.

$$P_{net} = \sigma A e (T^4 - T_0^{\,4}) \qquad (11.13)$$

SUGGESTIONS, SKILLS, AND STRATEGIES

If you are having difficulty with calorimetry problems, one or more of the following factors should be considered:

1. Be sure your units are consistent throughout. That is, if you are using specific heats measured in cal/g·°C, be sure that masses are in grams and temperatures are in Celsius units throughout.

2. Losses and gains in heat are found by using $Q = mc\Delta T$ only for those intervals in which no phase changes occur. Likewise, the equations $Q = mL_f$ and $Q = mL_v$ are to be used only when phase changes *are* taking place.

3. Often sign errors occur in heat loss = heat gain equations. One way to determine whether your equation is correct is to examine the sign of all ΔT's that appear in your equation. Every ΔT that appears should be a positive number.

REVIEW CHECKLIST

▷ Understand the concepts of heat, internal energy, and thermodynamic processes.

▷ Define and discuss the calorie, Btu, specific heat, and latent heat. Convert between calories, Btu's, and joules.

▷ Use equations for specific heat, latent heat, temperature change and energy gain (loss) to solve calorimetry problems.

▷ Discuss the possible mechanisms which can give rise to heat transfer between a system and its surroundings; that is, heat conduction, convection and radiation; and give a realistic example of each heat transfer mechanism.

▷ Apply the basic law of heat conduction, and Stefan's law for heat transfer by radiation.

SOLUTIONS TO SELECTED END-OF-CHAPTER PROBLEMS

9. Water at the top of Niagara Falls has a temperature of 10°C. If it falls a distance of 50 m and all of its potential energy goes into heating the water, calculate the temperature of the water at the bottom of the falls.

Solution Consider a 1-kg mass of water:

$$Q = \Delta PE = mgh = (1.0\text{ kg})(9.80\text{ m/s}^2)(50\text{ m}) = 490\text{ J}$$

Also, $$\Delta T = \frac{Q}{mc} = \frac{490\text{ J}}{(1.0\text{ kg})(4.19\times 10^3\text{ J/kg}\cdot°\text{C})} = 0.117°\text{C} \quad \text{so} \quad T_f = 10.1\text{ °C} \quad \lozenge$$

---

13. A 0.4-kg iron horseshoe that is initially at 500°C is dropped into a bucket containing 20 kg of water at 22°C. What is the final equilibrium temperature? Neglect any heat transfer to or from the surroundings.

Solution Our heat loss = heat gain equation becomes

$$m_{\text{iron}}c_{\text{iron}}\Delta T_{\text{iron}} = m_w c_w \Delta T_w$$

or $$(0.40\text{ kg})(448\text{ J/kg}\cdot°\text{C})(500°\text{C} - T) = (20\text{ kg})(4186\text{ J/kg}\cdot°\text{C})(T - 22°\text{C})$$

This becomes

$$T = \frac{(1.84\times 10^6 + 8.96\times 10^4)\text{J}\cdot°\text{C}^2}{(8.37\times 10^4 + 179)\text{J}\cdot°\text{C}} = 23°\text{C} \quad \lozenge$$

---

17. An aluminum cup contains 225 g of water at 27°C. A 400-g sample of silver at an initial temperature of 87°C is placed in the water. A 40-g copper stirrer is used to stir the mixture until it reaches its final equilibrium temperature of 32°C. Calculate the mass of the aluminum cup.

Solution Heat gain, H_{gain}, (by cup, water and stirrer) equals heat loss, H_{loss} (by silver). In equation form:

$$Q_{\text{gain}} = m_c c_c\,\Delta T_c + m_w c_w\,\Delta T_w + m_{cu}c_{cu}\,\Delta T_{cu}$$

$$Q_{\text{gain}} = m_c\left(900\frac{\text{J}}{\text{kg}\cdot°\text{C}}\right)(5°\text{C}) + (0.225\text{ kg})\left(4186\frac{\text{J}}{\text{kg}\cdot°\text{C}}\right)(55\text{ °C}) + (0.04\text{ kg})\left(387\frac{\text{J}}{\text{kg}\cdot°\text{C}}\right)(5°\text{C})$$

$$Q_{\text{gain}} = \left(4500\frac{\text{J}}{\text{kg}}\right)m_c + 4786\text{ J}$$

$$Q_{\text{loss}} = m_{Ag}c_{Ag}\,\Delta T_{Ag} = (0.4\text{ kg})\left(234\frac{\text{J}}{\text{kg}\cdot{}^\circ\text{C}}\right)(5^\circ\text{C}) = 5148\text{ J}$$

and setting $Q_{\text{gain}} = Q_{\text{loss}}$, we find $\quad m_c = \dfrac{362\text{ J}}{4500\text{ J/kg}} = 0.080\text{ kg} = 80\text{ g}$ ◊

---

21. A 100-g aluminum calorimeter contains 250 g of water. The two substances are in thermal equilibrium at 10°C. Two metallic blocks are placed in the water. One is a 50-g piece of copper at 80°C. The other sample has a mass of 70 g and is originally at a temperature of 100°C. The entire system stabilizes at a final temperature of 20°C. Determine the specific heat of the unknown second sample.

Solution Two substances (aluminum and water) experience heat gain and increase in temperature, while the two blocks (copper and unknown) undergo decrease in temperature and a corresponding heat loss. Requiring the heat loss to be equal to the heat gain:

$$Q_{\text{gain}} = Q_{\text{loss}}$$

$$m_{Al}c_{Al}\,\Delta T_{Al} + m_w c_w\,\Delta T_w = m_{cu}c_{cu}\,\Delta T_{cu} + m_x c_x\,\Delta T_x$$

The specific heat of the unknown material is $\quad c_x = \dfrac{m_{Al}c_{Al}\,\Delta T_{Al} + m_w c_w\,\Delta T_w - m_{cu}c_{cu}\,\Delta T_{cu}}{m_x\,\Delta T_x}$

Since $\quad \Delta T_{Al} = \Delta T_w = 10\text{ C}^\circ, \quad$ we have

$$c_x = \frac{\left[(0.10\text{ kg})\left(900\frac{\text{J}}{\text{kg}\cdot{}^\circ\text{C}}\right) + (0.25\text{ kg})\left(4186\frac{\text{J}}{\text{kg}\cdot{}^\circ\text{C}}\right)\right]10\ {}^\circ\text{C} - \left[(0.05\text{ kg})\left(387\frac{\text{J}}{\text{kg}\cdot{}^\circ\text{C}}\right)(60\ {}^\circ\text{C})\right]}{(0.07\text{ kg})(80\ {}^\circ\text{C})}$$

and $\quad c_x = 1800\dfrac{\text{J}}{\text{kg}\cdot{}^\circ\text{C}} \quad$ or $\quad 0.44\dfrac{\text{cal}}{\text{g}\cdot{}^\circ\text{C}}$ ◊

---

25. A 50-g ice cube at 0°C is heated until 45 g has become water at 100°C and 5 g has become steam at 100°C. How much heat was added to accomplish this?

Solution Total heat $Q = Q_{\text{melt}} + Q_{\text{warm}} + Q_{\text{vaporize}}$

$$Q_{\text{melt}} = m_i L_f = (0.05\ \text{kg})\left(3.33\times10^5\ \frac{\text{J}}{\text{kg}}\right) = 16650\ \text{J}$$

$$Q_{\text{warm}} = m_w c_w \Delta T_w = (0.05\ \text{kg})\left(4186\frac{\text{J}}{\text{kg·°C}}\right)(100\ \text{°C}) = 20930\ \text{J}$$

$$Q_{\text{vaporize}} = m_v L_v = (5\times10^{-3}\ \text{kg})\left(2.26\times10^6\ \frac{\text{J}}{\text{kg}}\right) = 11300\ \text{J}$$

and $Q_{\text{total}} = 4.9\times10^{-4}\ \text{J}$ ◊

---

29. A 100-g ice cube at 0°C is placed in 650 g of water at 25°C. What is the final temperature of the mixture?

Solution heat loss = heat gain

(loss as water cools to T_f) = (heat to melt ice) + (heat to warm melted ice to T_f)

$$m_w c_w \Delta T_w = m_i L_f + m_i c_w T_f$$

Note that the change in temperature of the water resulting from the melted ice is T_f since that water has an initial temperature of 0°.

Substituting into the calorimetry equation above:

$$(0.65\ \text{kg})\left(4186\frac{\text{J}}{\text{kg.°C}}\right)(25\text{°C} - T_f) = (0.10\ \text{kg})\left(3.33\times10^5\frac{\text{J}}{\text{kg}}\right) + (0.10\ \text{kg})\left(4186\frac{\text{J}}{\text{kg·°C}}\right)T_f$$

This becomes $\left(3140\frac{\text{J}}{\text{°C}}\right)T_f = 34720\ \text{J}$ and $T_f = 11\text{°C}$ ◊

---

33. A beaker of water sits in the Sun until it reaches an equilibrium temperature of 30°C. The beaker is made of 100 g of aluminum and contains 180 g of water. In an attempt to cool this system down, 100 g of ice at 0°C is added to the water. (a) Determine the final temperature. If T = 0°C, determine how much ice remains. (b) Repeat this for 50 g of ice.

Solution (a) First determine whether or not all the ice will melt. Heat energy necessary to reduce the temperature of the beaker and water from 30°C to 0°C is ΔQ_1.

$$\Delta Q_1 = m_{\text{cup}} c_{al} (\Delta T)_{\text{cup}} + m_{\text{water}} c_{\text{water}} (\Delta T)_{\text{water}}$$

$$\Delta Q_1 = 0.10 \text{ kg})(900 \text{ J/kg}\cdot°\text{C})(30°\text{C}) + (0.18 \text{ kg})(4186 \text{ J/kg}\cdot°\text{C})(30°\text{C})$$

$$\Delta Q_1 = 25304 \text{ J}$$

Now calculate the mass of ice which must melt in order to absorb a quantity of heat equal to ΔQ_1.

$$\Delta Q_1 = m_{\text{ice}} L_f \quad \text{or} \quad m_{\text{ice}} = \frac{\Delta Q_1}{L_f} = \frac{25304 \text{ J}}{3.34 \times 10^5 \ \frac{\text{J}}{\text{kg}}} \quad \text{or} \quad m_{\text{ice}} = 7.58 \times 10^{-2} \text{ kg} = 75.8 \text{ g}$$

Therefore, not all of the ice will melt. The final temperature will be 0°C and 24 g of ice will be left over. ◊

(b) If only 50 g of ice is used, we know from part (a) that all of it will melt and the final temperature T_f will be greater than 0°C. Compute T_f using the equation of heat loss equals heat gained.

$$Q_{\text{gain}} = m_{\text{ice}} L_f + m_{\text{ice}} c_{\text{water}} (T_f - 0°\text{C}) =$$

$$Q_{\text{gain}} = (0.05 \text{ kg})\left(3.34 \times 10^5 \frac{\text{J}}{\text{kg}}\right) + (0.05 \text{ kg})\left(4186 \frac{\text{J}}{\text{kg}\cdot°\text{C}}\right) T_f$$

$$Q_{\text{loss}} = m_{\text{water}} c_{\text{water}} (T_f - 30°\text{C}) + m_{\text{cup}} c_{\text{cup}} (T_f - 30°)$$

$$Q_{\text{loss}} = (0.18 \text{ kg})\left(4186 \frac{\text{J}}{\text{kg}\cdot°\text{C}}\right)(T_f - 30°\text{C}) + (0.10 \text{ kg})\left(900 \frac{\text{J}}{\text{kg}\cdot°\text{C}}\right)(T_f - 30°\text{C})$$

Get $Q_{\text{gain}} = Q_{\text{loss}}$ and solve for $T_f = 8.2°\text{C}$ ◊

37. Determine the R value for a wall constructed as follows: The outside of the house consists of lapped wood shingles placed over 0.5-in.-thick sheathing, over 3 in. of cellulose fiber, over 0.5 in. of dry wall.

Solution $R_{\text{total}} = \sum R_i = R_{\text{wood}} + R_{\text{sheathing}} + R_{\text{cellulose}} + R_{\text{dry wall}}$

$$= [0.17 + 0.87 + 1.32 + (3)(3.70) + (0.45) + (0.17)]$$

$$R_{\text{total}} = 14\frac{\text{ft}^2\cdot{}^\circ\text{F}\cdot\text{h}}{\text{Btu}} \quad \Diamond$$

---

41. A Thermopane window consists of two glass panes, each 0.5 cm thick, with a 1.0-cm-thick sealed layer of air between. If the inside temperature is 23.0°C and the outside temperature is 0.0°C, determine the rate of heat transfer through 1.0 m^2 of the window. Compare this with the rate of heat transfer through 1.0 m^2 of a single 1.0-cm-thick pane of glass.

Solution For the Thermopane, the rate of heat transfer (or heat current) is

$$H = \frac{\Delta Q}{\Delta t} = \frac{A(\Delta T)}{\Sigma R_i} = \frac{A(\Delta T)}{\sum (L_i / k_i)}$$

$$H = \frac{(1.0\ \text{m}^2)(23^\circ\text{C} - 0^\circ\text{C})}{\dfrac{10^{-2}\ \text{m}}{0.0234\ \text{J}/(\text{s}\cdot\text{m}\cdot{}^\circ\text{C})} + \dfrac{5\times 10^{-3}\ \text{m}}{0.80\ \text{J}/(\text{s}\cdot\text{m}\cdot{}^\circ\text{C})} + \dfrac{5\times 10^{-3}\ \text{m}}{0.80\ \text{J}/(\text{s}\cdot\text{m}\cdot{}^\circ\text{C})}}$$

$$H = 52\ \text{J/s} \quad \Diamond$$

For a single pane 1 cm thick,

$$H = \frac{A(\Delta T)}{L/k} = \frac{Ak(\Delta T)}{L} = \frac{(1\ \text{m}^2)\left(0.80\dfrac{\text{J}}{\text{s}\cdot\text{m}\cdot{}^\circ\text{C}}\right)(23^\circ\text{C})}{0.01\ \text{m}} = 1840\ \text{J/s} \quad \Diamond$$

---

45. A Styrofoam box has a surface area of 0.80 m^2 and a thickness of 2 cm. The temperature inside is 5°C, and that outside is 25°C. If it takes 8 h for 5 kg of ice to melt in the container, determine the thermal conductivity of the Styrofoam.

Solution Quantity of heat required to melt the ice,

$$\Delta Q = mL_f = (5.0 \text{ kg})\left(3.33 \times 10^5 \frac{\text{J}}{\text{kg}}\right) = 1.67 \times 10^6 \text{ J}$$

and the energy flow rate,

$$H = \frac{\Delta Q}{\Delta t} = \frac{1.67 \times 10^6 \text{ J}}{(8 \text{ h})(3600 \text{ s/h})} = 58 \text{ J/s}$$

But $H = \frac{kA(\Delta T)}{L}$ where L = wall thickness

so $k = \frac{HL}{A(\Delta T)} = \frac{(58 \text{ J/s})(0.02 \text{ m})}{(0.80 \text{ m}^2)(20°\text{C})} = 7.30 \times 10^{-2} \frac{\text{J}}{\text{s} \cdot \text{m} \cdot °\text{C}}$ ◊

---

47. A copper rod and an aluminum rod of equal diameter are joined end to end in good thermal contact. The temperature of the free end of the copper rod is held constant at 100°C, and that of the far end of the aluminum rod is held at 0°C. If the copper rod is 0.15 m long, what must be the length of the aluminum rod so that the temperature at the junction is 50°C?

Solution Let us call the copper rod object 1 and the aluminum rod object 2. At equilibrium, the flow rate through each must be the same. We have

$$\frac{k_2 A_2 \Delta T}{L_2} = \frac{k_1 A_1 \Delta T}{L_1}$$

Since the cross-sectional areas of the rods are the same and the temperature difference across the rods are also equal, we have

$$\frac{k_2}{L_2} = \frac{k_1}{L_1} \quad \text{and} \quad L_2 = \left(\frac{k_2}{k_1}\right)L_1 = \left(\frac{238}{397}\right)(0.15 \text{ m}) = 9 \times 10^{-2} \text{ m} \quad ◊$$

---

49. A 1-m^2 solar collector collects radiation from the Sun and focuses it on 250 g of water that is initially at 23°C. The average thermal energy arriving from the Sun at the surface of the Earth at this location is 550 W/m^2, and we assume that this is collected with 100% efficiency. Find the time required for the collector to raise the temperature of the water to 100°C.

Solution To raise the temperature of water to 100°C, we need an amount of heat equal to $\Delta Q = m_w c_w \Delta T_w$.

$$\Delta Q = (0.25 \text{ kg})(4186 \text{ J/kg}\cdot{}^\circ\text{C})(77^\circ\text{C}) = 8.06 \times 10^4 \text{ J}$$

The collection rate is (550 W/m^2)(1 m^2) = 550 W = 550 J/s. Therefore, the time required is

$$t = \frac{\Delta Q}{\text{rate}} = \frac{8.06 \times 10^4 \text{ J}}{550 \text{ J/s}} = 146 \text{ s} = 2.4 \text{ min} \quad \lozenge$$

---

53. A brass statue (60.0% copper, 40.0% zinc) has a mass of 50.0 kg. If its temperature increases by 20.0°C, what is the change of internal energy of the statue? (Specific heat of brass = 380 J / kg·°C = 0.092 cal / g·°C.)

Solution Change in internal energy equals $\Delta Q = mc\Delta T$

$$\Delta Q = (50 \text{ kg})(380 \text{ J/kg}\cdot{}^\circ\text{C})(20^\circ\text{C}) = 3.8 \times 10^5 \text{ J} = 9.08 \times 10^4 \text{ cal} \quad \lozenge$$

---

57. A class of 10 students taking an exam has a power output per student of about 200 W. Assume that the initial temperature of the room is 20°C and that its dimensions are 6 m by 15 m by 3 m. What is the temperature of the room at the end of 1 h if all the heat remains in the air in the room and none is added by an outside source? The specific heat of air is 837 J / kg·°C, and its density is about 1.3×10^{-3} g/cm^3.

Solution The heat added to the air in one hour is $Q = 10(200 \text{ J/s})(3600 \text{ s}) = 7.2 \times 10^6 \text{ J}$

The mass of the air in the room is $m = \rho V = (6 \text{ m})(15 \text{ m})(3 \text{ m})(1.30 \text{ kg/m}^3) = 351 \text{ kg}$

Thus, the change in temperature of the room is

$$\Delta T = \frac{Q}{mc} = \frac{7.2 \times 10^6 \text{ J}}{(351 \text{ kg})(837 \text{ J/kg} \cdot °\text{C})} = 24.5°\text{C}$$

Thus, the final temperature is 44.5°C. ◊

---

61. A 1-m-long aluminum rod of cross-sectional area 2 cm^2 is inserted vertically into a thermally insulated vessel containing liquid helium at 4.2 K. The rod is initially at 300 K. If half of the rod is inserted into the helium, how many liters of helium boil off by the time the inserted half cools to 4. 2 K?

Solution During any time interval, the heat energy lost by the rod equals the heat energy gained by the helium. Therefore,

$$(mL)_{He} = (mc\,\Delta T)_{Al} \quad \text{or} \quad (\rho VL)_{He} = (\rho Vc\,\Delta T)_{Al} \quad \text{so that}$$

$$V_{He} = \frac{(\rho Vc\,\Delta T)_{Al}}{(\rho L)_{He}} = \frac{(2.7 \text{ g/cm}^3)(100 \text{ cm}^3)(0.21 \text{ cal/g} \cdot °\text{C})(295.8°\text{C})}{(0.125 \text{ g/cm}^3)(4.99 \text{ cal/g})}$$

$$= 2.69 \times 10^4 \text{ cm}^3 = 27 \text{ liters} \quad ◊$$

---

67. Water is being boiled in an open kettle that has a 0.5-cm-thick circular aluminum bottom with a radius of 12 cm. If the water boils away at rate of 0.5 kg/min, what is the temperature of the lower surface of the bottom of the kettle? Assume that the top surface of the bottom of the kettle is at 100°C.

Solution The heat required to vaporize 0.5 kg of water at 100°C is

$Q = (0.5 \text{ kg})(2.26 \times 10^6 \text{ J/kg}) = 1.13 \times 10^6 \text{ J}$ Thus, the rate of heat transfer is

$\dfrac{\Delta Q}{\Delta t} = \dfrac{1.13 \times 10^6 \text{ J}}{60 \text{ s}} = 1.88 \times 10^4 \text{ J/s}$ From $\dfrac{\Delta Q}{\Delta t} = \dfrac{kA\,\Delta T}{L}$ we have

$$\Delta T = \frac{\Delta Q}{\Delta t}\frac{L}{kA} = (1.88 \times 10^4 \text{ J/s})\frac{5 \times 10^{-3} \text{ m}}{(238 \text{ J/s} \cdot \text{m} \cdot °\text{C})\pi(0.12 \text{ m})^2} = 8.75°\text{C}$$

Thus, $T = 108.75°\text{C} = 109°\text{C}$ ◊

CHAPTER SELF-QUIZ

1. In winter, light-colored clothes will keep you warmer than dark-colored clothes if
   a. you are warmer than your surroundings
   b. you are at the same temperature as your surroundings
   c. you are cooler than your surroundings
   d. you are standing in sunlight

2. A 2-kg block is made of a metal that has higher specific heat than the metal out of which a 1-kg block is made. The 2-kg block was originally at 50°C and the 1-kg block at 20°C. When the two blocks are placed in contact, what will the final equilibrium temperature be?
   a. below 35°C
   b. 35°C
   c. between 35 and 40°C
   d. above 40°C

3. There are four identical blocks, all at 10°C. The first block is heated to 70°C while the others are left at 10°C. The first block is then placed in contact with the second block until an equilibrium temperature is reached. Then the first block is placed in contact with the third block until an equilibrium temperature is reached. Then the first block is placed in contact with the fourth block until an equilibrium temperature is reached. What is the final temperature of the first block?
   a. 25°C
   b. 20°C
   d. 17.5°C
   d. 15°C

4. The primary reason why a sandy beach gets so hot on a sunny day is because
   a. sand absorbs sunlight so well
   b sand reflects sunlight so well
   c. sand has a large specific heat
   d. sand has a small specific heat

5. A silver bar of length 30 cm and cross-sectional area 1 cm^2 is used to transfer heat from a 100°C reservoir to a 0°C reservoir. How much heat is transferred per second? (k_{silver} = 427 W/m – °C)
   a. 7.1 W
   b. 9.2 W
   c. 11.7 W
   d. 14.2 W

6. A 3-gram copper penny at 20°C is dropped from a height of 300 m and strikes the ground. If 60% of the energy goes into increasing the internal energy of the coin, determine its final temperature. (The specific heat of copper is 387 J/kg · °C).
   a. 22.22°C
   b 24.55°C
   c. 26.78°C
   d. 29.03°C

7. A solar heating system has a 25% conversion efficiency; the solar radiation incident on the panels is 500 W/m^2. What is the increase in temperature of 30 kg of water in a 1.0 h period by a 4.0 m^2 area collector?
   a. 14°C
   b. 22°C
   c. 29°C
   d. 44°C

8. If the absolute temperature of a spherical object were tripled, by what factor would the rate of radiated energy emitted from its surface be changed?
   a. 3.0
   b 9.0
   c. 27.0
   d. 81.0

9. On a cold day, a piece of metal feels much colder to the touch than a piece of wood. This is attributed to a difference in which property with respect to the two objects?
   a. mass density
   b. specific heat
   c. temperature
   d. thermal conductivity

10. A glass pane 0.4 cm thick has an area of 2×10^4 cm^2. On a winter day the temperature difference between the inside and outside surfaces of the pane is 25°C. What is the rate of heat flow through this window? (Thermal conductivity for glass is $0.837 \text{ J/s} \cdot \text{m} \cdot °\text{C}$)
    a. 837 J/s
    b. 1×10^5 J/s
    c. 16.7 J/s
    d. 34.3 J/s

11. A 1-kg block of copper at 20°C is dropped into a large vessel of liquid nitrogen at 77 K. How many kilograms of nitrogen boil away by the time the copper reaches 77 K? (The specific heat of copper is $385 \text{ J/kg} \cdot °\text{C}$. The heat of vaporization of nitrogen is 2×10^5 J/kg.)
    a. 0.212 kg
    b. 0.414 kg
    c. 0.636 kg
    d. 0.898 kg

12. How much heat energy is required to melt a 20-gram block of silver at 20°C? The melting point of silver is 960°C, its specific heat is $2.34 \times 10^2 \text{ J/kg} \cdot °\text{C}$, and its heat of fusion is 8.83×10^4 J/kg.
    a 4395 J
    b. 6175 J
    c. 7518 J
    d. 8795 J

12
The Laws of Thermodynamics

Chapter 12

THE LAWS OF THERMODYNAMICS

The first law of thermodynamics is essentially the principle of conservation of energy generalized to include heat as a form of energy. It tells us that an increase in one form of energy must be accompanied by a decrease in some other form of energy. The law considers both heat and work and places no restrictions on the types of energy conversions that occur. According to the first law, the internal energy of an object can be increased either by heat added to the object or by work done on it.

The second law of thermodynamics, which can be stated in many equivalent ways, establishes which processes can occur in nature and which cannot. For example, the second law tells us that heat never flows spontaneously from a cold body to a hot body. One important application of this law is in the study of heat engines, such as the internal combustion engine, and the principles that limit their efficiency.

NOTES FROM SELECTED CHAPTER SECTIONS

12.2 Work and Heat

The work done in the expansion from the initial state to the final state is the area under the curve in a *PV* diagram, as shown in the figure.

If the gas is compressed, $V_f < V_i$, and the work is negative. That is, work is done *on* the gas. If the gas expands, $V_f > V_i$, the work is positive, and the gas does work on the piston. If the gas expands at *constant pressure,* called an *isobaric process,* then $W = P(V_f - V_i)$.

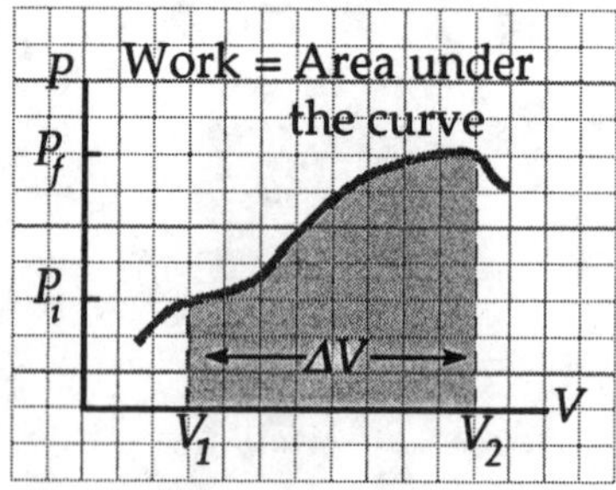

The work done by a system depends on the process by which the system goes from the initial to the final state. In other words, the work done depends on the initial, final, and intermediate states of the system.

The amount of heat gained or lost by a system depends on the initial, final, and intermediate states of the system.

12.3 The First Law of Thermodynamics

In the first law of thermodynamics, $\Delta U = Q - W$, Q is the heat added to the system and W is the work done by the system. Note that by convention, Q is *positive* when heat enters the system and *negative* when heat is removed from the system. Likewise, W can be positive or negative as mentioned earlier. The initial and final states must be *equilibrium* states; however, the intermediate states are, in general, nonequilibrium states since the thermodynamic coordinates undergo finite changes during the thermodynamic process.

A process that occurs at constant temperature is called an *isothermal process,* and a plot of P versus V at constant temperature for an ideal gas yields a hyperbolic curve called an *isotherm.* The internal energy of an ideal gas is a function of temperature only.

An *isolated system* is one which does not interact with its surroundings. In such a system, $Q = W = 0$. That is, the internal energy of an isolated system cannot change.

A *cyclic process* is one that originates and ends up at the same state. The work done per cycle equals the heat added to the system per cycle. This is important to remember when dealing with heat engines in the next section.

An *adiabatic process* is a process in which no heat enters or leaves the system; that is, $Q = 0$. A system may undergo an adiabatic process if it is thermally insulated from its surroundings.

An *isobaric process* is a process which occurs at constant pressure. For such a process, the heat transferred and the work done are nonzero.

An *isovolumetric process* is one which occurs at constant volume. By definition, $W = 0$ for such a process (since the volume does not change). All of the heat added to the system kept at constant volume goes into increasing the internal energy of the system.

12.4 Heat Engines and the Second Law of Thermodynamics

A heat engine is a device that converts thermal energy to other useful forms, such as electrical and mechanical energy.

A heat engine carries some working substance through a cyclic process during which (1) heat is absorbed from a source at a high temperature, (2) work is done by the engine, and (3) heat is expelled by the engine to a source at a lower temperature.

The engine absorbs a quantity of heat, Q_h, from a hot reservoir, does work W, and then gives up heat Q_c to a cold reservoir. Because the working substance goes through a cycle, its initial and final internal energies are equal, so $\Delta U = 0$. Hence, from the first equation we see that *the net work, W, done by a heat engine equals the net heat flowing into it.*

If the working substance is a gas, *the net work done for a cyclic process is the area enclosed by the curve representing the process on a PV diagram.*

The **thermal efficiency,** e, of a heat engine is the ratio of the net work done to the heat absorbed at the higher temperature during one cycle.

The second law of thermodynamics can be stated as follows: *It is impossible to construct a heat engine that, operating in a cycle, produces no other effect than the absorption of heat from a reservoir and the performance of an equal amount of work.*

12.5 Reversible and Irreversible Processes

A process is *irreversible* if the system and its surroundings cannot be returned to their initial states. A process is *reversible* if the system passes from the initial to the final state through a succession of equilibrium states.

12.6 The Carnot Engine

The most efficient cyclic process is called the *Carnot cycle,* described in the *PV* diagram shown in the figure. The Carnot cycle consists of two adiabatic and two isothermal processes, all being reversible.

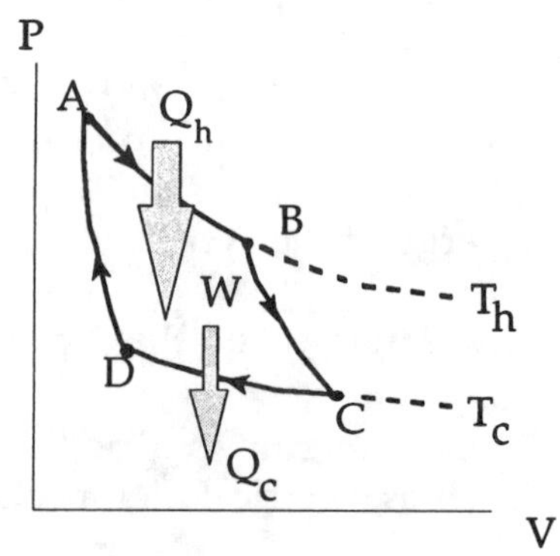

The Carnot Cycle

1. The process $A \rightarrow B$ is an isotherm (constant T), during which time the gas expands at constant temperature T_h, and absorbs heat Q_h from the hot reservoir.

2. The process $B \rightarrow C$ is an adiabatic expansion ($Q = 0$), during which time the gas expands and cools to a temperature T_c.

3. The process $C \rightarrow D$ is a second isotherm, during which time the gas is compressed at constant temperature T_c, and gives up heat Q_c to the cold reservoir.

4. The final process $D \rightarrow A$ is an adiabatic compression in which the gas temperature increases to a final temperature of T_h.

In practice, no working engine is 100% efficient, even when losses such as friction are neglected. One can obtain some theoretical limits on the efficiency of a real engine by comparison with the ideal Carnot engine. A *reversible engine* is one which will operate with the same efficiency in the forward and reverse directions. The Carnot engine is one example of a reversible engine.

> All Carnot engines operating reversibly between T_h and T_c have the *same* efficiency given by Equation 12.5.
>
> No real (irreversible) engine can have an efficiency greater than that of a reversible engine operating between the same two temperatures.

A *heat engine* is a device that converts thermal energy into other forms of energy such as mechanical and electrical energy. During its operation, a heat engine carries some working substance through a *cyclic process,* which is a process which begins and ends at the same state.

12.8 Entropy
12.9 Entropy and Disorder

Entropy is a quantity used to measure the degree of *disorder* in a system. For example, the molecules of a gas in a container at a high temperature are in a more disordered state (higher entropy) than the same molecules at a lower temperature.

When heat is added to a system, the entropy *increases*. When heat is removed, the entropy *decreases*. Note that only *changes* in entropy are defined by Equation 12.8; therefore, the concept of entropy is most useful when a system undergoes a *change in its state*.

The second law of thermodynamics can be stated in terms of entropy as follows: *The total entropy of an isolated system always increases in time if the system undergoes an irreversible process.* If an isolated system undergoes a *reversible* process, the total entropy *remains constant*.

EQUATIONS AND CONCEPTS

This equation can be used to calculate the work done on or by a gas sample, if the pressure of the gas remains constant during a compression or an expansion. When ΔV is positive, the work done is positive (work is done by the gas); when ΔV is negative, the work done is negative (work is done on the gas). The work done is equal to the area under the pressure-volume curve.

$$W = P\Delta V \quad (12.1)$$

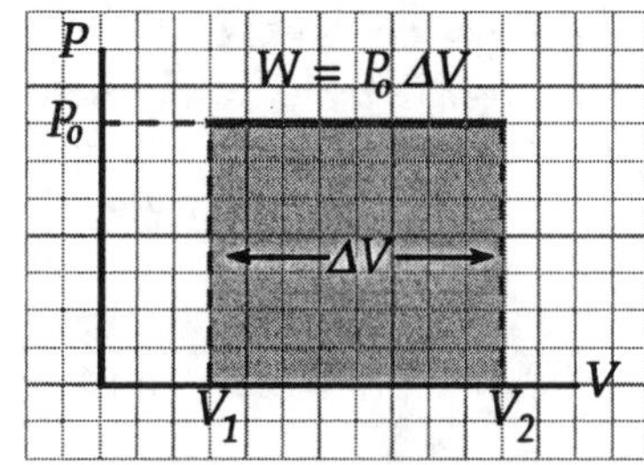

This is a statement of the first law of thermodynamics in equation form. This law is a generalization of the law of conservation of energy.

$$\Delta U = Q - W \quad (12.2)$$

Q = heat added to system

W = work done by system

ΔU = change in the internal energy of the system

Q is positive when heat energy is added to the system. W is positive when work is done by the system on its surroundings.

Comment on the first law of thermodynamics.

The values of both Q and W depend on the path or sequence of processes by which a system changes from an initial to a final state. The internal energy U depends only on the initial and final states of the system.

The following definitions will be important in describing the remaining equations in this chapter:

Isolated system

A system which does not interact with its surroundings. In such a system $Q = W = 0$, so that $\Delta U = 0$. The internal energy of an isolated system cannot change.

Cyclic process

A process for which the initial and final states are the same. For a cyclic process, $\Delta U = 0$ and $Q = W$. The net work done per cycle equals the heat energy added to the system per cycle.

Adiabatic process

A process in which no heat enters or leaves the system. $Q = 0$ and, therefore, $\Delta U = -W$. A system can undergo an adiabatic process if it is thermally insulated from its surroundings.

Isobaric process

A process which occurs at constant pressure. In an isobaric process, the work done and the heat transferred are both nonzero.

Isovolumetric process

A process which occurs at constant volume. Since at constant volume, $\Delta V = 0$, $W = 0$; and therefore, $\Delta U = Q$. All of the heat added at constant volume goes to increasing the internal energy.

Isothermal process

A process which occurs at constant temperature. The change in internal energy during an isothermal process results both from heat added and work done.

Chapter 12

A device that converts thermal energy into other forms of energy by carrying a substance through a cycle: (1) heat is absorbed from a source at a high temperature, (2) work is done by the engine, and (3) heat is expelled by the engine to a source at a low temperature.

Heat engine

A measure of the degree of disorder in a system.

Entropy

The net work done by a heat engine equals the net heat flowing into it. Q_h is the quantity of heat absorbed from the high temperature reservoir and Q_L is the quantity of heat expelled to the low (cold) temperature reservoir. Both Q_h and Q_L are taken to be positive quantities.

$$W = Q_h - Q_c \qquad (12.3)$$

The thermal efficiency of a heat engine is the ratio of the net work done to the heat absorbed during one cycle of the process.

$$e \equiv \frac{Q_h - Q_c}{Q_h} = 1 - \frac{Q_c}{Q_h} \qquad (12.4)$$

It is impossible to construct a heat engine that, operating in a cycle, produces no other effect than the absorption of thermal energy from a reservoir and the performance of an equal amount of work.

Statement on the second law of thermodynamics.

The Carnot cycle is the most efficient cyclic process and the thermal efficiency of an ideal Carnot engine depends on the temperatures of the hot and cold reservoirs.

$$e_c = \frac{T_h - T_c}{T_h} = 1 - \frac{T_c}{T_h} \qquad (12.5)$$

Entropy. S is a thermodynamic variable which characterizes the degree of disorder in a system. All physical processes tend toward a state of increasing entropy; and in going from an initial to a final state, the change in entropy. ΔS is the ratio of the heat energy added to the system to the absolute temperature of the system.

$$\Delta S = \frac{\Delta Q_r}{T} \tag{12.8}$$

The entropy of the Universe increases in all natural processes. This is an alternate way of expressing the second law of thermodynamics.

REVIEW CHECKLIST

▷ Understand how work is defined when a system undergoes a change in state, and the fact that work (like heat) depends on the path taken by the system. You should also know how to sketch processes on a *PV* diagram, and calculate work using these diagrams.

▷ State the first law of thermodynamics ($\Delta U = Q - W$), and explain the meaning of the three forms of energy contained in this statement. Discuss the implications of the first law of thermodynamics as applied to (i) an isolated system, (ii) a cyclic process, (iii) an adiabatic process, and (iv) an isothermal process.

▷ Describe the processes via which an ideal heat engine goes through a *Carnot cycle.* Express the efficiency of an ideal heat engine (Carnot engine) as a function of work and heat exchange with its environment. Express the maximum efficiency of an ideal heat engine as a function of its input and output temperatures.

▷ Understand the concept of entropy. Define *entropy* for a system in terms of its heat energy gain or loss, and its temperature. State the *second law of thermodynamics* as it applies to entropy changes in a thermodynamic system.

SOLUTIONS TO SELECTED END-OF-CHAPTER PROBLEMS

3. Sketch a PV diagram of the following processes. (a) A gas expands at constant pressure P_1 from volume V_1 to volume V_2. It is then kept at constant volume while the pressure is reduced to P_2. (b) A gas is reduced in pressure from P_1 to P_2 while its volume is held constant at V_1. It is then expanded at constant pressure P_2 to a final volume, V_2. (c) In which of the processes is more work done? Why?

Solution (a and b) The sketches of PV diagrams for the two processes are shown below.

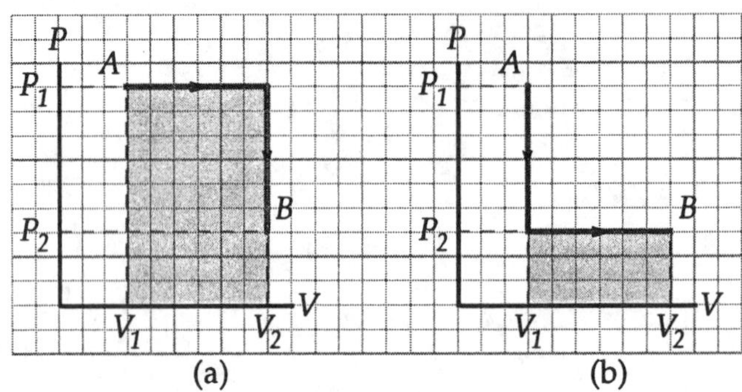

(c) There is more work done in process (a). We recognize this from the figures because there is more area under the PV curve in (a). Physically, more work is done because of the higher pressure during the expansion part of the process.

---

7. An ideal gas is enclosed under a movable piston in a cylinder. The piston has a mass of 8.00 kg and an area of 5.00 cm^2; it is free to slide up and down, keeping the pressure of the gas constant. How much work is done as the temperature of 0.2 mol of the gas is raised from 20.0°C to 300°C?

Solution For an ideal gas $PV = nRT$, and $V = \dfrac{nRT}{P}$

so
$$\Delta V = V_f - V_i = \frac{nR}{P}\left(T_f - T_i\right)$$

This gives
$$W = P\Delta V = nR\Delta T = (0.200 \text{ mol})(8.31 \text{ J/K mol})(280 \text{ K}) = 465 \text{ J} \quad \Diamond$$

---

11. A gas expands from I to F in Figure 12.15. The heat added to the gas is 418 J when the gas goes from I to F along the diagonal path. (a) What is the change in internal energy of the gas? (b) How much heat must be added to the gas for the indirect path IAF to give the same change internal energy?

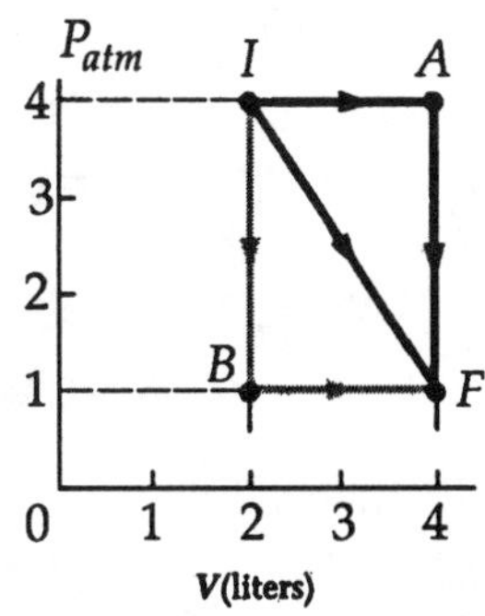

Figure 12.15

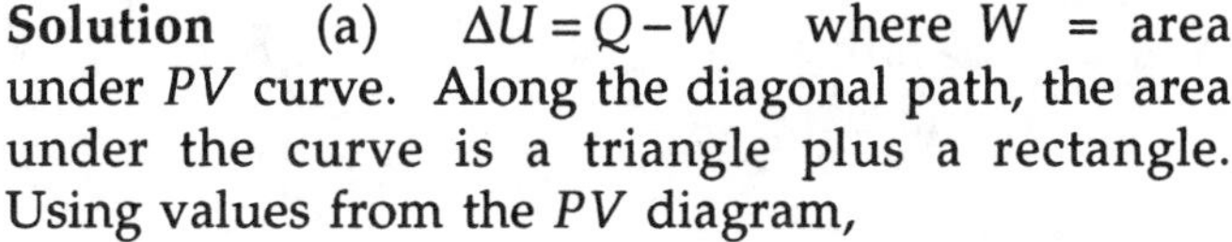

Solution (a) $\Delta U = Q - W$ where W = area under PV curve. Along the diagonal path, the area under the curve is a triangle plus a rectangle. Using values from the PV diagram,

$$W = \frac{1}{2}(2\times10^{-3}\text{ m}^3)(3.04\times10^5\text{ Pa}) + (2\times10^{-3}\text{ m}^3)(1.013\times10^5\text{ Pa})$$

$$W = 507\text{ J}$$

Therefore, $\Delta U = 418\text{ J} - 507\text{ J} = -89.0\text{ J}$ ◊

(b) Along path IAF, $W_{IAF} = W_{IA} + W_{AF}$

The volume is constant along AF, so $W_{AF} = 0$

$$W_{IAF} = W_{IA} = P_I(V_A - V_I) = (4.05\times10^5\text{ Pa})(2\times10^{-3}\text{ m}^3) = 810\text{ J}$$

ΔU has the same value as in part (a).

$$Q = \Delta U + W = -89.0\text{ J} + 810\text{ J} = 721\text{ J}$$ ◊

---

15. Two cm^3 of water is boiled at atmospheric pressure to become 3342 cm^3 of steam, also at atmospheric pressure. (a) Calculate the work done by the gas during this process. (b) Find the amount of heat added to the water to accomplish this process. (c) From (a) and (b), find the change in internal energy.

Solution

(a) $W = P\Delta V = (1.013\times10^5\text{ Pa})(3.34\times10^{-3}\text{ m}^3) = 338\text{ J}$ ◊

(b) The heat added is

$$Q = mL_v = (2.00\times10^{-3}\ \text{kg})(2.26\times10^{6}\ \text{J/kg}) = 4520\ \text{J} \quad \lozenge$$

(c) $\Delta U = Q - W = 4520\ \text{J} - 338\ \text{J} = 4180\ \text{J} \quad \lozenge$

---

17. Consider the cyclic process described by Figure 12.16. If Q is negative for the process BC, and ΔU is negative for the process CA, determine the signs of Q, W, and ΔU associated with each process.

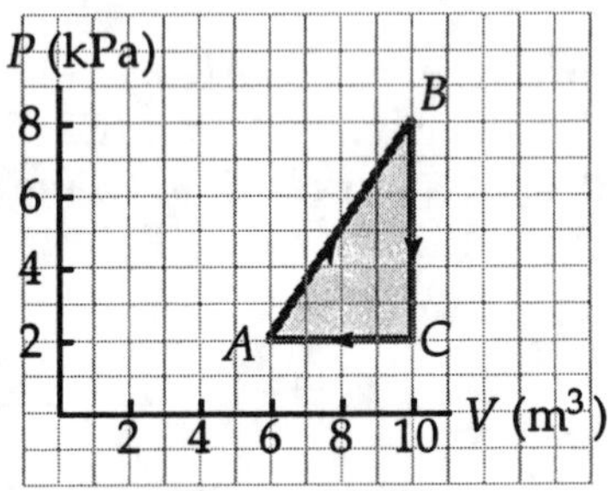

Figure 12.16

Solution In the table below, the signs of Q_{BC} and ΔU_{CA} are given in the statement of the problem. Determine and enter the remaining seven values as outlined in the sequence of steps following the table.

Process	Q	ΔU	W
BC	<0	<0	0
CA	<0	<0	<0
AB	>0	>0	>0

(1) $W_{BC} = 0$ because $\Delta V_{BC} = 0$

(2) $W_{CA} < 0$ because $\Delta V_{CA} < 0$ at constant pressure

(3) $W_{AB} > 0$ because both ΔV_{AB} and ΔP_{AB} are positive

(4) $\Delta U_{BC} = Q_{BC} - W_{BC} < 0$

(5) For any cycle process, $\Delta U_{\text{cycle}} = 0$ so $\Delta U_{AB} + \Delta U_{BC} + \Delta U_{CA} = 0$

Therefore, since ΔU_{BC} and ΔU_{CA} are negative, $\Delta U_{AB} > 0$

(6) $Q_{CA} = \Delta U_{CA} + W_{CA}$ and both ΔU_{CA} and W_{CA} are negative. Therefore, $Q_{CA} < 0$

(7) $Q_{AB} = \Delta U_{AB} + W_{AB}$ and $Q_{AB} > 0$ because both ΔU_{AB} and W_{AB} are positive

---

19. A 100-kg steel support rod in a building has a length of 2.00 m at a temperature of 20.0°C. The rod supports a load of 6000 kg. Find (a) the work done by the rod as the temperature increases to 40.0°C, (b) the heat added to the rod (assume the specific heat of steel is the same as that for iron), and (c) the change in internal energy of the rod.

Solution (a) The change in length of the rod is

$$\Delta L = \alpha L_0(\Delta T) = [11.0\times10^{-6}(^\circ\text{C})^{-1}](2.00\text{ m})(20.0^\circ\text{C}) = 4.40\times10^{-4}\text{ m}$$

The force exerted by the rod equals the weight of the load. Thus, $F = 5.88\times10^{4}\text{ N}$

so $$W = F(\Delta L) = (5.88\times10^{4}\text{ N})(4.40\times10^{-4}\text{ m}) = 25.9\text{ J} \quad \lozenge$$

(b) $Q = mc\Delta T = (10^2\text{ kg})(448\text{ J/kg}\cdot{}^\circ\text{C})(20.0^\circ\text{C}) = 8.96\times10^{5}\text{ J} \quad \lozenge$

(c) $\Delta U = Q - W = 8.96\times10^{5}\text{ J} - 25.9\text{ J} = 8.96\times10^{5}\text{ J} \quad \lozenge$

---

23. A container is placed in a water bath and held at constant volume as a mixture of fuel and oxygen is burned inside it. The temperature of the water is observed to rise during the burning (the water is also held at constant volume). (a) Consider the burning mixture to be the system. What are the signs of W, Q, and ΔU ? (b) What are the signs of these quantities if the water bath is considered to be the system?

Solution (a) $W = 0$ because the volume equals a constant. ◊

The system gives off heat. Thus, $Q < 0$ ◊

$\Delta U = Q - W$ and since $W = 0$, we have $\Delta U = Q$.

Therefore, $\Delta U < 0$ ◊

(b) Again, $W = 0$ because the volume remains constant. ◊

$Q > 0$ because the water receives heat. ◊

Again, $\Delta U = Q$ so $\Delta U > 0$ ◊

---

27. A heat engine performs 200 J of work in each cycle and has an efficiency of 30%. For each cycle of operation, (a) how much heat is absorbed and (b) how much heat is expelled?

Solution $$W = Q_h - Q_c = 200 \text{ J} \quad (1)$$

$$e = \frac{W}{Q_h} = 1 - \frac{Q_c}{Q_h} = 0.300 \quad (2)$$

From (2), $$Q_c = 0.700\, Q_h \quad (3)$$

Solving (3) and (1) simultaneously, we have

$$Q_h = 667 \text{ J absorbed} \quad \text{and} \quad Q_c = 467 \text{ J expelled} \quad \Diamond$$

---

29. A particular engine has a power output of 5.00 kW and an efficiency of 25%. If the engine expels 8000 J of heat in each cycle, find (a) the heat absorbed in each cycle and (b) the time required for each cycle.

Solution (a) $e = \frac{W}{Q_h} = \frac{Q_h - Q_c}{Q_h} = 1 - \frac{Q_c}{Q_h} = 0.250$

With $Q_c = 8000$ J, we have $Q_h = 10667 \text{ J} = 10.7 \text{ kJ}$ ◊

(b) $W = Q_h - Q_c = 2667$ J and from $P = \frac{W}{\Delta t}$, we have

$$\Delta t = \frac{W}{P} = \frac{2667 \text{ J}}{5000 \text{ J/s}} = 0.533 \text{ s} \quad \Diamond$$

---

33. The exhaust temperature of a Carnot heat engine is 300°C. What is the intake temperature if the efficiency of the engine is 30%?

Solution For Carnot efficiency, $e_c = 1 - \frac{T_c}{T_h}$

or $$T_h = \frac{T_c}{1 - e_c} = \frac{573 \text{ K}}{1 - 0.30} = 819 \text{ K} = 546°\text{C} \quad \Diamond$$

---

37. A power plant that uses the temperature gradient in the ocean has been proposed. The system is to operate between 20.0°C (surface water temperature) and 5.00°C (water temperature at a depth of about 1.00 km). (a) What is the maximum efficiency of such a system? (b) If the power output of the plant is 75.0 MW, how much thermal energy is absorbed per hour? (c) In view of your results to (a) do you think such a system is worthwhile?

Solution

(a) $e_{\text{max}} = 1 - \frac{T_c}{T_h} = 1 - \frac{278}{293} = 5.12 \times 10^{-2}$ (or 5.12%) ◊

(b) $P = \frac{W}{\Delta t} = 75.0 \times 10^6 \text{ J/s}$

Therefore, $W = (75.0 \times 10^6 \text{ J/s})(3600 \text{ s/h}) = 2.70 \times 10^{11} \text{ J/h}$

From $e = \frac{W}{Q_h}$ we find

$$Q_h = \frac{W}{e} = \frac{2.70 \times 10^{11} \text{ J/h}}{5.12 \times 10^{-2}} = 5.27 \times 10^{12} \text{ J/h} \quad ◊$$

---

41. A 70.0-kg log falls from a height of 25.0 m into a lake. If the log, the lake, and the air are all at 300 K, find the change in entropy of the Universe for this process.

Solution The heat generated equals the potential energy given up by the log.

$$Q = mgh = (70.0 \text{ kg})(9.80 \text{ m/s}^2)(25.0 \text{ m}) = 1.72 \times 10^4 \text{ J}$$

Thus,

$$\Delta S = \frac{Q}{T} = \frac{1.72 \times 10^4 \text{ J}}{300 \text{ K}} = 57.2 \text{ J/K} \quad ◊$$

---

45. Repeat the procedure used to construct Table 12.1 (a) for the case in which you draw three marbles from your bag rather than four and (b) for the case in which you draw five rather than four.

Solution

(a)

Result	Possible Combinations	Total
all red	RRR	1
2R, 1G	RRG, RGR, GRR	3
1R, 2G	RGG, GRG, GGR	3
all green	GGG	1

(b)

Result	Possible Combinations	Total
all red	RRRRR	1
4R, 1G	RRRRG, RRRGR, RRGRR, RGRRR, GRRRR	5
3R, 2G	RRRGG, RRGRG, RGRRG, GRRRG, RRGGR, RGRGR, GRRGR, RGGRR, GRGRR, GGRRR	10
2R, 3G	GGGRR, GGRGR, GRGGR, RGGGR, GGRRG, GRGRG, RGGRG, GRRGG, RGRGG, RRGGG	10
1R, 4G	RGGGG, GRGGG, GGRGG, GGGRG, GGGGR	5
all green	GGGGG	1

49. A gas follows path 123 on the PV diagram in Figure 12.19, and 418 J of heat flows into the system; also, 167 J of work is done. (a) What is the internal energy change of the system? (b) How much heat flows into the system if the process follows path 143? The work done by the gas along this path is 63 J. What net work would be done on or by the system if the system followed (c) path 12341? (d) path 14321? (e) What is the change in internal energy of the system in the processes described in parts (c) and (d)?

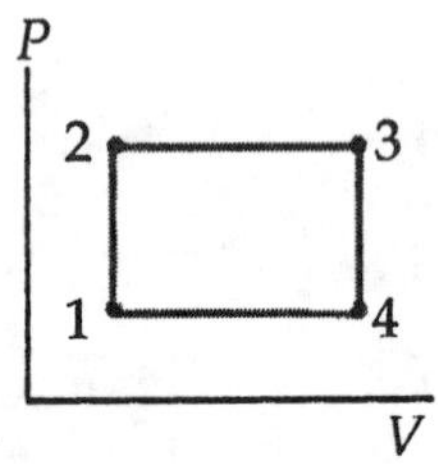

Figure 12.19

Solution

(a) $\Delta U = Q - W = 418\text{ J} - 167\text{ J} = 251\text{ J}$ ◊

(b) Along path (143), $Q = \Delta U + W = 251\text{ J} + 63\text{ J} = 314\text{ J}$ ◊

(c) Work equals area under the *PV* diagram,

$$W(12341) = W(123) - W(143) = 167\text{ J} - 63\text{ J} = 104\text{ J} \quad ◊$$

(d) $W(14321) = W(143) - W(123) = 63\text{ J} - 167\text{ J} = -104\text{ J}$ ◊

(e) For *any cyclic process,* $\Delta U = 0$ ◊

---

53. What is the minimum amount of work that must be done to extract 400 J of heat from a massive object at 0°C while rejecting heat to a hot reservoir at 20.0°C?

Solution The minimum quantity of work will be done by a refrigerator operating at maximum COP. The maximum COP for the refrigerator = $\text{COP}_{\text{Carnot}}$, or

$$\text{COP}_{\max} = \left(\frac{T_c}{T_h - T_c}\right) = \frac{273\text{ K}}{20\text{ K}} = 13.7$$

For a given Q_c, $\text{COP}_{\max} = \dfrac{Q_c}{W_{\min}}$ so $W_{\min} = \dfrac{Q_c}{\text{COP}_{\max}} = \dfrac{400\text{ J}}{13.7} = 29.3\text{ J}$ ◊

---

57. A 1500-kW heat engine operates at 25% efficiency. The heat energy expelled at the low temperature is absorbed by a stream of water that enters the cooling coils at 20.0°C. If 60.0 L flows across the coils per second, determine the increase in temperature of the water.

Solution We have $e = \dfrac{W}{Q_h} = 0.25$

Thus, $Q_h = \dfrac{W}{0.25} = \dfrac{1500 \times 10^3\text{ J/s}}{0.25} = 6.00 \times 10^6\text{ J/s}$

and $$Q_c = Q_h - W = 6.00\times10^6\ \text{J/s} - 1.50\times10^6\ \text{J/s} = 4.50\times10^6\ \text{J/s}$$

The coolant flow is $60\ \text{L/s} = 6.00\times10^{-2}\ \text{m}^3\text{/s}$. Thus, the mass flow is

$$\text{mass/s} = (6.00\times10^{-2}\ \text{m}^3\text{/s})(1000\ \text{kg/m}^3) = 60.0\ \text{kg/s}$$

Therefore, we see that the 60.0 kg of water must absorb 4.50×10^6 J.

From $Q = mc\Delta T$,

$$\Delta T = \frac{Q}{mc} = \frac{4.50\times10^6\ \text{J}}{(60.0\ \text{kg})(4184\ \text{J/kg}\cdot{}^\circ\text{C})} = 17.9^\circ\text{C} \quad \Diamond$$

---

61. An ideal gas initially at pressure P_0, volume V_0, and temperature T_0 is taken through the cycle described in Figure 12.22. (a) Find the net work done by the gas per cycle in terms of P_0 and V_0. (b) What is the net heat added to the system per cycle? (c) Obtain a numerical value for the net work done per cycle for 1 mol of gas initially at 0°C. (See the hint for Problem 60.)

Solution (a) The net work done in the entire cycle equals the area enclosed within the cycle on a *PV* diagram. Thus,

$$W_{\text{net}} = (3P_0 - P_0)(3V_0 - V_0) = 4P_0V_0 \quad \Diamond$$

(b) $\Delta U = 0$ for any complete cycle. Thus, from the first law,

$$Q_{\text{net}} = W_{\text{net}} = 4P_0V_0 \quad \Diamond$$

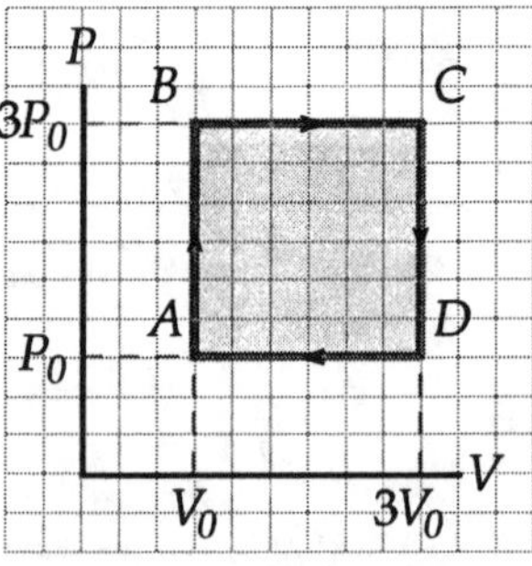

Figure 12.22

(c) From the ideal gas law, we have

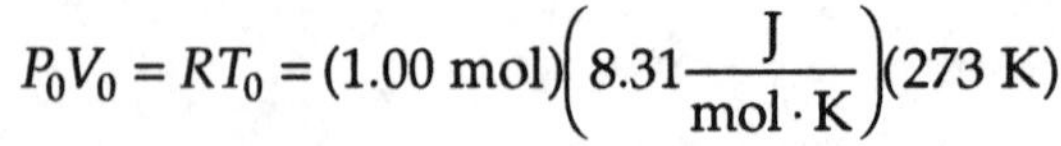

$$P_0V_0 = RT_0 = (1.00\ \text{mol})\left(8.31\frac{\text{J}}{\text{mol}\cdot\text{K}}\right)(273\ \text{K})$$

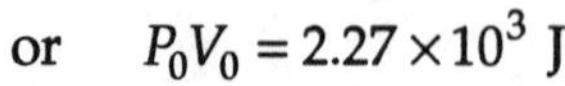

or $P_0V_0 = 2.27\times10^3$ J

Thus, $$W_{\text{net}} = 4P_0V_0 = 4(2.27\times10^3\ \text{J}) = 9.07\times10^3\ \text{J} \quad \Diamond$$

CHAPTER SELF-QUIZ

1. An ideal gas undergoes an adiabatic process in doing 25 J of work on its environment. What is its change in internal energy?
   a. 50 J
   b. 25 J
   c. Zero
   d. −25 J

2. A 3-mol ideal gas system is maintained at a constant volume of 4 liters; if 20 J of heat are added, what is the work done by the system?
   a. Zero
   b. 5.0 J
   c. 6.7 J
   d. 20 J

3. An ideal gas is maintained at a constant pressure of 200 N/m^2 during an isobaric process while its volume decreases by 0.2 m^3. What work is done by the system on its environment?
   a. 40 J
   b. 1000 J
   c. −40 J
   d. −1000 J

4. What is the work done by the gas as it expands from pressure P_1 and volume V_1 to pressure P_2 and volume V_2 along the indicated straight line?
   a. $(P_1 + P_2)(V_2 - V_1)/2$
   b. $(P_1 + P_2)(V_2 - V_1)$
   c. $(P_1 - P_2)(V_2 - V_1)/2$
   d. $(P_1 - P_2)(V_2 + V_1)$

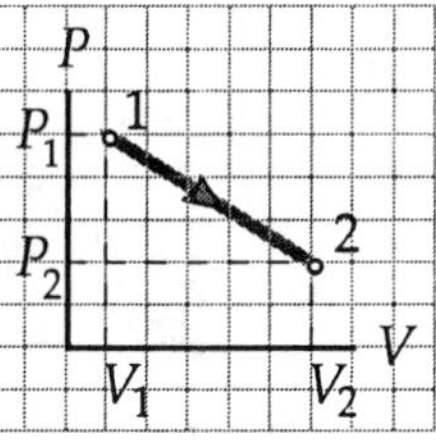

5. A 3-mol ideal gas system is maintained at a constant volume of 4 liters; if 20 J of heat are added, what is the change in internal energy of the system?
   a. Zero
   b. 5.0 J
   c. 6.7 J
   d. 20 J

6. A boulder of mass 600 kg tumbles down a mountainside and stops 160 m below. If the temperature of the boulder, mountain, and surrounding air are all at 300 K, what is the change in entropy of the Universe?
   a. 320 J/K
   b. 3100 J/K
   c. 1200 J/K
   d. 5200 J/K

7. An ideal gas at pressure, volume, and temperature, P_0, V_0, and T_0, respectively, is heated to point A, allowed to expand to point B at constant temperature $2T_0$, and then returned to the original condition. The internal energy increases by $3P_0V_0/2$ going from point B to Point T_0. How much heat left the gas from point B to point T_0?
   a. 0
   b. $P_0V_0/2$
   c. $3P_0V_0/2$
   d. $5P_0V_0/2$

8. The efficiency of a 1000-MW nuclear power plant is 33%. That is, 2000 MW of heat energy per second is rejected for every 1000 MW of electrical power produced. If a river of flow rate 10^6 kg/s is used to transport the excess heat away, what is the average temperature increase of the river?
   a. 1.44°C
   b. 0.96°C
   c. 0.48°C
   d. 0.24°C

9. A bottle containing an ideal gas has a volume of 2 m^3 and a pressure of 1×10^5 N/m^2 at a temperature of 300°K. The bottle is placed against a metal block that is maintained at 900°K and the gas expands as the pressure remains constant until the temperature of the gas reaches 900°K. How much work is done by the gas?
   a. 0
   b. 2×10^5 J
   c. 4×10^5 J
   d. 6×10^5 J

10. A gasoline engine absorbs 2500 J of heat energy and performs 500 J of mechanical work in each cycle. The efficiency of the engine is
    a. 80%
    b. 40%
    c. 60%
    d. 20%

11. What is the coefficient of performance of a refrigerator that operates with Carnot efficiency between temperatures −3°C and +27°C?
    a. 8
    b. 9
    c. 10
    d. 11

12. Calculate the entropy change when 1 mol (18 grams) of water at 100°C is converted into steam. (The heat of vaporization of water is 540 cal/gm).
    a. 26.1 cal/K
    b. 35.6 cal/K
    c. 97.2 cal/K
    d. 174.9 cal/K

13
Vibrations and Waves

Chapter 13

VIBRATIONS AND WAVES

During your study of this chapter, you will have an opportunity to use many of the concepts which were developed in the chapter on mechanics. You will examine various forms of *periodic motion,* concentrating especially on motion that occurs when the force on an object is proportional to the displacement of the object from its equilibrium position. When such a force acts only toward the equilibrium position, and has a magnitude which is proportional to the displacement from equilibrium, the result is a back-and-forth motion called simple harmonic motion--oscillation, or vibration, between two extreme positions for an indefinite period of time with no loss of energy. The terms *harmonic motion* and *periodic motion* are used interchangeably in this chapter. Both refer to back-and-forth motion.

Since vibrations can move through a medium, we also study wave motion in this chapter. Many kinds of waves occur in nature, including sound waves, seismic waves, and electromagnetic waves. We end this chapter with a brief discussion of some terms and concepts that are common to all types of waves, and in later chapters we shall focus our attention on specific categories of waves.

NOTES FROM SELECTED CHAPTER SECTIONS

13.1 Hooke's Law
13.2 Elastic Potential Energy

Simple harmonic motion occurs when the net force along the direction of motion is a Hooke's law type of force; that is, when the net force is proportional to the displacement and in the opposite direction.

It is necessary to define a few terms relative to harmonic motion.

1. The amplitude, A, is the *maximum distance that an object moves away from its equilibrium position.* In the absence of friction, an object will continue in simple harmonic motion and reach a maximum displacement equal to the amplitude on each side of the equilibrium position during each cycle.

2. The period, T, *is the time it takes the object to execute one complete cycle of the motion.*

3. The frequency, f, *is the number of cycles or vibrations per unit of time.*

Oscillatory motions are exhibited by many physical systems such as a mass attached to a spring, a pendulum, atoms in a solid, stringed musical instruments, and electrical circuits driven by a source of alternating current. *Simple harmonic motion* of a mechanical system corresponds to the oscillation of an object between two points for an indefinite period of time, with no loss in mechanical energy.

An object exhibits simple harmonic motion if the net external force acting on it is a *linear restoring force.*

13.3 Velocity as a Function of Position
13.5 Position as a Function of Time

The position (x), velocity (v), and acceleration (a) of an object moving with simple harmonic motion are shown in the three graphs below. In this particular case, the object was released from rest when it was a maximum distance (amplitude) from the equilibrium position.

$$x = A \cos(2\pi ft)$$

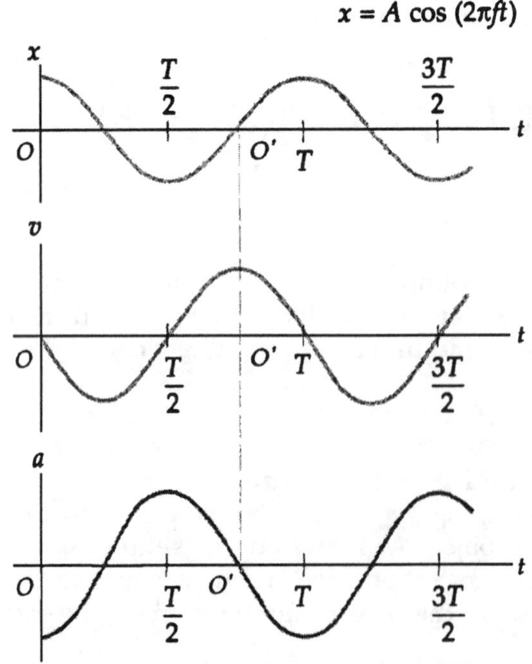

(a) Displacement, (b) velocity, and (c) acceleration versus time for an object moving with simple harmonic motion under the initial conditions that $x_0 = A$ and $v_0 = 0$ at $t=0$.

The most common system which undergoes simple harmonic motion is the mass-spring system shown in the figure below. The mass is assumed to move on a horizontal, *frictionless* surface. The point $x = 0$ is the equilibrium position of the mass; that is, the point where the mass would reside if left undisturbed. In this position, there is no horizontal force on the mass. When the mass is displaced a distance x from its equilibrium position, the spring produces a linear restoring force given by Hooke's law, $F = -kx$, where k is the force constant of the spring, and has SI units of N/m. The minus sign means that F is to the *left* when the displacement x is positive, whereas F is to the *right* when x is negative. In other words, the direction of the force F is *always* towards the equilibrium position.

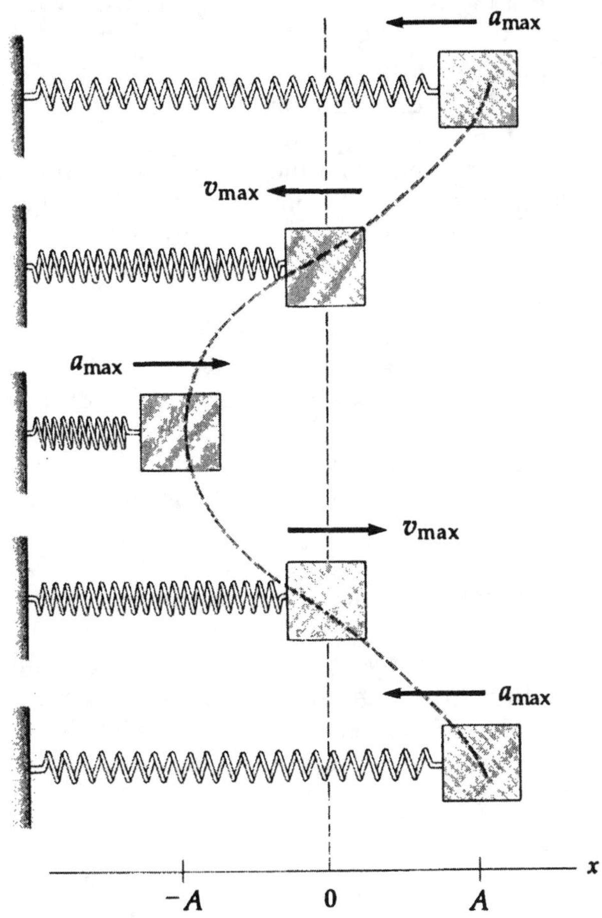

13.6 Motion of a Pendulum

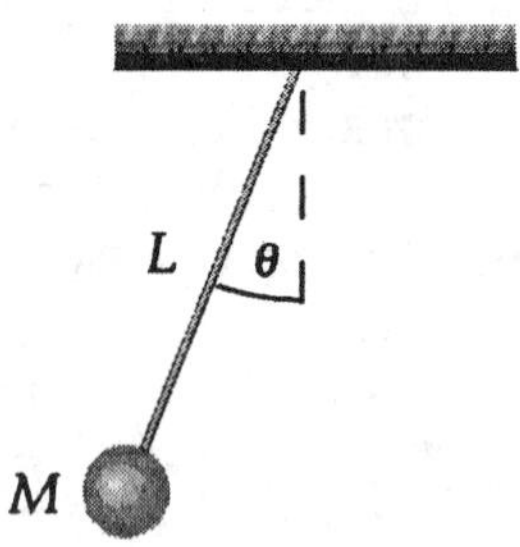

A *simple pendulum consists* of a mass m attached to a light string of length L as shown in the figure. When the angular displacement θ is small during the entire motion (less than about 15°), the pendulum exhibits simple harmonic motion. In this case, the resultant force acting on the mass m equals the component of weight *tangent* to the circle, and has a magnitude $mg\sin\theta$. Since this force is always directed towards $\theta = 0$, it corresponds to a restoring force. The period depends only on the length of the pendulum and the acceleration of gravity. The period *does not* depend on mass, so we conclude that *all* simple pendula of equal length oscillate with the same frequency and period.

13.7 Damped Oscillations

Damped oscillations occur in realistic systems in which retarding forces such as friction are present. These forces will reduce the amplitudes of the oscillations with time, since mechanical energy is continually lost by the system. When the retarding force is assumed to be proportional to the velocity, but small compared to the restoring force, the system will still oscillate, but the amplitude will decrease exponentially with time.

It is possible to compensate for the energy lost in a damped oscillator by adding an additional driving force that does positive work on the system. This additional energy supplied to the system must at least equal the energy lost due to friction to maintain constant amplitude. The energy transferred to the system is a maximum when the driving force is in phase with the velocity of the system. The amplitude is a maximum when the frequency of the driving force matches the natural (resonance) frequency of the system.

13.8 Wave Motion

The production of *mechanical waves* require: (1) an *elastic medium* which can be disturbed, (2) an *energy source* to provide a disturbance or deformation in the medium, and (3) a physical mechanism by way of which adjacent portions on the medium can *influence* each other. The three parameters important in characterizing waves are (1) wavelength, (2) frequency, and (3) wave velocity.

13.9 Types of Waves

Transverse waves are those in which particles of the disturbed medium move along a direction which is perpendicular to the direction of the wave velocity. For *longitudinal waves,* the particles of the medium undergo displacements which are parallel to the direction of wave motion.

13.11 The Speed of Waves on Strings

For linear waves, the *velocity* of *mechanical waves* depends only on the physical properties of the medium through which the disturbance travels. In the case of waves on a *string,* the velocity depends on the tension in the string and the mass per unit length (linear mass density).

13.12 Superposition and Interference of Waves

If two or more waves are moving through a medium, the *resultant wave function* is the *algebraic sum* of the wave functions of the individual waves. Two traveling waves can pass through each other without being destroyed or altered.

EQUATIONS AND CONCEPTS

The force exerted by a spring on a mass attached to the spring and displaced a distance x from the unstretched position is given by Hooke's law. The force constant, k, is always positive and has a value which corresponds to the relative stiffness of the spring. The negative sign means that the force exerted on the mass is always directed opposite the displacement--the force is a restoring force, always directed toward the equilibrium position.

(13.1) $$F_s = -kx$$

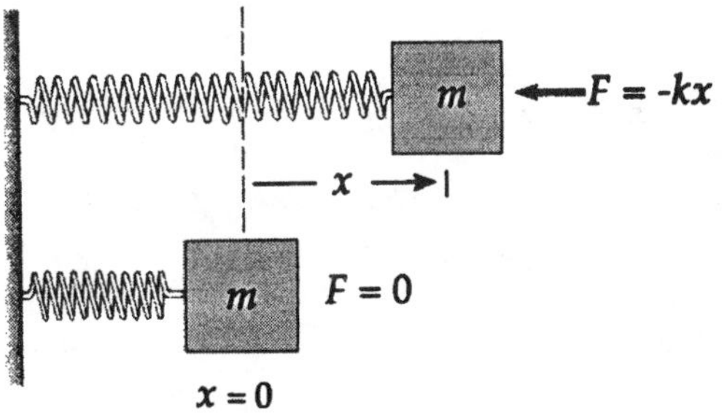

m ←$F = -kx$

x

m $F = 0$

$x = 0$

An object exhibits simple harmonic motion when the net force along the direction of motion is proportional to the displacement and oppositely directed.

Comment on simple harmonic motion.

This equation gives the acceleration of an object in simple harmonic motion as a function of position. Note that when the oscillating mass is at the equilibrium position ($x = 0$), the acceleration $a = 0$. The acceleration has its maximum magnitude when the displacement of the mass is maximum, $x = A$ (amplitude).

(13.2) $$a = -\left(\frac{k}{m}\right)x$$

Work must be done by an external applied force in order to stretch or compress a spring. This work results in energy, called elastic potential energy, being stored in the spring.

(13.3) $$PE_s \equiv \frac{1}{2}kx^2$$

The spring force is conservative; hence, in the absence of friction or other nonconservative forces, the total mechanical energy of the spring-mass system remains constant.

(13.4)

$$(KE + PE_g + PE_s)_i = (KE + PE_g + PE_s)_f$$

The speed of an object in simple harmonic motion is a maximum at $x = 0$; the speed is zero when the mass is at the points of maximum displacement ($x = \pm A$).

(13.6) $$v = \pm\sqrt{\frac{k}{m}(A^2 - x^2)}$$

The period of an object in simple harmonic motion is the time required to complete a full cycle of its motion.

(13.8) $$T = 2\pi\sqrt{\frac{m}{k}}$$

The frequency, the number of complete cycles per unit time, is the reciprocal of the period. The units of frequency are hertz (Hz).

(13.9) $$f = \frac{1}{T}$$

(13.10) $$f = \frac{1}{2\pi}\sqrt{\frac{k}{m}}$$

These equations represent the position of an object moving in simple harmonic motion as a function of time.

(13.11) $x = A\cos(\omega t)$

(13.13) $x = A\cos(2\pi f t)$

Notice that since cos(0) = 1, when $t = 0$, $x = A$. Therefore, the particular form of the position equations shown here assumes that the vibrating object was at the point of maximum displacement when $t = 0$.

Comment on Equation 13.11 and Equation 13.13.

$t = 0$, $v_0 = 0$ } Initial Conditions

$x = 0$

m

A

ω is the angular frequency (rad/s) of the object in simple harmonic motion and f is number of oscillations completed per unit time measured in hertz (Hz).

(13.12) $\omega = 2\pi f$

The period of oscillation of a simple pendulum depends only on its length, L, and the acceleration due to gravity, g. The period does not depend on the mass; and to a good approximation, the period does not depend on the amplitude, θ, within the range of small amplitudes.

(13.14) $T = 2\pi\sqrt{\frac{L}{g}}$

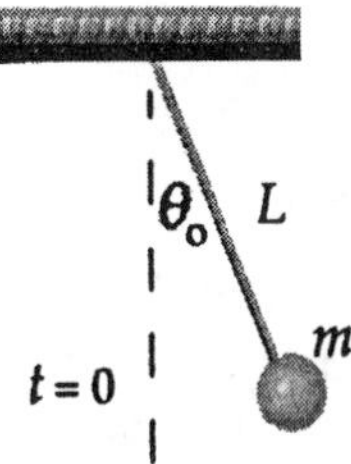

The following definitions of terms used to describe waves and wave motion will be important in discussing the remaining equations in this chapter:

Essential requirements of mechanical waves

(1) A source of disturbance.

(2) A material or medium which can undergo a disturbance.

(3) Some physical mechanism via which adjacent parts of the medium can influence each other. This allows the disturbance (pulse) to be propagated (travel along the medium).

A transverse wave is one in which the particles of the disturbed medium oscillate back and forth along a direction perpendicular to the direction of the wave velocity.

transverse wave

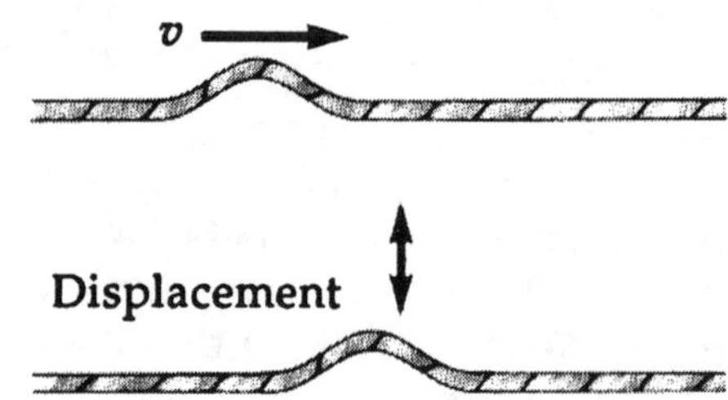

A longitudinal wave is one in which the particles of the medium undergo a displacement (oscillate back and forth) along a direction parallel to the direction of the wave velocity (direction along which the pulse travels).

longitudinal wave

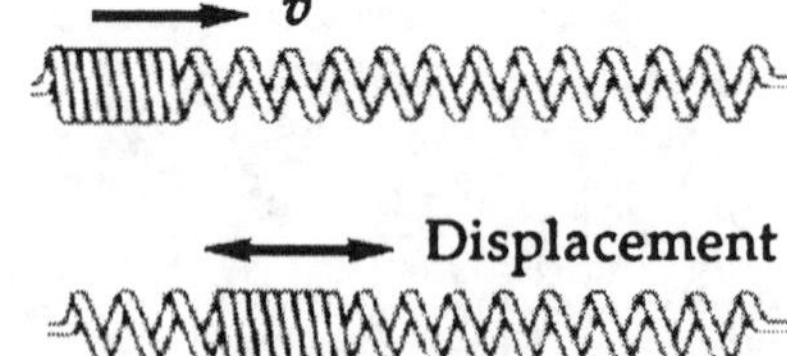

The wave amplitude, A, is the maximum possible value of the displacement of a particle of the medium away from its equilibrium position.

Amplitude, A

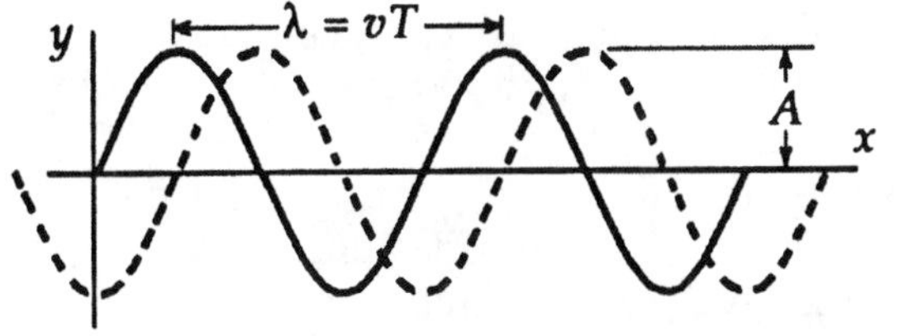

The wavelength, λ, is the minimum distance between two points which are the same distance from their equilibrium positions and are moving in the same direction (such a pair of points are said to be in phase).

wave length, λ

The period of the wave is the time required for the disturbance (or pulse) to travel along the direction of propagation a distance equal to the wavelength. The period is also the time required for any point in the medium to complete one complete cycle in its harmonic motion about its equilibrium point.

period, T

The wave speed, v, is the rate at which the disturbance or pulse moves along the direction of travel of the wave.

(13.15) $$v = f\lambda$$

The wave speed in a stretched string depends on the tension in the string and the linear density (mass per unit length). For any mechanical wave, the speed depends only on the properties of the medium through which the wave travels.

(13.16) $$v = \sqrt{\frac{F}{\mu}}$$

REVIEW CHECKLIST

▷ Describe the general characteristics of a system in simple harmonic motion; and define amplitude, period, frequency, and displacement.

▷ Define the following terms relating to wave motion: frequency, wavelength, velocity, and amplitude; and express a given harmonic wave function in several alternative forms involving different combinations of the wave parameters: wavelength, period, phase velocity, angular frequency, and harmonic frequency.

▷ Given a specific wave function for a harmonic wave, obtain values for the characteristic wave parameters: A, λ, k, ω, and f.

▷ Make calculations which involve the relationships between wave speed and the inertial and elastic characteristics of a string through which the disturbance is propagating.

▷ Define and describe the following wave associated phenomena: superposition, phase, interference, and reflection.

SOLUTIONS TO SELECTED END-OF-CHAPTER PROBLEMS

3. A load of 50.0 N attached to a spring hanging vertically will stretch the spring 5.00 cm. The spring is now placed horizontally on a table and stretched 11.0 cm. (a) What force is required to stretch the spring by this amount? (b) Plot a graph of force (on the y axis) versus spring displacement from the equilibrium position along the x axis.

Solution (a) The spring constant is

$$k = \frac{mg}{x} = \frac{50.0 \text{ N}}{5.00 \times 10^{-2} \text{ m}} = 1000 \text{ N/m}$$

and the force required to stretch the spring 11.0 cm (0.110 m) is

$$F = kx = (1000 \text{ N/m})(0.110 \text{ m}) = 110 \text{ N} \lozenge$$

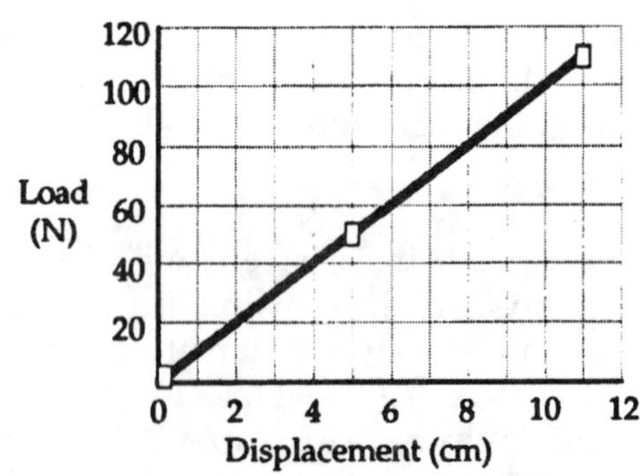

(b) The graph will be a straight line passing through the origin and having a slope of 1000 N/m (10 N/cm). ◊

11. A simple harmonic oscillator has a total energy E. (a) Determine the kinetic and potential energies when the displacement is one half the amplitude. (b) For what value of the displacement does the kinetic energy equal the potential energy?

Solution

(a) At $x = \frac{A}{2}$, $PE = \frac{1}{2}k\left(\frac{A}{2}\right)^2 = \frac{1}{4}E$ where $E = \frac{1}{2}kA^2$

Thus, $KE + \frac{1}{4}E = E$ so $KE = \frac{3}{4}E$ ◊

(b) If $KE = PE$, then $PE + PE = E$ or $PE = \frac{1}{2}E$

Thus, $\frac{1}{2}kx^2 = \frac{1}{2}\left(\frac{1}{2}kA^2\right)$ and $x = \frac{A}{\pm\sqrt{2}}$ ◊

---

19. A mass-spring system oscillates with an amplitude of 3.50 cm. If the spring constant is 250 N/m and the mass is 0.500 kg, determine (a) the mechanical energy of the system, (b) the maximum speed of the mass, and (c) the maximum acceleration.

Solution

(a) $E = \frac{1}{2}kA^2 = \frac{1}{2}(250 \text{ N/m})(3.50 \times 10^{-2} \text{ m})^2 = 0.153 \text{ J}$ ◊

(b) $v = \sqrt{\frac{k}{m}(A^2 - x^2)}$ becomes, with $v = v_{max}$ at $x = 0$,

$$v_{max} = \sqrt{\frac{k}{m}}A = \sqrt{\frac{250 \text{ N/m}}{0.500 \text{ kg}}}(3.50 \times 10^{-2} \text{ m}) = 0.783 \text{ m/s} \; ◊$$

(c) From $F = kx = ma$, we have $a = \frac{k}{m}x$. Thus, $a = a_{max}$ at $x = x_{max} = A$

or, $a_{max} = \frac{k}{m}A = \frac{250 \text{ N/m}}{0.500 \text{ kg}}(3.50 \times 10^{-2} \text{ m}) = 17.5 \text{ m/s}^2$ ◊

---

21. A ball moves with constant speed of 5 m/s in a circular path of radius 0.4 m (see Fig. 13.8). Find the x component of the velocity of the ball when θ equals (a) 0°, (b) 60°, (c) 90°, (d) 180°, (e) 270°.

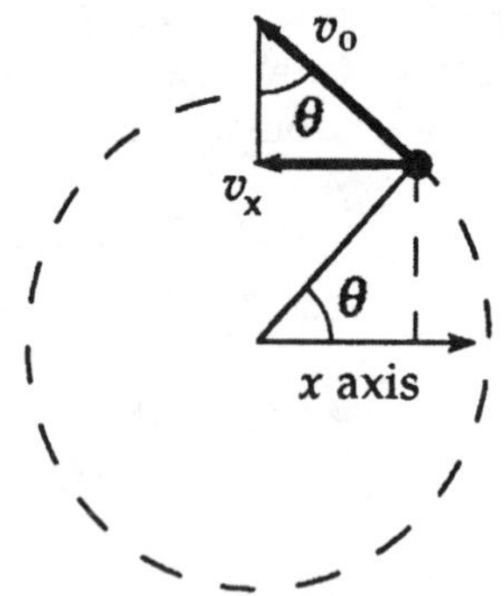

Figure 13.8

Solution We know that $v_x = -v\sin\theta$.

a) at $\theta = 0$, $v_x = 0$ ◊

b) at $\theta = 60°$, $v_x = -4.33 \text{ m/s}$ ◊

(c) at $\theta = 90°$, $v_x = -5.00 \text{ m/s}$ ◊

(d) at $\theta = 180°$, $v_x = 0$ ◊

(e) at $\theta = 270°$, $v_x = +5.00 \text{ m}$ ◊

---

27. The frequency of vibration of a mass-spring system is 5.00 Hz when a 4.00-g mass is attached to the spring. What is the force constant of the spring?

Solution

$$T = \frac{1}{f} = \frac{1}{5.00 \text{ Hz}} = 0.200 \text{ s} \quad \text{and} \quad T = 2\pi\sqrt{\frac{m}{k}}$$

or $$k = \frac{4\pi^2 m}{T^2} = \frac{4\pi^2(4.00\times10^{-3}\text{ kg})}{0.200 \text{ s}}$$

from which $k = 3.95 \text{ N/m}$ ◊

---

33. A spring of negligible mass stretches 3.00 cm from its relaxed length when a force of 7.50 N is applied. A 0.500-kg particle rests on a frictionless horizontal surface and is attached to the free end of the spring. The particle is pulled horizontally so that it stretches the spring 5.00 cm and is then released from rest at $t = 0$. (a) What is the force constant of the spring? (b) What are the period, frequency (f), and the angular frequency (ω) of the motion? (c) What is the total energy of the system? (d) What is the amplitude of the motion? (e) What are the maximum velocity and the maximum acceleration of the particle? (f) Determine the displacement x of the particle from the equilibrium position at $t = 0.500$ s.

Solution

(a) $k = \frac{F}{x} = \frac{7.50 \text{ N}}{3.00 \times 10^{-2} \text{ m}} = 250 \text{ N/m}$ ◊

(b) $T = 2\pi\sqrt{\frac{m}{k}} = 2\pi\sqrt{\frac{0.500 \text{ kg}}{250 \text{ N/m}}} = 0.281 \text{ s}$ ◊ and $f = \frac{1}{T} = 3.56 \text{ Hz}$ ◊

$\omega = 2\pi f = 2\pi(3.56 \text{ Hz}) = 22.4 \text{ rad/s}$ ◊

(c) $E = \frac{1}{2}kA^2 = \frac{1}{2}(250 \text{ N/m})(5.00 \times 10^{-2} \text{ m})^2 = 0.313 \text{ J}$ ◊

(d) $E = \text{const} = \frac{1}{2}mv_0^2 + \frac{1}{2}kx_0^2 = \frac{1}{2}kA^2$

When $v_0 = 0$, $x_0 = A = 5.00 \text{ cm}$ ◊

(e) $v_{max} = \omega A = (22.4 \text{ rad/s})(0.050 \text{ m}) = 1.12 \text{ m/s}$ ◊

$a_{max} = \omega^2 A = (22.5 \text{ rad/s})^2(0.050 \text{ m}) = 25.0 \text{ m/s}^2$ ◊

(f) $x = A\cos\omega t = (5.00 \text{ cm})\cos[(22.4 \text{ rad/s})(0.500 \text{ s})] = 0.919 \text{ cm}$ ◊

39. A visitor to a lighthouse wishes to determine the height of the tower. She ties a spool of thread to a small rock to make a simple pendulum, which she hangs down the center of a spiral staircase of the tower. The period of oscillation is 9.40 s. What is the height of the tower?

Solution $T = 2\pi\sqrt{\frac{L}{g}}$; $L = \left(\frac{T^2}{4\pi^2}\right)g$

$$L = \frac{(9.40\text{ s})^2}{4\pi^2} 9.80\text{ m/s}^2 = 21.9\text{ m} \quad \lozenge$$

---

45. A wave traveling in the positive x direction is pictured in Figure 13.38. Find (a) the amplitude, (b) wavelength, (c) period, and (d) speed of the wave if it has a frequency of 25.0 Hz.

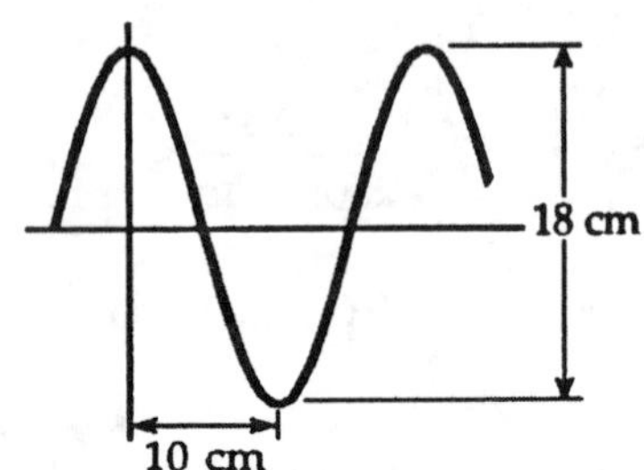

Figure 13.38

Solution (a) The vertical distance from the center line to the top of a crest or the bottom of a trough is the amplitude. Thus, by inspection, we see

$A = 9.00\text{ cm} \quad \lozenge$

(b) The wavelength is the horizontal distance along the wave between two points behaving identically. Again by inspection, we see that

$\lambda = 20.0\text{ cm} \quad \lozenge$

(c) $T = \frac{1}{25.0\text{ Hz}} = 0.040\text{ s} \quad \lozenge$

(d) $v = \lambda f = (20.0\text{ cm})(25.0\text{ Hz}) = 5.00\text{ cm/s} = 5.00\text{ m/s} \quad \lozenge$

---

49. A sound wave, traveling at 343 m/s, is emitted by the foghorn of a tugboat. An echo is heard 2.60 s later. How far away is the reflecting object?

Solution If d is the distance to the reflecting object, the total distance traveled by the wave is $2d$. Thus,

$$v = \frac{2d}{t} \quad \text{or} \quad d = \frac{vt}{2} = \frac{(343\text{ m/s})(2.60\text{ s})}{2} = 446\text{ m} \;\lozenge$$

---

55. Transverse waves travel at 20.0 m/s on a string that is under a tension of 6.00 N. What tension is required for a wave speed of 30.0 m/s in the same string?

Solution We know that the mass per unit length of the string is the same in both instances. Thus,

$$\frac{F_1}{v_1^2} = \frac{F_2}{v_2^2}$$

$$\frac{6.00\text{ N}}{(20.0\text{ m/s})^2} = \frac{F_2}{(30.0\text{ m/s})^2} \quad \text{which gives} \quad F_2 = 13.5\text{ N} \;\lozenge$$

---

57. A series of pulses of amplitude 0.150 m are sent down a string that is attached to a post at one end. The pulses are reflected at the post and travel back along the string without loss of amplitude. What is the amplitude at a point on the string where two pulses are crossing, (a) if the string is rigidly attached to the post? (b) if the end at which reflection occurs is free to slide up and down?

Solution

(a) If the end is fixed, there is inversion of the pulse upon reflection. Thus, when they meet, they cancel and the amplitude is zero. ◊

(b) If the end is free, there is no inversion on reflection. When they meet, the amplitude is $2A = 2(0.150\text{ m}) = 0.300\text{ m}$. ◊

---

65. A 500-g block is released from rest and slides down a frictionless track that begins 2.00 m above the horizontal, as shown in Figure 13.41. At the bottom of the track, where the surface is horizontal, the block strikes and sticks to a light spring with a spring constant of 20.0 N/m. Find the maximum distance the spring is compressed.

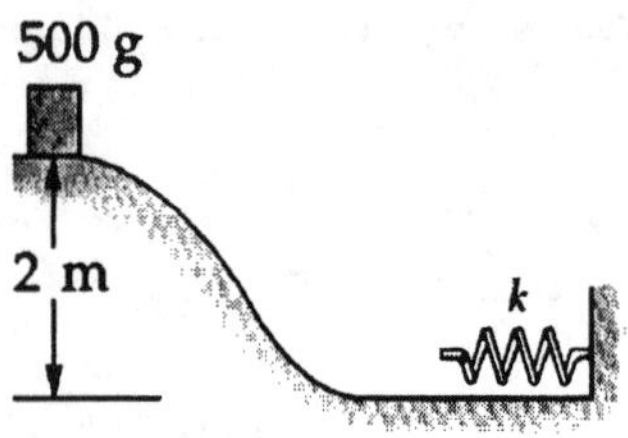

Figure 13.41

Solution

From conservation of mechanical energy, we have

$$mgh = \frac{1}{2}kx^2$$

$$(0.500\ \text{kg})(9.80\ \text{m/s}^2)(2.00\ \text{m}) = \frac{1}{2}(20.0\ \text{N/m})x^2$$

From which $x = 0.990$ m ◊

CHAPTER SELF-QUIZ

1. A 0.2-kg object, suspended from a spring with a spring constant of $k = 10$ N/m, is moving in simple harmonic motion and has an amplitude of 0.08 m. What is its kinetic energy at the instant when its displacement is 0.04 m?
   a. 0.8×10^{-2} J
   b. zero
   c. 2.4×10^{-2} J
   d. 31.25 J

2. The tension in a guitar string is increased by a factor of four. By what factor does the wave velocity change?
   a. 4.0
   b. 2.0
   c. 0.5
   d. 0.25

3. A mass of 0.4 kg, hanging from a spring with a spring constant of 80 N/m, is set into an up-and-down simple harmonic motion. What is the potential energy stored *in the spring alone* when the mass is displaced 0.1 m?
   a. zero
   b. 0.2 J
   c. 0.4 J
   d. 0.8 J

4. A runaway railroad car, with mass 30×10^4 kg, coasts across a level track at 1.5 m/s when it collides with a spring-loaded bumper at the end of the track. If the spring constant of the bumper is 2×10^6 N/m, what is the maximum displacement of the spring after the collision? (Assume collision is elastic.)
   a. 0.58 m
   b. 0.34 m
   c. 3.8 m
   d. 1.7 m

5. A 0.20-kg block rests on a frictionless level surface and is attached to a horizontally aligned spring with a spring constant of 40 N/m. The block is initially displaced 4 cm from the equilibrium point and then released to set up a simple harmonic motion. What is the frequency?
   a. 10.0 Hz
   b. 2.3 Hz
   c. 88.0 Hz
   d. 0.9 Hz

6. A mass on a spring vibrates in simple harmonic motion at a frequency of 4.0 Hz and an amplitude of 8.0 cm. If a timer is started when its displacement is a maximum (hence $x = 4$ cm when $t = 0$), what is the magnitude of its displacement when $t = 3$ s?
   a. zero
   b. 3.0 cm
   c. 6.7 cm
   d. 8.0 cm

7. A radio wave has a speed of 3×10^8 m/s and a frequency of 101 MHz. What is its wavelength?
   a. 3.0 m
   b. 45 m
   c. 0.10 m
   d. 2.97 m

8. Suppose a mass m is on a spring that has been compressed a distance h. How much further must the spring be compressed to triple the elastic potential energy?
   a. $2mgh$
   b. $2h$
   c. $1.7h$
   d. $0.73h$

9. Ocean waves with a wavelength of 120 m are coming in at a rate of 8 per minute. What is their speed?
   a. 8 m/s
   b. 16 m/s
   c. 24 m/s
   d. 30 m/s

10. A piano string of mass 0.005 kg/m is under a tension of 1350 N. Find the velocity with which a wave travels on this string.
   a. 260 m/s
   b. 520 m/s
   c. 1040 m/s
   d. 2080 m/s

11. An earthquake emits both P-waves and S-waves which travel at different speeds through the Earth. A P-wave travels at 9000 ms and an S-wave at 5000 m/s. If P-waves are received at a seismic station 1 minute before an S-wave arrives, how far is it to the earthquake center?
   a. 2420 km
   b. 1210 km
   c. 680 km
   d. 240 km

12. The speed of a sound wave in sea water is 1500 m/s. If this wave is transmitted at frequency 10 kHz, what is its wavelength?
   a. 5 cm
   b. 10 cm
   c. 15 cm
   d. 20 cm

14
Sound

Chapter 14

SOUND

Sound waves are the most important example of longitudinal waves. In this chapter we discuss the characteristics of sound waves—how they are produced, what they are, and how they travel through matter. We then investigate what happens when sound waves interfere with each other. The insights gained in this chapter will help you understand why we hear what we hear.

NOTES FROM SELECTED CHAPTER SECTIONS

14.2 Characteristics of Sound Waves

The motion of the medium particles is *back and forth along the direction in which the wave travels.* This is in contrast to a transverse wave, in which the vibrations of the medium are *at right angles to the direction of travel of the wave.*

14.4 Energy and Intensity of Sound Waves

The *intensity* of a wave is the rate at which sound energy flows through a unit area perpendicular to the direction of travel of the wave.

14.5 Spherical and Plane Waves

The intensity of a *spherical wave* produced by a point source is proportional to the average power emitted and inversely proportional to the square of the distance from the source.

14.6 The Doppler Effect

In general, a Doppler effect is experienced whenever there is relative motion between source and observer. When the source and observer are moving toward each other, the frequency heard by the observer is higher than the frequency of the source. When the source and observer are moving away from each other, the observer hears a frequency lower than the source frequency.

14.8 Standing Waves

Standing waves can be set up in a string by a continuous superposition of waves incident on and reflected from the ends of the string. The string has a number of natural patterns of vibration, called *normal modes.* Each normal mode has a *characteristic frequency.* The lowest of these frequencies is called the *fundamental frequency,* which together with the higher frequencies form a *harmonic series.*

The figure below is a schematic representation of the first three normal modes of vibration of string fixed at both ends.

14.10 Standing Waves in Air Columns

Standing waves are produced in strings by interfering *transverse* waves. Sound sources can be used to produce *longitudinal* standing waves in air columns. The phase relationship between incident and reflected waves depends on whether or not the reflecting end of the air column is open or closed. This gives rise to two sets of possible standing wave conditions:

> In *a pipe open at both ends,* the natural frequencies of vibration form a series in which all harmonics are present and are equal to integral multiples of the fundamental.
>
> In *a pipe closed at one end and open at the other,* only odd harmonics are present.

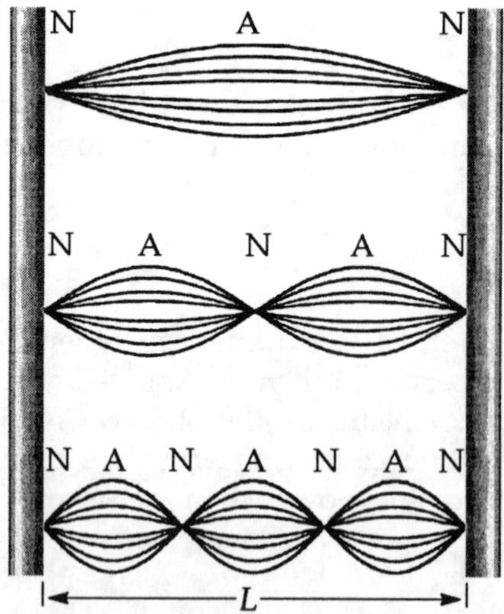

Schematic representation of standing waves on a stretched string of length L, where the envelope represents many successive vibrations. The points of zero displacement are called *nodes;* the points of maximum displacement are called *antinodes.*

EQUATIONS AND CONCEPTS

A sound wave propagates through an elastic medium as a compressional wave. The speed of the sound wave depends on value of the bulk modulus, B (an elastic property), and the equilibrium density, ρ (an inertial property), of the material through which it is traveling.

$$v = \sqrt{\frac{B}{\rho}} \tag{14.1}$$

$$B = -\frac{\Delta P}{\Delta V / V} \tag{14.2}$$

The speed of sound (or any longitudinal wave) in a solid depends on the value of Young's modulus and the density of the material.

$$v = \sqrt{\frac{Y}{\rho}} \tag{14.3}$$

The velocity of sound depends on the temperature of the medium. Equation 14.4 shows the temperature dependence in air where T is the temperature in degrees Celsius and v_0 is the speed of sound in air at 0°C.

$$v = (331\text{ m/s})\sqrt{1 + \frac{T}{273}} \tag{14.4}$$

where $v_0 = 331$ m/s

The intensity of a wave is the rate at which energy flows across a unit area, A, in a plane perpendicular to the direction of travel of the wave. The SI units of intensity, I, are watts per square meter, W/m^2.

$$I \equiv \frac{\text{power}}{\text{area}} = \frac{P}{A} \tag{14.6}$$

At a frequency of 1000 Hz, the faintest sound detectable by the human ear (threshold of hearing) has an intensity of 10^{-12} W/m^2. An intensity of 1 W/m^2 is the greatest intensity which the ear can tolerate (the threshold of pain).

Comment on sound intensity.

The decibel scale is a logarithmic intensity scale. On this scale, the unit of sound intensity is the decibel, dB. The constant I_0 is a reference intensity.

$$\beta \equiv 10 \log\left(\frac{I}{I_0}\right) \tag{14.7}$$

$$I_0 = 10^{-12}\text{ W/m}^2$$

In order to determine the decibel level corresponding to two different sources sounded simultaneously, first find the individual intensities I_1 and I_2 in W/m^2 and add these values to obtain the combined intensity $I = I_1 + I_2$. Finally, use Equation 14.7 to convert the intensity I to the decibel scale.

Comment on the decibel scale.

The intensity of a spherical wave is inversely proportional to the square of the distance from the source.

$$I = \frac{P_{av}}{4\pi r^2} \qquad (14.8)$$

The apparent change in frequency heard by an observer whenever there is relative motion between the source and the observer is called the Doppler effect. In Equation 14.15, the upper signs ($+v_o$ in numerator and $-v_o$ in the denominator) are used when there is motion of source or observer toward the other. The lower signs ($-v_o$ in numerator and $+v_o$ in denominator) are used when there is motion of source or observer away from the other. Also, it is important to remember that v_o (velocity of the observer) and v_s (velocity of the source) are *each measured relative to the medium in which the sound travels.*

$$f' = f\left(\frac{v \pm v_o}{v \mp v_s}\right) \qquad (14.15)$$

(See Table, following page)

A series of standing wave patterns called normal modes can be excited in a stretched string. Each mode corresponds to a characteristic frequency and wavelength.

$$\lambda_n = \frac{2L}{n}$$

$$f_n = \frac{nv}{2L} = \frac{n}{2L}\sqrt{\frac{F}{\mu}} \qquad (4.20)$$

where n = 1, 2, 3, . . .

EXAMPLES OF DOPPLER EFFECT WITH OBSERVER/SOURCE IN MOTION

Observer	Source	Equation	Remark
O ⟶	S	$f' = f\left(\dfrac{v + v_o}{v}\right)$	Observer moving toward stationary source
⟵ O	S	$f' = f\left(\dfrac{v - v_o}{v}\right)$	Observer moving away from stationary source
O	⟵ S	$f' = f\left(\dfrac{v}{v - v_s}\right)$	Source moving toward stationary observer
O	S ⟶	$f' = f\left(\dfrac{v}{v + v_s}\right)$	Source moving away from stationary observer
O ⟶	S ⟶	$f' = f\left(\dfrac{v + v_o}{v + v_s}\right)$	Observer following moving source
⟵ O	⟵ S	$f' = f\left(\dfrac{v - v_o}{v - v_s}\right)$	Source following moving observer
⟵ O	S ⟶	$f' = f\left(\dfrac{v - v_o}{v + v_s}\right)$	Observer and source moving away from each other along opposite directions
O ⟶	⟵ S	$f' = f\left(\dfrac{v + v_o}{v - v_s}\right)$	Observer and source moving toward each other
O	S	$f' = f$	Observer and source both stationary

The frequency f_1 corresponding to $n = 1$, is the fundamental frequency and is the lowest frequency for which a standing wave is possible.

Comment on standing waves in strings.

The fundamental together with the other frequencies ($n = 1, 2, 3, \ldots$) form a harmonic series where f_2 is the second harmonic, f_3 is the third harmonic, and so on. The second harmonic is also called the first overtone; the third harmonic, the second overtone; and so on.

(See Figure below)

Standing waves are produced in strings by interfering transverse waves. Sound sources can also be used to produce longitudinal standing waves in air columns. The phase relationship between incident and reflected waves depends on whether the reflecting end of the air column is open or closed. This gives rise to two possible sets of standing wave conditions: those produced in an "open" pipe (open at both ends) and those produced in a "closed" pipe (open at only one end).

Comment on longitudinal standing waves in air columns.

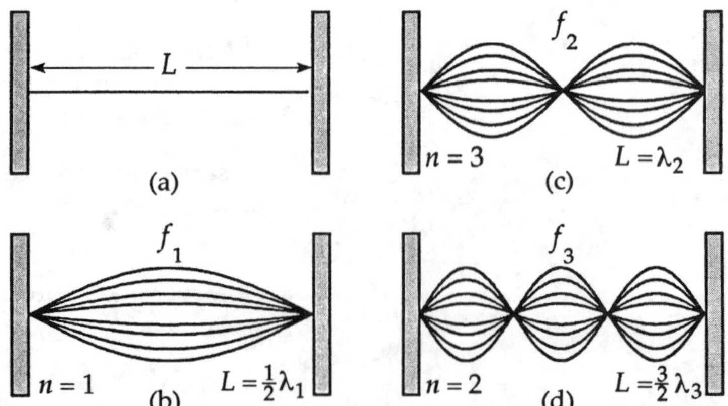

Standing waves in a stretched string of length L fixed at both ends. The normal frequencies of vibrations form a harmonic series: (b) the fundamental frequency, or first harmonic, (c) the second harmonic, and (d) the third harmonic.

In a pipe open at both ends, the natural frequencies of vibration form a series in which all harmonics (integer multiples of the fundamental) are present.

$$f_n = n\frac{v}{2L} \tag{14.21}$$

$$n = 1, 2, 3, \ldots$$

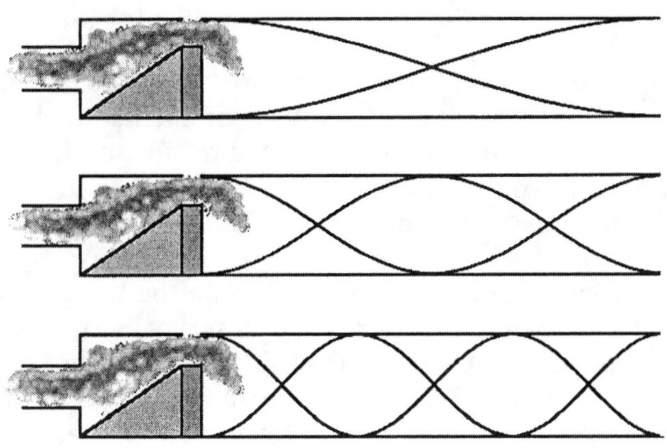

In a pipe open at one end, only the odd harmonics (odd multiples of the fundamental) are possible.

$$f_n = n\frac{v}{4L} \tag{14.22}$$

$$n = 1, 3, 5, \ldots$$

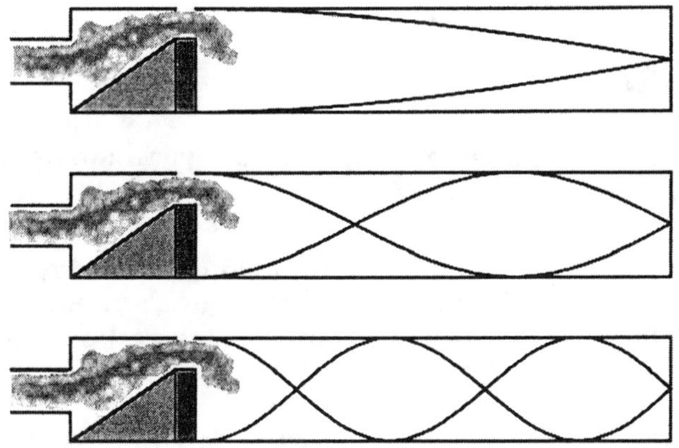

SUGGESTIONS, SKILLS, AND STRATEGIES

When making calculations using Equation 14.7 which defines the intensity of a sound wave on the decibel scale, the properties of logarithms must be kept clearly in mind.

In order to determine the decibel level corresponding to two sources sounded simultaneously, you must first find the intensity, I, of each source in W/m^2, add these values, and then convert the resulting intensity to the decibel scale. As an illustration of this technique, note that if two sounds of intensity 40 dB and 45 dB are sounded together, the intensity level of the combined sources *is* 46.2 dB (*NOT 85 dB*).

The most likely error in using Equation 14.15 to calculate the Doppler frequency shift due to relative motion between a sound source and an observer is due to using the incorrect algebraic sign for the velocity of either the observer or the source. These sign conventions are illustrated in the chart in the section on Equations and Concepts in which the directions of motion of the observer O and source S are indicated by arrows; and the correct choice of signs used in Equation 14.15,

$$f' = f\left(\frac{v \pm v_o}{v \mp v_s}\right)$$

where f is the true frequency of the source and f' is the apparent frequency as measured by the observer.

REVIEW CHECKLIST

▷ Describe the harmonic displacement and pressure variation as functions of time and position for a harmonic sound wave.

▷ Calculate the speed of sound in various media in terms of appropriate elastic properties (these can include bulk modulus, Young's modulus, and the pressure-volume relationships of an ideal gas) and the corresponding inertial properties (usually the mass density).

▷ Understand the basis of the logarithmic intensity scale (decibel scale) and convert intensity values (given in W/m^2) to loudness levels on the dB scale. Determine the intensity ratio for two sound sources whose decibel levels are known. Calculate the intensity of a point source wave at a given distance from the source.

- ▷ Describe the various situations under which a Doppler shifted frequency is produced. Calculate the apparent frequency for a given actual frequency for each of the various possible relative motions between source and observer.

- ▷ Describe in both qualitative and quantitative terms the conditions which produce standing waves in a stretched string and in an open or closed air column pipe.

SOLUTIONS TO SELECTED END-OF-CHAPTER PROBLEMS

3. A group of hikers hear an echo 3.00 s after they shout. If the temperature is 22.0°C, how far away is the mountain that reflected the sound wave?

Solution $$v = (331 \text{ m/s})\sqrt{1+\frac{22.0}{273}} = 344 \text{ m/s}$$

If d is the distance to the object, the total distance traveled by the sound is $2d$. Thus,

$$d = v\left(\frac{t}{2}\right) = (344 \text{ m/s})\left(\frac{3.00 \text{ s}}{2}\right) = 516 \text{ m} \quad \Diamond$$

---

13. Intensity is defined as power per unit area. As a result, show that the decibel level of two sounds can be related to the ratio of two corresponding powers as

$$dB = 10 \log\left(\frac{P_1}{P_0}\right)$$

Solution $I = \dfrac{P}{A}$

Therefore, $$\beta = 10 \log\left(\frac{I}{I_0}\right) = 10 \log\left(\frac{P/A}{P_0/A}\right) = 10 \log\left(\frac{P}{P_0}\right) \quad \Diamond$$

---

17. A stereo speaker (considered a small source) emits sound waves with a power output of 100 W. (a) Find the intensity 10.0 m from the source. (b) Find the intensity level in dB at this distance. (c) At what distance would you experience the sound at the threshold of pain, 120 dB?

Solution

(a) $I = \frac{P}{4\pi r^2} = \frac{100 \text{ W}}{4\pi(10.0 \text{ m})^2} = 7.96 \times 10^{-2} \text{ W/m}^2$ ◊

(b) $\beta = 10 \log\left(\frac{7.96 \times 10^{-2}}{10^{-12}}\right) = 10 \log(7.96 \times 10^{10}) = 109 \text{ dB}$ ◊

(c) If $\beta = 120$ dB (the threshold of pain), $I = 1.00$ W/m^2; and from $I = \frac{P}{4\pi r^2}$,

$$r^2 = \frac{P}{4\pi I} = \frac{100 \text{ W}}{4\pi(1.00 \text{ W/m}^2)} = 7.96 \text{ m}^2 \quad \text{giving} \quad r = 2.82 \text{ m} \quad \lozenge$$

---

23. At rest, a car's horn sounds the note A (440 Hz). While the car is moving down the street, the horn is sounded. A bicyclist moving in the same direction with one-third the car's speed hears the note A (415). How fast is the car moving? Is the cyclist ahead of or behind the car?

Solution Since the observer hears an apparent frequency which is lower than the actual frequency, the source must be moving away from the observer. Also, we know both source and observer are moving in the same direction. Therefore, the cyclist must be behind the car. ◊

Since both source and observer are in motion, we use $f' = f\left(\frac{v \pm v_o}{v \mp v_s}\right)$

Since the observer moves toward the source, use upper sign in numerator. Since source moves away from observer, use plus sign in denominator.

Then with $v_o = \frac{1}{3}v_s$, we have

$$415 \text{ Hz} = 440 \text{ Hz}\frac{345 \text{ m/s} + \frac{1}{3}v_s}{345 \text{ m/s} + v_s}$$

$$f' = f\left[\frac{v + \frac{v_s}{3}}{v + v_s}\right] \quad \text{or} \quad v_s = v\left[\frac{f - f'}{f' - \frac{f'}{3}}\right]$$

so
$$v_s = 345 \text{ m/s}\left[\frac{440 - 415}{415 - \frac{440}{3}}\right] = 32.1 \text{ m/s} \quad \Diamond$$

---

27. Two trains on separate tracks move toward one another. Train 1 has a speed of 130 km/h and train 2 has a speed of 90.0 km/h. Train 2 blows its horn, emitting a frequency of 500 Hz. What is the frequency heard by the engineer on train 1?

Solution Both observer and source are moving toward each other.

$$f' = f\left(\frac{v \pm v_o}{v \mp v_s}\right) \quad \text{becomes} \quad f' = f\frac{v + v_o}{v - v_o}$$

or
$$f = 500 \text{ Hz}\left(\frac{345 \text{ m/s} + 38.4 \text{ m/s}}{345 \text{ m/s} - 25.0 \text{ m/s}}\right) = 599 \text{ Hz} \quad \Diamond$$

---

31. Two loudspeakers are placed above and below one another, as in Figure 14.13, and driven by the same source at a frequency of 500 Hz. (a) What minimum distance should the top speaker be moved back in order to create destructive interference between the two speakers? (b) If the top speaker is moved back twice the distance calculated in part (a), will constructive or destructive interference occur?

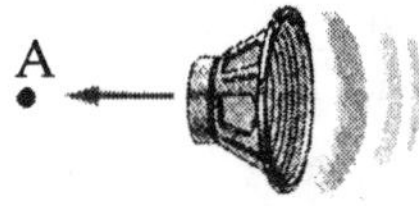

Figure 14.13

Solution (a) The sound emitted by the speakers has a wavelength of

$$\lambda = \frac{v}{f} = \frac{345 \text{ m/s}}{500 \text{ Hz}} = 0.690 \text{ m}$$

To produce destructive interference, the path difference should be $\frac{\lambda}{2}$.
Therefore, the top speaker must be moved back 34.5 cm. ◊

(b) If the path difference is λ, constructive interference will occur. ◊

---

35. Take the speed of sound to be 344 m/s. Two speakers are driven by a common oscillator at 800 Hz and face each other at a distance of 1.25 m. Locate the points along a line joining the two speakers where relative minima would be expected.

Solution The wavelength is

$$\lambda = \frac{v}{f} = \frac{343 \text{ m/s}}{800 \text{ Hz}} = 0.423 \text{ m}$$

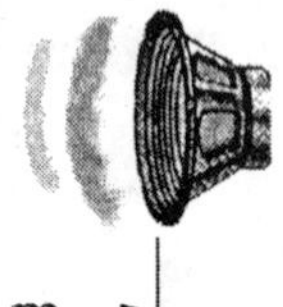

For minima, the difference in path length must be an odd number of half wavelengths. So, if x is the distance from the source to the speaker on the left, then the path difference will be

$$(1.25 \text{ m} - x) - x = (2n+1)\frac{\lambda}{2}$$

where n is any integer. Use $\lambda = 0.430 \text{ m}$ and substitute allowed values of n to calculate corresponding values of x. Solving, gives the following results:

$n = 0,\ x = 0.518$ m	$n = -1,\ x = 0.732$ m
$n = 1,\ x = 0.3035$ m	$n = -2,\ x = 0.947$ m
$n = 2,\ x = 0.089$ m	$n = -3,\ x = 1.161$ m

Thus, at distances of 0.089 m, 0.304 m, 0.518 m, 0.732 m, 0.947 m, and 1.161 m from either speaker. ◊

---

43. Two pieces of steel wire with identical cross sections have lengths of L and $2L$. Each of the wires is fixed at both ends and stretched so that the tension in the longer wire is four times greater than that in the shorter wire. If the fundamental frequency in the shorter wire is 60 Hz, what is the frequency of the second harmonic in the longer wire?

Solution μ is the same for both wires. We have $F_{\text{long wire}} = 4F_{\text{short wire}}$

Since $v = \sqrt{\dfrac{F}{\mu}}$ then $v_{\text{long wire}} = 2v_{\text{short wire}}$

For the fundamental frequency, the wavelength will be equal to twice the length of the wire.

$$v_{\text{short wire}} = \lambda f = (2L_{\text{short}})(60\text{ Hz}) = (120\text{ Hz})L_{\text{short}}$$

so $$v_{\text{long wire}} = 2v_{\text{short wire}} = (240\text{ Hz})L_{\text{short}}$$

The second harmonic in the long wire will have a wavelength equal to the length of the long wire which is twice the length of the short wire.

$$f_{\text{long wire}} = \frac{v_{\text{long wire}}}{\lambda_{\text{long wire}}} = \frac{(240\text{ Hz})L_{\text{short}}}{2L_{\text{short}}} = 120\text{ Hz}$$

= frequency of 2nd harmonic in long wire ◊

---

47. The fundamental frequency of an open organ pipe corresponds to middle C (261.6 Hz on the chromatic musical scale). The third resonance of a closed organ pipe has the same frequency. What are the lengths of the two pipes?

Solution For the open pipe (in the fundamental mode),

$$\lambda = \frac{v}{f} = \frac{345\text{ m/s}}{261.6\text{ Hz}} = 1.319\text{ m}$$

and $$L = \frac{\lambda}{2} = 0.659\text{ m} = 65.9\text{ cm} \quad ◊$$

For the closed pipe (in the third harmonic),

$$\lambda = \frac{v}{f} = \frac{345\text{ m/s}}{261.6\text{ Hz}} = 1.319\text{ m}$$

and $$L = \frac{3}{4}(1.319\text{ m}) = 0.989\text{ m} = 98.9\text{ cm} \quad ◊$$

---

51. A pipe open at both ends has a fundamental frequency of 300 Hz when the temperature is 0°C. (a) What is the length of the pipe? (b) What is the fundamental frequency at a temperature of 30°C?

Solution (a) At 0°C, the speed of sound is 331 m/s. The wavelength of the wave is

$$\lambda = \frac{v}{f} = \frac{331\text{ m/s}}{300\text{ Hz}} = 1.103\text{ m}$$

The wavelength (for the fundamental frequency) in a pipe open at both ends is equal to twice the length of the pipe, so

$$L = \frac{\lambda}{2} = 0.552\text{ m} \quad \Diamond$$

(b) At $T = 30°\text{C}$, the speed of sound is

$$v = (331\text{ m/s})\sqrt{1 + \frac{T_C}{273}} = (331\text{ m/s})\sqrt{1 + \frac{30}{273}} = 349\text{ m/s}$$

Thus,
$$f = \frac{v}{\lambda} = \frac{v}{2L} = \frac{349\text{ m/s}}{2(0.552\text{ m})} = 317\text{ Hz} \quad \Diamond$$

---

57. The G string on a violin has a fundamental frequency of 196 Hz. It is 30.0 cm long and has a mass of 0.500 g. As it is being plucked, a nearby violinist fingers a string on her identical violin until a beat frequency of 2.00 Hz is heard between the two. What is the length of the violin string being plucked by the nearby violinist?

Solution In the fundamental mode, the wavelength of the sound emitted by the violin of the first player is

$$\lambda = 2L = 2(30.0\text{ cm}) = 60.0\text{ cm}$$

As the second player shortens the length of her string, the frequency it emits will increase. Thus, the frequency her instrument emits is

$$f' = f + 2.00\text{ Hz} = 198\text{ Hz}$$

The velocity of the waves is the same for each instrument. Therefore,

$$\lambda f = \lambda' f' \qquad \text{and} \qquad \lambda' = \lambda\left(\frac{f}{f'}\right) = \frac{(60.0\text{ cm})(196\text{ Hz})}{198\text{ Hz}} = 59.4\text{ cm} \quad \Diamond$$

The length of the shortened string is

$$L' = \frac{0.594\text{ m}}{2} = 0.297\text{ m} = 29.7\text{ cm} \quad \Diamond$$

---

65. Two point sound sources have measured intensities of $I_1 = 100\text{ W/m}^2$ and $I_2 = 200\text{ W/m}^2$. By how many decibels is the level of source 1 lower than the level of source 2? (Assume that the observer is the same distance from both sources.)

Solution Since power equals intensity times area, we have for the two sources

$$P_1 = I_1 A \qquad \text{and} \qquad P_2 = I_2 A$$

Divide the former by the latter and solve for I_1.

$$I_1 = \frac{P_1}{P_2}(I_2) = \frac{100}{200}(I_2) \qquad \text{or} \qquad I_2 = 2I_1$$

Therefore,

$$\beta_2 = 10\log\left(\frac{I_2}{I_0}\right) = 10\log\left(\frac{2I_1}{I_0}\right) = 10\log\left[(2)\left(\frac{I_1}{I_0}\right)\right]$$

or

$$\beta_2 = 10\left[\log 2 + \log\left(\frac{I_1}{I_0}\right)\right] \qquad \text{but} \qquad \beta_1 = 10\log\left(\frac{I_1}{I_0}\right)$$

Therefore,

$$\beta_2 = 10\log 2 + \beta_1 \qquad \text{or} \qquad \beta_2 - \beta_1 = 10\log 2 = 3.01\text{ dB} \quad \Diamond$$

---

69. A variable-length air column is placed just below a vibrating wire that is fixed at both ends. The length of the air column is gradually increased from zero until the first position of resonance is observed at L= 34.0 cm. The wire is 120 cm long and is vibrating in its third harmonic. If the speed of sound in air is 340 m/s, what is the speed of transverse waves in the wire?

Solution For a closed column, the first resonance will occur for

$$L_{air} = \frac{\lambda_{air}}{4} = 0.340 \text{ m}$$

Therefore, $\lambda_{air} = 1.36 \text{ m}$ and $f = \frac{v_{air}}{\lambda_{air}} = 250 \text{ Hz}$ is the frequency of both the sound wave and the vibrating wire. Since the wire is vibrating in its third harmonic,

$$L_{wire} = \frac{3\lambda_{wire}}{2} \quad \text{or} \quad \lambda_{wire} = \frac{2}{3}L_{wire} = 0.800 \text{ m}$$

The wave velocity in the wire is then

$$v_{wire} = f\lambda_{wire} = (250 \text{ Hz})(0.800 \text{ m}) = 200 \text{ m/s} \quad \Diamond$$

---

73. A speaker at the front of a room and an identical speaker at the rear of the room are being driven by the same sound source at 456 Hz. A student walks at a uniform rate of 1.50 m/s along the length of the room. How many beats does the student hear per second?

Solution To find the apparent frequency as heard by the student we use $f' = f\left(\frac{v \pm v_o}{v \mp v_s}\right)$

with f_1' = frequency of the speaker in front of the student
and f_2' = frequency of the speaker behind the student

$$f_1' = (456 \text{ Hz})\frac{(331 \text{ m/s} + 1.50 \text{ m/s})}{(331 \text{ m/s} - 0)} = 458 \text{ Hz}$$

and

$$f_2' = (456 \text{ Hz})\frac{(331 \text{ m/s} - 1.50 \text{ m/s})}{(331 \text{ m/s} - 0)} = 454 \text{ Hz}$$

Therefore, the beat frequency is 4.00 Hz $\Diamond$

---

75. Two ships are moving along a line due east. The trailing vessel has a speed of 64.0 km/h relative to a land-based observation point, and the leading ship has a speed of 45.0 km/h relative to the same station. The two ships are in a region of the ocean where the current is moving uniformly due west at 10.0 km/h. The trailing ship transmits a sonar signal at a frequency of 1200 Hz. What frequency is monitored by the leading ship? (Use 1520 m/s as the speed of sound in ocean water.)

Solution When the observer is moving in front of and in the same direction as the source,

$$f_o = f_s \frac{v - v_o}{v - v_s}$$

where v_o and v_s are measured *relative to the medium* in which the sound is propagated. In this case the ocean current is opposite the direction of travel of the ships and

$$v_o = 45.0 \text{ km/h} - (-10.0 \text{ km/h}) = 55.0 \text{ km/h} = 15.3 \text{ m/s}$$
$$v_s = 64.0 \text{ km/h} - (-10.0 \text{ km/h}) = 74.0 \text{ km/h} = 20.55 \text{ m/s}$$

Therefore, $$f_o = (1200 \text{ Hz})\frac{1520 \text{ m/s} - 15.3 \text{ m/s}}{1520 \text{ m/s} - 20.55 \text{ m/s}} = 1200 \text{ Hz} \quad \Diamond$$

CHAPTER SELF-QUIZ

1. A series of ocean waves, each 6.0 m from crest to crest, moving past the observer at a rate of 2 waves per second, have what velocity?
   a. 1/3 m/s
   b. 3.0 m/s
   c. 8.0 m/s
   d. 12.0 m/s

2. A standing wave is set up in a 2.0-m length string fixed at both ends. The string vibrates in 5 distinct segments when driven by a 120 Hz source. What is the wave velocity in this string?
   a. 96 m/s
   b. 48 m/s
   c. 24 m/s
   d. 12 m/s

3. If a sound source with a 1000 Hz frequency is at rest, and a listener moves at a speed of 30.0 m/s away from the source, what is the apparent frequency heard by the listener? (The velocity of sound = 340 m/s.)
   a. 919 Hz
   b. 912 Hz
   c. 1090 Hz
   d. 1097 Hz

4. A low C (f = 65 Hz) is sounded on a piano. If the length of the piano wire is 2.0 m and its mass density is 5.0 g/m, what is the tension of the wire?
   a. 84 N
   b. 168 N
   c. 338 N
   d. 677 N

5. When I stand half way between two speakers, with one on my left and one on my right, a musical note from the speaker gives me constructive interference. How far to my left should I move to obtain destructive interference?
   a. one-fourth of a wavelength
   b. half a wavelength
   c. one wavelength
   d. one and a half wavelengths

6. For a standing wave on a string, the wavelength must equal
   a. the distance between adjacent nodes
   b. the distance between adjacent antinodes
   c. twice the distance between adjacent antinodes
   d. the distance between supports

7. An organ pipe, closed at one end, and a guitar string have the same fundamental frequency The frequency of the second overtone of the pipe corresponds to which overtone of the guitar string?
   a. first
   b. second
   c. third
   d. fourth

8. The sound level 5.0 m from a point source is 95 dB. At what distance will it be 75 dB?
   a. 50 m
   b. 75 m
   c. 225 m
   d. 500 m

9. A fireworks rocket explodes at a height of 100 m above the ground. An observer on the ground directly under the explosion experiences an average sound intensity of 7×10^{-2} W/m^2. What is the sound level in dB heard by the observer? ($I_0 = 10^{-12}$ W/m^2)
   a. 94.4 dB
   b. 100.0 dB
   c. 108.4 dB
   d. 119.4 dB

10. A clarinet behaves like a tube closed at one end. If its length is 80 cm, and the velocity of sound is 340 m/s, what is its fundamental frequency in Hz?
   a. 106 Hz
   b. 159 Hz
   c. 212 Hz
   d. 265 Hz

11. A bat, flying at 5.0 m/s, emits a chirp at 40 kHz. If this sound pulse is reflected by a wall, what is the frequency of the echo received by the bat?
   a. 41.2 kHz
   b. 40.9 kHz
   c. 40.6 kHz
   d. 40.3 kHz

12. Shortening a guitar string to one-third its initial length will change its natural frequency by what factor?
   a. 0.58
   b. 1.0
   c. 1.7
   d. 3.0

15
Electric Forces and Electric Fields

Chapter 15

ELECTRIC FORCES AND ELECTRIC FIELDS

In this chapter we use the effect, charging by friction, to begin an investigation of electric forces. We then discuss Coulomb's law, which is the fundamental law of force between any two charged particles. The concept of an electric field associated with charges is then introduced, and its effects on other charged particles described. We end with brief discussions of the Van de Graaff generator and the oscilloscope.

NOTES FROM SELECTED CHAPTER SECTIONS

15.1 Properties of Electric Charges

Electric charge has the following important properties:

1. There are two kinds of charges in nature, with the property that unlike charges attract one another and like charges repel one another.

2. The force between charges varies as the inverse square of their separation.

3. Charge is conserved.

4. Charge is quantized.

15.2 Insulators and Conductors

Conductors are materials in which electric charges move freely under the influence of an electric field; *insulators* are materials that do not readily transport charge.

15.3 Coulomb's Law

Experiments show that an *electric force* has the following properties:

1. It is inversely proportional to the square of the separation, r, between the two particles and is along the line joining them.

2. It is proportional to the product of the magnitudes of the charges, $|q_1|$ and $|q_2|$, on the two particles.

3. It is attractive if the charges are of opposite sign and repulsive if the charges have the same sign.

15.5 The Electric Field

The electric field vector **E** at some point in space is defined as the electric force **F** acting on a positive test charge placed at that point divided by the magnitude of the test charge q_0.

An electric field exists at some point if a test charge at rest placed at that point experiences an electrical force.

The total electric field due to a group of charges equals the *vector sum* of the electric fields of all the charges at some point.

15.6 Electric Field Lines

A convenient aid for visualizing electric field patterns is to draw lines pointing in the same direction as the electric field vector at any point. These lines, called electric field lines, are related to the electric field in any region of space in the following manner:

1. The electric field vector **E** is *tangent* to the electric field line at each point.

2. The number of lines per unit area through a surface perpendicular to the lines is proportional to the strength of the electric field in that region. Thus, **E** is large when the field lines are close together and small when they are far apart.

The rules for drawing electric field lines for any charge distribution are as follows:

1. The lines must begin on positive charges and terminate on negative charges, or at infinity in the case of an excess of charge.

2. The number of lines drawn leaving a positive charge or approaching a negative charge is proportional to the magnitude of the charge.

3. No two field lines can cross.

15.7 Conductors in Electrostatic Equilibrium

A conductor in *electrostatic equilibrium* has the following properties:

1. The electric field is zero everywhere inside the conductor.

2. Any excess charge on an isolated conductor resides entirely on its surface.

3. The electric field just outside a charged conductor is perpendicular to the conductor's surface.

4. On an irregularly shaped conductor, the charge per unit area is greatest at locations where the curvature of the surface is greatest, that is, at sharp points.

EQUATIONS AND CONCEPTS

The magnitude of the electrostatic force between two stationary point charges, q_1 and q_2, separated by a distance, r, is given by Coulomb's law. In calculations the approximate value of the Coulomb constant, k, may be used.

$$F = k\frac{|q_1||q_2|}{r^2} \quad (15.1)$$

$$k = 8.99 \times 10^9\ \text{N} \cdot \text{m}^2/\text{C}^2 \quad (15.2)$$

The smallest unit of electric charge known in nature is the charge on the electron or on the proton, represented by the symbol, e.

$$e = 1.6 \times 10^{-19}\ \text{C}$$

The direction of the electrostatic force on each charge is determined from the experimental observation that like sign charges experience forces of mutual repulsion and unlike sign charges attract each other. By virtue of Newton's third law, the magnitude of the force on each of the two charges is the same regardless of the relative magnitude of the values of q_1 and q_2.

Comment on Coulomb's law.

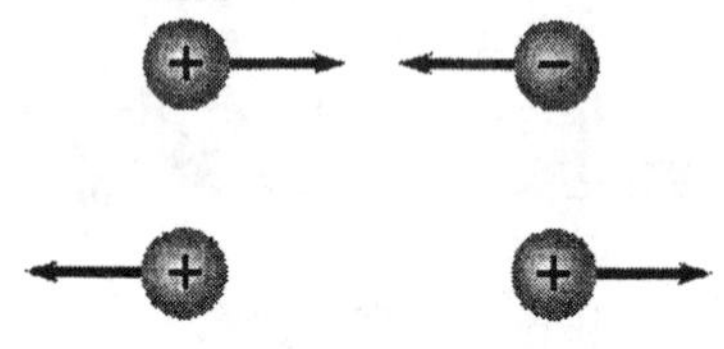

In cases where there are more than two charges present, the resultant force on any one charge is the vector sum of the forces exerted on that charge by the remaining individual charges present.

Comment on the principle of superposition.

The electric field, **E**, at any point in space is defined as the ratio of electric force per unit charge exerted on a small positive test charge, q_0, placed at the point where the field is to be determined.

$$E \equiv \frac{|\mathbf{F}|}{|q_0|} \qquad (15.4)$$

The direction of **E** at any point is defined to be the direction of the electric force that would be exerted on small positive charge if placed at the point in question. The SI units of the electric field are newtons per coulomb (N/C).

Comment on direction and units of the electric field.

The definition of the electric field combined with Coulomb's law leads to an expression for calculating the electric field a distance, r, from a point charge, q_0. The direction of the electric field is radially outward from a positive point charge and radially inward toward a negative point charge. The superposition principle holds when the electric field at a point is due to a number of point charges.

$$E = k\frac{|q|}{r^2} \qquad (15.5)$$

Electric field lines are a convenient graphical representation of electric field patterns. These lines are drawn so that the electric field vector, **E**, is tangent to the electric field lines at each point. Also, the number of lines per unit area through a surface perpendicular to the lines is proportional to the strength or magnitude of the electric field over the region. In every case, electric field lines must begin on positive charges and terminate on negative charges or at infinity; the number of lines leaving or approaching a charge is proportional to the magnitude of the charge; and no two field lines can cross.

Comment on electric field lines.

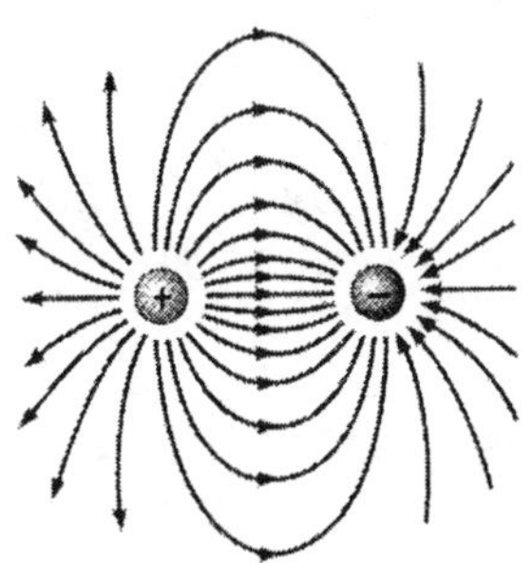

A conductor in electrostatic equilibrium has the following properties: (1) The electric field is zero everywhere inside the conductor; (2) Any excess charge on an isolated conductor resides entirely on its surface; (3) The electric field just outside a charged conductor is perpendicular to the surface of the conductor; and (4) On an irregularly shaped conductor, the surface charge density is greatest at points where the curvature of the surface is greatest (i.e. when the radius is smallest).

Comment on conductors in electrostatic equilibrium.

SUGGESTIONS, SKILLS, AND STRATEGIES

Here is a problem-solving strategy for electric forces and fields:

1. Units: When performing calculations that involve the use of the Coulomb constant k that appears in Coulomb's law, charges must be in coulombs and distances in meters. If they are given in other units, you must convert them to SI.

2. Applying Coulomb's law to point charges: It is important to use the superposition principle properly when dealing with a collection of interacting point charges. When several charges are present, the resultant force on any one of them is found by finding the individual force that every other charge exerts on it and then finding the vector sum of all these forces. The magnitude of the force that any charged object exerts on another is given by Coulomb's law, and the direction of the force is found by noting that the forces are repulsive between like charges and attractive between unlike charges.

3. Calculating the electric field of point charges: Remember that the superposition principle can also be applied to electric fields, which are also vector quantities. To find the total electric field at a given point, first calculate the electric field at the point due to each individual charge. The resultant field at the point is the vector sum of the fields due to the individual charges.

REVIEW CHECKLIST

▷ Use Coulomb's law to determine the net electrostatic force on a point electric charge due to a known distribution of a finite number of point charges.

▷ Calculate the electric field **E** (magnitude and direction) at a specified location in the vicinity of a group of point charges.

▷ Describe the configuration of electric field lines as they are associated with various patterns of charge distribution such as (i) point charges, (ii) dipole, (iii) charged metallic sphere, (iv) etc.

▷ State and justify the conditions for charge distribution on conductors in electrostatic equilibrium.

SOLUTIONS TO SELECTED END-OF-CHAPTER PROBLEMS

7. The Moon and Earth are bound together by gravity. If, instead, the force of attraction were the result of each having a charge of the same magnitude but opposite in sign, find the quantity of charge that would have to be placed on each to produce the required force.

Solution If $F_e = F_g$, we have $\frac{kQ^2}{r^2} = \frac{GMm}{r^2}$, where M is the mass of the Earth; m, the mass of the moon; and Q is the charge that would have to be on each. We solve for Q to find $Q = \sqrt{\frac{GMm}{k}}$ or

$$Q = \sqrt{\frac{(6.67 \times 10^{-11}\ \mathrm{N \cdot m^2/kg^2})(5.98 \times 10^{24}\ \mathrm{kg})(7.36 \times 10^{22}\ \mathrm{kg})}{(8.99 \times 10^{9}\ \mathrm{N \cdot m^2/C^2})}}$$

$$Q = 5.71 \times 10^{13}\ \mathrm{C} \quad \lozenge$$

---

9. A 2.20×10^{-9} C charge is located on the x axis at $x = -1.50$ m, a 5.40×10^{-9} C charge is located on the x axis at $x = 2.00$ m, and a 3.50×10^{-9} C charge is located at the origin. Find the net force on the 3.50×10^{-9} C charge.

Solution The force exerted on the 3.50×10^{-9} C charge by the 2.20×10^{-9} C charge is in the positive x direction (a force of repulsion) and has a magnitude of

$$F_1 = \frac{kq_1q_3}{r^2} = \left(8.99\times10^9\,\frac{\text{N}\cdot\text{m}^2}{\text{C}^2}\right)\frac{(2.20\times10^{-9}\text{ C})(3.50\times10^{-9}\text{ C})}{(1.50\text{ m})^2} = 3.08\times10^{-8}\text{ N}$$

and the force exerted on the 3.50×10^{-9} C charge by the 5.40×10^{-9} C charge is in the negative x direction (also a force of repulsion). Its magnitude is

$$F_2 = \frac{kq_2q_3}{r^2} = \left(8.99\times10^9\,\frac{\text{N}\cdot\text{m}^2}{\text{C}^2}\right)\frac{(5.40\times10^{-9}\text{ C})(3.50\times10^{-9}\text{ C})}{(2.00\text{ m})^2} = 4.25\times10^{-8}\text{ N}$$

The net force is

$$F_{\text{net}} = -4.25\times10^{-8}\text{ N} + 3.08\times10^{-8}\text{ N} = -1.17\times10^{-8}\text{ N (negative } x \text{ direction)} \quad \Diamond$$

---

13. Three charges are arranged as shown in Figure 15.28. Find the magnitude and direction of the electrostatic force on the 6-nC charge.

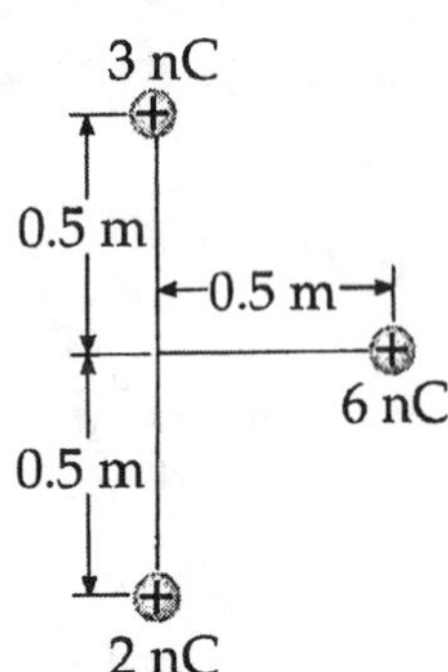

Figure 15.28

Solution The force $\mathbf{F}_1$ exerted on the 6.00×10^{-9} C charge by the 2.00×10^{-9} C is repulsive and in the direction indicated in the sketch (following page). Its magnitude is

$$F_1 = (8.99\times10^9\text{ N}\cdot\text{m}^2/\text{C}^2)\frac{(2.00\times10^{-9}\text{ C})(6.00\times10^{-9}\text{C})}{2(0.500\text{ m})^2}$$

$$F_1 = 2.16\times10^{-7}\text{ N}$$

The force $\mathbf{F}_2$ on the 6.00×10^{-9} C charge by the 3.00×10^{-9} C is repulsive and in the direction shown. The magnitude is

$$F_2 = (8.99\times10^9\text{ N}\cdot\text{m}^2/\text{C}^2)\frac{(3.00\times10^{-9}\text{ C})(6.00\times10^{-9}\text{ C})}{2(0.500\text{ m})^2} = 3.24\times10^{-7}\text{ N}$$

Now, resolve the forces into their x and y components as:

Force	x-component	y-component
F_1	$+1.53\times10^{-7}$ N	$+1.53\times10^{-7}$ N
F_2	$+2.29\times10^{-7}$ N	-2.29×10^{-7} N
F_R	3.82×10^{-7} N	-7.61×10^{-8} N

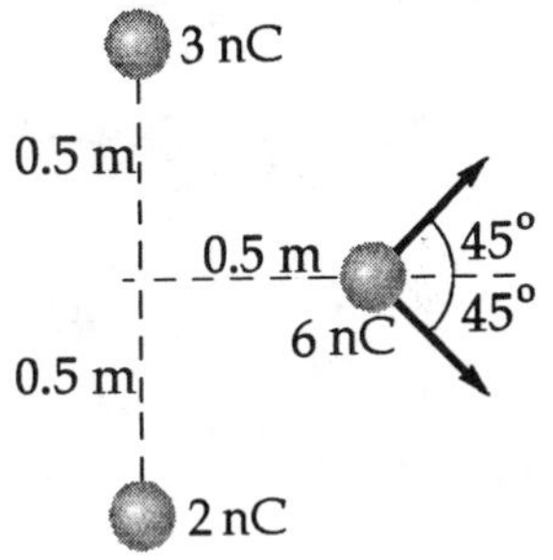

3 nC
0.5 m
0.5 m
45°
45°
6 nC
0.5 m
2 nC

$$F_R=\sqrt{(F_{Rx})^2+(F_{Ry})^2}=3.90\times10^{-7}\text{ N} \quad \Diamond$$

and $\tan\theta=\dfrac{F_{Ry}}{F_{Rx}}=0.199$ or $\theta=11.3°$ below the positive x-axis. ◊

---

17. A charge of 2.00×10^{-9} C is placed at the origin, and a charge of 4.00×10^{-9} C is placed at x = 1.50 m. Locate the point between the two charges where a charge of 3.00×10^{-9} C should be placed so that the net electric force on it is zero.

Solution If the net force is zero, $F_1=F_2$ (See the sketch.). We see that

$$\frac{k(2.00\times10^{-9}\text{ C})q}{x^2}=\frac{k(4.00\times10^{-9}\text{ C})q}{(1.50\text{ m}-x)^2}$$

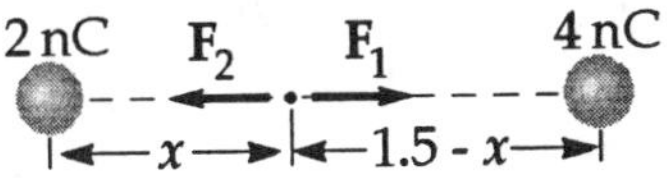

2 nC F₂ F₁ 4 nC
x 1.5 - x

which reduces to $(1.50\text{ m}-x)^2=2x^2$

Now take square root of both sides of the equation to yield

$$(1.50\text{ m}-x)=\sqrt{2}x \quad \text{or} \quad x=\frac{1.50\text{ m}}{\sqrt{2}+1}$$

Note that the square root of 2 can be either plus or minus. We select the plus sign so that the position of equilibrium will be between the two charges. We find x = 0.621 m. ◊

---

19. A charge of 6.00×10^{-9} C and a charge of -3.00×10^{-9} C are separated by a distance of 60 cm. Find the position at which a third charge of 12.0×10^{-9} C can be placed so that the net electrostatic force on it is zero.

Solution The required position for q is indicated in the sketch. We find

$$\frac{k(6.00\times10^{-9}\text{ C})q}{(x+0.600\text{ m})^2}=\frac{k(3.00\times10^{-9}\text{ C})q}{(x)^2}$$

6 nC -3 nC F_2 F_1
0.6 m x

which gives

$$2x^2=(x+0.600\text{ m})^2 \qquad \text{or} \qquad x=\frac{0.600\text{ m}}{\sqrt{2}-1}$$

We must choose the signs as shown in order to have $x>0$ so that the forces on the charge q will be in opposite directions. We solve for x, obtaining

$$x=1.45\text{ m}\quad(\text{beyond the }-3\times10^{-9}\text{ C charge})\quad\Diamond$$

---

27. (a) Determine the electric field strength at a point 1.00 cm to the left of the middle charge shown in Figure 15.26. (b) If a charge of $-2.00\ \mu$C is placed at this point, what are the magnitude and direction of the force on it?

6 μC 1.5 μC -2 μC
3 cm 2 cm

Figure 15.26

Solution

6 μC E_2 E_1 1.5 μC -2 μC
1 2 3
E_3
0.02 m 0.01 m 0.02 m

a) Note from the sketch that

$$\mathbf{E}=\mathbf{E}_1-\mathbf{E}_2+\mathbf{E}_3$$

or

$$E=\left(8.99\times10^9\frac{\text{N}\cdot\text{m}^2}{\text{C}^2}\right)\left(\frac{6.00\times10^{-6}\text{ C}}{(2.00\times10^{-2}\text{ m})^2}-\frac{1.50\times10^{-6}\text{ C}}{(1.00\times10^{-2}\text{ m})^2}+\frac{2.00\times10^{-6}\text{ C}}{(3.00\times10^{-2}\text{ m})^2}\right)$$

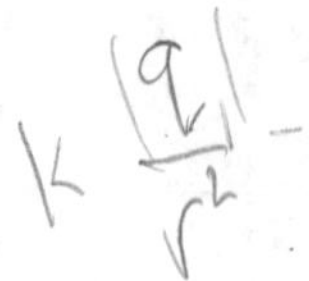

Giving $E = 2.00 \times 10^7$ N/C (directed toward the right) ◊

(b) $F = |q|E = (2.00 \times 10^{-6}\ \text{N})(2.00 \times 10^7\ \text{N/C}) = 40.0\ \text{N}$ (toward the left) ◊

The direction of **F** is toward the left because q is negative and **E** is directed toward the right.

---

29. A piece of aluminum foil of mass 5.00×10^{-2} kg is suspended by a string in an electric field directed vertically upward. If the charge on the foil is 3.00 μC, find the strength of the field that will reduce the tension in the string to zero.

Solution If there is zero tension in the string, the magnitudes of the electric and gravitational forces will be equal, so that

$$qE = mg \qquad \text{or} \qquad E = \frac{mg}{q}$$

Thus,

$$E = \frac{(5.00 \times 10^{-2}\ \text{kg})(9.80\ \text{m/s}^2)}{3.00 \times 10^{-6}\ \text{C}} = 1.63 \times 10^5\ \text{N} \quad ◊$$

---

33. A proton accelerates from rest in a uniform electric field of 640 N/C. At some later time, its speed is 1.20×10^6 m/s. (a) Find the acceleration of the proton. (b) How long does it take the proton to reach this velocity? (c) How far has it moved in this interval? (d) What is its kinetic energy at the later time?

Solution The force on the proton is

$$F = qE = (1.60 \times 10^{-19}\ \text{C})(640\ \text{N/C}) = 1.02 \times 10^{-16}\ \text{N}$$

(a) The acceleration is found from Newton's second law:

$$a = \frac{F}{m} = \frac{1.02 \times 10^{-16}\ \text{N}}{1.67 \times 10^{-27}\ \text{kg}} = 6.13 \times 10^{10}\ \text{m/s}^2 \quad ◊$$

(b) $v = v_0 + at$ or $t = \frac{v}{a}$ when $v_0 = 0$; so

$$t = \frac{v}{a} = \frac{1.20 \times 10^6 \text{ m/s}}{6.13 \times 10^{10} \text{ m/s}^2} = 1.96 \times 10^{-5} \text{ s} = 19.6 \ \mu\text{s} \quad \lozenge$$

(c) $v^2 = v_0^2 + 2as$ or $s = \frac{v^2}{2a}$ when $v_0 = 0$; so

$$s = \frac{v^2}{2a} = \frac{(1.20 \times 10^6 \text{ m/s})^2}{2(6.13 \times 10^{10} \text{ m/s}^2)} = 11.7 \text{ m} \quad \lozenge$$

(d) $KE = \frac{1}{2}mv^2 = \frac{1}{2}(1.67 \times 10^{-27} \text{ s})(1.20 \times 10^6 \text{ m/s})^2 = 1.20 \times 10^{-15} \text{ J} \quad \lozenge$

---

37. Positive charges are situated at three corners of a rectangle, as shown in Figure 15.30. Find the electric field at the fourth corner.

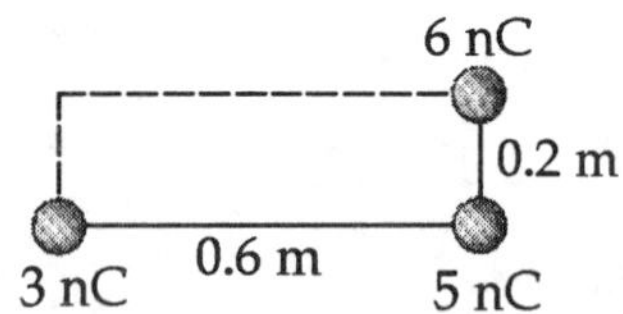

Figure 15.30

Solution The length of the diagonal can be shown, from the Pythagorean theorem, to be 0.630 m. Also, the angle ϕ (See figure below) is found from the tangent function to be 18.4°.

The magnitude of the field $\mathbf{E}_1$, due to the 3.00×10^{-9} C charge (direction shown on sketch) is

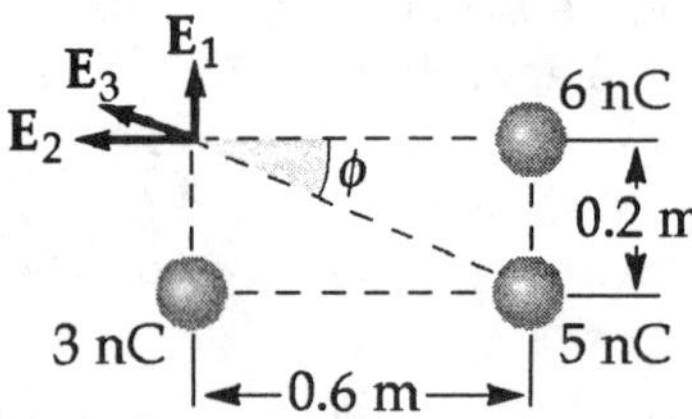

$$E_1 = \frac{kq}{r^2} = \frac{\left(8.99 \times 10^9 \frac{\text{N} \cdot \text{m}^2}{\text{C}^2}\right)(3.00 \times 10^{-9} \text{ C})}{(0.200 \text{ m})^2} = 675 \text{ N/C}$$

Likewise, the magnitude of $\mathbf{E}_2$ due to the 6.00×10^{-9} C charge is

$$E_2 = \frac{kq}{r^2} = \frac{\left(8.99 \times 10^9 \frac{\text{N} \cdot \text{m}^2}{\text{C}^2}\right)(6.00 \times 10^{-9} \text{ C})}{(0.600 \text{ m})^2} = 150 \text{ N/C}$$

and the magnitude of $\mathbf{E}_3$ due to the 5.00×10^{-9} C charge is

$$E_3 = \frac{kq}{r^2} = \frac{\left(8.99 \times 10^9 \frac{\text{N·m}^2}{\text{C}^2}\right)(5.00 \times 10^{-9}\text{ C})}{(0.630\text{ m})^2} = 112.5\text{ N/C}$$

Now, resolve the fields into their x and y components as

Field	x-comp	y-comp
$\mathbf{E}_1$	0	675 N/C
$\mathbf{E}_2$	−150 N/C	0
$\mathbf{E}_3$	−107 N/C	35.6 N/C
$\mathbf{E}_R$	−257 N/C	711 N/C

$$E_R = \sqrt{(E_{Rx})^2 + (E_{Ry})^2} = 756\text{ N/C} \quad \Diamond$$

and
$$\theta = \tan^{-1}\left(\frac{E_{Ry}}{E_{Rx}}\right) = 70.1° \text{ (above the negative } x \text{ axis)} \quad \Diamond$$

---

39. Three identical charges ($q = -5.00\ \mu$C) are along a circle of 2.00-m radius at angles of 30.0°, 150°, and 270°, as shown in Figure 15.31. What is the resultant electric field at the center of the circle?

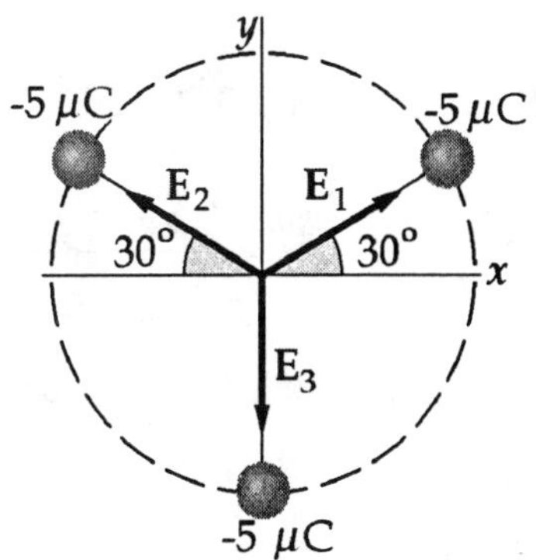

Figure 15.31

Solution You should recognize that $E_{net} = 0$ at the center from the symmetry of the arrangement. However, this result can be verified by calculation. $\mathbf{E}_1 = \mathbf{E}_2 = \mathbf{E}_3$, where $\mathbf{E}_1$ is the field due to the $-5\ \mu$C charge in the upper right quadrant, $\mathbf{E}_2$ is the field due to the $-5\ \mu$C in the upper left quadrant, and $\mathbf{E}_3$ is the field due to the $-5\ \mu$C charge directly below the point.

$$E_x = E_{1x} - E_{2x} = E_1 \cos 30° - E_2 \cos 30° = 0$$

and

$$E_y = E_{1y} - E_{2y} - E_{3y} = 2E_1 \sin 30° - E_2 = E_1 - E_2 = 0$$

Therefore, the net force at the center, $E = 0$. ◊

---

41. Three charges are at the corners of an equilateral triangle as shown in Figure 15.33. Calculate the electric field at a point midway between the two charges on the x axis.

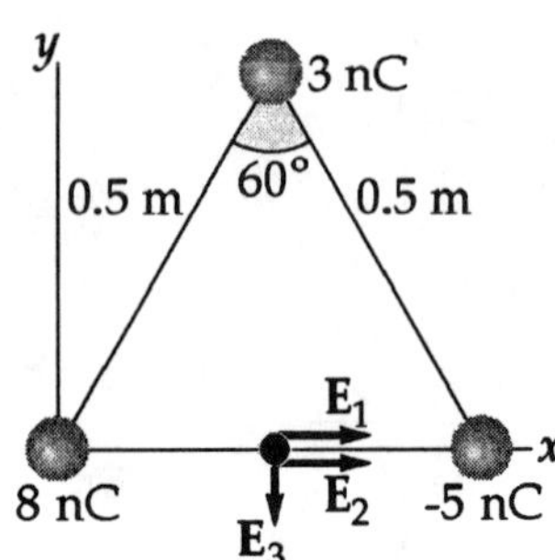

Solution The magnitude of field $\mathbf{E}_1$, due to the 8.00×10^{-9} C charge is

$$E_1 = \frac{kq}{r^2} = \frac{(8.99 \times 10^9 \text{ N} \cdot \text{m}^2/\text{C}^2)(8.00 \times 10^{-9} \text{ C})}{(0.250 \text{ m})^2} = 1152 \text{ N/C}$$

(direction shown in sketch.)

Likewise, the magnitude of $\mathbf{E}_2$, due to the -5.00×10^{-9} C charge is

$$E_2 = \frac{kq}{r^2} = \frac{(8.99 \times 10^9 \text{ N} \cdot \text{m}^2/\text{C}^2)(5.00 \times 10^{-9} \text{ C})}{(0.250 \text{ m})^2} = 720 \text{ N/C}$$

and that of $\mathbf{E}_3$, due to the 3.00×10^{-9} C charge is

$$E_3 = \frac{kq}{r^2} = \frac{(8.99 \times 10^9 \text{ N} \cdot \text{m}^2/\text{C}^2)(3.00 \times 10^{-9} \text{ C})}{(0.433 \text{ m})^2} = 144 \text{ N/C}$$

$E_{Rx} = \Sigma E_x = E_1 + E_2 = 1872 \text{ N/C}$ and $E_{Ry} = \Sigma E_y = -E_3 = -144 \text{ N/C}$

$E_R = \sqrt{(E_{Rx})^2 + (E_{Ry})^2} = 1880 \text{ N/C}$ ◊ and $\theta = \tan^{-1}\left(\frac{E_{Ry}}{E_{Rx}}\right) = -4.40°$ (below x axis) ◊

---

43. In Figure 15.34, determine the point (other than infinity) at which the total electric field is zero.

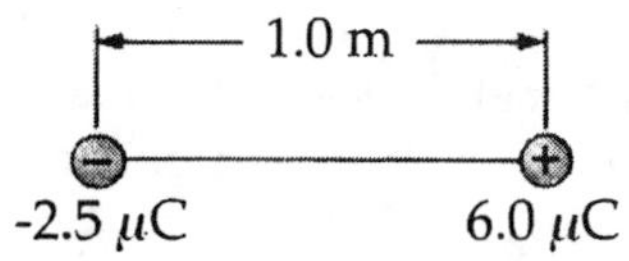

Figure 15.34

Solution At the point x designated in the sketch, the electric field, $\mathbf{E}_1$, due to the -2.50×10^{-6} C charge, is

$$E_1 = \frac{kq}{r^2} = \frac{(8.99\times10^9\ \text{N}\cdot\text{m}^2/\text{C}^2)(2.50\times10^{-6}\ \text{C})}{d^2}$$

or $$E_1 = \frac{2.25\times10^4}{d^2}\text{N}\cdot\text{m}^2/\text{C}$$

and E_2 due to the 6.00×10^{-6} C charge is

$$E_2 = \frac{kq}{r^2} = \frac{(8.99\times10^9\ \text{N}\cdot\text{m}^2/\text{C}^2)(6.00\times10^{-6}\ \text{C})}{(d+1.00\ \text{m})^2}$$

or $$E_2 = \frac{5.40\times10^4}{(d+1.00\ \text{m})^2}\text{N}\cdot\text{m}^2/\text{C}$$

For net zero, field at point x must have $E_1 = E_2$.

Therefore, $(d+1.00\ \text{m})^2 = 2.4d^2$ so $d(\pm1.55-1) = 1.00\ \text{m}$

The negative value for d is unsatisfactory because that locates a point between the charges where both fields are in the same direction. Thus,

$$d = 1.82\ \text{m to the left of the} -2.50\times10^{-6}\ \text{C charge} \quad \lozenge$$

---

49. Consider a rod of finite length having a uniform negative charge. Sketch the pattern of the electric field lines in a plane containing the rod.

Solution Note in the sketch at the right that the lines strike the surface at right angles. The pattern is symmetrical about the axis of the rod and symmetrical about a line perpendicular to the rod and passing through the center of the rod.

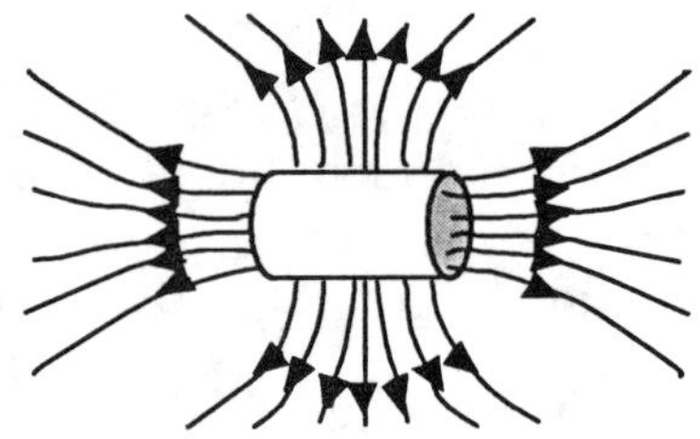

---

53. The dome of a Van de Graaff generator receives a charge of 2.00×10^{-4} C. Find the strength of the electric field (a) inside the dome; (b) at the surface of the dome, assuming it has a radius of 1.00 m; (c) 4.00 m from the center of the dome. (*Hint:* See Section 15.7 to review properties of conductors in electrostatic equilibrium. Also use the fact that the points on the surface are outside a spherically symmetric charge distribution; the total charge may be considered as located at the center of the sphere.)

Solution (a) The dome is a closed conducting surface. Therefore, the electric field inside the dome is zero. ◊

(b) $$E = \frac{kq}{r^2} = \frac{(8.99 \times 10^9 \text{ N}\cdot\text{m}^2/\text{C}^2)(2.00 \times 10^{-4}\text{ C})}{(1.00\text{ m})^2} = 1.80 \times 10^6 \text{ N/C} \quad ◊$$

(c) $$E = \frac{kq}{r^2} = \frac{(8.99 \times 10^9 \text{ N}\cdot\text{m}^2/\text{C}^2)(2.00 \times 10^{-4}\text{ C})}{(4.00\text{ m})^2} = 1.12 \times 10^5 \text{ N/C} \quad ◊$$

---

55. Air breaks down (loses its insulating quality) and sparking results if the field strength is increased to about 3.00×10^6 N/C. (a) What acceleration does an electron experience in such a field? (b) If the electron starts from rest, in what distance does it acquire a speed equal to 10% of the speed of light?

Solution (a) $$a = \frac{F}{m} = \frac{qE}{m} = \frac{(1.60 \times 10^{-19}\text{ C})(3.00 \times 10^6\text{ N/C})}{9.11 \times 10^{-31}\text{ kg}} = 5.27 \times 10^{17}\text{ m/s}^2 \quad ◊$$

(b) Anticipating that this distance is very small, we assume the field is uniform over this short distance.

$$\text{work done} = \Delta KE = KE_f - KE_i = KE_f - 0 \qquad \text{so} \qquad Fs = \frac{1}{2}\,\text{mv}^2$$

and $$s = \frac{mv^2}{2qE} = \frac{(9.11\times10^{-31}\text{ kg})(3.00\times10^7)^2}{2(1.60\times10^{-19}\text{ C})(3.00\times10^6\text{ N/C})} = 8.54\times10^{-4}\text{ m} = 0.854\text{ mm} \quad \lozenge$$

---

59. Three point charges are aligned along the x axis as shown in Figure 15.37. Find the electric field at (a) the position (2, 0) and (b) the position (0, 2).

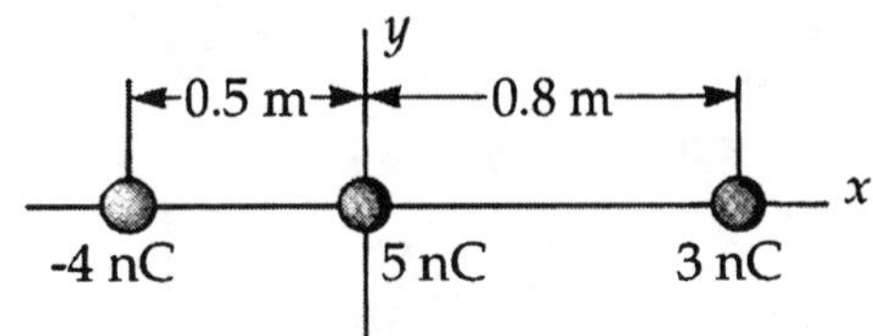

Figure 15.37

Solution (a) Refer to figure (a) below.

The field, $\mathbf{E}_1$, due to the 4.00×10^{-9} C charge is in the negative x direction (see sketch).

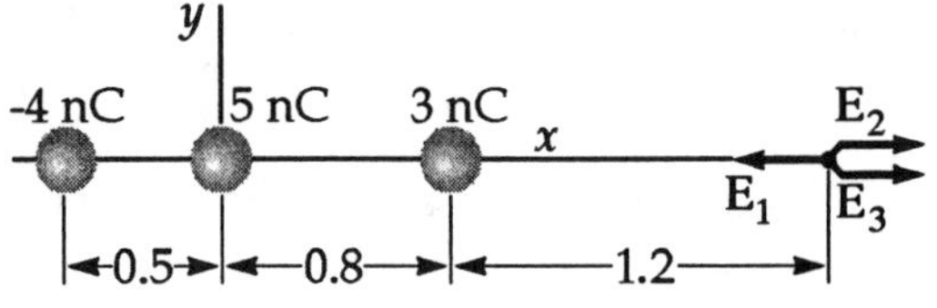

(a)

$$E_1 = \frac{kq}{r^2} = \frac{(8.99\times10^9\text{ N}\cdot\text{m}^2/\text{C}^2)(4.00\times10^{-9}\text{ C})}{(2.50\text{ m})^2} = 5.76\text{ N/C}$$

Likewise, $\mathbf{E}_2$, due to the 5.00×10^{-9} C charge is

$$E_2 = \frac{kq}{r^2} = \frac{(8.99\times10^9\text{ N}\cdot\text{m}^2/\text{C}^2)(5.00\times10^{-9}\text{ C})}{(2.00\text{ m})^2} = 11.3\text{ N/C}$$

and $\mathbf{E}_3$, due to the 3.00×10^{-9} C charge is

$$E_3 = \frac{(8.99\times10^9\text{ N}\cdot\text{m}^2/\text{C}^2)(3.00\times10^{-9}\text{ C})}{(1.20\text{ m})^2} = 18.8\text{ N/C}$$

$$E_R = -E_1 + E_2 + E_3 = 24.2\text{ N/C in } +x\text{ direction} \quad \lozenge$$

(b) The directions of the various fields are shown in figure (b) below. Using the equation, $E = \frac{kq}{r^2}$, the magnitudes are calculated to be:

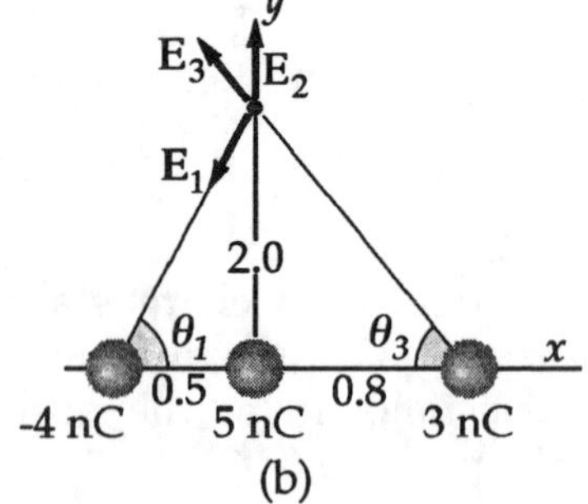

$E_1 = 8.47$ N/C; $E_2 = 11.3$ N/C; and $E_3 = 5.82$ N/C

The resultant x component is

$$\Sigma E_x = -E_{1x} - E_{3x} = -E_1 \cos\theta_1 - E_3 \cos\theta_3 = -4.21 \text{ N/C}$$

The resultant y component is

$$\Sigma E_y = -E_{1y} - E_{2y} + E_{3y} = -E_1 \sin\theta_1 + E_2 + E_3 \sin\theta_3 = 8.44 \text{ N/C}$$

and the magnitude of the field is

$$E = \sqrt{(\Sigma E_x)^2 + (\Sigma E_y)^2} = 9.43 \text{ N/C} \quad \Diamond$$

$$\phi = \tan^{-1}\left(\frac{\Sigma E_y}{\Sigma E_x}\right) = -63.5°$$

or the direction of the resultant field is 63.5° above the negative x axis. ◊

---

65. Two 2.00-g spheres are suspended by 10.0-cm-long light strings (Fig. 15.40). A uniform electric field is applied in the x direction. If the spheres have charges of -5.00×10^{-8} C and $+5.00 \times 10^{-8}$ C, determine the electric field intensity that enables the spheres to be in equilibrium at $\theta = 10.0°$.

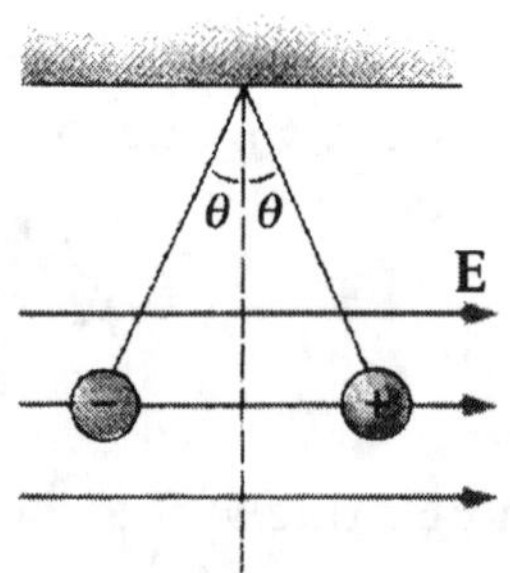

Figure 15.40

Solution The distance between the two charges at equilibrium is

$$r = 2l\sin\theta \quad \text{or} \quad r = 2(0.100 \text{ m})\sin 10° = 3.47 \times 10^{-2} \text{ m}$$

Now consider the forces on the sphere with charge $+q$ [see sketch (a)], and use $\Sigma F_y = 0$ to obtain

$$T\cos\theta = mg \quad \text{or} \quad T = \frac{mg}{\cos\theta} \qquad (1)$$

(a)

Now use $\Sigma F_x = 0$, $\quad F_{\text{net}} = T\sin\theta \qquad (2)$

F_{net} is the net electrical force on the charged sphere. Eliminate T from (2) by use of (1).

$$F_{net} = \frac{mg\sin\theta}{\cos\theta} = mg\tan\theta \quad \text{or}$$

$$F_{\text{net}} = (2.00\times10^{-3}\ \text{kg})(9.80\ \text{m/s}^2)\tan 10° = 3.46\times10^{-3}\ \text{N}$$

F_{net} is the resultant of two forces, $\mathbf{F}_1$ and $\mathbf{F}_2$ [see sketch (b)].

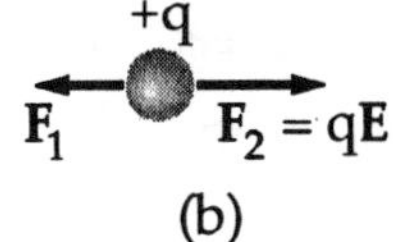

(b)

F_1 is the attractive force on $+q$ exerted by $-q$, and $\mathbf{F}_2$ is the force exerted on $+q$ by the *external electric field.*

$F_{\text{net}} = F_2 - F_1 \quad$ or $\quad F_2 = F_{\text{net}} + F_1 \quad$ and

$$F_1 = (8.99\times10^9\ \text{N}\cdot\text{m}^2/\text{C}^2)\frac{(5.00\times10^{-8}\ \text{C})(5.00\times10^{-8}\ \text{C})}{(3.47\times10^{-3}\ \text{m})^2} = 1.87\times10^{-2}\ \text{N}$$

Thus, $F_2 = F_{\text{net}} + F_1$ yields

$$F_2 = 3.46\times10^{-3}\ \text{N} + 1.87\times10^{-2}\ \text{N} = 2.21\times10^{-2}\ \text{N}$$

and

$$E = \frac{F_2}{q} = \frac{2.21\times10^{-2}\ \text{N}}{5.00\times10^{-8}\ \text{C}} = 4.43\times10^5\ \text{N/C} \quad \Diamond$$

CHAPTER SELF-QUIZ

1. A charge of +3 C is at the origin. When charge Q is placed at 2.00 m along the positive x axis, the electric field at 2.00 m along the negative x axis becomes zero. What is the value of Q?

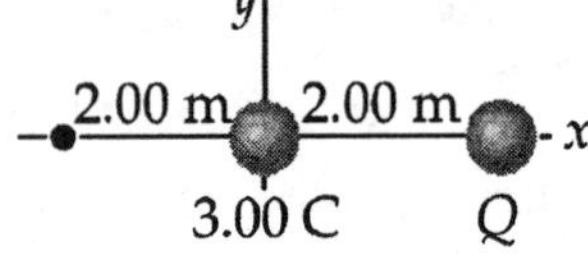

   a. −3 C
   b. −6 C
   c. −9 C
   d. −12 C

2. A charge, $+Q$, is placed inside a balloon and the balloon is blown up. As the radius, r, of the balloon increases, the number of field lines going through the surface of the balloon
   a. increases proportional to r^2
   b. increases proportional to r
   c. stays the same
   d. decreases as $1/r$

3. At what point is the electrical field associated with a uniformly-charged, hollow, metallic sphere greatest?
   a. center of the sphere
   b. at the sphere's inner surface
   c. at infinity
   d. at the sphere's outer surface

4. Two point charges are placed along a horizontal axis with the following values and positions: +3 μC at x = 0 cm and −7 μC at x = 20 cm. What is the magnitude of the electric field at the point midway between the two charges (at x = 10 cm)?
   a. 9.00×10^6 N/C
   b. 3.60×10^6 N/C
   c. 4.50×10^6 N/C
   d. 1.80×10^6 N/C

5. Initially, a net charge of –5.0 μC is placed on a 5.0-cm radius metallic hollow sphere. Next, a +10 μC charge is carefully inserted at the center through a hole in the latter's surface. What electric field is present at a point 10 cm from the center of the sphere? $(k = 8.99 \times 10^9\ \text{N} \cdot \text{m}^2 / \text{C}^2)$
   a. 2.30×10^6 N/C
   b. 4.50×10^6 N/C
   c. 9.00×10^6 N/C
   d. 18.0×10^6 N/C

6. Two point charges are separated by 4 cm and have charges of +2.0 μC and –2.0 μC, respectively. What is the electric field at a point midway between the two charges?
   a. 18.0×10^7 N/C
   b. 9.00×10^7 N/C
   c. 4.50×10^7 N/C
   d. Zero

7. The number of electric field lines passing through a unit cross-sectional area is indicative of which of the following?
   a. field direction
   b. charge density
   c. field strength
   d. charge motion

8. The electric field in a cathode ray tube is supposed to accelerate electrons from 0 to 1.60×10^7 m/s in a distance of 2 cm. What electric field is required? $(m_e = 9.10 \times 10^{-31}$ kg and $e = 1.60 \times 10^{-19}$ C)
   a. 9,000 N/m
   b. 18,200 N/m
   c. 36,400 N/m
   d. 72,800 N/m

9. A Van De Graaff generator has a spherical dome of radius 30 cm. Operating in dry air, where "atmospheric breakdown" is $E_{max} = 3.00 \times 10^6$ N/C what is the maximum charge that can be held on the dome? $(k = 8.99 \times 10^9 \text{ N} \cdot \text{m}^2/\text{C}^2)$
   a. 1.50×10^{-5} C
   b. 3.00×10^{-5} C
   c. 1.50×10^{-6} C
   d. 7.50×10^{-6} C

10. Two charges, $+Q$ and $-Q$, are located two meters apart and there is a point along the line that is equidistant from the two charges as indicated. Which vector best represents the direction of the electric field at that point?
   a. Vector $\mathbf{E_A}$
   b. Vector $\mathbf{E_B}$
   c. Vector $\mathbf{E_C}$
   d. The electric field at that point is zero

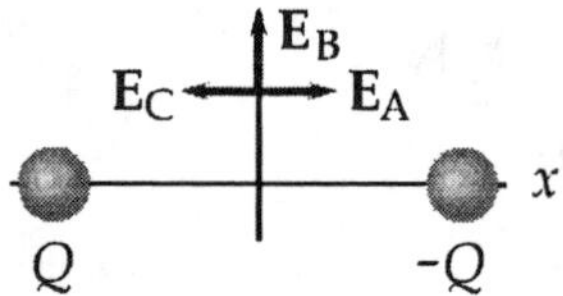

11. A proton (mass = 1.67×10^{-27} kg, charge = $+1.60 \times 10^{-19}$ C) is accelerated from rest by an electric field of 400 N/C. Find its velocity after 10^{-6} s.
   a. 3.83×10^4 m/s
   b. 7.66×10^4 m/s
   c. 15.32×10^4 m/s
   d. 30.6×10^4 m/s

12. A glass rod rubbed with a silk cloth acquires a charge of $+7.40 \times 10^{-9}$ C. How many electrons were transferred from the glass to the silk?
   a. 3.90×10^{10}
   b. 2.10×10^{10}
   c. 0.90×10^9
   d. 1.30×10^9

16
Electrical Energy and Capacitance

Chapter 16

ELECTRICAL ENERGY AND CAPACITANCE

The concept of potential energy was first introduced in Chapter 5. A potential energy function can be defined for any conservative force, such as the force of gravity. By using the principle of conservation of energy, we were often able to avoid working directly with forces when solving problems. In this chapter we discover that the energy concept is also useful in the study of electricity. Because the Coulomb force is conservative, we can define an electrical potential energy corresponding to the Coulomb force. This concept of potential energy is of value, but perhaps even more valuable is a quantity called electric potential, defined as potential energy per unit charge.

We take our first steps toward circuits with a discussion of electric potential, carried forward by an investigation of a common circuit element called a capacitor.

NOTES FROM SELECTED CHAPTER SECTIONS

16.3 Potentials and Charged Conductors

No work is required to move a charge between two points that are at the same potential. That is, $W = 0$ when $V_B = V_A$.

The electric potential is a constant everywhere on the surface of a charged conductor in equilibrium.

The electric potential is constant everywhere inside a conductor and equal to its value at the surface.

The electron volt is defined as the energy that an electron (or proton) gains when accelerated through a potential difference of 1 V.

16.7 Combinations of Capacitors

The potential difference across each capacitor in the parallel circuit is the same and is equal to the voltage of the battery, V.

The equivalent capacitance of a parallel combination of capacitors is larger than any of the individual capacitances.

For a series combination of capacitors, the magnitude of the charge must be the same on all the plates.

In general, the potential difference across any number of capacitors (or other circuit elements) in series is equal to the sum of the potential differences across the individual capacitors.

EQUATIONS AND CONCEPTS

The change in electric potential energy of an electric charge in moving between two points in an electric field is equal to the negative of the work done by the electric force. The change in electric potential energy can be expressed in terms of the charge, the magnitude of the field, and the distance moved parallel to the direction of the field.

$$\Delta PE = -W = -qEd \qquad (16.1)$$

The potential difference between points A and B is defined as the change in potential energy per unit charge as a positive charge is moved from point A to point B.

$$\Delta V \equiv V_B - V_A = \frac{\Delta PE}{q} \qquad (16.2)$$

The SI unit of potential is the volt, V. One joule of work must be done by an external force in order to move 1 coulomb of charge from point A to a second point B where the potential is 1 volt greater. Since the electric force is a conservative force, the work done is independent of the path taken from A to B.

$$1\text{ V} \equiv 1\text{ J/C} \qquad (16.3)$$

$$1\text{ N/C} = 1\text{ V/m}$$

The electric potential in the vicinity of a point charge, q, is inversely proportional to the distance from the charge. This equation assumes that the potential at infinity is zero. The potential can be positive or negative depending on the sign of q.

$$V = k\frac{q}{r} \qquad (16.5)$$

The total electric potential at some point, P, due to several point charges is the algebraic sum of the potentials due to the individual charges.

Comment on a potential due to several charges.

The potential energy of a pair of charges separated by a distance, r, represents the minimum work required to assemble the charges from an infinite separation. The potential energy of the two charges is positive if the two charges have the same sign; and it is negative if the two charges are of opposite sign.

$$PE = k\frac{q_1 q_2}{r} \quad (16.6)$$

No work is required to move a charge between two points that are at the same potential.

$$W = -q(V_B - V_A) \quad (16.7)$$

The net charge on a conductor in electrostatic equilibrium resides entirely on the surface. The electric potential is constant everywhere on the surface. The electric potential is constant everywhere inside and equal to its value at the surface. Also, recall that the electric field inside a charged conductor is zero.

Comment on charged conductors in electrostatic equilibrium.

The electron volt is defined as that quantity of energy which an electron or proton gains when accelerated through a potential difference of 1 volt.

$$1\text{ eV} = 1.60 \times 10^{-19}\text{ J} \quad (16.8)$$

The capacitance, C, of a capacitor is defined as the ratio of the charge on either plate (conductor) to the potential difference between the plates. (C is always positive.)

$$C \equiv \frac{Q}{V} \quad (16.9)$$

The SI unit of capacitance is the farad (F). The farad is a very large unit of capacitance and in practice typical devices have capacitances ranging from picofarads (10^{-9} F) to microfarads (10^{-6} F).

$$1\text{ F} \equiv 1\text{ C/V}$$

Comment on capacitors.

The capacitance of a capacitor depends on the physical characteristics of the device (size, shape, plate separation, and the nature of the dielectric medium filling the region between the plates).

The capacitance of an air-filled parallel plate capacitor is proportional to the area of the plates and inversely proportional to the separation of the plates.

$$C = \varepsilon_0 \frac{A}{d} \qquad (16.10)$$

The value for the permittivity constant for free space.

$$\varepsilon_0 = 8.85 \times 10^{-12}\ \mathrm{C^2/N \cdot m^2}$$

The equivalent capacitance of a parallel combination of capacitors is the sum of the values of the individual capacitors in the parallel group. The total charge stored by a group of capacitors connected in parallel is the sum of the charges stored on the individual capacitors.

$$C_{eq} = C_1 + C_2 + C_3 + \ldots \qquad (16.13)$$

$$Q_{total} = Q_1 + Q_2 + Q_3 + \ldots$$

(parallel combination)

The equivalent capacitance of a series combination of capacitors is smaller than the smallest individual value of capacitance in the group. The potential difference across the series group equals the sum of the values of potential difference across the individual capacitors.

$$\frac{1}{C_{eq}} = \frac{1}{C_1} + \frac{1}{C_2} + \frac{1}{C_3} + \ldots \qquad (16.16)$$

$$V = V_1 + V_2 + V_3 + \ldots$$

(series combination)

The electrostatic energy stored in the electric field of a charged capacitor is equal to the work done by a battery (or other source of emf) in charging the capacitor from $q = 0$ to $q = Q$.

$$W = \frac{1}{2} QV \qquad (16.17)$$

$$\text{Energy stored} = \frac{1}{2} CV^2 = \frac{Q^2}{2C}$$

When a dielectric (insulating material) is inserted between the plates of a capacitor, the capacitance of the device increases.

$$C = \kappa \varepsilon_0 \left(\frac{A}{d}\right) \qquad (16.20)$$

The diffusion of ions through the walls of living cells is regulated by an equilibrium potential difference across the cell wall called the Nernst potential. The equilibrium potential is dependent on the absolute temperature, ion concentration within and outside the cell, and the charge on each ion.

$$V_N = \frac{kT}{q}\ln\left(\frac{c_i}{c_o}\right) \quad (16.21)$$

Boltzmann's constant.

$$k_B = 1.38 \times 10^{-23} \text{ J/K}$$

SUGGESTIONS, SKILLS, AND STRATEGIES

A strategy for problems involving electric potential:

1. When working problems involving electric potential, remember that potential is a *scalar quantity* (rather than a vector quantity like the electric field), so there are no components to worry about. Therefore, when using the superposition principle to evaluate the electric potential at a point due to a system of point charges, you simply take the algebraic sum of the potentials due to each charge. However, you must keep track of signs. The potential for each positive charge is positive, while the potential for each negative charge is negative. The basic equation to use is $V = kq/r$.

2. As in mechanics, only changes in potential are significant; hence, the point where you choose the potential to be zero is arbitrary.

A problem-solving strategy for capacitors:

1. Be careful with your choice of units. To calculate the capacitance of a device in farads, make sure that distances are in meters and use the SI value of ε_0.

2. When two or more unequal capacitors are connected in *series,* they carry the same charge, but the potential differences across them are not the same. Their capacitances add as reciprocals, and the equivalent capacitance of the combination is always *less* than the smallest individual capacitor.

3. When two or more capacitors are connected in *parallel,* the potential difference across each is the same. The charge on each capacitor is proportional to its capacitance; hence, the capacitances add directly to give the equivalent capacitance of the parallel combination.

4. A complicated circuit consisting of capacitors can often be reduced to a simple circuit containing only one capacitor. To do so, examine your initial circuit and replace any capacitors in series or any in parallel using the rules of Steps 2 and 3 above. Draw a sketch of your new circuit after these changes have been made. Examine this new circuit and replace any series or parallel combinations. Continue this process until a single, equivalent capacitor is found.

5. If the charge on, or the potential difference across, one of the capacitors in the complicated circuit is to be found, start with the final circuit found in Step 4 and gradually work your way back through the circuits using $C = Q/V$ and the rules given in Steps 2 and 3 above.

REVIEW CHECKLIST

▷ Understand that each point in the vicinity of a charge distribution can be characterized by a scalar quantity called the electric potential, V; and define the quantity, *electrical potential difference.*

▷ Calculate the electric potential difference between any two points in a uniform *electric field* and calculate the electric potential difference between any two points in the vicinity of a *group of point charges.*

▷ Calculate the electric *potential energy* associated with a group of point charges and define the unit of energy, *electron volt.*

Justify the claims that (i) all points on the surface and within a charged conductor are at the same potential and (ii) the electric field within a charged conductor is zero.

▷ Define the quantity, *capacitance;* and evaluate the capacitance of a parallel plate capacitor of given area and plate separation.

▷ Determine the equivalent capacitance of a network of capacitors in series-parallel combination and calculate the final charge on each capacitor and the potential difference across each when a known potential is applied across the combination.

SOLUTIONS TO SELECTED END-OF-CHAPTER PROBLEMS

3. A uniform electric field of magnitude 250 V/m is directed in the positive x direction. A +12.0 μC charge moves from the origin to the point (x, y) = (20.0 cm, 50.0 cm).

(a) What was the change in the potential energy of this charge? (b) Through what potential difference did the charge move?

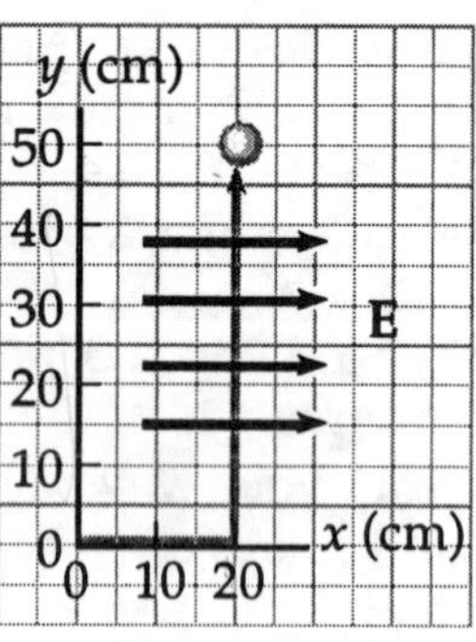

Solution (a) The electrostatic force is *conservative*. Therefore, the work done is independent of the path. Consider a path parallel to the x axis from the origin at (0, 0) to the point (20.0 cm, 0); and then parallel to the y axis from (20.0 cm, 0) to (20.0 cm, 50.0 cm).

$\Delta PE = -W$, the work done.

Note that along the second segment of the path, the electric field (and force) is perpendicular to the direction of displacement of the charge; and hence, along that segment, $W = 0$. So, $\Delta PE = -F(\Delta x)$ where Δx is the path from (0, 0) to (20.0 cm, 0).

$$\Delta PE = -qE(\Delta x) = -(12.0\times10^{-6}\text{ C})(250\text{ V/m})(0.200\text{ m})$$

$$\Delta PE = -6.00\times10^{-4}\text{ J} \quad \Diamond$$

(b) $$\Delta V = \frac{\Delta PE}{q} = \frac{-6.00\times10^{-4}\text{ J}}{12.0\times10^{-6}\text{ C}} = -50.0\text{ V} \quad \Diamond$$

---

7. How much work is done (by a battery, generator, or some other source of electrical energy) in moving Avogadro's number of electrons from a point where the electric potential is 6.00 V to a point where the electric potential is −10.0 V?

Solution $\Delta V = \dfrac{W}{Q}$; $\quad W = (\Delta V)(Q) = (\Delta V)(N_a e)$

$$W = (-10.0\text{ V} - 6.00\text{ V})(6.02\times10^{23}\text{ electrons})(-1.60\times10^{-19}\text{ C/electrons})$$

$$W = 1.54\times10^{6}\text{ J} \quad \Diamond$$

---

11. (a) Calculate the speed of a proton that is accelerated from rest through a potential difference of 120 V. (b) Calculate the speed of an electron that is accelerated through the same potential difference.

Solution $W = \Delta KE = q\Delta V$ or $\frac{1}{2}mv^2 = q(\Delta V)$

so $$v = \sqrt{\frac{2q(\Delta V)}{m}} = \sqrt{\frac{(2)(1.60\times10^{-19}\ \text{C})(120\ \text{V})}{m}}$$

(a) For a proton, $m = 1.67\times10^{-27}$ kg

$$v = \sqrt{\frac{3.84\times10^{-17}\ \text{J}}{1.67\times10^{-27}\ \text{kg}}} = 1.52\times10^5\ \text{m/s} \quad \Diamond$$

(b) For an electron, $m = 9.11\times10^{-31}$ kg

$$v = \sqrt{\frac{3.84\times10^{-17}\ \text{J}}{9.11\times10^{-31}\ \text{kg}}} = 6.49\times10^6\ \text{m/s} \quad \Diamond$$

This speed is only 2% of the speed of light; and relativistic effects are not important in this case.

---

13. (a) Through what potential difference would an electron need to accelerate to achieve a speed of 60% of the speed of light, starting from rest? The speed of light is 3.00×10^8 m/s. (b) Repeat this calculation for a proton. (Do not consider relativistic effects.)

Solution Sixty percent of the speed of light $(3.00\times10^8\ \text{m/s})$ is $1.80\times10^8\ \text{m/s}$. From conservation of energy, we have

$$\frac{1}{2}mv^2 = qV \quad \text{or} \quad V = \frac{mv^2}{2q}$$

(a) For an electron,

$$|V| = \frac{mv^2}{2q} = \frac{(9.11\times10^{-31}\text{ kg})(1.80\times10^{8}\text{ m/s})^2}{2(1.60\times10^{-19}\text{ C})} = 9.22\times10^{4}\text{ V} \quad \lozenge$$

(b) For a proton,

$$|V| = \frac{mv^2}{2q} = \frac{(1.67\times10^{-27}\text{ kg})(1.80\times10^{8}\text{ m/s})^2}{2(1.60\times10^{-19}\text{ C})} = 1.69\times10^{8}\text{ V} \quad \lozenge$$

---

23. Three charges are situated at corners of a rectangle as in Figure 16.28. How much energy would be expended in moving the 8.00 μC to infinity?

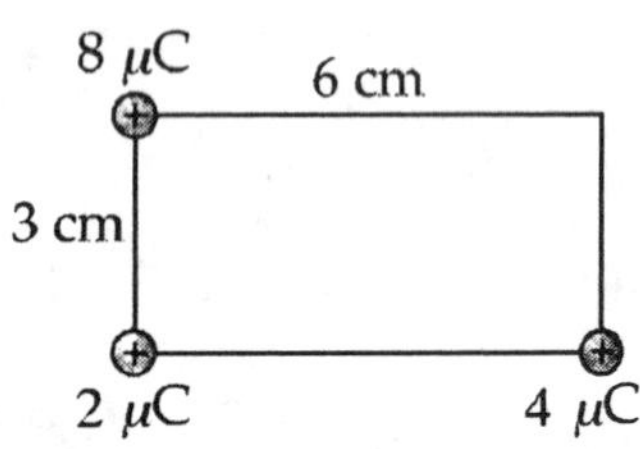

Figure 16.28

Solution Work done on 8.00 μC charge equals $\Delta PE = PE_f - PE_i = 0 - PE_i$. ($PE_f = 0$ because, at the end, the charge is at an infinite distance from all other charges.)

So,

$$W = -\left(PE_{\text{due to presence of 2.00 }\mu\text{C charge}} + PE_{\text{due to presence of 4.00 }\mu\text{C charge}}\right)$$

In each case use $PE = \frac{kq}{r}$.

$$W = -\frac{(8.99\times10^{9}\text{ N}\cdot\text{m}^2/\text{C}^2)(8.00\times10^{-6}\text{ C})(2.00\times10^{-6}\text{ C})}{3.00\times10^{-2}\text{ m}}$$

$$+\frac{(8.99\times10^{9}\text{ N}\cdot\text{m}^2/\text{C}^2)(8.00\times10^{-6}\text{ C})(4.00\times10^{-6}\text{ C})}{\sqrt{(0.030\text{ m})^2 + (0.060\text{ m})^2}}$$

or $\quad W = -9.09\text{ J} \quad \lozenge$

---

25. Calculate the speed of (a) an electron that has a kinetic energy of 1.00 eV and (b) a proton that has a kinetic energy of 1.00 eV.

Solution (a) 1.00 eV = 1.60×10^{-19} J. Thus, this is the kinetic energy of the electron. We have

$$KE=\frac{1}{2}mv^2 \quad \text{and} \quad v=\sqrt{\frac{2KE}{m}}$$

For an electron,

$$v=\sqrt{\frac{2(1.60\times10^{-19}\text{ J})}{9.11\times10^{-31}\text{ kg}}}=5.93\times10^{5}\text{ m/s} \quad \lozenge$$

(b) For a proton of the same energy,

$$v=\sqrt{\frac{2(1.60\times10^{-19}\text{ J})}{1.67\times10^{-27}\text{ kg}}}=1.38\times10^{4}\text{ m/s} \quad \lozenge$$

---

31. The plates of a parallel-plate capacitor are separated by 0.100 mm. If the material between the plates is air, what plate area is required to provide a capacitance of 2.00 pF?

Solution $C=\frac{\varepsilon_0 A}{d}$ gives $A=\frac{Cd}{\varepsilon_0}=\frac{(2.00\times10^{-12}\text{ F})(1.00\times10^{-4}\text{ m})}{8.85\times10^{-12}\text{ F/m}}=2.26\times10^{-5}\text{ m}^2$ ◊

---

35. A parallel-plate capacitor has an area of 5.00 cm^2, and the plates are separated by 1.00 mm with air between them. It stores a charge of 400 pC. (a) What is the potential difference across the plates of the capacitor? (b) What is the magnitude of the uniform electric field in the region between the plates?

Solution (a) $C=\frac{\varepsilon_0 A}{d}=\frac{(8.85\times10^{-12}\text{ F/m})(5.00\times10^{-4}\text{ m}^2)}{1.00\times10^{-3}\text{ m}}=4.43\times10^{-12}\text{ F}$

and $V=\frac{Q}{C}=\frac{400\times10^{-12}\text{ C}}{4.43\times10^{-12}\text{ F}}=90.4\text{ V}$ ◊

(b) $E = \frac{V}{d} = \frac{90.4 \text{ V}}{10^{-3} \text{ m}} = 9.04 \times 10^4 \text{ V/m}$ ◊

---

41. Three capacitors, $C_1 = 5.00\ \mu\text{F}$, $C_2 = 4.00\ \mu\text{F}$, and $C_3 = 9.00\ \mu\text{F}$, are connected together. (a) Find the effective capacitance of the group if they are all in parallel. (b) Find the effective capacitance of the group if they are all in series.

Solution (a) $C_{eq} = C_1 + C_2 + C_3 = 5.00\ \mu\text{F} + 4.00\ \mu\text{F} + 9.00\ \mu\text{F} = 18.0\ \mu\text{F}$ ◊

(b) $$\frac{1}{C_{eq}} = \frac{1}{C_1} + \frac{1}{C_2} + \frac{1}{C_3} = \frac{1}{5.00\ \mu\text{F}} + \frac{1}{4.00\ \mu\text{F}} + \frac{1}{9.00\ \mu\text{F}}$$

$$\frac{1}{C_{eq}} = \frac{36 + 45 + 20}{180\ \mu\text{F}} = \frac{101}{180\ \mu\text{F}}$$

$$C_{eq} = \frac{180\ \mu\text{F}}{101} = 1.78\ \mu\text{F} \quad ◊$$

---

43. Find the charge on each of the capacitors in Figure 16.30.

Solution

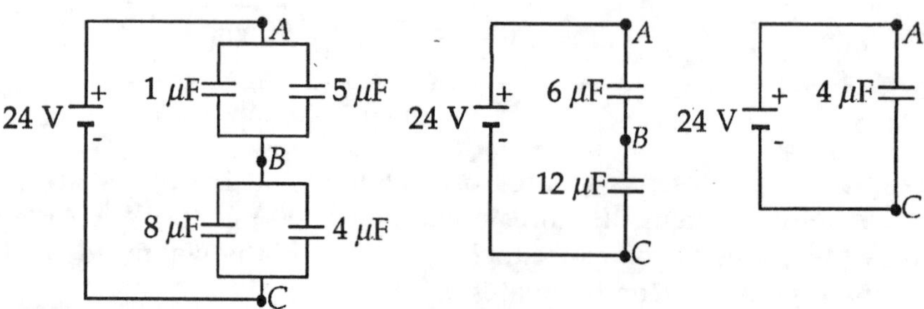

Figure 16.30

We reduce the circuit in steps as shown above. Using the equivalent circuit, we find

$$Q_{total} = CV = (4.00\ \mu\text{F})(24.0\ \text{V}) = 96.0\ \mu\text{C}$$

Then in the second circuit,

$$V_{AB} = \frac{Q_{\text{total}}}{C_{AB}} = \frac{96.0\ \mu\text{C}}{6.00\ \mu\text{F}} = 16.0\ \text{V} \quad \text{and} \quad V_{BC} = \frac{Q_{\text{total}}}{C_{BC}} = \frac{96.0\ \mu\text{C}}{12.0\ \mu\text{F}} = 8.00\ \text{V}$$

Finally, using the original circuit,

$$Q_1 = C_1 V_{AB} = (1.00\ \mu\text{F})(16.0\ \text{V}) = 16.0\ \mu\text{C} \quad \Diamond$$

$$Q_5 = C_5 V_{AB} = (5.00\ \mu\text{F})(16.0\ \text{V}) = 80.0\ \mu\text{C} \quad \Diamond$$

$$Q_8 = C_8 V_{BC} = (8.00\ \mu\text{F})(8.00\ \text{V}) = 64.0\ \mu\text{C} \quad \Diamond$$

and

$$Q_4 = C_4 V_{BC} = (4.00\ \mu\text{F})(8.00\ \text{V}) = 32.0\ \mu\text{C} \quad \Diamond$$

---

47. A 10.0-μF capacitor is fully charged across a 12.0-V battery. The capacitor is then disconnected from the battery and connected across an initially uncharged capacitor, C. The voltage across each capacitor is 3.00 V. What is the capacitance, C?

Solution For the first capacitor, initially,

$$Q_{1i} = C_1 V_{1i} = (10.0 \times 10^{-6}\ \text{F})(12.0\ \text{V}) = 1.20 \times 10^{-4}\ \text{C}$$

When connected across the second capacitor,

$$Q_{1f} = C_1 V_{1f} = (10.0 \times 10^{-6}\ \text{F})(3.00\ \text{V}) = 3.00 \times 10^{-5}\ \text{C}$$

The charge on the second capacitor is

$$Q_2 = Q_{1i} - Q_{1f} = 1.20 \times 10^{-4}\ \text{C} - 3.00 \times 10^{-5}\ \text{C} = 9.00 \times 10^{-5}\ \text{C}$$

and

$$C_2 = \frac{Q_2}{V_2} = \frac{9.00 \times 10^{-5}\ \text{C}}{3.00\ \text{V}} = 3.00 \times 10^{-5}\ \text{F} = 30.0\ \mu\text{F} \quad \Diamond$$

---

51. A 1.00-μF capacitor is first charged by being connected across a 10.0-V battery. It is then disconnected from the battery and connected across an uncharged 2.00-μF capacitor. Determine the charge on each capacitor.

Solution When the capacitor is connected across the battery, it receives a charge of

$$Q = CV = (1.00\ \mu\text{F})(10.0\ \text{V}) = 10.0\ \mu\text{C}$$

When it is connected across the 2.00 μF, the charges move about until Q_1 is on the 1.00 μF and Q_2 is on the 2.00 μF. But, we know that

$$Q_1 + Q_2 = 10.0\ \mu\text{C} \qquad (1)$$

Also, since they are in parallel, the voltage across each must be the same, so $V_1 = V_2$

or $$\frac{Q_1}{C_1} = \frac{Q_2}{C_2} = \frac{Q_1}{1.00\ \mu\text{F}} = \frac{Q_2}{2.00\ \mu\text{F}} \quad \text{yielding} \quad Q_2 = 2Q_1 \qquad (2)$$

Solving (1) and (2) simultaneously, we find $Q_1 = \frac{10.0}{3.00}\ \mu\text{C}$ and $Q_2 = \frac{20.0}{3.00}\ \mu\text{C}$ ◊

---

55. A parallel-plate capacitor has 2.00-cm² plates that are separated by 5.00 mm with air between them. If a 12.0-V battery is connected to this capacitor, how much energy does it store?

Solution $$C = \frac{\varepsilon_0 A}{d} = \frac{(8.85\times10^{-12}\ \text{F/m})(2.00\times10^{-4}\ \text{m}^2)}{5.00\times10^{-3}\ \text{m}} = 3.54\times10^{-13}\ \text{F}$$

and $$W = \frac{1}{2}CV^2 = \frac{1}{2}(3.54\times10^{-13}\ \text{F})(12.0\ \text{V})^2 = 2.55\times10^{-11}\ \text{J} \quad ◊$$

---

61. Determine (a) the capacitance and (b) the maximum voltage that can be applied to a Teflon-filled parallel-plate capacitor having a plate area of 175 cm² and insulation thickness of 0.040 mm.

Solution

(a) $$C = \frac{k\varepsilon_0 A}{d} = \frac{2.10(8.85\times10^{-12}\ \text{F/m})(1.75\times10^{-2}\ \text{m}^2)}{4.00\times10^{-5}\ \text{m}} \quad \text{or} \quad C = 8.13\times10^{-9}\ \text{F} = 8.13\ \text{nF} \quad ◊$$

(b) $$V_{max} = E_{max}d = (60.0\times10^{6}\ \text{V/m})(4.00\times10^{-5}\ \text{m}) = 2.40\ \text{kV} \quad ◊$$

---

65. Find the potential at point P for the rectangular grouping of charges shown in Figure 16.36.

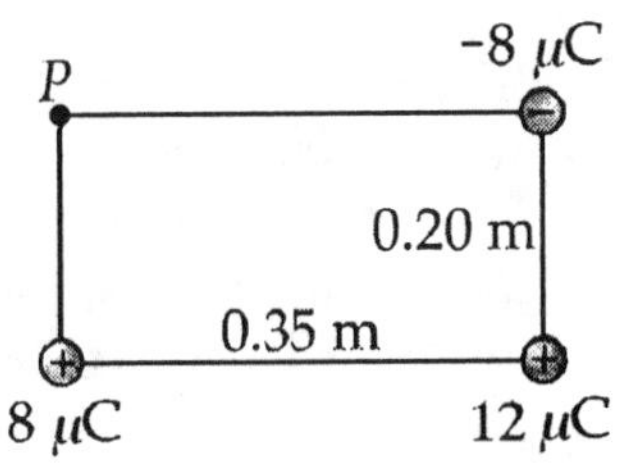

Figure 16.36

Solution The distance across the diagonal (from the Pythagorean theorem) is 0.403 m.

$$V = k\Sigma\frac{q}{r} = (8.99\times10^9\ \text{N}\cdot\text{m}^2/\text{C}^2)\left(\frac{8.00\times10^{-6}\ \text{C}}{0.200\ \text{m}} + \frac{12.0\times10^{-6}\ \text{C}}{0.403\ \text{m}} - \frac{8.00\times10^{-6}\ \text{C}}{0.350\ \text{m}}\right)$$

or $\quad V = 4.22\times10^5\ \text{V}$ ◊

---

73. It is possible to create large potential differences by first charging a group of capacitors connected in parallel and then activating a switching arrangement that in effect disconnects the capacitors from the charging source and reconnects them in series. The group of charged capacitors is then discharged in series. What is the maximum potential difference that can be achieved in this manner, using ten 500-μF capacitors and a charging source of 800 V?

Solution In parallel, the charge on each capacitor is $Q = CV_P$; and for a group of N capacitors, each of equal capacitance, $Q_{\text{total}} = NQ = NCV_p$.

In series, the total capacitance will be

$$\frac{1}{C_{\text{eq}}} = \frac{1}{C_1} + \frac{1}{C_2} + \ldots + \frac{1}{C_N} = N\left(\frac{1}{C}\right); \qquad C_{\text{eq}} = \frac{C}{N}$$

The potential difference in series will be

$$V_s = \frac{Q_{\text{total}}}{C_s} = \frac{NCV_p}{C/N} = N^2V_p = (10)^2(800\ \text{V}) = 80\ \text{kV}$$ ◊

CHAPTER SELF-QUIZ

1. At which location will the magnitude of the electric field between the two parallel plates of a charged capacitor be the greatest?
   a. near the positive plate
   b. near the negative plate
   c. midway between the two plates
   d. electric field is constant throughout space between plates

2. When moving an electrical charge from one point to another in the presence of an electrical field, which quantity depends on the size of the charge that is moved?
   a. the electric field
   b. the work done
   c. the potential difference
   d. the distance moved

3. Two capacitors with charges of 1.00 and 0.50 μF, respectively, are connected in parallel. The system is connected to a 100-V battery. What charge accumulates on the 1.00-μF capacitor?
   a. 150 μC
   b. 100 μC
   c. 50 μC
   d. 33 μC

4. A 200-V battery is connected to a 0.50 μF, parallel plate, air-filled capacitor. Now, the battery is disconnected, with care taken not to discharge the plates. Some Pyrex glass is next inserted between the two plates, completely filling up the space. What is the final potential difference between the plates? (For Pyrex, dielectric constant = 5.60.)
   a. 36 V
   b. 200 V
   c. 560 V
   d. 1120 V

5. A parallel-plate capacitor has dimensions 2.00 cm × 3.00 cm. The plates are separated by a 1.00 mm thickness of paper (dielectric constant κ= 3.70). What is the charge that can be stored on this capacitor, when connected to a 9.00-volt battery? ($\varepsilon_0 = 8.85 \times 10^{-12}\ C^2/N \cdot m^2$)
   a. 19.6×10^{-12} C
   b. 4.75×10^{-12} C
   c. 4.75×10^{-11} C
   d. 1.76×10^{-10} C

6. If $C_1 = 15.0\ \mu F$, $C_2 = 10.0\ \mu F$, $C_3 = 20.0\ \mu F$, and $V_0 = 18.0$ V, determine the energy stored by C_2.
   a. 0.72 mJ
   b. 0.32 mJ
   c. 0.50 mJ
   d. 0.18 mJ

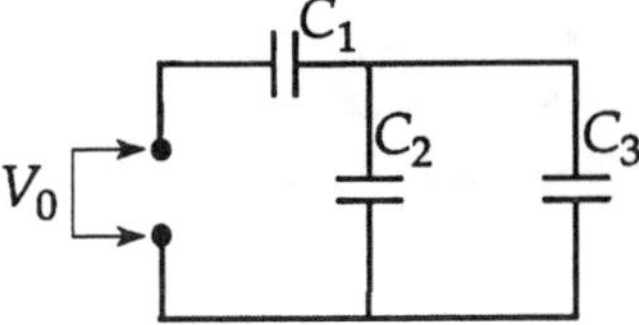

7. Very large capacitors have been considered as a means for storing electrical energy. If we constructed a very large parallel plate capacitor of plate area 1.00 m^2 using Pyrex (k = 5.60) of thickness 2.00 mm as a dielectric, how much electrical energy would it store at a plate voltage of 6000 V? ($\varepsilon_0 = 8.85 \times 10^{-12}\ C^2/N \cdot m$)
   a. 0.45 J
   b. 90 J
   c. 9,000 J
   d. 45, 000 J

8. There is a hollow, conducting, uncharged sphere with a charge $+Q$ inside the sphere. Consider the potential and the electrical field at a point P_1 inside the metal of the sphere. At this point,
   a. only the electrical field is zero
   b. only the electrical potential is zero
   c. both the electrical field and electrical potential are zero
   d. neither the electrical field nor the electrical potential are zero

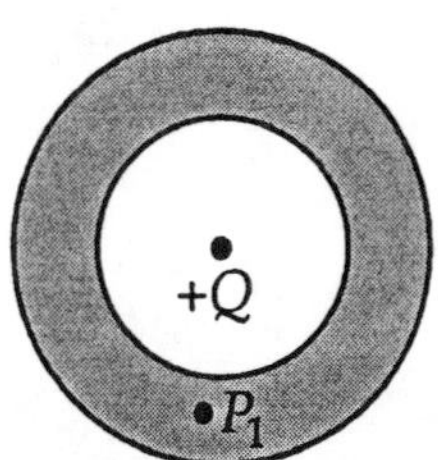

9. What is the equivalent capacitance of the combination shown?
   a. 20.0 μF
   b. 90.0 μF
   c. 22.0 μF
   d. 4.60 μF

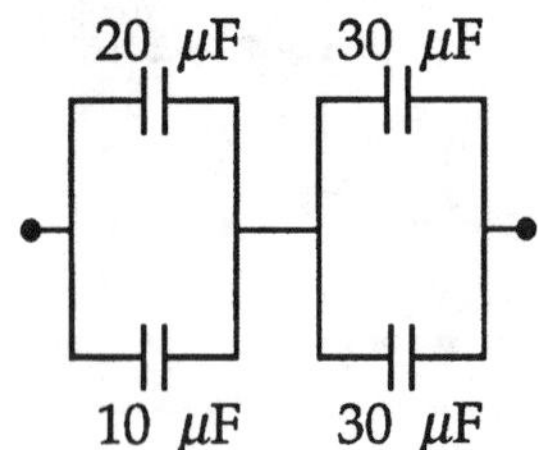

10. Inserting a dielectric material between two charged parallel conducting plates, originally separated by air and disconnected from a battery, will produce what effect on the capacitor?
    a. increase charge
    b. increase voltage
    c. increase capacitance
    d. decrease capacitance

11. A uniform electric field, with a magnitude of 5×10^2 N/C, is directed parallel to the positive x axis. If the potential at x = 5.00 m is 2500 V, what is the potential at x = 2.00 m?
    a. 1000 V
    b. 2000 V
    c. 4000 V
    d. 1500 V

12. If C = 45.0 μF, determine the equivalent capacitance for the combination shown.
    a. 36.0 μF
    b. 32.0 μF
    c. 34.0 μF
    d. 30.0 μF

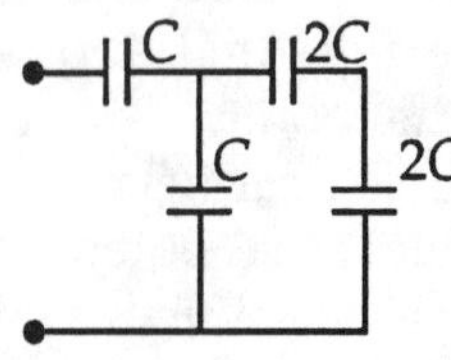

17
Current and Resistance

Chapter 17

CURRENT AND RESISTANCE

Many practical applications and devices are based on the principles of static electricity, but electricity truly became an inseparable part of our daily lives when scientists learned how to control the *flow of electric charges.*

In this chapter we define current and discuss some of the factors that contribute to the resistance to flow of charge in conductors. We also discuss energy transformations in electric circuits. (These topics will be the foundation for additional work with circuits in later chapters.) Finally, an interesting essay on the exciting topic of superconductivity follows this chapter.

NOTES FROM SELECTED CHAPTER SECTIONS

17.1 Electric Current

The direction of conventional current is designated as the direction of motion of positive charge. In an ordinary metal conductor, the direction of current will be *opposite* the *direction of flow of electrons* (which are the charge carriers in this case).

17.2 Current and Drift Speed

In the classical model of electronic conduction in a metal, electrons are treated like molecules in a gas and, in the absence of an electric field, have a *zero average velocity*.

Under the influence of an electric field, the electrons move along a direction opposite the direction of the applied field with a *drift velocity* which is proportional to the average time between collisions with atoms of the metal and inversely proportional to the number of free electrons per unit volume.

17.3 Resistance and Ohm's Law

When a voltage (potential difference), V, is applied across the ends of a metallic conductor, the current in the conductor is found to be proportional to the applied voltage. If the proportionality is exact, we can write $V = IR$, where the proportionality constant R is called the resistance of the conductor. In fact, we define this resistance as the ratio of the voltage across the conductor to the current it carries.

For many materials, including most metals, experiments show that *the resistance is constant over a wide range of applied voltages.* This statement is known as Ohm's law.

Ohm's law is not a fundamental law of nature, but an empirical relationship that is valid only for certain materials. Materials that obey Ohm's law, and hence have a constant resistance over a wide range of voltages, are said to be *ohmic.* Materials that do not obey Ohm's law are *nonohmic.*

Resistance has the SI units volts per ampere, called ohms (Ω). Thus, if a potential difference of 1 V across a conductor produces a current of 1 A, the resistance of the conductor is 1 Ω.

17.4 Resistivity

For ohmic materials, the ratio of the ratio of the current density and the electric field which gives rise to the current is equal to a constant, σ, which is the conductivity of the material. The reciprocal of the conductivity is called the resistivity, ρ. Each ohmic material has a characteristic resistivity which depends only on the properties of the specific material and is a function of temperature.

17.5 Temperature Variation of Resistance

The resistivity, and hence the resistance, of a conductor depends on a number of factors. One of the most important is the temperature of the metal. For most metals, resistivity increases with increasing temperature. For most metals, resistivity increases approximately linearly with temperature over a limited temperature range.

EQUATIONS AND CONCEPTS

Under the influence of an electric field, electric charges will move through conducting gases, liquids, and solids. The electric current is defined as the rate at which charge flows through a cross section of a conductor.

$$I \equiv \frac{\Delta Q}{\Delta t} \tag{17.1}$$

The SI unit of current is the ampere (A).

$$1\,\text{A} = 1\,\text{C/s} \tag{17.2}$$

It is conventional to choose the direction of current to be in the direction of flow of positive charge. In a solid conductor, the current is due to the motion of negatively charged electrons. In such conductors, the direction of the current will be opposite the direction of flow of electrons.

Comment on direction of current.

The velocity of the charge carriers in a conductor is actually an average value of the individual velocities and is called the drift velocity, v_d. The current can be expressed in terms of the drift velocity and the number of mobile charge carriers per unit volume of conductor.

$$I = nqv_d A \qquad (17.3)$$

For many practical applications, the resistance of a given conductor can be more conveniently stated as the ratio of the potential difference across a conductor to the value of the current in the conductor. This equation is usually referred to as Ohm's law.

$$V = IR \qquad (17.5)$$

The resistance of a given conductor made of a homogeneous material of uniform cross section can be expressed in terms of the dimensions of the conductor and an intrinsic property of the material of which the conductor is made called its resistivity. The value of the resistivity of a given material depends on the electronic structure of the material and on the temperature.

$$R = \rho \frac{L}{A} \qquad (17.6)$$

The symbol ρ used for resistivity should not be confused with the same symbol used earlier in the book for density. Very often, a single symbol is used to represent different quantities.

The SI unit of resistance is the ohm. If a potential difference of 1 V across a conductor produces a current of 1 A, the resistance of the conductor is 1 ohm.

$$1\,\Omega = 1\,\text{V/A}$$

The resistivity, and therefore the resistance, of a conductor varies with temperature. Over a limited range of temperatures, this variation is approximately linear. In these equations, the parameter, α, is the temperature coefficient of resistivity, and T_0 is a reference temperature usually taken to be 20.0°C. Some materials (for example, carbon) have a negative temperature of resistivity and in these cases the resistance decreases as the temperature increases.

$$\rho = \rho_0[1+\alpha(T-T_0)] \qquad (17.7)$$

$$R = R_0[1+\alpha(T-T_0)] \qquad (17.8)$$

Joule's law can be used to calculate the power delivered to a resistor or other device carrying a current, I, and having a potential difference, V, between its terminals.

$$P = IV \qquad (17.9)$$

When the device obeys Ohm's law (i.e. a resistor), the power can be expressed in either of two alternative forms.

$$P = I^2R = \frac{V^2}{R} \qquad (17.10)$$

The SI unit of power is the watt (W).

$$1\text{ W} = 1\text{ J/s} = 1\text{ V·A}$$

The kilowatt-hour is the quantity of energy consumed in one hour at a constant use rate (or power) of 1 kW.

$$1\text{ kWh} = 3.60\times 10^6\text{ J} \qquad (17.11)$$

REVIEW CHECKLIST

▷ Define the term, electric current, in terms of rate of charge flow, and its corresponding unit of measure, the ampere. Calculate electron drift velocity, and quantity of charge passing a point in a given time interval in a specified current-carrying conductor.

▷ Determine the resistance of a conductor using Ohm's law. Also, calculate the resistance based on the physical characteristics of a conductor. Distinguish between ohmic and nonohmic conductors.

▷ Make calculations of the variation of resistance with temperature, which involves the concept of the temperature coefficient of resistivity.

▷ Sketch a simple single-loop circuit to illustrate the use of basic circuit element symbols and direction of conventional current.

▷ Use Joule's law to calculate the power dissipated in a resistor.

SOLUTIONS TO SELECTED END-OF-CHAPTER PROBLEMS

5. In a particular television picture tube, the measured beam current is 60.0 μA. How many electrons strike the screen every second?

Solution $I = ne = 60.0 \times 10^{-6}\,\text{A}$

Thus, $$n = \frac{I}{e} = \frac{60.0 \times 10^{-6}\,\text{A}}{1.60 \times 10^{-19}\,\text{C}} = 3.75 \times 10^{14}\ \text{electrons/s} \quad \lozenge$$

---

7. If 3.25×10^{-3} kg of gold is deposited on the negative electrode of an electrolytic cell in a period of 2.78 h, what is the current through the cell in this period? Assume that the gold ions carry one elementary unit of positive charge.

Solution The atomic weight of gold = 197, so the mass of one atom is

$$\text{Mass of atom} = \frac{197\ \text{g}}{6.02 \times 10^{23}\ \text{atoms}} = 3.27 \times 10^{-22}\ \text{g} = 3.27 \times 10^{-25}\ \text{kg}$$

The number of gold atoms deposited is

$$N = \frac{3.25 \times 10^{-3}\ \text{kg}}{3.27 \times 10^{-25}\ \text{kg/atom}} = 9.93 \times 10^{21}\ \text{ions}$$

and the charge deposited is

$$Q = (9.93 \times 10^{21} \text{ ions})(1.60 \times 10^{-19} \text{ C/ion}) = 1.59 \times 10^{3} \text{ C}$$

The elapsed time = 2.78 h = 1.00×10^4 s, so

$$I = \frac{\Delta Q}{\Delta t} = \frac{1.59 \times 10^3 \text{ C}}{1.00 \times 10^4 \text{ s}} = 1.59 \times 10^{-1} \text{ C/s} = 0.159 \text{ A} \quad \Diamond$$

---

9. Calculate the number of free electrons per cubic meter for gold, assuming one free electron per atom. (Density of gold = 19.3×10^3 kg/m^3)

Solution The atomic weight of gold = 197, and its density = 19.3×10^3 kg/m^3. Thus, the mass of 1.00 m^3 = 19.3×10^3 kg. Let N equal the number of gold atoms in 1 m^3.

$$N = 19.3 \times 10^3 \text{ kg}\left(\frac{6.02 \times 10^{26} \text{ atoms / kg} \cdot \text{mol}}{197 \text{ kg / kg} \cdot \text{mol}}\right) = 5.90 \times 10^{28} \text{ atoms / m}^3$$

Let n equal the number of free electrons per m^3.

$$n = \text{(number of atoms/m}^3\text{)(number of free electrons/atom)}$$

or

$$n = (5.90 \times 10^{28} \text{ atoms/m}^3)(1 \text{ electron/atom}) = 5.90 \times 10^{28} \text{ electrons/m}^3 \quad \Diamond$$

---

17. A person notices a mild shock if the current along a path through the thumb and index finger exceeds 80.0 μA. Compare the maximum allowable voltage without shock across the thumb and index finger with a dry-skin resistance of 4.00×10^5 Ω and a wet-skin resistance of 2000 Ω.

Solution $V_{max} = I_{max}R$ and for dry skin,

$$V_{max} = (8.00 \times 10^{-5} \text{ A})(4.00 \times 10^5 \ \Omega) = 32.0 \text{ V} \quad \Diamond$$

For wet skin,

$$V_{max} = (8.00 \times 10^{-5} \text{ A})(2.00 \times 10^3 \ \Omega) = 0.160 \text{ V} \quad \Diamond$$

---

27. A rectangular block of copper has sides of length 10.0 cm, 20.0 cm, and 40.0 cm. If the block is connected to a 6.00-V source across opposite faces of the rectangular block, what are (a) the maximum current and (b) minimum current that can be carried?

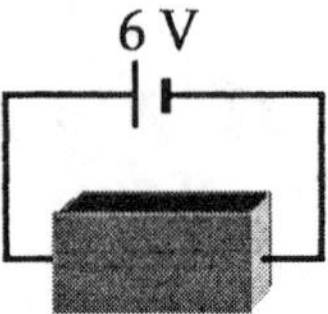

Solution (a) From $R = \rho \frac{L}{A}$, we see that the minimum resistance, and therefore maximum current, occurs for the minimum L and the maximum A. The maximum cross-sectional area is

$$A_{max} = (0.200\text{ m})(0.400\text{ m}) = 8.00 \times 10^{-2}\text{ m}^2$$

with a corresponding value of $L = 0.100\text{ m}$.

so $$R_{min} = (1.78 \times 10^{-8}\ \Omega \cdot \text{m})\left(\frac{0.100\text{ m}}{8.00\ \times 10^{-2}\text{ m}^2}\right) = 2.23 \times 10^{-8}\ \Omega$$

and $$I_{max} = \frac{V}{R_{min}} = \frac{6.00\text{ V}}{2.23 \times 10^{-8}\ \Omega} = 2.69 \times 10^{8}\text{ A} \quad \Diamond$$

(b) For minimum current, use the arrangement that will result in maximum resistance.

$$R_{max} = (1.78 \times 10^{-8}\ \Omega \cdot \text{m})\left(\frac{0.400\text{ m}}{(0.0200\text{ m})(0.0100\text{ m})}\right) = 3.56 \times 10^{-7}\ \Omega$$

and $$I_{min} = \frac{V}{R_{max}} = \frac{6.00\text{ V}}{3.56 \times 10^{-7}\ \Omega} = 1.69 \times 10^{7}\text{ A} \quad \Diamond$$

---

31. At 40.0°C, the resistance of a segment of gold wire is 100 Ω. When the wire is placed in a liquid bath, the resistance decreases to 97.0 Ω. What is the temperature of the bath?

Solution Use $R = R_0(1+\alpha\,\Delta T)$ to calculate the resistance, R_0, at 20.0°C.

$$R_0 = \frac{R}{1+\alpha\,\Delta T} = \frac{100\ \Omega}{1+(3.40\times10^{-3}/°\text{C})(20.0\ °\text{C})} = 93.63\ \Omega$$

Now, solve the general equation for $\Delta T = \frac{1}{\alpha}\left(\frac{R}{R_0}-1\right)$ where R is the resistance at the unknown temperature.

$$\Delta T = \frac{1}{(3.40\times10^{-3}/°\text{C})}\left(\frac{97.0\ \Omega}{93.63\ \Omega}-1\right) = 10.58\ °\text{C}$$

The temperature, T, of the bath is

$$T = 20.0°\text{C} + \Delta T = 30.6°\text{C} \quad \Diamond$$

---

33. A wire 3.00 m long and 0.450 mm² in cross-sectional area has a resistance of 41.0 Ω at 20.0°C. If its resistance increases to 41.4 Ω at 29.0°C, what is the temperature coefficient of resistivity?

Solution $R = R_0(1+\alpha\,\Delta T)$ and $\alpha = \frac{1}{\Delta T}\left(\frac{R}{R_0}-1\right)$

$$\alpha = \frac{1}{9.00\ \text{C}°}\left(\frac{41.4\ \Omega}{41.0\ \Omega}-1\right) = 1.08\times10^{-3}(°\text{C})^{-1} \quad \Diamond$$

---

37. A 100-cm-long copper wire 0.500 cm in radius has a potential difference across it sufficient to produce a current of 3.00 A at 20.0°C. (a) What is the potential difference? (b) If the temperature of the wire is increased to 200°C, what potential difference is now required to produce a current of 3.00 A?

Solution (a) The resistance of the wire is

$$R = \frac{\rho L}{A} = \frac{(1.70\times10^{-8}\ \Omega\cdot\text{m})(1.00\ \text{m})}{\pi(0.500\times10^{-2}\ \text{m})^2} = 2.165\times10^{-4}\ \Omega$$

Then, $V = IR = (3.00\ \text{A})(2.165\times10^{-4}\ \Omega) = 6.49\times10^{-4}\ \text{V} = 0.649\ \text{mV} \quad \Diamond$

(b) At T = 200°C, the resistance is

$$R = R_0[1+\alpha(\Delta T)] = 2.165\times10^{-4}\ \Omega[1+(3.90\times10^{-3}(°\text{C})^{-1})(180°\text{C})]$$

or $$R = 3.69\times10^{-4}\ \Omega$$

Thus, $$V = IR = (3.00\text{ A})(3.69\times10^{-4}\ \Omega) = 1.11\times10^{-3}\text{ V} = 1.11\text{ mV}\quad\Diamond$$

---

43. The tungsten heating element in a 1500-W heater is 3.00 m long, and the resistor is to be connected to a 120-V source. What is the cross-sectional area of the wire?

Solution Combine Ohm's law, $V = IR$, and Joule's law, $P = VI$, to get $P = \frac{V^2}{R}$.

So, $R = \frac{V^2}{P} = \frac{(120\text{ V})^2}{1500\text{ W}} = 9.60\ \Omega$ and from $R = \rho\frac{L}{A}$,

$$A = \frac{\rho L}{R} = \frac{(5.60\times10^{-8}\ \Omega\cdot\text{m})(3.00\text{ m})}{9.60\ \Omega} = 1.75\times10^{-8}\text{ m}^2\quad\Diamond$$

---

45. Suppose that a voltage surge produces 140 V for a moment. By what percentage will the output of a 120-V, 100-W lightbulb increase, assuming its resistance does not change?

Solution

$$\%\text{ change} = \left(\frac{P_f - P_i}{P_i}\right)100\% = \left(\frac{\frac{V_f^2}{R} - \frac{V_i^2}{R}}{\frac{V_i^2}{R}}\right)100\% = \left(\frac{V^2{}_f - V^2{}_i}{V^2{}_i}\right)100\%$$

$$\%\text{ change} = \left(\frac{(140\text{ V})^2 - (120\text{ V})^2}{(120\text{ V})^2}\right)100\% = 36.1\%\quad\Diamond$$

---

49. In a hydroelectric installation, a turbine delivers 2000 hp to a generator, which in turn converts 90% of the mechanical energy to electrical energy. Under these conditions, what current does the generator deliver at a potential difference of 3000 V?

Solution At 90% conversion efficiency, the power appearing in electrical form is

$$P = (0.90)(2000\text{ hp})(746\text{ W/hp}) = 1.34\times10^6\text{ W}$$

Thus, $$I = \frac{P}{V} = \frac{1.34\times10^6\text{ W}}{3000\text{ V}} = 448\text{ A} \quad \Diamond$$

---

51. How much does it cost to watch a complete 21-hour-long World Series on a 90.0-W black-and-white television set? Assume that electricity costs $0.070/kWh.

Solution $$W = Pt = (90.0\text{ W})(21.0\text{ h}) = 1890\text{ Wh} = 1.89\text{ kWh}$$

Thus, the cost is Cost = (7.00 cents/kWh) (1.89 kWh) = 13.2 cents ◊

---

55. An electric resistance heater is to deliver 1500 kcal/h to a room using 110-V (effective dc) electricity. If fuses come in 10.0-A, 20.0-A, and 30.0-A sizes, what is the smallest fuse that can safely be used in the heater circuit?

Solution $P = 1500\text{ kcal/h} = 1744\text{ W}$

and $$I = \frac{P}{V} = \frac{1744\text{ W}}{110\text{ V}} = 15.9\text{ A}$$ (A 20-Ampere fuse is required.) ◊

---

59. Storage batteries are often rated in terms of the amounts of charge they can deliver. How much charge can a 90.0-amp-hour battery deliver?

Solution $$Q = IA = (90.0\text{ A})(1.00\text{ h}) = (90.0\text{ C/s})(3600\text{ s}) = 3.24\times10^5\text{ C} \quad \Diamond$$

---

63. A particular wire has a resistivity of $3.00 \times 10^{-8}\ \Omega \cdot \text{m}$ and a cross-sectional area of $4.00 \times 10^{-6}\ \text{m}^2$. A length of this wire is to be used as a resistor that will develop 48.0 W of power when connected across a 20.0-V battery. What length of wire is required?

Solution The current drawn by the wire is

$$I = \frac{P}{V} = \frac{48.0\ \text{W}}{20.0\ \text{V}} = 2.40\ \text{A}$$

and the resistance of the wire is

$$R = \frac{V}{I} = \frac{20.0\ \text{V}}{2.40\ \text{A}} = 8.33\ \Omega$$

Thus, from $R = \frac{\rho L}{A}$, we have

$$L = \frac{RA}{\rho} = \frac{(8.33\ \Omega)4.00 \times 10^{-6}\ \text{m}^2}{3.00 \times 10^{-8}\ \Omega \cdot \text{m}} = 1110\ \text{m} \quad \Diamond$$

---

65. A length of metal wire has a radius of 5.00×10^{-3} m and a resistance of $0.100\ \Omega$. When the potential difference across the wire is 15.0 V, the electron drift speed is found to be 3.17×10^{-4} m/s. Based on these data, calculate the density of free electrons in the wire.

Solution The area of the wire is $A = \pi r^2 = 7.85 \times 10^{-5}\ \text{m}^2$

The current in the wire is $I = \frac{V}{R} = \frac{15.0\ \text{V}}{0.100\ \Omega} = 150\ \text{A}$

We now can find the density of free electrons from $I = nqAv_d$.

$$n = \frac{I}{qv_dA} = \frac{150\ \text{A}}{(1.60 \times 10^{-19}\ \text{C})(3.17 \times 10^{-4}\ \text{m/s})(7.85 \times 10^{-5}\ \text{m}^2)}$$

or $$n = 3.77 \times 10^{28}\ \text{electrons/m}^3 \quad \Diamond$$

---

67. A small sphere that carries a charge of 8.00 nC is whirled in a circle at the end of an insulating string. The rotation frequency is 100π rad/s. What average current does this rotating charge represent?

Solution $I = \dfrac{\Delta Q}{\Delta t}$ and the sphere of charge, q, completes the circular path in a time equal to the period. Therefore, $\Delta t = T = \dfrac{1}{f}$ and from $\omega = 2\pi f$, $f = \dfrac{\omega}{2\pi}$ so

$$I = \frac{q}{1/f} = qf$$

$$I = (8.00 \times 10^{-9}\ \text{C})\left(\frac{w}{2\pi}\right)$$

$$I = (8.00 \times 10^{-9}\ \text{C})\left(\frac{100\pi/\text{s}}{2\pi/\text{s}}\right) = 4.00 \times 10^{-7}\ \text{A} \quad \text{or} \quad I = 0.400\ \mu\text{A} \quad \Diamond$$

---

69. (a) A 115-g mass of aluminum is formed into a right circular cylinder, shaped so that its diameter equals it height. Calculate the resistance between the top and bottom faces of the cylinder at 20.0°C. (b) Calculate the resistance between opposite faces if the same mass of aluminum is formed into the cube.

Solution (a) $R = \dfrac{\rho L}{A}$ and since the radius equals $\dfrac{L}{2}$, we have $R = \dfrac{\rho L}{\pi r^2} = \dfrac{4\rho}{\pi L}$.

Density, $D = \dfrac{m}{V} = \dfrac{m}{\pi\left(\dfrac{L}{2}\right)^2 L} = \dfrac{4m}{\pi L^3}$, so

$$L = \left(\frac{4m}{\pi D}\right)^{1/3} = \left(\frac{4(0.115\ \text{kg})}{\pi(2.70 \times 10^3\ \text{kg/m}^3)}\right)^{1/3} = 3.79 \times 10^{-2}\ \text{m}$$

Using this value for L and the resistivity of Al in the equation for R, we get

$$R = \frac{4(2.80 \times 10^{-8}\ \Omega \cdot \text{m})}{\pi(3.79 \times 10^{-2}\ \text{m})} = 9.40 \times 10^{-7}\ \Omega \quad \Diamond$$

(b) For the cube, $D = \frac{m}{V} = \frac{m}{L^3}$ so $L = \left(\frac{m}{D}\right)^{1/3} = 3.49 \times 10^{-2}\ \text{m}$ and

$$R = \frac{\rho L}{A} = \frac{\rho L}{L^2} = \frac{\rho}{L} = \frac{2.80 \times 10^{-8}\ \Omega \cdot \text{m}}{3.49 \times 10^{-2}\ \text{m}} = 8.02 \times 10^{-7}\ \Omega \quad \Diamond$$

---

71. The current in a conductor varies in time as shown in Figure 17.10. (a) How many coulombs of charge pass through a cross section of the conductor in the interval $t = 0$ to $t = 5.00$ s? (b) What constant current would transport the same total charge during the 5.00-s interval as does the actual current?

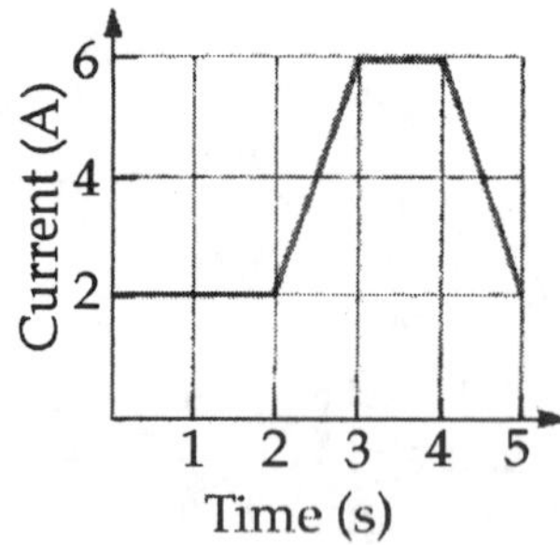

Figure 17.10

Solution (a) From the definition of current, we have $\Delta Q = I(\Delta t)$. From this, we see that the charge can be found by finding the total area under a curve of I versus t. Thus,

$$Q = (\text{area of rectangle } A_1) + 2(\text{area of triangle } A_2) + (\text{rectangular area } A_3)$$

$$Q = (2.00\ \text{A})(5.00\ \text{s}) + 2\left(\frac{(1.00\ \text{s})(4.00\ \text{A})}{2}\right) + (1.00\ \text{s})(4.00\ \text{A}) = 10.0\ \text{C} + 4.00\ \text{C} + 4.00\ \text{C} = 18.0\ \text{C} \quad \Diamond$$

(b) The constant current would be $I = \frac{\Delta Q}{\Delta t} = \frac{18.0\ \text{C}}{5.00\ \text{s}} = 3.60\ \text{A} \quad \Diamond$

---

73. A 50.0-g sample of a conducting material is all that is available. The resistivity of the material is measured to be $11.0 \times 10^{-8}\ \Omega \cdot \text{m}$, and the density is 7.86 g/cm^3. The material is to be shaped into a wire that has a total resistance of 1.50 Ω. (a) What length is required? (b) What must be the diameter of the wire?

Solution (a) The volume of the wire is

$$\text{Vol} = \frac{50.0\ \text{g}}{7.86\ \text{g/cm}^3} = 6.36\ \text{cm}^3$$

We also know that Volume $= AL$ or

$$A = \frac{\text{Vol}}{L} = \frac{6.36\ \text{cm}^3}{L} = \frac{6.36 \times 10^{-6}\ \text{m}^3}{L} \qquad (1)$$

From the definition of resistance, we have

$$R = \frac{\rho L}{A} \quad \text{which becomes} \quad 1.50\ \Omega = (11.0 \times 10^{-8}\ \Omega \cdot \text{m})\frac{L}{A} \quad \text{or}$$

$$\frac{L}{A} = 1.36 \times 10^{7}\ \text{m}^{-1} \qquad (2)$$

Solve (1) and (2) simultaneously to find

$$L = 9.31\ \text{m} \quad \text{and} \quad A = 6.83 \times 10^{-7}\ \text{m}^2 \quad \Diamond$$

(b) From the area of the wire, we can find its diameter to be

$$d = 9.33 \times 10^{-4}\ \text{m} \quad \Diamond$$

CHAPTER SELF-QUIZ

1. If a 6.00-V battery, with negligible internal resistance, and a 12.0-ohm resistor are connected in series, what is the amount of electrical energy transformed to heat per coul of charge that flows through the circuit?
   a. 0.50 J
   b. 3.00 J
   c. 6.00 J
   d. 72.0 J

2. When the potential difference between the ends of a conductor is tripled, the average drift velocity of mobile charge carriers is changed by what factor?
   a. 0.33
   b. 1.00
   c. 3.00
   d. 9.00

3. The heating coil of a hot water heater has a resistance of 20.0 ohm and operates at 210 V. What is its power rating?
   a. 8.80×10^5 W
   b. 2205 W
   c. 10.5 W
   d. 2940 W

4. If a certain resistor obeys Ohm's law, its resistance will change
   a. as the voltage across the resistor changes
   b. as the current through the resistor changes
   c. as the energy given off by the electrons in their collisions changes
   d. none of the above, since resistance is a constant for a given resistor

5. A certain material is in a room at 27.0°C. If the absolute temperature (°K) of the material is doubled, its resistance also doubles. (Water freezes at 273.0°K.) What is the value for α, the temperature coefficient of resistivity?
   a. 1/°C
   b. 2/°C
   c. 0.0033/°C
   d. 0.038/°C

6. A metal wire has a radius of 0.30 mm, is 1.00 m long, and has a resistance of 20.0 ohms. What must be the length of a second wire made of the same type of metal if its radius is 0.10 mm and also has a resistance of 20.0 ohms?
   a. 0.11 m
   b. 0.33 m
   c. 3.00 m
   d. 9.00 m

7. A platinum wire is utilized to determine the melting point of indium. The resistance of the platinum wire is 2.00 Ω at 20.0°C and increases to 3.072 Ω as the indium starts to melt. $\alpha_{platinum} = 3.92 \times 10^{-3}/°C$. What is the melting temperature of indium?
   a. 137°C
   b. 157°C
   c. 351°C
   d. 731°C

8. Two wires made of the same metal and of the same length are connected across the same voltage. If one wire has three times the radius of the other, what is the ratio of currents, large wire to small wire?
   a. 1/9
   b. 1/3
   c. 3/1
   d. 9/1

9. A copper cable is to be designed to carry a current of 300 A with a power loss of only 2 watts per meter. What is the required radius of the copper cable? (The resistivity of copper is $1.70 \times 10^{-8}\ \Omega \cdot \text{m}$.)
   a. 0.80 cm
   b. 1.60 cm
   c. 3.20 cm
   d. 4.00 cm

10. A color television set draws about 2.50 A when connected to 120 V. What is the cost (with electrical energy at 6 cents/kWh) of running the color TV for 8 hours?
    a. 1.4 cents
    b. 3.0 cents
    c. 14.4 cents
    d. 30.0 cents

11. An electric car is designed to run off a bank of 12.0-V batteries with total energy storage of 2.00×10^{7} J. If the electric motor draws 8000 W in moving the car at a steady speed of 20 m/s, how far will the car go before it is "out of juice"?
    a. 25 km
    b. 50 km
    c. 100 km
    d. 150 km

12. Suppose you have a 10-gram copper cylinder. It is to be drawn down into a wire that will have a resistance of 1.00 Ω. What is the length of such a wire? (The resistivity of copper is $1.70 \times 10^{-8}\ \Omega \cdot \text{m}$.) (The density of copper is $8.90 \times 10^{3}\ \text{kg/m}^{3}$.)
    a. 2.00 m
    b. 4.00 m
    c. 8.00 m
    d. 16.0 m

18
Direct Current Circuits

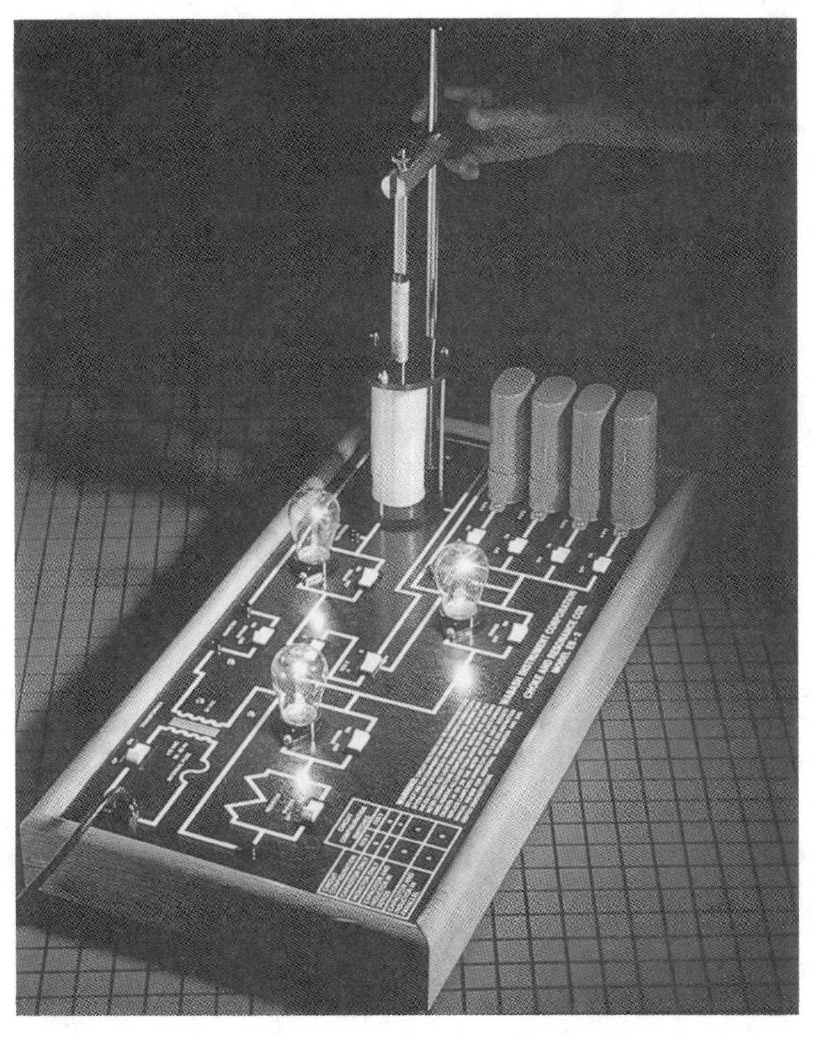

DIRECT CURRENT CIRCUITS

This chapter analyzes some simple circuits whose elements include batteries, resistors, and capacitors in varied combinations. Such analysis is simplified by the use of two rules known as Kirchhoff's rules, which follow from the principles of conservation of energy and the law of conservation of charge. Most of the circuits are assumed to be in *steady state*, which means that the currents are constant in magnitude and direction. We close the chapter with a discussion of circuits containing resistors and capacitors, in which current varies with time.

NOTES FROM SELECTED CHAPTER SECTIONS

18.2 Resistors in Series
18.3 Resistors in Parallel

The *current* must be the same for each of a group of resistors connected in *series*.

The *potential difference* must be the same across each of a group of resistors in *parallel*.

18.4 Kirchhoff's Rules and Simple DC Circuits

1. The sum of the currents entering any junction must equal the sum of the currents leaving that junction. (A junction is any point in the circuit where the current can split.)

2. The algebraic sum of the changes in potential around any closed circuit loop must be *zero*.

The first rule is a statement of *conservation of charge*; the second rule follows from the *conservation of energy*.

18.5 RC Circuits

Consider an uncharged capacitor in series with a resistor, a battery, and a switch. In the charging process, charges are transferred from one plate of the capacitor to the other moving along a path *through the resistor, battery, and switch*. The charges *do not move across the gap between the plates of the capacitor*.

The battery does work on the charges to increase their electrostatic potential energy as they move from one plate to the other.

EQUATIONS AND CONCEPTS

When a battery is providing a current to an external circuit, the *terminal voltage* of the battery will be less than the emf due to *internal resistance* of the battery.

(18.1) $V = \mathcal{E} - Ir$

The total or equivalent resistance of a series combination of resistors is equal to the sum of the resistances of the individual resistors.

(18.4) $R_{eq} = R_1 + R_2 + R_3 + \ldots$

(series combination)

A group of resistors connected in parallel has an equivalent resistance which is less than the smallest individual value of resistance in the group.

(18.6) $\dfrac{1}{R_{eq}} = \dfrac{1}{R_1} + \dfrac{1}{R_2} + \dfrac{1}{R_3} + \ldots$

Resistors in series are connected so that they have only one common circuit point per pair; there is a common current through each resistor in the group.

Comment on series and parallel connection of resistors.

Resistors in parallel are connected so that each resistor in the group has two circuit points in common with each of the other resistors; there is a common potential difference across each resistor in the group.

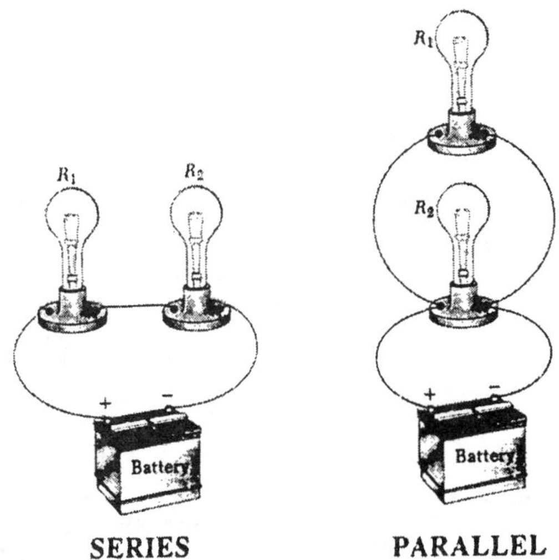

Many circuits which contain several resistors can be reduced to an equivalent single-loop circuit by successive step-by-step combinations of groups of resistors in series and parallel. However, in the most general case, such a reduction is not possible and you must solve a true multiloop circuit by use of Kirchhoff's rules. Review the procedure suggested in the previous section to apply Kirchhoff's rules.

When a battery is used to charge a capacitor in series with a resistor, a quantity τ, called the time constant of the circuit, is used to describe the manner in which the charge on the capacitor varies with time. The charge on the capacitor increases from zero to 63 percent of its maximum value in a time interval equal to one time constant. Also, during one time constant, the charging current decreases from its initial maximum value of $I_0 = \frac{\varepsilon}{R}$ to 37 percent of I_0.

(18.7) $$q = Q(1 - e^{-t/RC})$$

(18.8) $$\tau = RC$$

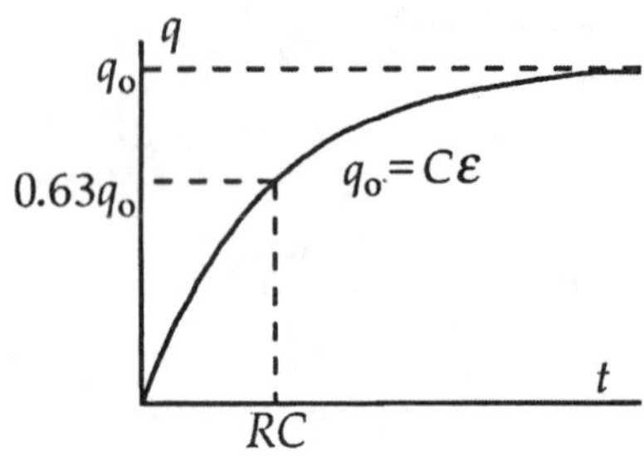

SUGGESTIONS, SKILLS, AND STRATEGIES

A problem-solving strategy for resistors:

1. Be careful with your choice of units. To calculate the resistance of a device in ohms, make sure that distances are in meters and use the SI value of ρ.

2. When two or more unequal resistors are connected in *series,* they carry the same current, but the potential differences across them are not the same. The resistors add directly to give the equivalent resistance of the series combination.

3. When two or more unequal resistors are connected in *parallel,* the potential differences across them are the same. Since the current is inversely proportional to the resistance, the currents through them are not the same. The equivalent resistance of a parallel combination of resistors is found through reciprocal addition, and the equivalent resistor is always *less* than the smallest individual resistor.

4. A complicated circuit consisting of resistors can often be reduced to a simple circuit containing only one resistor. To do so, examine the initial circuit and replace any resistors in series or any in parallel using the procedures outlined in Steps 2 and 3. Sketch the new circuit after these changes have been made. Examine the new circuit and replace any series or parallel combinations. Continue this process until a single equivalent resistance is found.

5. If the current through, or the potential difference across, a resistor in the complicated circuit is to be identified, start with the final circuit found in Step 4 and gradually work your way back through the circuits, using $V = IR$ and the rules of Steps 2 and 3.

A strategy for using Kirchhoff's rules:

1. First, draw the circuit diagram and assign labels and symbols to all the known and unknown quantities. You must assign *directions* to the currents in each part of the circuit. Do not be alarmed if you guess the direction of a current incorrectly; the resulting value will be negative, but *its magnitude will be correct.* Although the assignment of current directions is arbitrary, you must stick with your guess throughout as you apply Kirchhoff's rules.

2. Apply the junction rule to any junction in the circuit. The junction rule may be applied as many times as a new current (one not used in a previous application) appears in the resulting equation. In general, the number of times the junction rule can be used is one fewer than the number of junction points in the circuit.

3. Now apply Kirchhoff's loop rule to as many loops in the circuit as are needed to solve for the unknowns. Remember you must have as many equations as there are unknowns (I's, R's, and $\mathcal{E}$'s). In order to apply this rule, you must correctly identify the change in potential as you cross each element in traversing the closed loop. Watch out for signs!

   Convenient "rules of thumb" which you may use to determine the increase or decrease in potential as you cross a resistor or seat of emf in traversing a circuit loop are illustrated in the following figure. Notice that the potential *decreases* (changes by $-IR$) when the resistor is traversed *in the direction of the current.* There is an *increase* in potential of $+IR$ if the direction of travel is *opposite* the direction of current. If a seat of emf is traversed *in* the direction of the emf (from − to + on the battery), the potential *increases* by $\mathcal{E}$. If the direction of travel is from + to −, the potential *decreases* by $\mathcal{E}$ (changes by $-\mathcal{E}$).

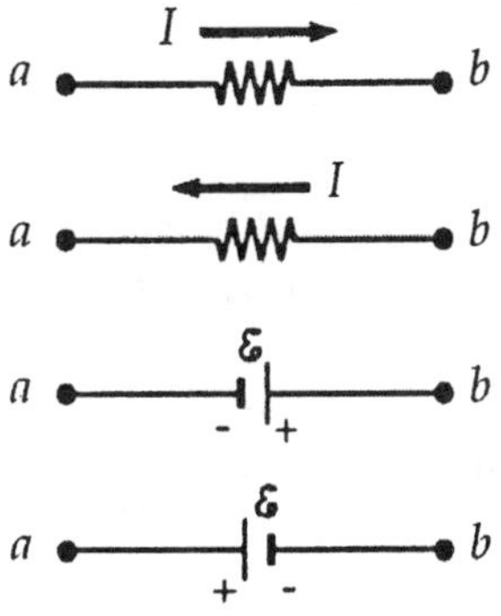

$$\Delta V = V_b - V_a = -IR$$

$$\Delta V = V_b - V_a = IR$$

$$\Delta V = V_b - V_a = \mathcal{E}$$

$$\Delta V = V_b - V_a = -\mathcal{E}$$

4. Finally, you must solve the equations simultaneously for the unknown quantities. Be careful in your algebraic steps, and check your numerical answers for consistency.

As an illustration of the use of Kirchhoff's rules, consider a three-loop circuit which has the *general form* shown in the following figure on the left. In this illustration, the actual circuit elements, R's and $\mathcal{E}$'s are not shown but assumed known. There are six possible different values of I in the circuit; therefore you will need six independent equations to solve for the six values of I. There are four junction points in the circuit (at points $a, d, f,$ and h). The first rule applied at *any three* of these points will yield three equations. The circuit can be thought of as a group of three "blocks" as shown in the following figure on the right. Kirchhoff's second law, when applied to each of these loops (*abcda*, *ahfga*, and *defhd*), will yield three additional equations. You can then solve the total of six equations simultaneously for the six values of I_1, I_2, I_3, I_4, I_5, and I_6. You can, of course, expect that the sum of the changes in potential difference around *any other closed loop* in the circuit will be zero (for example, *abcdefga* or *ahfedcba*); however the equations found by applying Kirchhoff's second rule to these additional loops *will not be independent* of the six equations found previously.

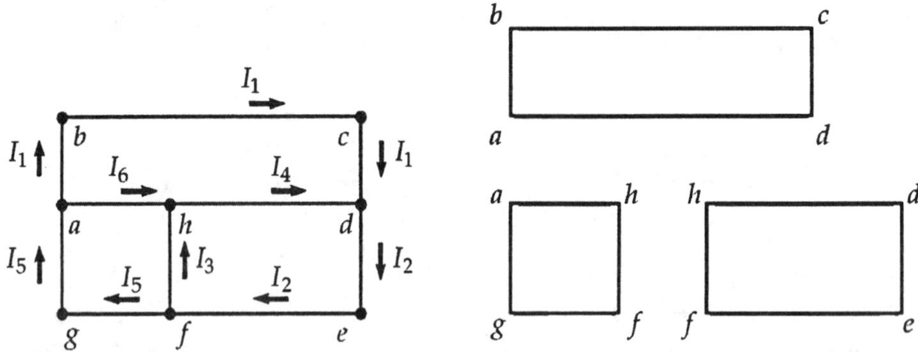

REVIEW CHECKLIST

▷ Calculate the current in a single-loop circuit and the potential difference between any two points in the circuit; and calculate the equivalent resistance of a group of resistors in parallel, series, or series-parallel combination.

▷ Use Ohm's law to calculate the current in a circuit and the potential difference between any two points in a circuit which can be reduced to an equivalent single-loop circuit.

- ▷ Use Joule's law to calculate the power dissipated by any resistor or group of resistors in a circuit.
- ▷ Apply Kirchhoff's rules to solve multiloop circuits; that is, find the currents at any point and the potential difference between any two points.
- ▷ Describe in qualitative terms the manner in which charge accumulates on a capacitor or current flow changes through a resistor with time in a series circuit with battery, capacitor, resistor, and switch.

SOLUTIONS TO SELECTED END-OF-CHAPTER PROBLEMS

3. The resistors of Problem 1 are connected in parallel across a 24-V battery. Find (a) the equivalent resistance and (b) the current in each resistor.

Solution

(a) The equivalent resistance for a parallel combination of resistors is given by Equation 18.6:

$$\frac{1}{R_{eq}} = \frac{1}{R_1} + \frac{1}{R_2} + \frac{1}{R_3} + \ldots$$

In this case $\frac{1}{R_{eq}} = \frac{1}{4\,\Omega} + \frac{1}{8\,\Omega} + \frac{1}{12\,\Omega} = \frac{11}{24\,\Omega}$ so $R_{eq} = \frac{24}{11}\,\Omega$ ◊

(b) In parallel, the potential difference across each element is the same and is equal to the battery terminal potential difference. Therefore,

$$I_4 = \frac{V}{R_4} = \frac{24\text{ V}}{4\,\Omega} = 6\text{ A} \quad ◊$$

$$I_8 = \frac{V}{R_8} = \frac{24\text{ V}}{8\,\Omega} = 3\text{ A} \quad ◊$$

$$I_{12} = \frac{V}{R_{12}} = \frac{24\text{ V}}{12\,\Omega} = 2\text{ A} \quad ◊$$

11. Find the equivalent resistance of the circuit in Figure 18.22.

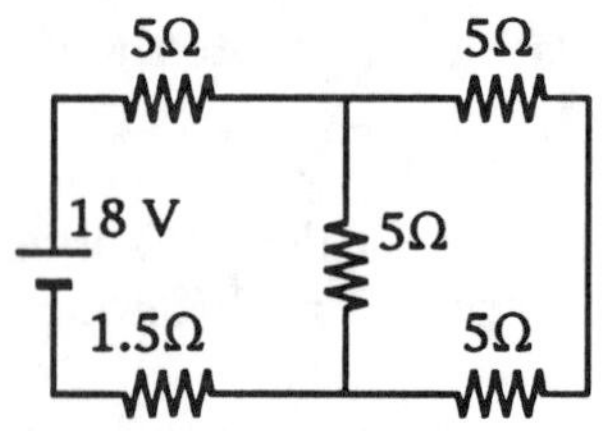

Figure 18.22

Solution The rules for combining resistors in series and parallel are used to reduce the circuit to an equivalent resistor according to the stages indicated below.

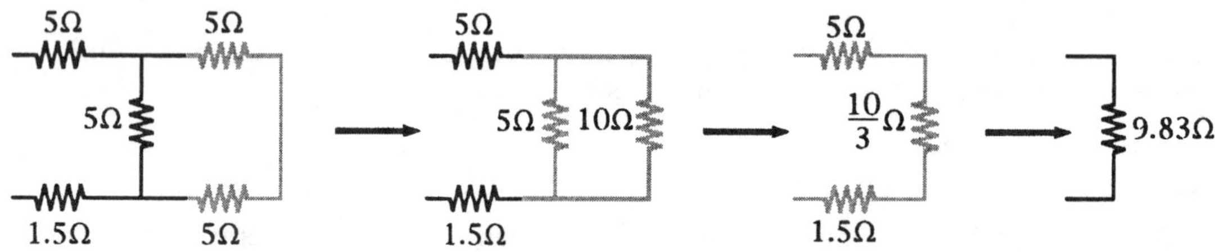

The resultant value for the equivalent resistance is 9.83 Ω. ◊

---

15. Find the current in the 12-Ω resistor in Figure 18.25.

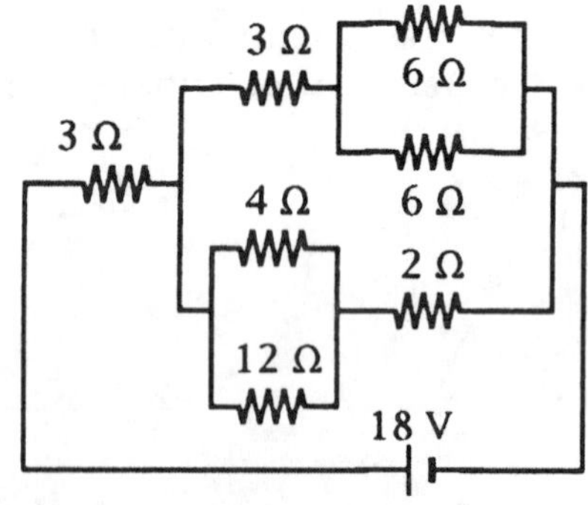

Figure 18.25

Solution The resistors in the circuit can be combined in the stages shown on the following page to yield an equivalent resistance of $\frac{63}{11}$ Ω.

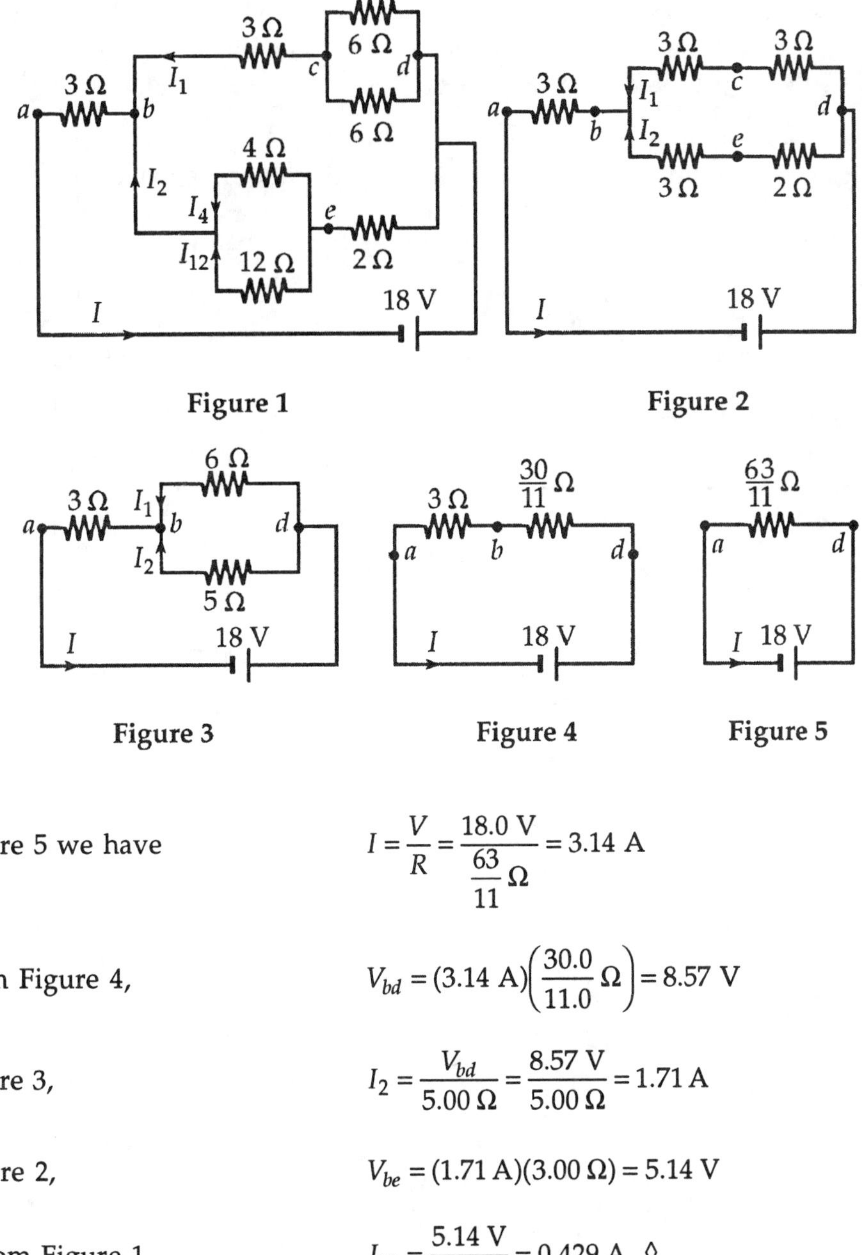

Figure 1

Figure 2

Figure 3

Figure 4

Figure 5

From Figure 5 we have $$I = \frac{V}{R} = \frac{18.0\text{ V}}{\frac{63}{11}\,\Omega} = 3.14\text{ A}$$

Then, from Figure 4, $$V_{bd} = (3.14\text{ A})\left(\frac{30.0}{11.0}\,\Omega\right) = 8.57\text{ V}$$

From Figure 3, $$I_2 = \frac{V_{bd}}{5.00\,\Omega} = \frac{8.57\text{ V}}{5.00\,\Omega} = 1.71\text{ A}$$

From Figure 2, $$V_{be} = (1.71\text{ A})(3.00\,\Omega) = 5.14\text{ V}$$

Finally, from Figure 1, $$I_{12} = \frac{5.14\text{ V}}{12.0\,\Omega} = 0.429\text{ A} \quad \Diamond$$

21. An unmarked battery has an unknown internal resistance. If the battery is connected to a fresh 5.60-V battery (negligible internal resistance) positive to positive and negative to negative, the current through the circuit is 10.0 mA. If the polarity of the unknown battery is reversed, the current increases to 25.0 mA. Determine the source voltage and internal resistance of the unknown battery. Assume that in each case the direction of the current is negative to positive in the 5.60-V battery. (See the note from Problem 8 below.)

Solution The two arrangements are shown in the figures below.

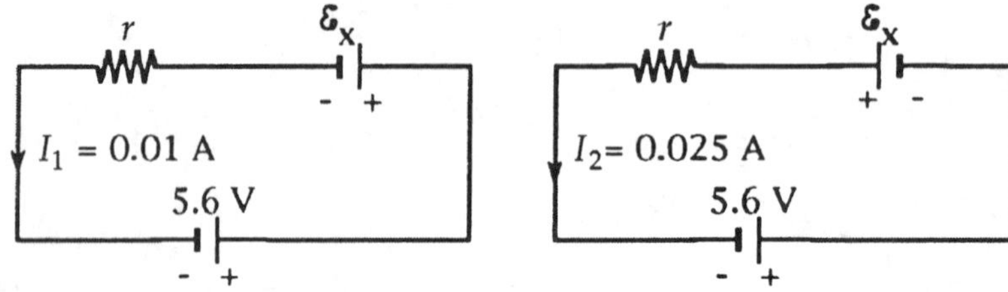

Using the loop rule for the case:

on the left, $-\mathcal{E}_x - 0.010r + 5.60 = 0 \text{ V}$

on the right, $+\mathcal{E}_x - 0.025r + 5.60 = 0 \text{ V}$

Add these two equations to get $-0.035r = 11.2 \text{ V}$

Then:

$r = 320\ \Omega$ ◊;

and

$\mathcal{E}_x = 2.40 \text{ V}$ ◊

(*Note :* A battery has a resistance to the flow of current which depends on the condition of the chemicals in it and increases as the battery ages. This so-called internal resistance may be represented in a simple circuit diagram as a resistor in series with the battery.)

29. What is the emf, $\mathcal{E}_1$, of the battery in the circuit of Figure 18.36?

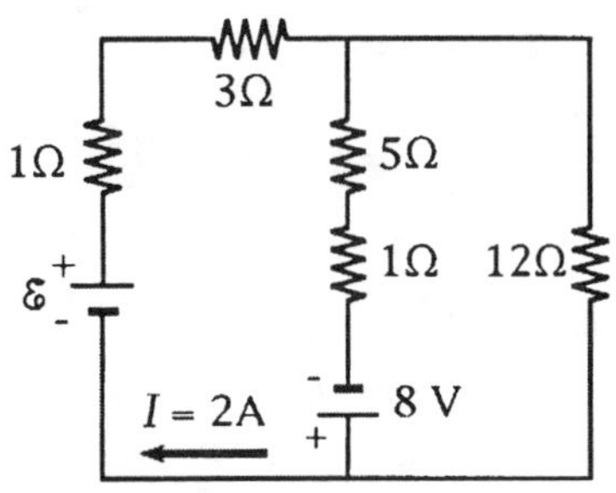

Figure 18.36

Solution

Apply the junction rule at point *a*. $I_1 + I_2 = 2.00$ or $I_2 = 2 - I_1$. Apply the loop rule to the right-hand loop moving clockwise from *a*.

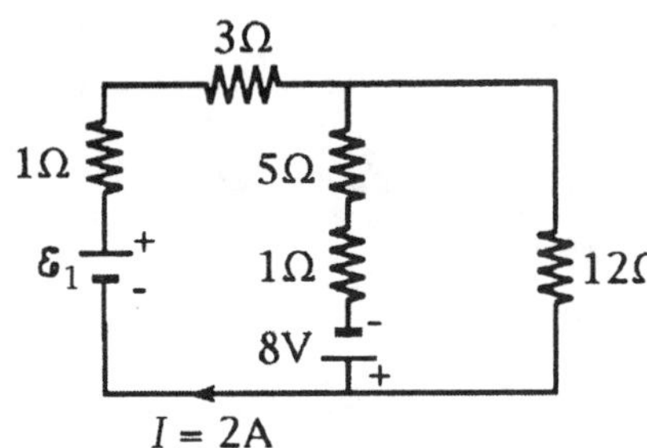

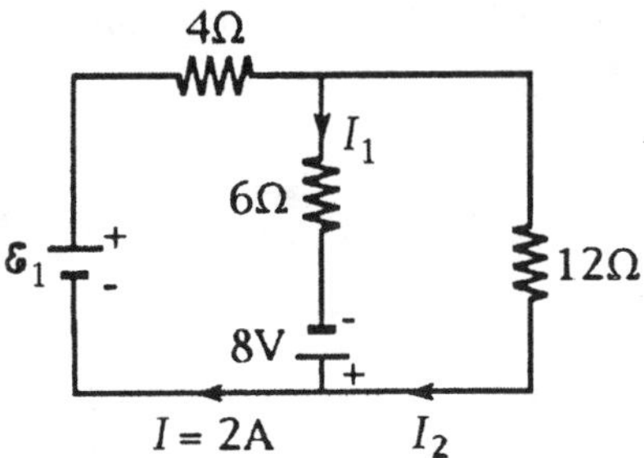

$$-8.00\text{ V} + 6.00\, I_1 - 12.0(2.00 - I_1) = 0.$$ From which, $I_1 = 1.78$ A.

Finally, use the loop rule on the left-hand loop moving clockwise from *a*.

$$+\mathcal{E}_1 - (4.00\ \Omega)(2.00\text{ A}) - (6.00\ \Omega)(1.78\text{ A}) + 8.00\text{ V} = 0$$

gives $\mathcal{E}_1$ = 10.67 V with the polarity indicated. ◊

---

33. Consider an *RC* circuit in which the capacitor is being charged by a battery connected in the circuit. In two time constants, what percentage of the *final* charge is on the capacitor?

Solution Use Equation 18.7, $q = Q(1 - e^{-t/RC})$, where Q is the final charge on the capacitor and $RC = \tau$ is the time constant. When $t = 2\tau$, the equation becomes $\frac{q}{Q} = (1 - e^{-2}) = 0.865$. So in two constants, the charge on the capacitor will be 86.5% of the *final value.*

It is useful to remember that in a time interval equal to one time constant, the capacitor *gains* an amount of charge equal to 63% of the difference between the charge at the beginning of the time interval and the eventual final charge. The accumulation of charge on the capacitor as a function of time is shown in the figure below.

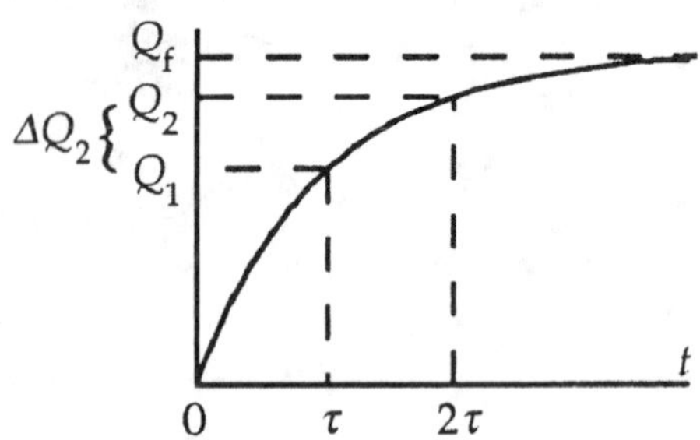

---

37. An electric heater is rated at 1300 W, a toaster is rated at 1000 W, and an electric grill is rated at 1500 W. The three appliances are connected in parallel to a common 120-V circuit. (a) How much current does each appliance draw? (b) Is a 30-A circuit sufficient in this situation? Explain.

Solution

(a) The current drawn by each appliance is found from Joule's law, $P = VI$ or $I = \frac{P}{V}$.

For the heater, $I = \frac{1300\text{ W}}{120\text{ V}} = 10.8\text{ A}$ ◊

For the toaster, $I = \frac{1000\text{ W}}{120\text{ V}} = 8.33\text{ A}$ ◊

For the grill, $I = \frac{1500\text{ W}}{120\text{ V}} = 12.5\text{ A}$ ◊

(b) The total current in that portion of the circuit leading directly to the power source will be the **sum** of the three currents, $I_{\text{total}} = 31.6$ A; thus a 30-A circuit is **not** sufficient. ◊

---

45. In Figure 18.39, $R_1 = 0.1\ \Omega$, $R_2 = 1.0\ \Omega$, and $R_3 = 10.0\ \Omega$. Find the equivalent resistance of the circuit and the current in each resistor when a 5-V power supply is connected between (a) points *a* and *b*; (b) points *a* and *c*; (c) points *a* and *d*.

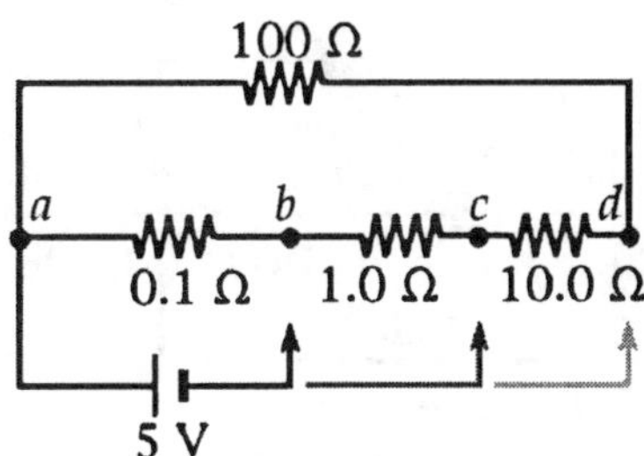

Figure 18.39

Solution

(a) When a 5.00-V power supply is connected between points *a* and *b*, the circuit looks like that shown below, and reduces as shown to an equivalent resistance of 0.0999 Ω.

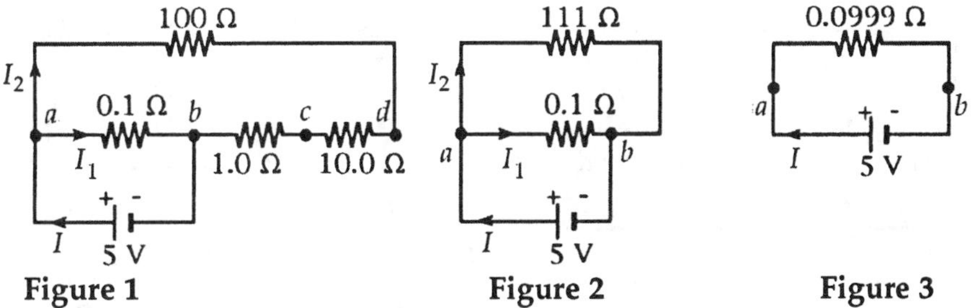

Figure 1 **Figure 2** **Figure 3**

From Figure 2,

$$I_1 = \frac{5.00\text{ V}}{0.100\ \Omega} = 50.0\text{ A} \quad \lozenge$$

and

$$I_2 = \frac{5.00\text{ V}}{111\ \Omega} = 0.045\text{ A} \quad \lozenge$$

(b) When the 5.00-V source is connected between points *a* and *c*, the circuit reduces to an equivalent resistance of 1.089 Ω as shown. ◊

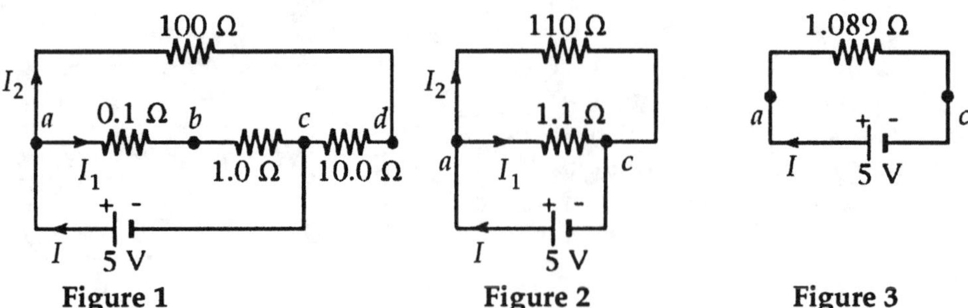

Figure 1 **Figure 2** **Figure 3**

From Figure 2,

$$I_1 = \frac{5.00 \text{ V}}{1.10 \ \Omega} = 4.55 \text{ A} \ \Diamond$$

and

$$I_2 = \frac{5.00 \text{ V}}{110 \ \Omega} = 0.0455 \text{ A} \ \Diamond$$

(c) Finally, when the 5.00-V source is connected between points *a* and *d*, we have an equivalent resistance of 9.99 Ω. ◊

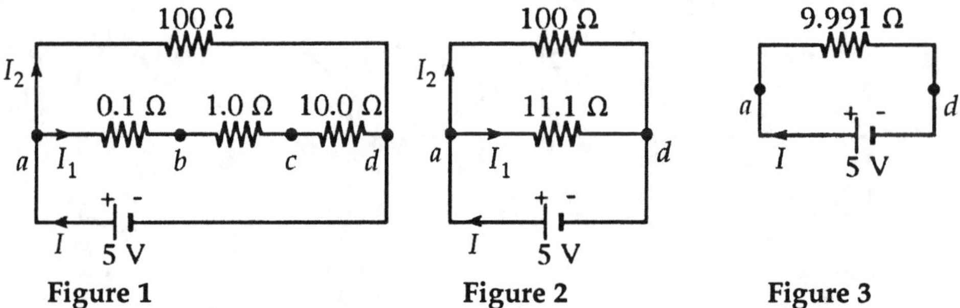

Figure 1 **Figure 2** **Figure 3**

From Figure 2,

$$I_1 = \frac{5.00 \text{ V}}{11.1 \ \Omega} = 0.455 \text{ A} \ \Diamond$$

and

$$I_2 = \frac{5.00 \text{ V}}{100 \ \Omega} = 0.0500 \text{ A} \ \Diamond$$

55. For the circuit in Figure 18.46, calculate (a) the equivalent resistance of the circuit and (b) the power dissipated by the entire circuit. (c) Find the current in the 5Ω resistor.

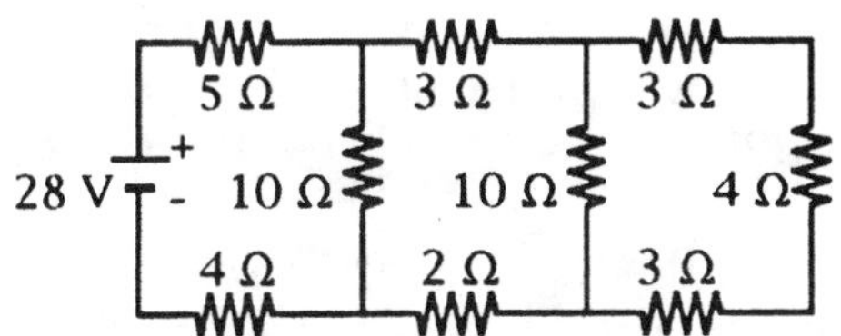

Figure 18.46

Solution

(a) The circuit is reduced as shown below to an equivalent resistance of $R_{eq} = 14\ \Omega$. ◊

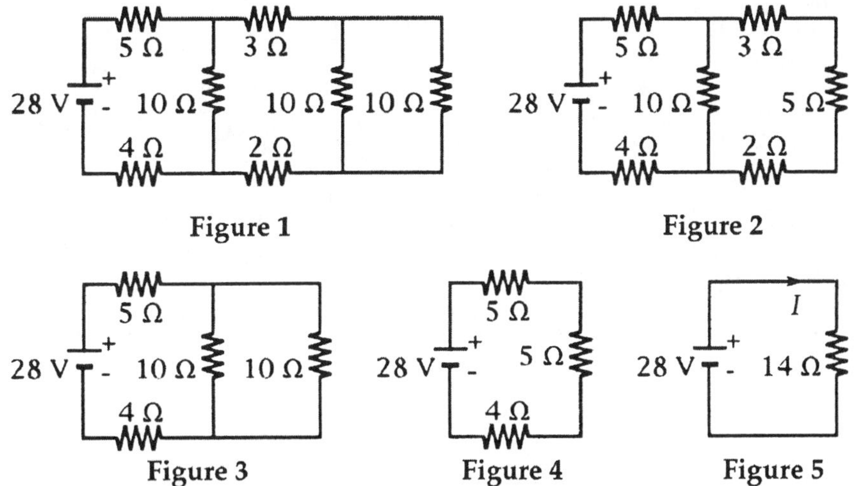

(b) The power dissipated in the entire circuit is

$$P = I^2 R_{eq} = (2.00\ \text{A})^2(14.0\ \Omega) = 56.0\ \text{W} \quad ◊$$

(c) From Figure 5 we have

$$I = \frac{28.0\ \text{V}}{14.0\ \Omega} = 2.00\ \text{A} \quad ◊$$

61. What are the expected readings of the ammeter and voltmeter for the circuit in Figure 18.48?

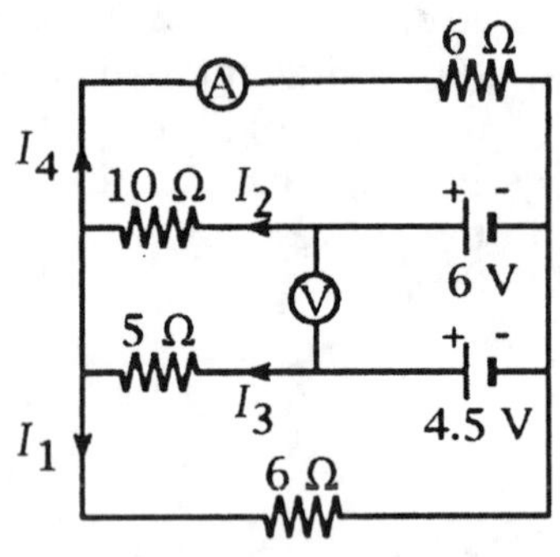

Figure 18.48

Solution Redrawing the circuit, and labeling currents and dirctions as shown to the right, we have from the new junction point:

$$I_1 + I_4 = I_2 + I_3$$

From the "loop rule" we set the following equations:

$$6.00I_1 - 6.00I_4 = 0$$

$$6.00I_1 + 5.00I_3 - 4.50 = 0$$

$$6.00I_1 + 10.0I_2 - 6.00 = 0$$

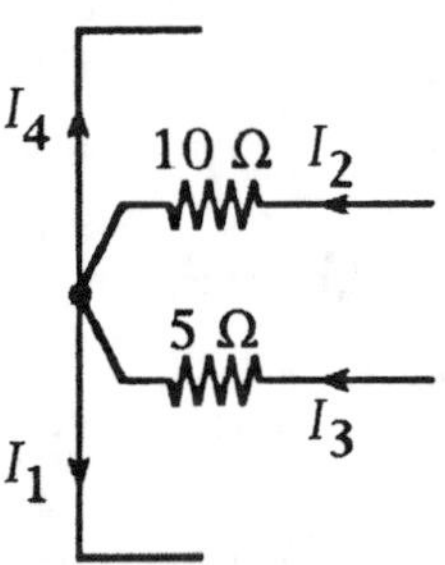

The solution of these equations yields an ammeter reading of:

$$I_4 = 0.390 \text{ A} \quad \Diamond$$

Going around the loop containing the 6.00 V and 4.50 emfs and the voltmeter, we have

$$-6.00 \text{ V} + 4.50 \text{ V} + V_m = 0$$

Thus, $V_m = 1.50 \text{ V}$ ◊

CHAPTER SELF-QUIZ

1. The following three appliances are connected to a 120-volt house circuit: (i) toaster, 1200 W; (ii) coffee pot, 750 W; and (iii) microwave, 600 W. If all were operated at the same time, what total current would they draw?
   a. 3.00 A
   b. 5.00 A
   c. 10.0 A
   d. 21.0 A

2. A circuit contains a 3.00-volt battery, a 2.00-ohm resistor, a 0.30-microfarad capacitor, an ammeter, and a switch, all in series. What will be the current reading 10 min after the switch is closed?
   a. zero
   b. 0.75 A
   c. 1.50 A
   d. 10^7 A

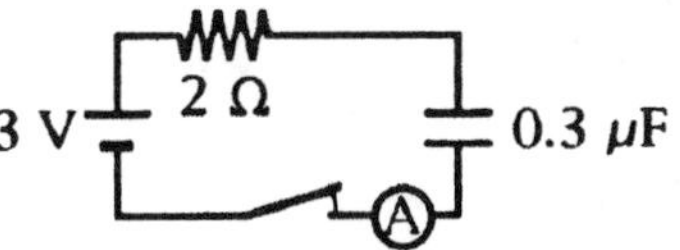

3. What is the current through the 4.00-ohm resistor?
   a. 1.00 A
   b. 0.50 A
   c. 1.50 A
   d. 2.00 A

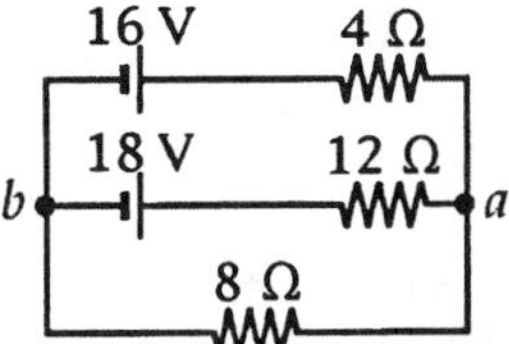

4. A circuit contains a 3.00-volt battery, a 2.00-ohm resistor, a 0.30-microfarad capacitor, an ammeter, and a switch in series. The sum of all the potential differences taken for all the components going completely around the loop is which of the following?
   a. zero
   b. 3.00 V
   c. -3.00 V
   d. 6.00 V

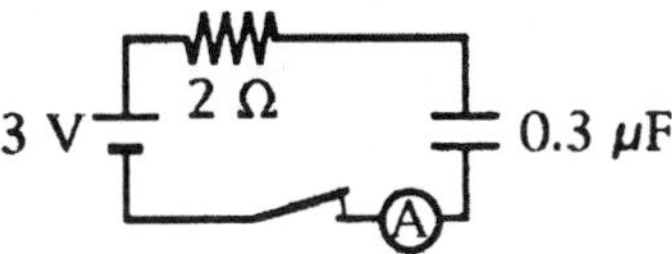

5. Three resistors, with values of 3.00, 6.00, and 12.0 ohms, respectively, are connected in parallel. What is the overall resistance of this combination?
   a. 0.58 ohms
   b. 1.70 ohms
   c. 7.00 ohms
   d. 21.0 ohms

6. Two resistors of values 6.00 and 12.00 ohms are connected in parallel. This combination in turn is hooked in series with a 3.00-ohm resistor and a 24.0-V battery. What is the current in the 6.00-ohm resistor?
   a. 2.00 A
   b. 3.00 A
   c. 6.00 A
   d. 12.0 A

7. Which two resistors are in parallel with each other?
   a. R and R_4
   b. R_2 and R_3
   c. R_2 and R_4
   d. R and R_1

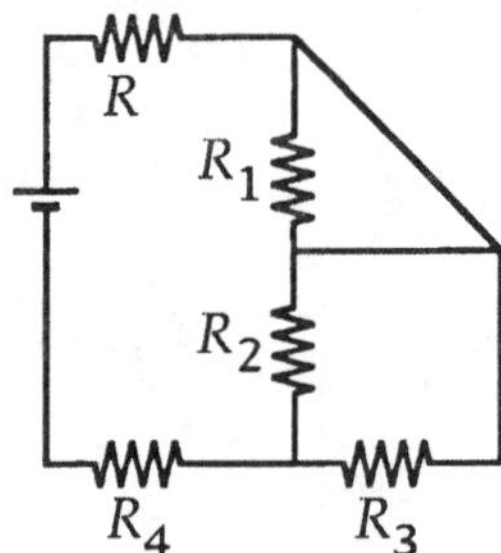

8. Which resistor in Question 7 above is in series with resistor R?
   a. $R1$
   b. R_2
   c. R_3
   d. R_4

9. What is the current flowing through the 8.00-ohm resistor?
   a. 1.00 A
   b. 2.00 A
   c. 3.00 A
   d. 6.00 A

8 V
2 Ω
16 V
8 Ω

10. What is the equivalent resistance for these resistors?
    a. 2.30 ohms
    b. 5.20 ohms
    c. 13.0 ohms
    d. 33.0 ohms

6 Ω
10 Ω
2 Ω
4 Ω

11. In a circuit, a current of 1.00 A is drawn from a battery. The current then divides and passes through two resistors in parallel. One of the resistors has a value of 64 Ω and the current through it is 0.20 A. What is the value of the other resistor?
    a. 8.00 Ω
    b. 16.0 Ω
    c. 24.0 Ω
    d. 32.0 Ω

12. What is the maximum number of 100-W light bulbs you can connect in parallel in a 120-volt home circuit without tripping the 20-amp circuit breaker?
    a. 11
    b. 17
    c. 23
    d. 29

VOLTSwagon towing a mobile OHM.

19
Magnetism

Chapter 19

MAGNETISM

As we investigate magnetism in this chapter, you will find that the subject cannot be divorced from electricity. For example, magnetic fields affect moving charges and moving charges produce magnetic fields. The ultimate source of all magnetic fields is electric current, whether it be the current in a wire or the current produced by the motion of charges within atoms or molecules.

NOTES FROM SELECTED CHAPTER SECTIONS

19.3 Magnetic Fields

Particles with charge q, moving with velocity **v** in a magnetic field **B**, experience a magnetic force **F**:

1. The magnetic force is proportional to the charge q and velocity **v** of the particle.
2. The magnitude and direction of the magnetic force depend on the velocity of the particle and on the magnitude and direction of the magnetic field.
3. When a charged particle moves in a direction *parallel* to the magnetic field vector, the magnetic force **F** on the charge is *zero*.
4. The magnetic force acts in a direction perpendicular to both **v** and **B**; that is, **F** is perpendicular to the plane formed by **v** and **B**.
5. The magnetic force on a positive charge is in the direction opposite to the force on a negative charge moving in the same direction.
6. If the velocity vector makes an angle θ with the magnetic field, the magnitude of the magnetic force is proportional to $\sin\theta$.

19.4 Magnetic Force on a Current-Carrying Conductor

The total magnetic force on a *closed* current loop in a *uniform* magnetic field is *zero*.

19.6 The Galvanometer and its Applications

The *galvanometer* is a device used in the construction of both ammeters and voltmeters. The basic operation of this instrument makes use of the fact that a *torque* acts on a current loop in the presence of a magnetic field.

In order to convert a galvanometer into an ammeter, a resistor (shunt) is placed in *parallel* with the galvanometer coil. For use as a voltmeter, a resistor is place in *series* with the galvanometer coil.

19.7 Motion of a Charged Particle in a Magnetic Field

When a charged particle moves in an external magnetic field, the work *done* by the magnetic force on the particle is *zero*. The magnetic force changes the direction of the velocity vector but does not change its magnitude.

When a charged particle enters an external magnetic field along a direction perpendicular to the field, the particle will move in a circular path in a plane perpendicular to the magnetic field.

19.8 Magnetic Field of a Long, Straight Wire and Ampère's Law

The direction of the magnetic field due to a current in a conductor is given by the right-hand rule:

> If the wire is grasped in the right hand with the thumb in the direction of the current, the fingers will wrap (or curl) in the direction of **B**.

Ampère's law is valid only for *steady* currents and is useful only in those cases where the current configuration has a *high degree* of *symmetry*.

19.9 Magnetic Force Between Two Parallel Conductors

Parallel conductors carrying currents in the *same direction attract* each other, whereas parallel conductors carrying currents in *opposite directions repel* each other.

The force between two parallel wires each carrying a current is used to define the ampere as follows:

> If two long, parallel wires 1.00 m apart carry the same current and the force per unit length on each wire is 2.00×10^{-7} N/m, then the current is defined to be 1.00 A.

EQUATIONS AND CONCEPTS

The magnitude of the magnetic field at a point in space is defined in terms of the magnetic force exerted on a moving electric charge. The magnetic force will be

$$F_{max} = qvB \qquad (19.4)$$

of maximum magnitude when the charge moves along a direction perpendicular to the direction of the magnetic field.

In general, the velocity vector may be directed along some direction other than 90.0° relative to the magnetic field. In this case, the magnetic force on the moving charge is less than its maximum value.

$$F = qvB\sin\theta \qquad (19.1)$$

The SI unit of magnetic field intensity is the tesla (T).

$$1\ \mathrm{T} = 1\frac{\mathrm{N \cdot s}}{\mathrm{C \cdot m}}$$

Equation 19.1 can be written in a form which serves to define the magnitude of the magnetic field.

$$B \equiv \frac{F}{qv\sin\theta} \qquad (19.2)$$

The cgs unit of magnetic field is the gauss (G).

$$1\ \mathrm{T} = 10^4\ \mathrm{G}$$

In order to determine the direction of the magnetic force, apply right-hand rule *A:* hold your open right hand with your fingers pointing in the direction of B and your thumb pointing in the direction of v. The magnetic force on a positive charge is then directed out of the palm of your hand. If the charge is negative, then the direction of the force is reversed.

Comment on the direction of the magnetic force on a moving charge.

A magnetic force is exerted on a straight conductor placed in a magnetic field. The magnitude of the magnetic force on the conductor depends on the angle between the direction of the conductor and the direction of the field.

$$F = BIL\sin\theta \qquad (19.6)$$

The magnetic force will be maximum when the conductor is directed perpendicular to the magnetic field.

$$F_{max} = BIL \qquad (19.5)$$

In order to determine the direction of the magnetic force on a current-carrying conductor, use right-hand rule *A* with your thumb in the direction of conventional current.

Comment on the direction of the magnetic force on a current-carrying conductor.

When a closed conducting loop is placed in a uniform external magnetic field, a net torque will be exerted on the loop. The magnitude of the torque will depend on the angle between the direction of the magnetic field and the direction of the normal (or perpendicular) to the plane of the loop.

$$\tau = BIA\sin\theta \qquad (19.8)$$

The magnitude of the torque will be maximum when the magnetic field is parallel to the plane of the loop.

$$\tau_{max} = BIA \qquad (19.7)$$

The direction of rotation of the loop is such that the normal to the plane of the loop turns into a direction parallel to the magnetic field.

Comment on the direction of rotation of the loop.

When a charged particle enters the region of a uniform magnetic field with the velocity of the particle initially perpendicular to the direction of the field, the particle will move in a circular path of constant radius. The circular path will have a radius which is proportional to the linear momentum of the moving charge and will be in a plane which is perpendicular to the direction of the magnetic field.

$$r = \frac{mv}{qB} \qquad (19.10)$$

The period of revolution in the circular path is independent of the value of the radius.

$$T = \frac{2\pi r}{v} = \frac{2\pi m}{qB}$$

The magnetic field at a distance r from a long straight current-carrying conductor.

$$B = \frac{\mu_0 I}{2\pi r} \qquad (19.11)$$

The magnetic field at the center of a current loop of radius R.

$$B = \frac{\mu_0 I}{2R}$$

The magnetic field inside a solenoid which has n turns of conductor per unit length.

$$B = \mu_0 n I \qquad (19.15)$$

The permeability of free space.

$$\mu_0 = 4\pi \times 10^{-7}\ \text{T} \cdot \text{m/A}$$

The direction of the magnetic field due to current in a long wire is determined by using right-hand rule B: hold the conductor in the right hand with the thumb pointing in the direction of the conventional current. The fingers will then wrap around the wire in the direction of the magnetic field lines. The magnetic field is tangent to the circular field lines at every point in the region around the conductor.

Comment on the direction of the magnetic field due to different current geometries.

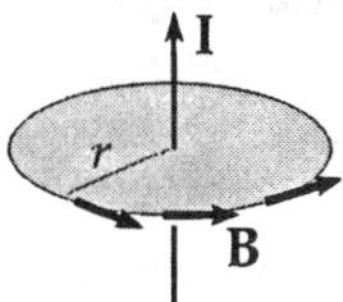

The direction of the magnetic field at the center of a current loop is perpendicular to the plane of the loop and directed in the sense given right-hand rule B.

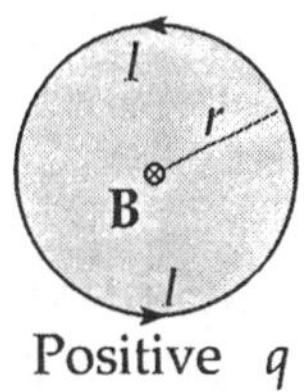

Positive q

Within a solenoid, the magnetic field is parallel to the axis of the solenoid and pointing in a sense determined by applying right-hand rule B to one of the coils.

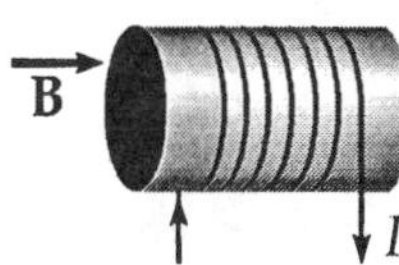

Ampère's circuital law is very useful in calculating the magnetic field due to current configurations which have a high degree of symmetry. Ampère's law states that the sum of the products of the components of the magnetic field parallel to each segment of a closed path is proportional to the net current which passes through the surface bounded by the closed path.

$$\sum B_{\parallel}\Delta l = \mu_0 I \tag{19.13}$$

The magnitude of the magnetic force per unit length between parallel conductors is proportional to the product of the two currents and inversely proportional to the distance between the two conductors.

$$\frac{F_1}{l} = \frac{\mu_0 I_1 I_2}{2\pi d} \tag{19.14}$$

Parallel currents in the same direction attract each other and parallel currents in opposite directions repel each other.

Comment on the direction of magnetic force between parallel conductors.

REVIEW CHECKLIST

▷ Use the defining equation for a magnetic field **B** and the right-hand rule to determine the magnitude and direction of the magnetic force exerted on an electric charge moving in a region where there is a magnetic field. You should understand clearly the important differences between the forces exerted on electric charges by electric fields and those forces exerted on moving charges by magnetic fields.

▷ Calculate the magnitude and direction of the magnetic force on a current-carrying conductor when placed in an external magnetic field. Determine the magnitude and direction of the torque exerted on a closed current loop in an external magnetic field.

▷ Describe how a moving coil galvanometer can be converted to either an ammeter or a voltmeter.

▷ Calculate the period and radius of the circular orbit of a charged particle moving in a uniform magnetic field. Understand the essential features of the velocity selector and mass spectrometer and make appropriate quantitative calculations regarding the operation of these instruments.

▷ Calculate the magnitude and determine the direction of the magnetic field for the following cases: a point in the vicinity of a long, straight current-carrying conductor at the center of a current loop, and at interior points of a solenoid. Correctly apply the right-hand rule for these situations.

SOLUTIONS TO SELECTED END-OF-CHAPTER PROBLEMS

3. A proton travels with a speed of 3.00×10^6 m/s at an angle of 37.0° with the direction of a magnetic field of 0.300 T in the $+y$ direction. What are (a) the magnitude of the magnetic force on the proton and (b) the proton's acceleration?

Solution

(a) $F = qvB\sin\theta = (1.60 \times 10^{-19}\text{ C})(3.00 \times 10^{6}\text{ m/s})(3.00 \times 10^{-1}\text{ T})\sin 37.0°$

or $F = 8.70 \times 10^{-14}\text{ N}$ ◊

(b) $a = \dfrac{F}{m} = \dfrac{8.67 \times 10^{-14}\text{ N}}{1.67 \times 10^{-27}\text{ kg}} = 5.20 \times 10^{13}\text{ m/s}^2$ ◊

---

7. Find the direction of the magnetic field on the positively charged particle moving in the various situations shown in Figure 19.35, if the direction of the magnetic force acting on it is as indicated.

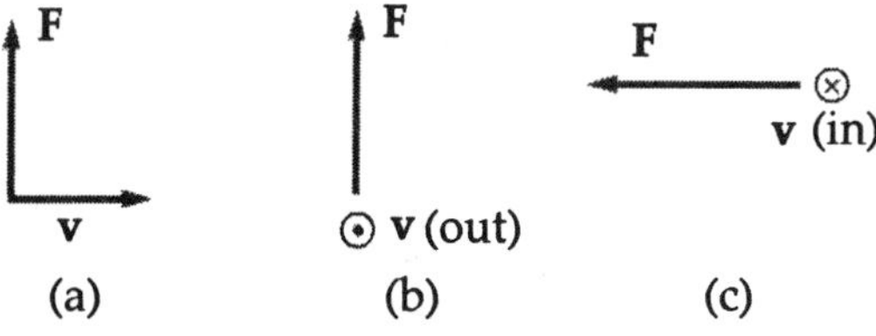

Figure 19.35

Solution $\mathbf{F} = q\mathbf{v} \times \mathbf{B}$.
Use the right-hand rule. Point the thumb along the direction of the force; the four fingers wrap around the force vector first in the direction of the velocity, and then in the direction of the magnetic field lines.

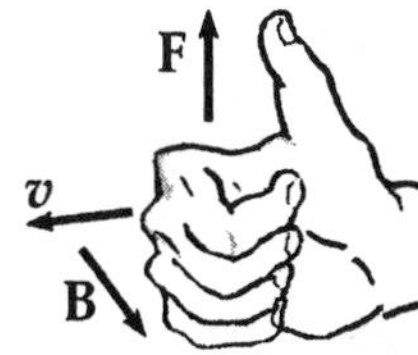

The answers are:

(a) into the page ◊
(b) toward the right ◊
(c) toward bottom of page ◊

---

11. A proton moves perpendicularly to a uniform magnetic field **B** with a speed of 10^7 m/s and experiences an acceleration of 2.00×10^{13} m/s^2 in the $+x$ direction when its velocity is in the +z direction. Determine the magnitude and direction of the field.

Solution $F = ma$

$$F = (1.67 \times 10^{-27}\text{ kg})(2.00 \times 10^{13}\text{ m/s}^2) = 3.34 \times 10^{-14}\text{ N}$$

$$B = \frac{F}{qv} = \frac{3.34 \times 10^{-14}\text{ N}}{(1.60 \times 10^{-19}\text{ C})(1.00 \times 10^7\text{ m/s})} = 2.10 \times 10^{-2}\text{ T} \quad \Diamond$$

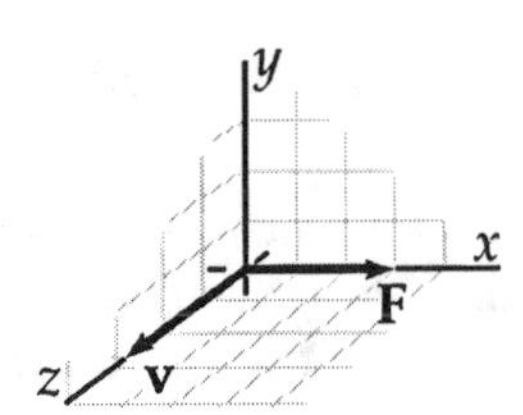

The right-hand rule shows that **B** must be in the negative y direction in order to yield a force in the positive x direction when **v** is in the z direction.

---

19. A wire with a mass of 1.00 g/cm is placed on a horizontal surface with a coefficient of friction of 0.200. The wire carries a current of 1.50 A eastward and moves horizontally to the north. What are the magnitude and the direction of the *smallest* magnetic field that enables the wire to move in this fashion?

Solution To find the minimum **B** field, B must be perpendicular to I, and from the right-hand rule, it must be directed downward.

The *friction force* per unit length = μN = μ(mass per unit length)(g), or

$$\frac{f}{L} = 0.200\left(\frac{10^{-3}\text{ kg}}{10^{-2}\text{ m}}\right)(9.80\text{ m/s}^2) = 0.196\text{ N/m}$$

Thus, the *magnetic force* per unit length must also be 0.196 N/m.

$$\frac{F}{L} = BI = 0.196\text{ N/m} \qquad \text{or} \qquad B = \frac{0.196\text{ N/m}}{1.50\text{ A}} = 0.131\text{ T} \quad \Diamond$$

---

21. A thin copper rod 1.00 m long has a mass of 50.0 g. What is the minimum current in the rod that will cause it to float in a magnetic field of 2.00 T?

Solution In order for the rod to float, the magnetic force must have the same magnitude as the weight of the rod. Thus,

$$F = BIL\sin\theta = mg \qquad \text{or} \qquad I = \frac{mg}{BL\sin\theta}$$

The minimum value of I occurs when $\sin\theta = 1$.

Thus, $$I_{min} = \frac{(5.00\times10^{-2}\text{ kg})(9.80\text{ m/s}^2)}{(2.00\text{ T})(1.00\text{ m})} = 0.245\text{ A} \quad \Diamond$$

---

23. The Earth has a magnetic field of 0.600×10^{-4} T, pointing 75.0° below the horizontal in a north-south plane. A 10.0-m-long straight wire carries a 15.0-A current. (a) If the current is directed horizontally toward the east, what are the magnitude and direction of the magnetic force on the wire? (b) What are the magnitude and direction of the force if the current is directed vertically upward?

Solution

(a) $F = BIL\sin\theta = (0.600\times10^{-4}\text{ T})(15.0\text{ A})(10.0\text{ m})\sin 90.0° = 9.00\times10^{-3}\text{ N} \quad \Diamond$

F is perpendicular to B and, by the right-hand rule, is directed at 15.0° above the horizontal in the northward direction. ◊

(b) $F = BIL\sin\theta = (0.600\times10^{-4}\text{ T})(15.0\text{ A})(10.0\text{ m})\sin 165° = 2.30\times10^{-3}\text{ N} \quad \Diamond$

The right-hand rule shows F is horizontal and directed due west. ◊

---

27. A 50.0-Ω, 10.0-mA galvanometer is to be converted to an ammeter that reads 3.00 A at full-scale deflection. What value of R_p should be placed in parallel with the coil?

Solution (See sketch.) The current through the parallel resistor is

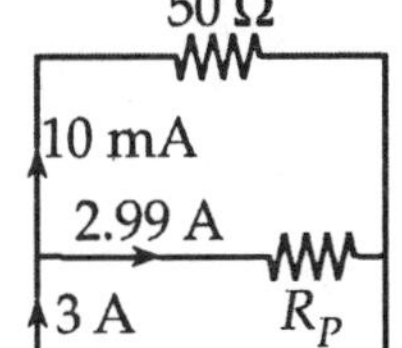

$$I = 3.00\text{ A} - 0.010\text{ A} = 2.99\text{ A}$$

Use the loop rule on the closed loop to find

$$-(0.010\text{ A})(50.0\ \Omega) + R_p(2.99\text{ A}) = 0$$

This yields $$R_p = 0.167\ \Omega \quad \Diamond$$

---

31. The galvanometer of Problem 32 is to be converted to a multi-range ammeter using the circuit shown in Figure 19.39. Find the values of R_1, R_2, and R_3 that will give the full-scale readings in the figure.

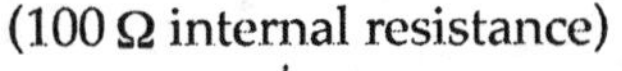

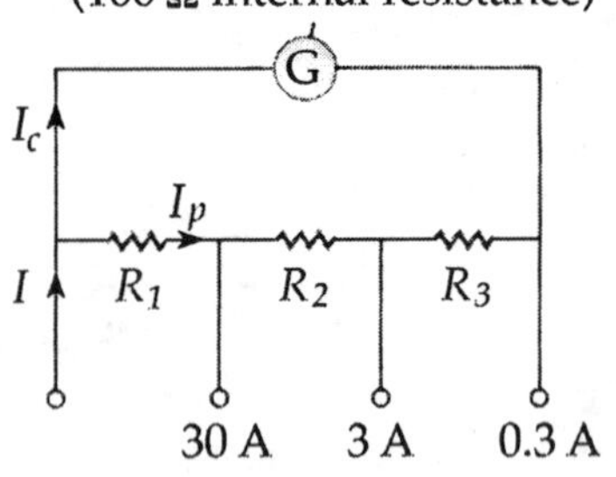

Figure 19.39
(with currents labeled)

Solution If $I = 0.300$ A, and a and d are the terminals,

$$I_p(R_1 + R_2 + R_3) = i_c(100\ \Omega)$$

In this case, $I_p = 0.2999$ A

and $i_c = 100 \times 10^{-6}$ A

Thus, $$R_1 + R_2 + R_3 = 3.33 \times 10^{-2}\ \Omega \qquad (1)$$

If $I = 3.00$ A, and a and c are the terminals,

$$I_p(R_1 + R_2) = i_c(100\ \Omega + R_3)$$

Here, $I_p = 2.9999$ A and $i_c = 100 \times 10^{-6}$ A

Thus, we have $$R_1 + R_2 - 3.33 \times 10^{-5} R_3 = 3.33 \times 10^{-3}\ \Omega \qquad (2)$$

Finally, if $I = 30.0$ A and a and b are the terminals, $I_p R_1 = i_c(100\ \Omega + R_2 + R_3)$

With $I_p = 29.9999$ A and $i_c = 100 \times 10^{-6}$ A, this reduces to

$$R_1 - 3.33 \times 10^{-6} R_2 - 3.33 \times 10^{-6} R_3 = 3.33 \times 10^{-4}\ \Omega \qquad (3)$$

Solving (1), (2), and (3) simultaneously, we find

$$R_1 = 3.33 \times 10^{-4}\ \Omega \quad \Diamond$$

$$R_2 = 2.998 \times 10^{-3}\ \Omega \quad \Diamond$$

and $$R_3 = 2.997 \times 10^{-2}\ \Omega \quad \Diamond$$

35. A proton moves in a circular orbit perpendicularly to a uniform magnetic field of 0.758 T. Find the time it takes the proton to make one pass around the orbit.

Solution $r = \dfrac{mv}{qB}$ gives $v = \dfrac{qrB}{m}$

and the time to complete one orbit (the period, T) is

$$T = \frac{\text{circumference}}{\text{speed}} = \frac{2\pi r}{v} = \frac{2\pi r}{\frac{qrB}{m}} = \frac{2\pi m}{qB}$$

or $$T = \frac{2\pi(1.67\times10^{-27}\text{ kg})}{(1.60\times10^{-19}\text{ C})(0.758\text{ T})} = 8.65\times10^{-8}\text{ s} \quad \Diamond$$

---

43. A mass spectrometer is used to examine the isotopes of uranium. Ions in the beam emerge from the velocity selector at a speed of 3.00×10^5 m/s and enter a uniform magnetic field of 0.600 T directed perpendicularly to the velocity of the ions. What is the distance between the impact points formed on the photographic plate by singly charged ions of ^{235}U and ^{238}U?

Solution $$r = \frac{mv}{qB} = m\left(\frac{v}{qB}\right) = m\left(\frac{3.00\times10^5\text{ m/s}}{(1.60\times10^{-19}\text{ C})(0.600\text{ T})}\right) = m\left(3.125\times10^{24}\,\frac{\text{m}}{\text{kg}}\right)$$

For U^{235}, $m = 235(1.67\times10^{-27}\text{ kg}) = 3.925\times10^{-25}\text{ kg}$

Thus, $$r = (3.125\times10^{24}\text{ m/kg})(3.925\times10^{-25}\text{ kg}) = 1.226\text{ m}$$

For U^{238}, $m = 238(1.67\times10^{-27}\text{ kg}) = 3.975\times10^{-25}\text{ kg}$

and $$r = (3.125\times10^{24}\text{ m/kg})(3.975\times10^{-25}\text{ kg}) = 1.242\text{ m}$$

The distance between the impact points will be equal to the difference between the *diameters* of the two arcs. Therefore,

$$\Delta d = 2\Delta r = 2(1.242 - 1.226) = 3.20\times10^{-2}\text{ m} = 3.20\text{ cm.} \quad \Diamond$$

---

47. Niobium metal becomes a superconductor (with electrical resistance equal to zero) when cooled below 9.00 K. If superconductivity is destroyed when the surface magnetic field exceeds 0.100 T, determine the maximum current a 2.00-mm diameter niobium wire can carry and remain superconducting.

Solution We use $B = \frac{\mu_0 I}{2\pi r}$ or $I = \frac{2\pi r B}{\mu_0}$

At the surface of the wire, this becomes $I = \frac{2\pi(0.100\text{ T})(10^{-3}\text{ m})}{4\pi \times 10^{-7}\text{ T}\cdot\text{m/A}} = 500\text{ A}$ ◊

---

49. The two wires shown in Figure 19.45 carry currents of 5.00 A in opposite directions and are separated by 10.0 cm. Find the direction and magnitude of the net magnetic field (a) at a point midway between the wires; (b) at point P_1, that is, 10.0 cm to the right of the wire on the right; and (c) at point P_2, that is, 20.0 cm to the left of the wire on the left.

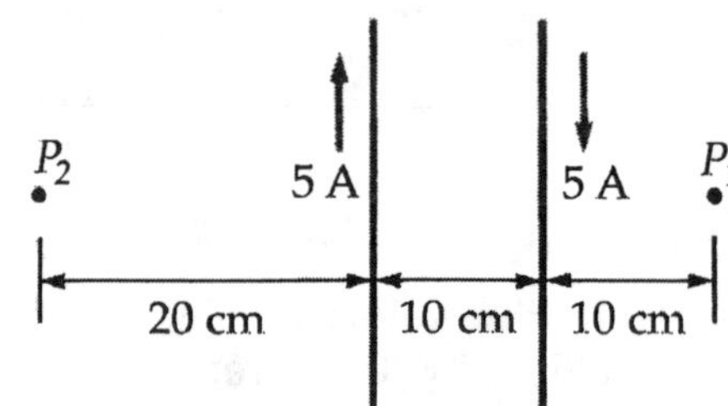

Figure 19.45

Solution Call the wire on the left wire 1 and the wire on the right wire 2.

The magnetic field a distance, r, from a long current-carrying conductor is

$$B = \frac{\mu_0 I}{2\pi r} = \frac{(4\pi \times 10^{-7}\text{ T}\cdot\text{m/A})(5.00\text{ A})}{2\pi r} = \frac{10^{-6}\text{ T}\cdot\text{m}}{r}$$

(a) At a point midway between the wires,

$$B_1 = B_2 = \frac{10^{-6}\text{ T}\cdot\text{m}}{0.050\text{ m}} = 2.00 \times 10^{-5}\text{ T}$$

Using the right-hand rule, we see that both fields are in the same direction (out of the page), which gives us a resultant field of

$$B_{net} = B_1 + B_2 = 4.00 \times 10^{-5}\text{ T} \quad \text{(directed into the page)} \; ◊$$

(b) At the point P_1, $B_1 = \dfrac{10^{-6}\ \text{T}\cdot\text{m}}{0.200\ \text{m}} = 5.00\times10^{-6}\ \text{T}$ (into the page)

and $B_2 = \dfrac{10^{-6}\ \text{T}\cdot\text{m}}{0.100\ \text{m}} = 10.0\times10^{-6}\ \text{T}$ (out of the page)

Thus $B_{net} = 5.00\times10^{-6}\ \text{T}$ (out of the page) ◊

(c) At P_2, $B_1 = \dfrac{10^{-6}\ \text{T}\cdot\text{m}}{0.200\ \text{m}} = 5.00\times10^{-6}\ \text{T}$ (out of the page)

and $B_2 = \dfrac{10^{-6}\ \text{T}\cdot\text{m}}{0.300\ \text{m}} = 3.33\times10^{-6}\ \text{T}$ (into the page)

for $B_{net} = 1.67\times10^{-6}\ \text{T}$ (out of the page) ◊

---

53. Two parallel wires are 10.0 cm apart, and each carries a current of 110 A. (a) If the currents are in the same direction, find the force per unit length exerted on one of the wires by the other. Are the wires attracted or repelled? (b) Repeat the problem with the currents in opposite directions.

Solution (a) We use $\dfrac{F_1}{L} = \dfrac{\mu_0 I_1 I_2}{2\pi d} = \dfrac{(4\pi\times10^{-7}\ \text{N/A}^2)(10.0\ \text{A})^2}{(2\pi)(0.100\ \text{m})} = 2.00\times10^{-4}\ \text{N/m}$ ◊

Consider one wire as a current-carrying conductor in a magnetic field which is due to current in the other wire. Use of the right-hand rule will show that the magnetic force between the two wires is one of attraction. ◊

(b) The magnitude remains the same as calculated in (a), but the wires are repelled. ◊

---

57. A single-turn square loop of wire, 2.00 cm on a side, carries a current of 0.200 A. The loop is inside a solenoid, with the plane of the loop perpendicular to the magnetic field of the solenoid. The solenoid has 30 turns per centimeter and carries a current of 15.0 A. Find the force on each side of the loop and the torque acting on it.

Solution The magnetic field at the center of the solenoid is

$$B = \mu_0 n I = (4\pi\times10^{-7}\ \text{T}\cdot\text{m/A})(3000\ \text{turns/m})(15.0\ \text{A}) = 5.66\times10^{-2}\ \text{T}$$

The force on one of the sides of the loop is

$$F = BIL = (5.66\times10^{-4}\ \text{T})(0.200\ \text{A})(0.020\ \text{m}) = 2.26\times10^{-4}\ \text{N} \quad \lozenge$$

With the loop aligned as in the problem, all the forces are directed such that they tend to stretch the loop. There is no tendency for the forces to cause rotation. Thus, $\tau = 0$. ◊

---

65. Two long, straight wires cross each other at right angles, as shown in Figure 19.52. (a) Find the direction and magnitude of the magnetic field at point *P*, which is in the same plane as the two wires. (b) Find the magnetic field at a point 30.0 cm above the point of intersection (30.0 cm out of the page, toward you).

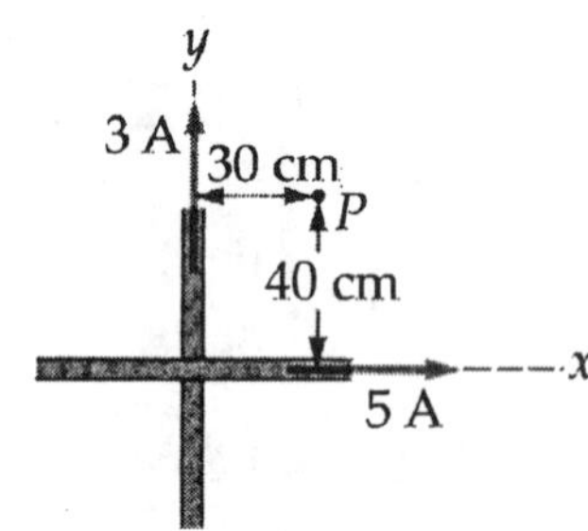

Figure 19.52

Solution (a) Let us call the wire carrying the 3.00 A current wire 1 and the wire with the 5.00 A current wire 2. The field at *P* produced by wire 1 is

$$B_1 = \frac{\mu_0 I}{2\pi r} = \frac{(4\pi\times10^{-7}\ \text{T}\cdot\text{m/A})(3.00\ \text{A})}{2\pi(0.300\ \text{m})} = 2.00\times10^{-6}\ \text{T} \quad \text{(into the page)}$$

and the field due to wire 2 is

$$B_2 = \frac{\mu_0 I}{2\pi r} = \frac{(4\pi\times10^{-7}\ \text{T}\cdot\text{m/A})(5.00\ \text{A})}{2\pi(0.400\ \text{m})} = 2.50\times10^{-6}\ \text{T} \quad \text{(out of the page)}$$

Thus, the net field is $\quad B_{net} = 5.00\times10^{-7}\ \text{T}$ (out of the page) ◊

(b) We use the same designation for discussing the wires as above. The field due to wire 1 is

$$B_1 = \frac{\mu_0 I}{2\pi r} = \frac{(4\pi\times10^{-7}\ \text{T}\cdot\text{m/A})(3.00\ \text{A})}{2\pi(0.300\ \text{m})} = 2.00\times10^{-6}\ \text{T}$$

The sketch indicates the direction of the field.

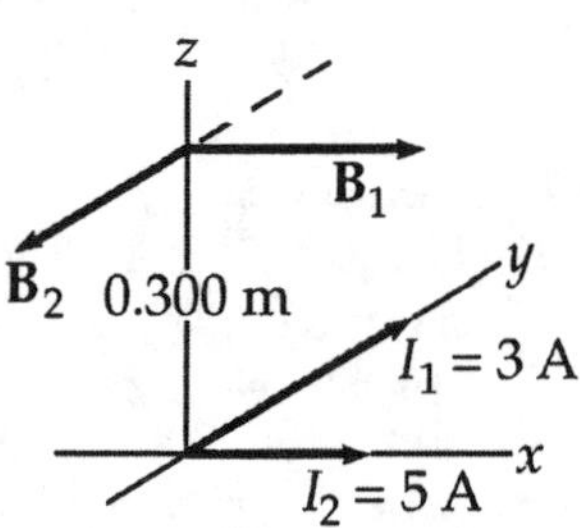

The field due to wire 2 is

$$B_2 = \frac{\mu_0 I}{2\pi r} = \frac{(4\pi \times 10^{-7}\ \text{T}\cdot\text{m/A})(5.00\ \text{A})}{2\pi(0.300\ \text{m})} = 3.33\times 10^{-6}\ \text{T}$$

(For direction, see sketch.)

The net field is found by using the Pythagorean theorem.

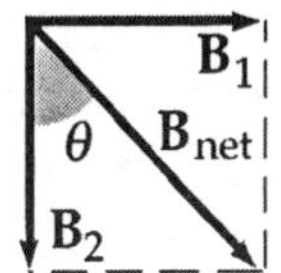

$$B_{net} = 3.89\times 10^{-6}\ \text{T} \quad \Diamond$$

The angle, as seen *from above the wires* is

$$\tan\theta = \frac{B_1}{B_2} = 0.600 \qquad \text{and} \qquad \theta = 31.0° \quad \Diamond$$

---

69. Two parallel conductors carry currents in opposite directions, as shown in Figure 19.53. One conductor carries a current of 10.0 A. Point A is the midpoint between the wires, and point C is 5.00 cm to the right of the 10.0-A current. I is adjusted so that the magnetic field at C is zero. Find (a) the value of the current I and (b) the value of the magnetic field at A.

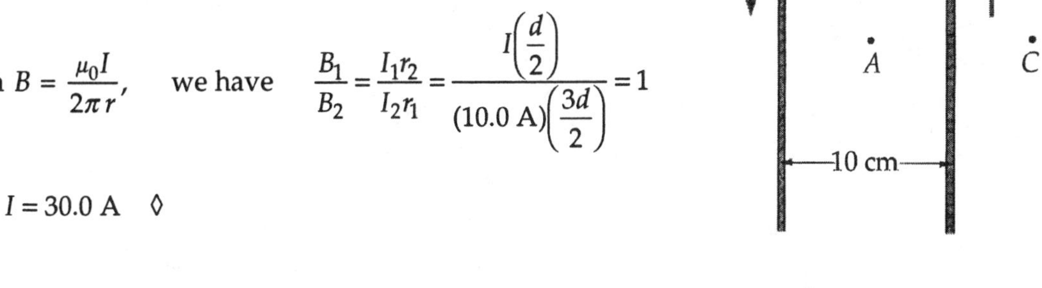

Solution (a) At point C, the two currents give rise to oppositely directed magnetic fields. Therefore, to have $B = 0$ at C, we require $B_1 = B_2$, where B_1 is due to the current I and B_2 is due to the 10.0-A current.

From $B = \frac{\mu_0 I}{2\pi r}$, we have $\frac{B_1}{B_2} = \frac{I_1 r_2}{I_2 r_1} = \frac{I\left(\frac{d}{2}\right)}{(10.0\ \text{A})\left(\frac{3d}{2}\right)} = 1$

or $\quad I = 30.0\ \text{A} \quad \Diamond$

(b) At point A, between the wires,

Figure 19.53

$$B_1 = \frac{\mu_0 I}{2\pi r} = \frac{(4\pi\times 10^{-7}\ \text{T}\cdot\text{m/A})(30.0\ \text{A})}{2\pi(5.00\times 10^{-2}\ \text{m})} = 1.20\times 10^{-4}\ \text{T} \quad \text{(out of the page)}$$

and $$B_2 = \frac{\mu_0 I}{2\pi r} = \frac{(4\pi \times 10^{-7}\ \text{T}\cdot\text{m/A})(10.0\ \text{A})}{2\pi(5.00\times 10^{-2}\ \text{m})} = 4.00\times 10^{-5}\ \text{T} \quad \text{(out of the page).}$$

Thus, $$B_{\text{total}} = B_1 + B_2 = 1.60\times 10^{-4}\ \text{T} \quad \text{(out of the page)} \ \lozenge$$

---

73. For the arrangement shown in Figure 19.56, the current in the straight conductor has the value $I_1 = 5.00$ A and lies in the plane of the rectangular loop, which carries a current $I_2 = 10.0$ A. The dimensions are $c = 0.100$ m, $a = 0.150$ m, and $L = 0.450$ m. Find the magnitude and direction of the *net force* exerted on the rectangle by the magnetic field of the straight current-carrying conductor.

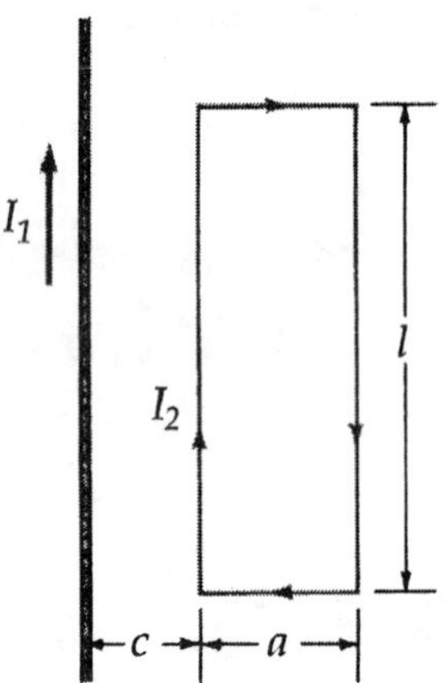

Figure 19.56

Solution The magnetic forces on the top and bottom segments of the rectangle cancel. The force on the left vertical segment is

$$F_L = \frac{\mu_0 I_1 I_2 L}{2\pi c} \quad \text{to the left}$$

and the force on the right vertical segment is

$$F_R = \frac{\mu_0 I_1 I_2 L}{2\pi(c+a)} \quad \text{(to the right)}$$

The net force on the rectangle, $F_{\text{net}} = F_L - F_R$

$$F_{\text{net}} = \frac{\mu_0 I_1 I_2 L}{2\pi}\left(\frac{1}{c} - \frac{1}{c+a}\right) = \frac{\mu_0 I_1 I_2 L}{2\pi}\left(\frac{a}{c(c+a)}\right)$$

$$F_{\text{net}} = \frac{\mu_0 (5.00\ \text{A})(10.0\ \text{A})(0.45\ \text{m})}{2\pi}\left(\frac{0.15\ \text{m}}{0.100\ \text{m}(0.250\ \text{m})}\right)$$

$$F_{\text{net}} = 2.70\times 10^{-5}\ \text{N} \quad \text{toward the left} \ \lozenge$$

CHAPTER SELF-QUIZ

1. When a magnetic field causes a charged particle to move in a circular path, the only quantity listed below which the magnetic force changes significantly as the particle goes around in a circle is the particle's
   a. energy
   b. momentum
   c. radius for the circle
   d. time to go around the circle once

2. A 10.0-ohm, 25.0-mA galvanometer is to be converted into a voltmeter which reads 20.0 V at full-scale deflection. What resistance should be placed in series with the galvanometer coil?
   a. 810 ohms
   b. 790 ohms
   c. 500 ohms
   d. 450 ohms

3. A 10.0-ohm, 15.0-mA galvanometer is converted into an ammeter by placing a 1.50×10^{-2} ohm shunt in parallel with the galvanometer coil. What is the maximum scale reading on the galvanometer?
   a. 25.0 A
   b. 15.0 A
   c. 10.0 A
   d. 5.00 A

4. A circular loop carrying a current of 2.00 A is oriented in a magnetic field of 3.50 T. The loop has an area of 0.12 m^2 and is mounted on an axis, perpendicular to the magnetic field, which allows the loop to rotate. What is the torque on the loop when its plane is oriented at a 37.0° angle to the field?
   a. 46.0 N·m
   b. 0.67 N·m
   c. 0.51 N·m
   d. 0.10 N·m

5. A current-carrying wire of length 0.300 m is positioned perpendicular to a uniform magnetic field. If the current is 5.00 A and it is determined that there is a resultant force of 2.25 N on the wire due to the interaction of the current and field, what is the magnetic field strength?
   a. 0.60 T
   b. 1.50 T
   c. 1.85×10^{3} T
   d. 6.70×10^{3} T

6. A proton, mass 1.67×10^{-27} kg and charge $+1.60 \times 10^{-19}$ C, moves in a circular orbit perpendicular to a uniform magnetic field of 0.75 T. Find the time for the proton to make one complete circular orbit.
   a. 4.30×10^{-8} s
   b. 8.70×10^{-8} s
   c. 4.90×10^{-7} s
   d. 9.80×10^{-7} s

7. An electron moves through a region of crossed electric and magnetic fields. The electric field **E** = 1000 V/m and is directed straight down. The magnetic field **B** = 0.40 T and is directed to the left. For what velocity v of the electron into the paper (perpendicular to the plane) will the electric force exactly cancel the magnetic force?
   a. 2500 m/s
   b. 4000 m/s
   c. 5000 m/s
   d. 8000 m/s

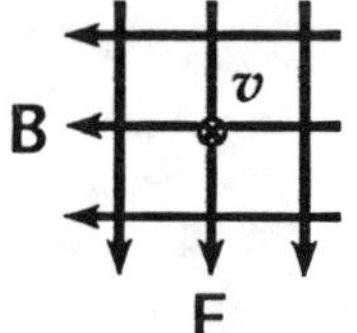

8. A proton is released such that its initial velocity is from left to right across this page. The proton's path, however, is deflected in a direction toward the bottom edge of the page due to the presence of a uniform magnetic field. What is the direction of this field?
   a. out of the page
   b. into the page
   c. from bottom edge to top edge of the page
   d. from right to left across the page

9. If a proton is released at the equator and falls toward the Earth under the influence of gravity, the magnetic force on the proton will be toward the
   a. north
   b. south
   c. east
   d. west

10. Two parallel cables of a high-voltage transmission line carry equal and opposite currents of 1500 A. The distance between the cables is 3.00 m. What is the magnetic force acting on a 40.0-meter length of each cable?
    a. 6 N
    b. 60 N
    c. 600 N
    d. 6000 N

11. A square loop (L = 0.200 m) consists of 50 closely-wrapped turns which each carry a current of 0.500 A. The loop is oriented as shown in a uniform magnetic field of 0.40 T directed in the positive y direction. What is the magnitude of the torque on the loop?
    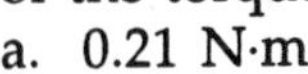
    a. 0.21 N·m
    b. 0.20 N·m
    c. 0.35 N·m
    d. 0.12 N·m

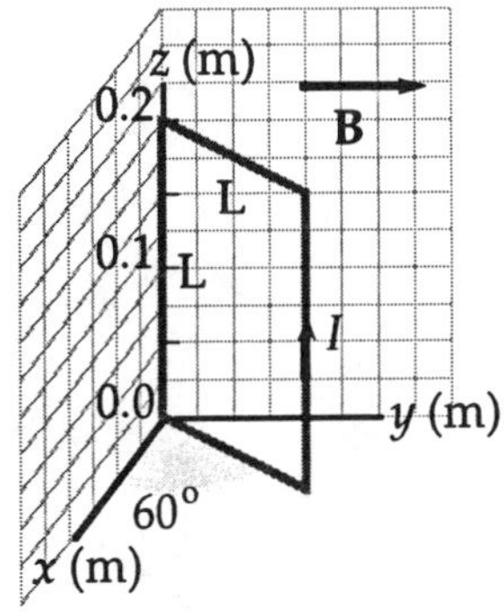

12. A superconducting solenoid is to be designed to generate a magnetic field of 10.0 T. If the solenoid winding has 2000 turns/meter, what is the required current? ($\mu_0 = 4\pi \times 10^{-7}$ A·m/T)
    a. 1000 A
    b. 1990 A
    c. 3980 A
    d. 5000 A

20
Induced Voltages and Inductance

Chapter 20

INDUCED VOLTAGES AND INDUCTANCE

Oersted (1819) discovered that a magnetic compass experiences a force in the vicinity of an electric current; this was the first evidence of a link between electricity and magnetism. Because nature is often symmetric, the discovery that electric currents produce magnetic fields led scientists to suspect that magnetic fields could produce electric currents. Indeed, experiments conducted by Michael Faraday in England and, independently, Joseph Henry in the United States in 1831 showed that a changing magnetic field could induce an electric current in a circuit. The results of these experiments led to a very basic and important law known as Faraday's law. In this chapter we discuss several practical applications of Faraday's law.

NOTES FROM SELECTED CHAPTER SECTIONS

20.2 Faraday's Law of Induction

The emf induced in a circuit is proportional to the time rate of change of magnetic flux through the circuit.

The polarity of the induced emf is such that it tends to produce a current that will create a magnetic flux to oppose the *change in flux* through the circuit.

20.3 Motional emf

A potential difference will be maintained across a conductor moving in a magnetic field as long as the direction of motion through the field is not parallel to the field direction. If the motion is reversed, the polarity of the potential difference will also be reversed.

20.7 Self-Inductance

The self-induced emf is always proportional to the *time rate of change* of current in the circuit.

The *inductance* of a device (an inductor) depends on its *geometry*.

20.8 *RL* Circuits

If a resistor and an inductor are connected in series to a battery, the current in the circuit will reach an *equilibrium* value ($\mathcal{E}/R$) after a time which is long compared to the *time constant* of the circuit, τ.

20.9 Energy Stored in a Magnetic Field

In an *RL* circuit, the rate at which energy is supplied by the battery equals the sum of the rate at which heat is dissipated in the resistor and the rate at which energy is stored in the inductor. The *energy density* is proportional to the *square of the magnetic field.*

EQUATIONS AND CONCEPTS

The total magnetic flux through a plane area, *A*, placed in a uniform magnetic field depends on the angle between the direction of the magnetic field and the direction perpendicular to the surface area.

$$\Phi \equiv B_{\perp} A = BA\cos\theta \qquad (20.1)$$

$$\Phi_{max} = BA$$

The maximum flux through the area occurs when the magnetic field is perpendicular to the plane of the surface area. When the magnetic field is parallel to the plane of the surface area, the flux through the area is zero. The unit of magnetic flux is the weber, Wb.

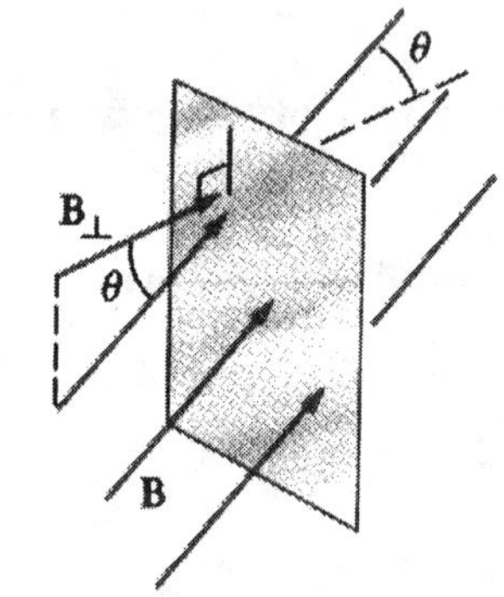

Faraday's law of induction states that the average emf induced in a circuit is proportional to the rate of change of magnetic flux through the circuit. The minus sign is included to indicate the polarity of the induced emf, which can be found by use of Lenz's law.

$$\mathcal{E} = -N\frac{\Delta\Phi}{\Delta t} \qquad (20.2)$$

Lenz's law states that the polarity of the induced emf (and the direction of the associated current in a closed circuit) produces a current whose magnetic field opposes the change in the flux through the loop. That is, the induced current tends to maintain the original flux through the circuit.

Comment on Lenz's law.

A "motional" emf is induced in a conductor of length, l, moving with speed, v, perpendicular to a magnetic field.

$$\mathcal{E} = Blv \qquad (20.4)$$

If the moving conductor is part of a complete circuit of resistance, R, a current will be induced in the circuit.

$$I = \frac{Blv}{R} \qquad (20.5)$$

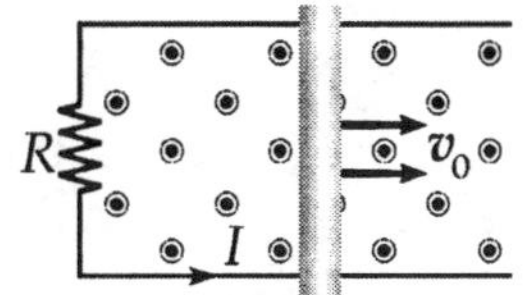

When a conducting loop of N turns and cross-sectional area, A, rotates with a constant angular velocity in a magnetic field, the emf induced in the loop will vary sinusoidally in time. For a given loop, the maximum value of the induced emf will be proportional to the angular velocity of the loop.

$$\mathcal{E} = NBA\omega \sin \omega t \qquad (20.7)$$

$$\mathcal{E}_{\text{max}} = NBA\omega \qquad (20.8)$$

A coil, solenoid, toroid, coaxial cable, or other conducting device is characterized by a parameter called its inductance, L. A change of current in the circuit will result in a self-induced emf which will be proportional to the rate of change of the current.

$$\mathcal{E} \equiv -L\frac{\Delta I}{\Delta t} \qquad (20.9)$$

The inductance of a given device, for example a coil, depends on its physical makeup—diameter, number of turns, type of material on which the wire is wound, and other geometric parameters. A circuit element which has a large inductance is called an inductor. The SI unit of inductance is the henry, H. A rate of change of current of 1 ampere per second in an inductor of 1 henry will produce a self-induced emf of 1 volt.

Comment on inductance.

$$1\,\text{H} = 1\frac{\text{V}\cdot\text{s}}{\text{A}}$$

The inductance of a coil can be calculated if the magnetic flux through the coil is known for a given current.

$$L = \frac{N\Phi}{I} \tag{20.10}$$

If the switch in a series circuit which contains a battery, resistor, and inductor is closed at time $t = 0$, the current in the circuit will increase in a characteristic fashion toward a maximum value of $I = \mathcal{E}/R$. The time required for the current to reach 63 percent of its final value is known as the time constant of the circuit.

$$\tau \equiv \frac{L}{R} \tag{20.15}$$

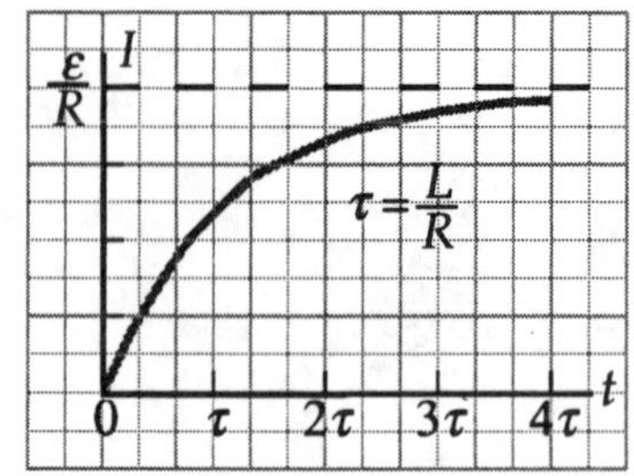

The energy stored in the magnetic field of an inductor is proportional to the square of the current; and the energy stored in the electric field of a charged capacitor is proportional to the square of the potential difference between the plates.

$$PE_L = \frac{1}{2}LI^2 \tag{20.16}$$

$$PE_C = \frac{1}{2}CV^2$$

REVIEW CHECKLIST

▷ Calculate the emf (or current) induced in a circuit when the magnetic flux through the circuit is changing in time. The variation in flux might be due to a change in (i) the area of the circuit, (ii) the magnitude of the magnetic field, (iii) the direction of the magnetic field, or (iv) the orientation/location of the circuit in the magnetic field.

▷ Apply Lenz's law to determine the direction of an induced emf or current. You should also understand that Lenz's law is a consequence of the law of conservation of energy.

▷ Calculate the emf induced between the ends of a conducting bar as it moves through a region where there is a constant magnetic field (motional emf).

▷ Define self-inductance, L, of a circuit in terms of appropriate circuit parameters. Calculate the total magnetic energy stored in a magnetic field if you are given the values of the inductance of the device with which the field is associated and the current in the circuit.

▷ Qualitatively describe the manner in which the instantaneous value of the current in an *LR* circuit changes while the current is either increasing or decreasing with time.

SOLUTIONS TO SELECTED END-OF-CHAPTER PROBLEMS

3. A square loop 2.00 m on a side is placed in a magnetic field of strength 0.300 T. If the field makes an angle of 50.0° with the normal to the plane of the loop, as in Figure 20.2, determine the magnetic flux through the loop.

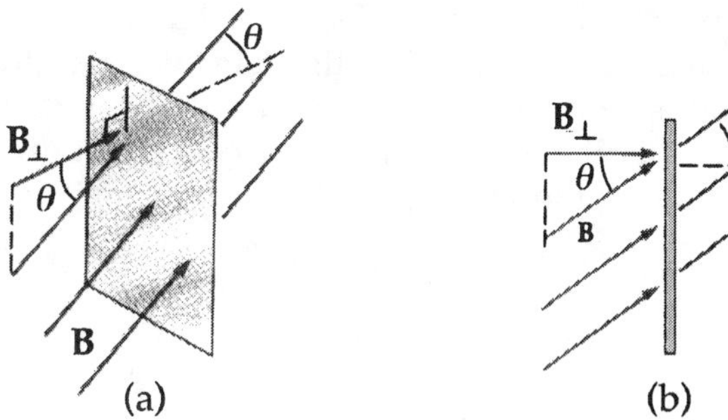

Figure 20.2

Solution $\quad \Phi = BA\cos\theta = (0.300\text{ T})(4.00\text{ m}^2)\cos 50.0° = 7.71\times10^{-1}\text{ T}\cdot\text{m}^2$ ◊

---

5. A solenoid 4.00 cm in diameter and 20.0 cm long has 250 turns and carries a current of 15.0 A. Calculate the magnetic flux through the circular cross-sectional area of the solenoid.

Solution $\quad$ We have $B = \mu_0 nI$, where $n = \dfrac{230\text{ turns}}{0.200\text{ m}} = 1250\text{ turns/m}$. Thus,

$$\Phi = BA\cos\theta = \mu_0 nIA \qquad \text{or}$$

$$\Phi = \left(4\pi\times10^{-7}\text{ T}\frac{\text{m}}{\text{A}}\right)\left(1250\frac{\text{turns}}{\text{m}}\right)(15.0\text{ A})\pi(2.00\times10^{-2}\text{ m})^2 = 2.96\times10^{-5}\text{ T}\cdot\text{m}^2 \quad ◊$$

---

15. A 500-turn circular-loop coil 15.0 cm in diameter is initially aligned so that its axis is parallel to the Earth's magnetic field. In 2.77 ms the coil is flipped so that its axis is perpendicular to the Earth's magnetic field. If a voltage of 0.166 V is thereby induced in the coil, what is the value of the Earth's magnetic field?

Solution $\Delta\Phi = \Phi_i - \Phi_f = B(A_f - A_i) = B\left[0 - \pi\left(\frac{d^2}{4}\right)\right]$ and $|\mathcal{E}_{\text{ave}}| = N\frac{|\Delta\Phi|}{\Delta t} = N\frac{B}{\Delta t}\frac{\pi d^2}{4}$

or $$B = \frac{4\,\Delta t|\mathcal{E}_{\text{ave}}|}{N\pi d^2} = \frac{4(2.77\times10^{-3}\text{ s})(0.166\text{ V})}{(500)\pi(15.0\times10^{-2}\text{ m})^2} = 5.20\times10^{-5}\text{ T} \quad \Diamond$$

---

17. The flexible loop in Figure 20.31 has a radius of 12.00 cm and is in a magnetic field of strength 0.150 T. The loop is grasped at points *A* and *B* and stretched until it closes. If it takes 0.200 s to close the loop, find the magnitude of the average induced emf in it during this time.

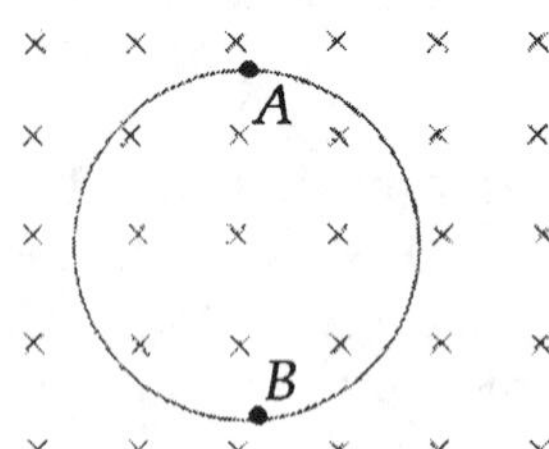

Figure 20.31

Solution

$$\Delta\Phi = B(A_f - A_i) = B(0 - A_i) = (0.150\text{ T})\left[-\pi(0.12\text{ m})^2\right] = -6.79\times10^{-3}\text{ T}\cdot\text{m}^2$$

The magnitude of the average value of the induced emf is

$$\mathcal{E}_{\text{ave}} = \left|\frac{\Delta\Phi}{\Delta t}\right| = \frac{6.79\times10^{-3}\text{ T}\cdot\text{m}^2}{0.200\text{ s}} = 3.39\times10^{-2}\text{ V} = 33.9\text{ mV} \quad \Diamond$$

---

21. A square, single-turn wire coil 1.00 cm on a side is placed inside a solenoid that has a circular cross-section of radius 3.00 cm, as shown in Figure 20.33. The solenoid is 20.0 cm long and wound with 100 turns of wire. (a) If the current in the solenoid is 3.00 A, find the flux through the coil. (b) If the current in the solenoid is reduced to zero in 3.00 s, find the magnitude of the average induced emf in the coil.

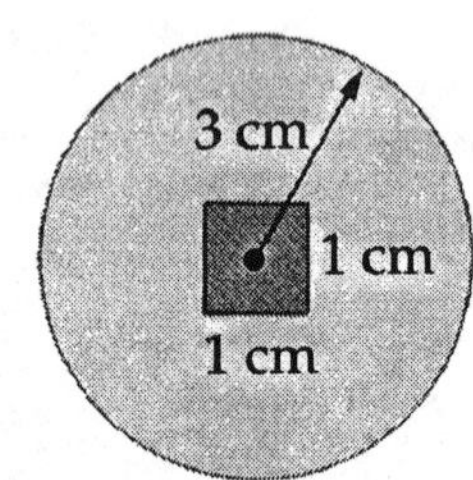

Figure 20.33

Solution (a) The magnetic field set up inside the solenoid is

$$B = \mu_0 nI = (4\pi \times 10^{-7}\ \text{T}\cdot\text{m/A})\frac{100\ \text{turns}}{0.200\ \text{m}}(3.00\ \text{A}) = 1.88 \times 10^{-3}\ \text{T}$$

$$\Phi_i = BA = (1.88 \times 10^{-3}\ \text{T})(10^{-2}\ \text{m})^2 = 1.88 \times 10^{-7}\ \text{T}\cdot\text{m}^2 \quad \Diamond$$

(b) When the current is reduced to zero, $\Phi_f = 0.$

Therefore, $\Delta\Phi = 1.88 \times 10^{-7}\ \text{T}\cdot\text{m}^2$

and $$\mathcal{E}_{\text{ave}} = N\frac{\Delta\Phi}{\Delta t} = \frac{1.88 \times 10^{-7}\ \text{T}\cdot\text{m}^2}{3.00\ \text{s}} = 6.28 \times 10^{-8}\ \text{V} \quad \Diamond$$

Note in the above equation, $N = 1$ (the number of turns in the square coil, not the turns in the solenoid).

---

25. Over a region where the *vertical* component of the Earth's magnetic field is 40.0 μT directed downward, a 5.00-m length of wire is held in an east-west direction and moved horizontally to the north with a speed of 10.0 m/s. Calculate the potential difference between the ends of the wire and determine which end is positive.

Solution $\mathcal{E} = Blv = (40.0 \times 10^{-6}\ \text{T})(5.00\ \text{m})(10.0\ \text{m/s}) = 2.00\ \times 10^{-3}\ \text{V} = 2.00\ \text{mV} \quad \Diamond$

Using the right-hand rule shows that the direction of the magnetic force on a positive charge in the wire is directed toward the west. Thus, a charge will drift to the western end of the wire, so the western end is *positive relative to the eastern end.* ◊

---

31. A bar magnet is positioned near a coil of wire as shown in Figure 20.37. What is the direction of the current through the resistor when the magnet is moved (a) to the left? (b) to the right?

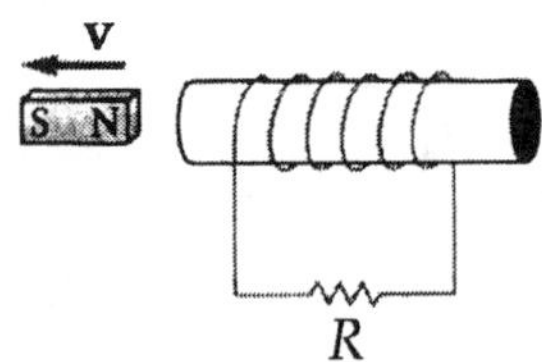

Figure 20.37

Solution Use Lenz's law to determine the direction of current in the coil and hence in the resistor.

(a) When the magnet is moved to left, the current is left to right. ◊

(b) When the magnet is moved to right, the current is right to left. ◊

---

35. What is the direction of the current induced in the resistor when the current in the long, straight wire in Figure 20.41 decreases rapidly to zero?

Solution Use Lenz's law to determine the direction of current in the loop when the current in the long wire decreases to zero. The result will be a current from left to right in the resistor. ◊

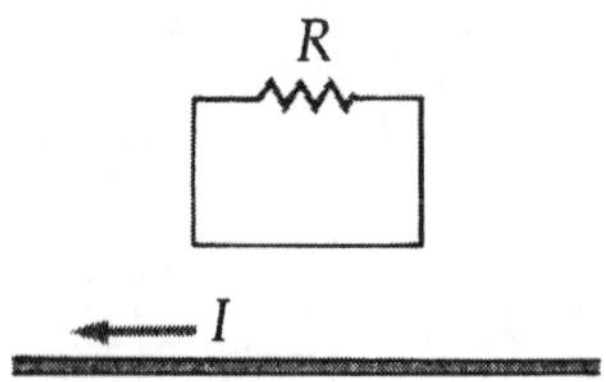

Figure 20.41

---

37. The magnetic field shown in Figure 20.43 has a uniform magnitude of 25.0 mT directed into the paper. The initial diameter of the kink is 2.00 cm. (a) The wire is quickly pulled taut, and the kink shrinks to a diameter of zero in 50.0 ms. Determine the average voltage induced between endpoints A and B. Include the polarity. (b) Suppose the kink is undisturbed, but the magnetic field increases to 100 mT in 4.00×10^{-3} s. Determine the average voltage across terminals A and B, including polarity, during this period.

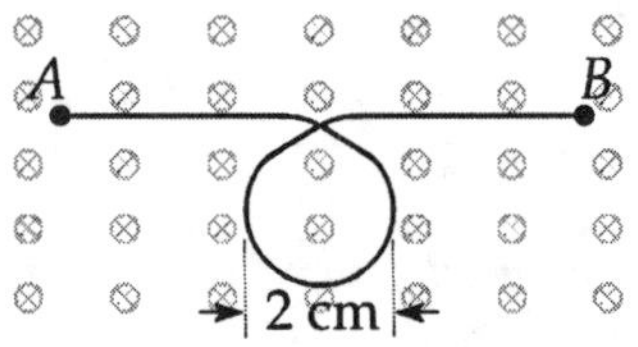

Figure 20.43

Solution (a) $A_i = \pi r_i^2 = \pi(10^{-2}\ \text{m})^2 = \pi 10^{-4}\ \text{m}^2$ and $A_f = 0$

Therefore,

$$\Delta\Phi = BA_i - BA_f = BA_i = (25.0 \times 10^{-3}\ \text{T})(\pi 10^{-4}\ \text{m}^2)$$

$$\Delta\Phi = 7.85 \times 10^{-6}\ \text{T} \cdot \text{m}^2$$

Thus, $$\mathcal{E}_{ave} = N\frac{\Delta\Phi}{\Delta t} = \frac{7.85\times10^{-6}\ \text{T}\cdot\text{m}^2}{50.0\times10^{-3}\ \text{s}} = 1.57\times10^{-4}\ \text{V} = 0.157\ \text{mV} \quad \Diamond$$

and Lenz's law shows that the induced current will flow from A to B. (End B will be positive.) ◊

(b) $\Delta\Phi = B_f A - B_i A = (100\ \text{m}\cdot\text{T})A - (25.0\ \text{m}\cdot\text{T})A = (75.0\times10^{-3}\ \text{T})(\pi 10^{-4}\ \text{m}^2)$ or

$\Delta\Phi = 2.36\times10^{-5}\ \text{T}\cdot\text{m}^2$ and $\mathcal{E}_{ave} = N\frac{\Delta\Phi}{\Delta t}$ gives

$$\mathcal{E}_{ave} = \frac{2.36\times10^{-5}\ \text{T}\cdot\text{m}^2}{4.00\times10^{-3}\ \text{s}} = 5.89\times10^{-3}\ \text{V} = 5.89\ \text{mV} \quad \Diamond$$

In this case, the magnetic force on a positive charge in the wire causes it to drift toward end A, so end A is positive. (Current flow is from B to A.) ◊

---

43. An ac generator with its terminals shorted together consists of 40.0 turns of wire with an area of 0.120 m^2 and a total resistance of 30.0 Ω. The loop rotates in a magnetic field of 0.100 T at a constant frequency of 60.0 Hz. (a) Find the maximum induced emf, (b) the maximum induced current, (c) an expression for the time variation of $\mathcal{E}$, and (d) an expression for the time variation of the induced current.

Solution (a) $\omega = 2\pi f = 2\pi(60.0\ \text{Hz}) = 377\ \text{s}^{-1}$ and

$$\mathcal{E}_{max} = NAB\omega = (40.0)(0.100\ \text{T})(0.120\ \text{m}^2)(377\ \text{s}^{-1}) = 181\ \text{V} \quad \Diamond$$

(b) $I_{max} = \frac{\mathcal{E}_{max}}{R} = \frac{181\ \text{V}}{30.0\ \Omega} = 6.03\ \text{A}$ ◊

(c) $\mathcal{E} = \mathcal{E}_{max}\sin\omega t = (181\ \text{V})\sin\left[(377\ \text{s}^{-1})t\right]$ ◊

(d) $i = \frac{\mathcal{E}}{R} = (6.03\ \text{A})\sin\left[(377\ \text{s}^{-1})t\right]$ ◊

---

47. In a model ac generator, a 500-turn rectangular coil of dimensions 8.00 cm by 20.0 cm rotates at 120 rev/min in a uniform magnetic field of 0.600 T. (a) What is the maximum emf induced in the coil? (b) What is the instantaneous value of the emf in the coil at $t = (\pi/32)$ s? Assume that the emf is zero at $t = 0$. (c) What is the smallest value of t for which the emf will have its maximum value?

Solution (a) $\omega = 120 \text{ rev/min} = 12.6 \text{ rad/s}$ and

$$\mathcal{E}_{max} = NAB\omega = (500)(0.600 \text{ T})(0.080 \text{ m})(0.200 \text{ m})(12.6 \text{ rad/s}) = 60.3 \text{ V} \quad \lozenge$$

(b) $\mathcal{E} = \mathcal{E}_{max} \sin\omega t = (60.3 \text{ V})\sin\left[12.6 \text{ s}^{-1}\left(\frac{\pi}{32}\right)\right] = 56.9 \text{ V} \quad \lozenge$

(c) The emf will be a maximum at

$$t = \frac{T}{4} = \frac{2\pi/\omega}{4} = \frac{\pi}{2\omega} = \frac{\pi}{2(12.6 \text{ rad/s})} = 0.125 \text{ s} \quad \lozenge$$

---

49. A Slinky toy spring has a radius of 4.00 cm and an inductance of 275 μH when extended to a length of 1.50 m. What is the total number of turns in the spring?

Solution From $L = \frac{\mu_0 N^2 A}{l}$, we find $N = \sqrt{\frac{Ll}{\mu_0 A}}$

$$N = \sqrt{\frac{(275 \times 10^{-6} \text{ H})(1.50 \text{ m})}{(4\pi \times 10^{-7} \text{ N/A}^2)\pi(0.040 \text{ m})^2}} = 256 \text{ turns} \quad \lozenge$$

---

55. A 6.00-V battery is connected in series with a resistor and an inductor. The series circuit has a time constant of 600 μs, and the maximum current is 300 mA. What is the value of the inductance?

Solution $R = \frac{V}{I_{max}} = \frac{6.00 \text{ V}}{3.00 \times 10^{-4} \text{ A}} = 20.0 \ \Omega$

Therefore, $L = \tau R = (6.00 \times 10^{-4} \text{ s})(20.0 \ \Omega) = 1.20 \times 10^{-2} \ \Omega\text{s} = 12.0 \text{ mH} \quad \lozenge$

---

59. How much energy is stored in a 70.0-mH inductor at the instant when the current is 2.00 A?

Solution $W = \frac{1}{2}LI^2 = \frac{1}{2}(70.0 \times 10^{-3}\text{ H})(2.00\text{ A})^2 = 0.140\text{ J}$ ◊

---

63. A tightly wound circular coil has 50.0 turns, each of radius 0.200 m. A uniform magnetic-field is introduced perpendicularly to the plane of the coil. If the field increases in strength from 0 to 0.300 T in 0.400 s, what average emf is induced in the windings of the coil?

Solution We use $\mathcal{E} = N\frac{\Delta\Phi}{\Delta t}$, with

$$\Delta\Phi = \Delta B(A) = (0.300\text{ T} - 0)\pi(0.200\text{ m})^2 = 3.77 \times 10^{-2}\text{ T}\cdot\text{m}^2$$

Thus, $$\mathcal{E} = \frac{50(3.77 \times 10^{-2}\text{ T}\cdot\text{m}^2)}{0.400\text{ s}} = 4.70\text{ V} \quad ◊$$

---

65. An automobile starter motor draws a current of 3.50 A from a 12.0-V battery when operating at normal speed. A broken pulley locks the armature in position, and the current increases to 18.0 A. What is the back emf of the motor?

Solution When not rotating, $\mathcal{E} = IR$, and from this, $R = \frac{\mathcal{E}}{I} = \frac{12.0\text{ V}}{18.0\text{ A}} = 0.667\ \Omega$.

When rotating, $\mathcal{E} - \mathcal{E}_{\text{back}} = IR$, or

$$\mathcal{E}_{\text{back}} = \mathcal{E} - IR = 12.0\text{ V} - (3.50\text{ A})(0.667\ \Omega) = 9.70\text{ V} \quad ◊$$

---

71. Shown in Figure 20.52 on the following page, is a graph of the induced emf versus time for a coil of N turns rotating with angular velocity ω in a uniform magnetic field directed perpendicularly to the axis of rotation of the coil. Copy this sketch (to a larger scale), and on the same set of axes show the graph of emf versus t when (a) the number of turns in the coil is doubled, (b) the angular velocity is doubled, and (c) the angular velocity is doubled while the number of turns in the coil is halved.

Solution

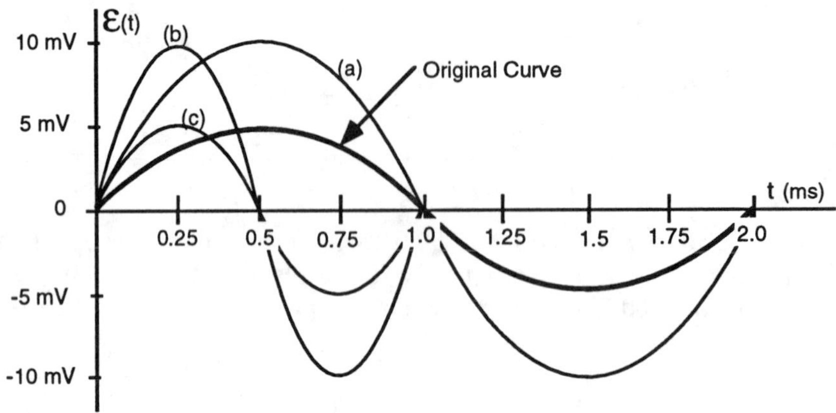

(a) Doubling the number of turns doubles the amplitude but does not alter the period. ◊

(b) Doubling the angular velocity doubles the amplitude and also cuts the period in half. ◊

(c) Doubling the angular velocity while reducing the number of turns to one half the original value leaves the amplitude unchanged but does cut the period in half. ◊

---

73. A single-turn circular loop of radius 0.200 m is coaxial with a long 1600-turn solenoid of radius 0.050 m and length 0.800 m, as in Figure 20.54 The variable resistor is changed so that the solenoid current decreases linearly from 6.00 A to 1.50 A in 0.200 s. Calculate the induced emf in the circular loop. (The field just outside the solenoid is small enough to be negligible.)

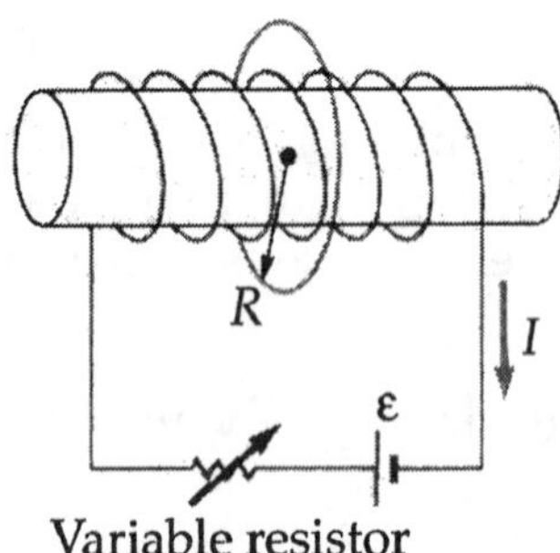

Figure 20.54

Solution The flux through the solenoid is given by $\Phi = BA = \mu_0 nIA$.

Thus, the change in flux is $\Delta\Phi = \mu_0 nA\,\Delta I$ and $\mathcal{E} = N\left(\frac{\Delta\Phi}{\Delta t}\right) = \mu_0 nA\left(\frac{\Delta I}{\Delta t}\right)$ or

$$\mathcal{E} = \left(4\pi\times10^{-7}\ \frac{\text{N}}{\text{A}^2}\right)\left(\frac{1600}{0.800\ \text{m}}\right)\pi(0.050\ \text{m})^2\left(\frac{6.00\ \text{A} - 1.50\ \text{A}}{0.200\ \text{s}}\right) = 4.44\times10^{-4}\ \text{V} = 440\ \mu\text{V} \quad ◊$$

CHAPTER SELF-QUIZ

1. In a circuit made up of inductor, resistance, ammeter, battery, and switch in series, at which one of the following time intervals after the switch is closed will the rate of current increase be greatest?
   a. zero
   b. one time constant
   c. reciprocal of one time constant
   d. ten time constants

2. A 0.20-m wire is moved perpendicular to a 0.50 magnetic field at a speed of 1.50 m/s. What emf is induced across the ends of the wire?
   a. 2.25 V
   b. 1.00 V
   c. 0.60 V
   d. 0.15 V

3. A coil with a self-inductance of 0.75 mH experiences a constant increase in current from zero to 5.00 A in 0.125 s. What is the induced emf during this interval?
   a. 0.045 V
   b. 0.030 V
   c. 0.470 V
   d. 0.019 V

4. What is the self-inductance in a coil which experiences a 1.50-V induced emf when the current is changing at a rate of 55 A/s?
   a. 83 mH
   b. 45 mH
   c. 37 mH
   d. 27 mH

5. An inductor, battery, resistor, ammeter, and switch are connected in series. If the switch, initially open, is now closed, what is the current's final value?
   a. zero
   b. battery voltage divided by inductance
   c. battery voltage times inductance
   d. battery voltage divided by resistance

6. A coil with a self-inductance of 1.50 mH will show what time rate of change of current when inducing an emf of 0.30 V?
   a. 2.00×10^2 A/s
   b. 0.45×10^{-4} A/s
   c. 5.00×10^{-2} A/s
   d. 0.30 A/s

7. The self-inductance of a solenoid increases under which of the following conditions?
   a. solenoid length is increased
   b. cross-sectional area is decreased
   c. number of coils per unit length is decreased
   d. number of coils is increased

8. A 12.0-V battery is connected in series with a switch, resistor, and coil. If the circuit's time constant is 4.00×10^{-4} s and the final steady current after the switch is closed becomes 2.00 A, what is the value of the inductance?
   a. 1.20 mH
   b. 2.40 mH
   c. 9.60 mH
   d. 48.0 mH

9. A bar magnet is falling through a loop of wire with constant velocity with the north pole entering first. As the north pole enters the wire, the induced current as viewed from above will be:
   a. clockwise
   b. counterclockwise
   c. zero
   d. along the length of the magnet

10. A uniform 1.50-T magnetic field passes perpendicularly through the plane of a wire loop 0.300 m^2 in area. What flux passes through the loop?
    a. 5.00 Wb
    b. 0.45 Wb
    c. 0.25 Wb
    d. 0.135 Wb

11. An *RL*-series circuit has the following components: 2.50-mH coil, 0.500-ohm resistor, 6.00-V battery, ammeter, and switch. What is the time constant of this circuit?
    a. 12.5×10^{-3} s
    b. 5.00×10^{-3} s
    c. 2.50×10^{-2} s
    d. 200 s

12. Electricity may be generated by rotating a loop of wire between the poles of a magnet. The induced current is greatest when:
    a. the plane of the loop is parallel to the magnetic field
    b. the plane of the loop is perpendicular to the magnetic field
    c. the magnetic flux through the loop is a maximum
    d. the plane of the loop makes an angle of 45.0° with the magnetic field

21
Alternating Current Circuits and Electromagnetic Waves

Chapter 21

ALTERNATING CURRENT CIRCUITS AND ELECTROMAGNETIC WAVES

It is important to understand the basic principles of alternating current (ac) circuits because they are so much a part of our everyday life. We begin our study of ac circuits by examining the characteristics of a circuit containing a source of emf and a single circuit element: a resistor, a capacitor, or an inductor. Then we examine what happens when these elements are connected in combination with each other. Our discussion is limited to situations in which the elements are arranged in simple series configurations.

We conclude this chapter with a discussion of electromagnetic waves, which are composed of fluctuating electric and magnetic fields.

NOTES FROM SELECTED CHAPTER SECTIONS

21.1 Resistors in an AC Circuit

If an ac circuit consists of a generator and a resistor, the *current in the circuit is in phase with the voltage.* That is, the current and voltage reach their maximum values at the same time. The *average* value of the current over *one complete cycle* is zero. The *rms* current refers to *root mean square,* which simply means the square root of the average value of the square of the current.

21.2 Capacitors in an AC Circuit

When an alternating voltage is applied across a capacitor, the voltage reaches its maximum value one quarter of a cycle after the current reaches its maximum value. In this situation, it is common to say that the *voltage always lags the current* through a capacitor by 90.0°.

21.3 Inductors in an AC Circuit

When a sinusoidal voltage is applied across an inductor, the voltage reaches its maximum value one quarter of an oscillation period before the current reaches its maximum value. In this situation, we say that the *voltage always leads the current* by 90.0°.

21.4 The *RLC* Series Circuit

The ac current at all points in a series ac circuit has the same amplitude and phase. Therefore, the voltage across each element will have *different* amplitudes and phases: the voltage across the resistor is in phase with the current, the voltage across the inductor leads the current by 90.0°, and the voltage across the capacitor lags behind the current by 90.0°.

21.5 Power in an AC Circuit

The average power delivered by the generator is dissipated as heat in the resistor. There is *no power* loss in an ideal inductor or capacitor.

21.6 Resonance in a Series *RLC* Circuit

The current in a series *RLC* circuit reaches its peak value when the frequency of the generator equals ω_0; that is, when the "driving" frequency matches the *resonance frequency*.

21.7 The Transformer

A transformer is a device designed to raise or lower an ac voltage and current without causing an appreciable change in the product IV. In its simplest form, it consists of a primary coil of N_1 turns and a secondary coil of N_2 turns, both wound on a common soft iron core. In an ideal transformer, the power delivered by the generator must equal the power dissipated in the load.

21.8 Maxwell's Predictions and
21.9 Hertz's Discoveries

Electromagnetic waves are generated by accelerating electric charges. The radiated waves consist of oscillating electric and magnetic fields, which are *at right angles to each other* and also *at right angles to the direction of wave propagation.*

The fundamental laws governing the behavior of electric and magnetic fields are Maxwell's equations. In this unified theory of electromagnetism, Maxwell showed that electromagnetic waves are a natural consequence of these fundamental laws. The theory he developed is based upon the following four pieces of information:

1. Electric fields originate on positive charges and terminate on negative charges. The electric field due to a point charge can be determined at a location by applying Coulomb's force law to a test charge placed at that location.

2. Magnetic field lines always form closed loops; that is, they do not begin or end anywhere.

3. A varying magnetic field induces an emf and hence an electric field. This is a statement of Faraday's law (Chapter 20).

4. Magnetic fields are generated by moving charges (or currents), as summarized in Ampère's law (Chapter 19).

21.11 Properties of Electromagnetic Waves

Following is a summary of the properties of electromagnetic waves:

1. The solutions of Maxwell's third and fourth equations are wavelike, where both **E** and **B** satisfy the same wave equation.

2. Electromagnetic waves travel through empty space with the speed of light, $c = 1/\sqrt{\varepsilon_0 \mu_0}$.

3. The electric and magnetic field components of plane electromagnetic waves are perpendicular to each other and also perpendicular to the direction of wave propagation. The latter property can be summarized by saying that electromagnetic waves are transverse waves.

4. The relative magnitudes of **E** and **B** in empty space are related by $E/B = c$.

5. Electromagnetic waves obey the principle of superposition.

EQUATIONS AND CONCEPTS

The output of an ac generator is sinusoidal where v is the instantaneous voltage and V_m is the maximum voltage.

$$v = V_m \sin 2\pi f t \tag{21.1}$$

These two equations relate the rms values of current and voltage to the maximum values of these quantities.

$$I = \frac{I_m}{\sqrt{2}} = 0.707\ I_m \tag{21.2}$$

$$V = \frac{V_m}{\sqrt{2}} = 0.707\ V_m \tag{21.3}$$

The rms voltage across a resistor is related to the rms current in the resistor by Ohm's law.

$$V_R = IR \qquad (21.4)$$

The impeding effect of a capacitor to the current in an ac circuit is expressed in terms of a factor called the *capacitive reactance.*

$$X_C \equiv \frac{1}{2\pi fC} \qquad (21.5)$$

$$V_C = IX_C \qquad (21.6)$$

The effective resistance of a coil in an ac circuit is measured by a quantity called the inductive reactance.

$$X_L \equiv 2\pi fL \qquad (21.8)$$

$$V_L = IX_L \qquad (21.9)$$

The instantaneous voltage across a resistor is in phase with the current.

$$i = I_m \sin 2\pi ft$$

$$V_R = V_R \sin 2\pi ft$$

The instantaneous voltage across an inductor leads the current by a quarter cycle.

$$V_L = V_L \cos 2\pi ft$$

The instantaneous voltage across the capacitor lags the current by a quarter cycle.

$$v_C = -v_C \cos 2\pi ft$$

The net instantaneous voltage across all three circuit elements is the sum of the instantaneous voltages across the separate elements; the net voltage is "out-of-step" with the instantaneous current by an amount called the phase angle, ϕ.

$$v = V\sin(2\pi ft + \phi)$$

The rms voltage across the combination of resistor, inductor, and capacitor can be determined in terms of the rms voltage values across the individual components.

$$V = \sqrt{V_R{}^2 + (V_L - V_C)^2} \qquad (21.11)$$

$$V = I\sqrt{R^2 + (X_L - X_C)^2} \qquad (21.13)$$

The resonance frequency of oscillation in an LC circuit depends on the values of the inductance and capacitance in the circuit.

$$f_0 = \frac{1}{2\pi\sqrt{LC}}$$

The speed of an electromagnetic wave is related to the permeability and permittivity of the medium through which it travels. Electromagnetic waves travel with the speed of light.

$$c = \frac{1}{\sqrt{\mu_0 \varepsilon_0}} \qquad (21.25)$$

Permeability constant of free space.

$$\mu_0 = 4\pi \times 10^{-7}\ \mathrm{T \cdot m/A}$$

Permittivity constant of free space.

$$\varepsilon_0 = 8.85 \times 10^{-12}\ \mathrm{C/N \cdot m^2}$$

Accurate value for the speed of light in a vacuum.

$$c = 2.99792 \times 10^{8}\ \mathrm{m/s} \qquad (21.26)$$

The ratio of the electric to the magnetic field in an electromagnetic wave is constant and equal to the speed of light.

$$\frac{E}{B} = c \qquad (21.27)$$

Electromagnetic waves carry energy as they travel through space. The rate of flow of energy, or power per unit area, transported perpendicular to a surface can be expressed in several alternate forms involving the maximum values of the electric and magnetic fields. The power per unit area given by these equations is the average power per unit area. Also, it can be shown that the energy carried by an electromagnetic wave is shared equally by the electric and magnetic fields.

Average power per unit area =

$$\frac{E_m B_m}{2\mu_0} \qquad (21.28)$$

Average power per unit area =

$$\frac{E_m{}^2}{2\mu_0 c} = \frac{c}{2\mu_0} B_m{}^2 \qquad (21.29)$$

The product of frequency and wavelength of an electromagnetic wave propagating in vacuum is constant and equal to c.

$$c = f\lambda \qquad (21.30)$$

The circuit rms voltage can also be calculated in terms of the common circuit current and the values of resistance, inductive reactance, and capacitive reactance.

By defining a quantity, Z, called the impedance, it is possible to relate the rms voltage and rms current in the form of a generalized Ohm's law.

$$Z \equiv \sqrt{R^2 + (X_L - X_C)^2} \qquad (21.14)$$

$$V = IZ \qquad (21.15)$$

The phase angle between rms voltage and rms current in the circuit can be determined from the impedance triangle or from the voltage triangle.

$$\tan\phi = \frac{X_L - X_C}{R} \qquad (21.16)$$

$$\tan\phi = \frac{V_L - V_C}{V_R} \qquad (21.12)$$

The only element in an ac circuit which dissipates energy is the resistor. The average power dissipated in an ac circuit can be expressed in terms of the power factor of the circuit, $\cos\phi$.

$$P_{av} = I^2 R \qquad (21.17)$$

$$P_{av} = IV\cos\phi \qquad (21.18)$$

When the frequency of an ac circuit is equal to the resonance frequency, the impedance of the circuit becomes equal to the circuit resistance and the current becomes maximum.

$$f_0 = \frac{1}{2\pi\sqrt{LC}} \qquad (21.20)$$

The primary voltage and current values in a transformer are related to the values of those quantities in the secondary in terms of the ratio of the number of turns in the primary and secondary coils. In a step-up transformer, N_2 is greater than N_1.

$$V_2 = \frac{N_2}{N_1} V_1 \qquad (21.23)$$

$$I_1 V_1 = I_2 V_2 \qquad (21.24)$$

Chapter 21

SUGGESTIONS, SKILLS, AND STRATEGIES

The following procedures are recommended when solving alternating current problems:

1. The first step in analyzing alternating current circuits is to calculate as many of the unknown quantities such as X_L and X_C as possible. (Note that when calculating X_C, the capacitance should be expressed in farads, rather than, say, microfarads.)

2. Apply the equation $V = IZ$ to that portion of the circuit of interest. That is, if you want to know the voltage drop across the combination of an inductor and a resistor, the equation reduces to $V = I\sqrt{R^2 + X_L^2}$.

REVIEW CHECKLIST

▷ Apply the formulas that give the reactance values in an ac circuit as a function of (i) capacitance, (ii) inductance, and (iii) frequency. Interpret the meaning of the terms *phase angle* and *power factor* in an ac circuit.

▷ Given an *RLC* series circuit in which values of resistance, inductance, capacitance, and the characteristics of the generator (source of emf) are known, calculate: (i) the instantaneous and rms voltage drop across each component, (ii) the instantaneous and rms current in the circuit, (iii) the phase angle by which the current leads or lags the voltage, (iv) the power expended in the circuit, and (v) the resonance frequency of the circuit.

▷ Understand the manner in which step-up and step-down transformers are used in the process of transmitting electrical power over large distances; and make calculations of primary to secondary voltage and current ratios for an ideal transformer.

▷ Describe the contribution by James Clerk Maxwell, properly relating the significance of the information available to him, to the theoretical understanding of the nature of electromagnetic radiation. Summarize the properties of electromagnetic waves.

▷ Relate the relative orientation of magnetic field, electric field, and direction of propagation in the corresponding electromagnetic wave. Justify the statement that electromagnetic waves carry both energy and momentum.

SOLUTIONS TO SELECTED END-OF-CHAPTER PROBLEMS

3. Figure 21.26 shows three lamps connected to the 110-V ac (rms) household supply voltage. Lamps 1 and 2 have 150-W bulbs; lamp 3 has a 100-W bulb (average power). Find the rms value of the current I and the resistance of each bulb.

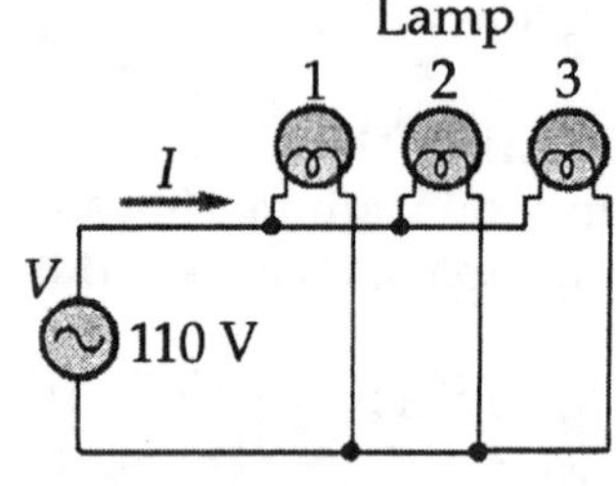

Figure 21.26

Solution $P = IV$ and $V = 110$ V for each bulb (parallel circuit), so

$$I_1 = I_2 = \frac{P_1}{V} = \frac{150 \text{ W}}{110 \text{ V}} = 1.36 \text{ A} \quad \text{and}$$

$$R_1 = R_2 = \frac{V}{I_1} = 80.7\ \Omega \quad \Diamond$$

$$I_3 = \frac{P_3}{V} = \frac{100 \text{ W}}{110 \text{ V}} = 0.909 \text{ A} \quad \text{and} \quad R_3 = \frac{V}{I_3} = 121\ \Omega$$

$$I_{\text{total}} = I_1 + I_2 + I_3 = 3.64 \text{ A} \quad \Diamond$$

---

5. An ac voltage source has an output of $v = 150 \sin 377t$. Find (a) the rms voltage output, (b) the frequency of the source, and (c) the voltage at $t = 1/120$ s. (d) Find the maximum current in the circuit when the generator is connected to a 50.0-Ω resistor.

Solution Compare the expression given for the voltage, $v = 150 \sin 377t.$, with the general expression for an ac voltage, $v = V_m \sin 2\pi f t$. By comparison, we see

(a) $V_m = 150$ V, and from this, $V = \frac{V_m}{\sqrt{2}} = \frac{150 \text{ V}}{\sqrt{2}} = 106 \text{ V} \quad \Diamond$

(b) We also see that $\omega = 2\pi f = 377$ rad/s. Thus, $f = \frac{377 \text{ rad/s}}{2\pi} = 60.0 \text{ Hz} \quad \Diamond$

(c) At $t = \frac{1}{120 \text{ s}}$, $v = 150 \text{ V} \sin\left(\frac{377}{120}\right) = 150 \text{ V} \sin \pi = 0 \quad \Diamond$

(d) $$I_m = \frac{V_m}{R} = \frac{150 \text{ V}}{50.0\ \Omega} = 3.00 \text{ A} \quad \Diamond$$

---

9. The generator in a purely capacitive ac circuit has an angular frequency of 120 π rad/s. If V_m = 140 V and C = 6.00 μF, what is the rms current in the circuit?

Solution $X_C = \dfrac{1}{\omega C}$, where $\omega = 2\pi f$ = the angular frequency.

Thus, $$X_C = \frac{1}{(120\pi \text{ rad/s})(6.00 \times 10^{-6} \text{ F})} = 442\ \Omega$$

Also, $V_m = 140 \text{ V} = \sqrt{2}\text{ V}$ and $V = \dfrac{140 \text{ V}}{\sqrt{2}} = 99.0 \text{ V}$

Thus, $$I = \frac{V}{X_C} = \frac{99.0 \text{ V}}{442\ \Omega} = 0.224 \text{ A} \quad \Diamond$$

---

13. An inductor is connected to a 20.0-Hz power supply that produces a 50.0-V rms voltage. What inductance is needed to keep the maximum current in the circuit below 80.0 mA?

Solution $X_L = 2\pi f L = \dfrac{V}{I}$ and $I = \dfrac{I_m}{\sqrt{2}}$

so $L = \dfrac{\sqrt{2}\, V}{2\pi f I_m}$ and for $I < 80.0$ mA,

$$L > \frac{\sqrt{2}\,(50.0 \text{ V})}{(2\pi)(20 \text{ s}^{-1})(8.00 \times 10^{-2} \text{ A})} \quad \text{or} \quad L > 7.03 \text{ H} \quad \Diamond$$

---

17. A 40.0-μF capacitor is connected to a 50.0-Ω resistor and a generator whose rms output is 30.0 V at 60.0 Hz. Find (a) the rms current in the circuit, (b) the voltage drop across the resistor, (c) the voltage drop across the capacitor, and (d) the phase angle for the circuit. (e) Sketch the phasor diagram for this circuit.

Solution We have $X_C = \dfrac{1}{2\pi f C} = \dfrac{1}{2\pi(60.0\text{ Hz})(40.0\times 10^{-6}\text{ F})} = 66.3\ \Omega$

Thus, $Z = \sqrt{R^2 + (X_L - X_C)^2}$ becomes

$$Z = \sqrt{R^2 + (X_C)^2} = \sqrt{(150\ \Omega)^2 + (66.3\ \Omega)^2} = 83.1\ \Omega$$

(a) $I = \dfrac{V}{Z} = \dfrac{30.0\text{ V}}{83.1\ \Omega} = 0.361\text{ A}$ ◊

(b) $V_R = IR = (0.361\text{ A})(50.0\ \Omega) = 18.1\text{ V}$ ◊

(c) $V_C = IX_C = (0.361\text{ A})(66.3\ \Omega) = 23.9\text{ V}$ ◊

(d) $\tan\phi = \dfrac{X_L - X_C}{R} = \dfrac{-X_C}{R} = \dfrac{-66.3\ \Omega}{50.0\ \Omega} = -1.33$

and $\phi = -53.0°$ ◊

(e) The phasor diagram is sketched at the right.

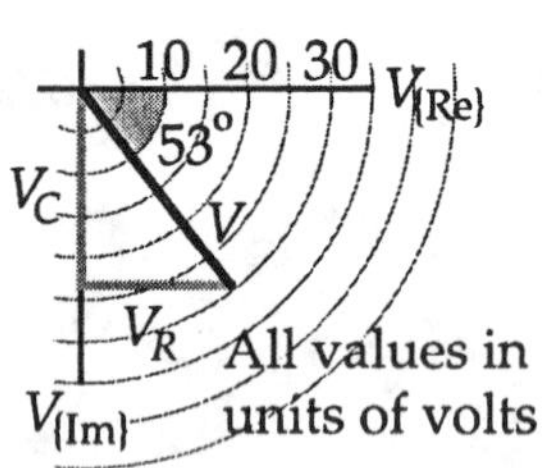

(e)

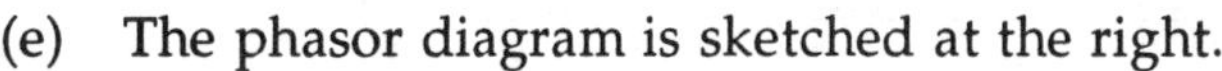

21. A 50.0-Ω resistor, a 0.100-H inductor, and a 10.0-μF capacitor are connected in series to a 60.0-Hz source. The rms current in the circuit is 2.75 A. Find the rms voltages across (a) the resistor, (b) the inductor, (c) the capacitor, and (d) the *RLC* combination. (e) Sketch the phasor diagram for this circuit.

Solution $X_C = \dfrac{1}{2\pi f C} = \dfrac{1}{2\pi(60.0\text{ Hz})(10.0\times 10^{-6}\text{ F})} = 265\ \Omega$

and $X_L = 2\pi f L = 2\pi(60.0\text{ Hz})(0.100\text{ H}) = 37.7\ \Omega$

(a) $V_R = IR = (2.75\text{ A})(50.0\ \Omega) = 138\text{ V}$ ◊

(b) $V_L = IX_L = (2.75\text{ A})(37.7\ \Omega) = 104\text{ V}$ ◊

(c) $V_C = IX_C = (2.75\text{ A})(265\ \Omega) = 729\text{ V}$ ◊

(d) $V = \sqrt{V_R^2 + (V_L - V_C)^2}$ or

$$V = \sqrt{(138\text{ V})^2 + (104\text{ V} - 729\text{ V})^2} = 640\text{ V}$$ ◊

(e) The phasor diagram is sketched at the right.

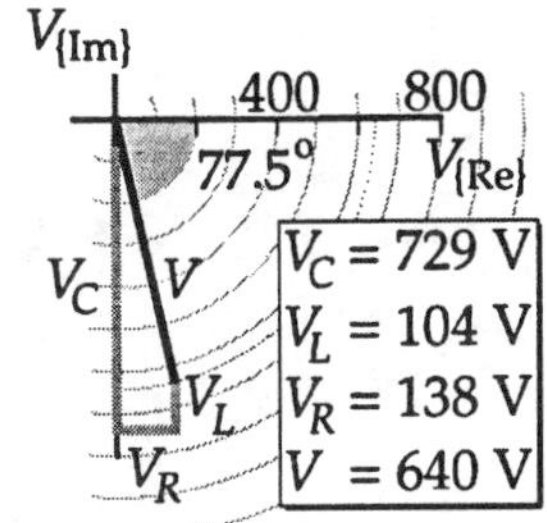

(e)

---

25. A 50.0-Ω resistor is connected in series with a 15.0-μF capacitor and a 60.0-Hz, 120-V source. (a) Find the current in the circuit. (b) What is the value of the inductor that must be inserted in the circuit to reduce the current to one-half that found in (a)?

Solution $X_C = \dfrac{1}{2\pi f C} = \dfrac{1}{2\pi(60.0\text{ Hz})(15.0\times 10^{-6}\text{ F})} = 177\ \Omega$

and $Z = \sqrt{R^2 + (X_L - X_C)^2} = \sqrt{R^2 + (X_C)^2} = 184\ \Omega$

(a) $I = \dfrac{V}{Z} = \dfrac{120\text{ V}}{184\ \Omega} = 0.653\text{ A} = 650\text{ mA}$ ◊

(b) If $I_2 = \dfrac{1}{2} I_1$, then $Z_2 = 2Z_1$ where Z_2 is the value of the impedance when an inductor is added to the circuit. Therefore,

$$Z_2^2 = 4Z_1^2 \quad \text{or} \quad R^2 + (X_L - X_C)^2 = 4\left[R^2 + (X_C)^2\right]$$

This reduces to the expression $X_L = X_C \pm 365\ \Omega = 177\ \Omega \pm 365\ \Omega$

We must use the + sign so that the inductive reactance will be greater than zero. Thus,

$$X_L = 542\ \Omega = 2\pi f L \quad \text{and} \quad L = \frac{542\ \Omega}{2\pi(60.0\text{ Hz})} = 1.44\text{ H}$$ ◊

---

27. A resistor (R = 900 Ω), a capacitor (C = 0.250 μF), and an inductor (L = 2.50 H) are connected in series across a 240-Hz ac source for which V_m = 140 V. Calculate the

(a) impedance of the circuit, (b) peak current delivered by the source, and (c) phase angle between the current and voltage. (d) Is the current leading or lagging behind the voltage?

Solution $X_L = 2\pi f L = (1.51\times10^3 \text{ rad/s})(2.50 \text{ H}) = 3770\ \Omega$

$$X_C = \frac{1}{2\pi f C} = \frac{1}{(1.51\times10^3 \text{ rad/s})(0.250\times10^{-6}\text{ F})} = 2653\ \Omega$$

(a) $Z = \sqrt{(900\ \Omega)^2 + (3770\ \Omega - 2653\ \Omega)^2} = 1440\ \Omega$ ◊

(b) $I_{max} = \frac{V_{max}}{Z} = \frac{140\text{ V}}{1435\ \Omega} = 9.76\times10^{-2}\text{ A} = 97.6\text{ mA}$ ◊

(c) $\tan\phi = \frac{X_L - X_C}{R} = \frac{3770\ \Omega - 2653\ \Omega}{900\ \Omega} = 1.241$ and $\phi = 51.1°$ ◊

(d) $X_L > X_C$ ϕ is greater than zero, so the *voltage leads the current.* ◊

---

29. A 50.0-Ω resistor is connected to a 30.0-μF capacitor and to a 60.0-Hz, 100-V rms source. (a) Find the power factor and the average power delivered to the circuit. (b) Repeat part (a) when the capacitor is replaced with a 0.300-H inductor.

Solution Use $X_C = \frac{1}{2\pi f C}$ and $Z = \sqrt{R^2 + (X_L - X_C)^2}$ to find

$$X_C = 88.4\ \Omega,\quad Z = 102\ \Omega,\quad \text{and}\quad I = \frac{V}{Z} = \frac{100\text{ V}}{102\ \Omega} = 0.980\text{ A}$$

(a) Then, $\tan\phi = \frac{X_L - X_C}{R} = \frac{-X_C}{R} = \frac{-88.4}{50} = -1.77$ and $\phi = -60.5°$

Thus, the power factor = $\cos\phi = \cos(-60.5°) = 0.492$ ◊

and $P_{ave} = IV\cos\phi = (0.980\text{ A})(100\text{ V})(0.492) = 48.3\text{ W}$ ◊

(b) For this case, $X_L = 113\ \Omega$, $Z = 124\ \Omega$, and $I = \frac{V}{Z} = \frac{100\text{ V}}{124\ \Omega} = 0.806\text{ A}$

$$\tan\phi = \frac{X_L - X_C}{R} = \frac{X_L}{R} = \frac{113}{50} = 2.26 \qquad \text{and} \qquad \phi = 66.1°$$

Thus, the power factor = $\cos\phi = \cos(66.1°) = 0.404$ ◊

and $P_{ave} = IV\cos\phi = (0.806 \text{ A})(100 \text{ V})(0.404) = 32.6 \text{ W}$ ◊

---

35. A resonant circuit in a radio receiver is tuned to a certain station when the inductor has a value of 0.200 mH and the capacitor has a value of 30.0 pF. Find the frequency of the radio station and the wavelength sent out by the station.

Solution The frequency of the station is equal to the resonant frequency of the tuning circuit:

$$f_0 = \frac{1}{2\pi\sqrt{LC}} = \frac{1}{2\pi\sqrt{(2.00\times10^{-4} \text{ H})(30.0\times10^{-12} \text{ F})}} \qquad \text{or}$$

$$f_0 = 2.05\times10^6 \text{ Hz} = 2.05 \text{ megahertz} \quad ◊$$

and

$$\lambda = \frac{c}{f} = \frac{3.00\times10^8 \text{ m/s}}{2.05\times10^6 \text{ Hz}} = 146 \text{ m} \quad ◊$$

---

39. The Q value of an RLC circuit is defined as the voltage drop across the inductor (or capacitor) at resonance, divided by the voltage drop across the resistor. The larger the Q value, the sharper, or narrower, is the curve of power versus frequency. Figure 21.32 shows such curves for small R and for large R. (a) Show that the Q value is given by the expression $Q = 2\pi f_0 L/R$, where f_0 is the resonance frequency. (b) Calculate the Q value for the circuit of Problem 38. (Problem 38 describes a series circuit which contains a 3.00-H inductor, a 3.00-μF capacitor, and a 30.0-Ω resistor connected to a 120-V rms source of variable frequency.)

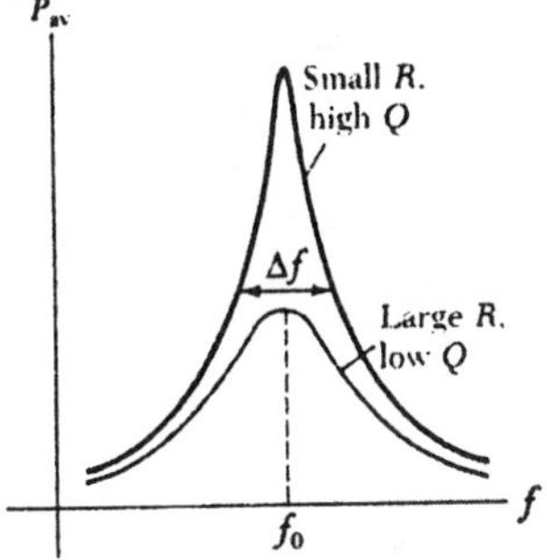

Figure 21.32

Solution (a) $Q = \frac{V_L}{V_R}$ (at the resonant frequency)

Thus, $Q = \frac{V_L}{V_R} = \frac{2\pi f_0 L I_{\text{max}}}{R I_{\text{max}}} = \frac{2\pi f_0 L}{R}$ ◊

(b) The resonant frequency is $f_0 = \frac{1}{2\pi\sqrt{LC}}$

Using this expression in the result of part (a), we have

$$Q = \frac{2\pi L}{R}\left(\frac{1}{2\pi\sqrt{LC}}\right) = \frac{1}{R}\sqrt{\frac{L}{C}}$$

$$Q = \frac{1}{30.0\ \Omega}\sqrt{\frac{3.00\ \text{H}}{3.00\times 10^{-6}\ \text{F}}} = 33.3\ \text{Hz} \quad \Diamond$$

---

43. At a given moment, every inhabitant of a city of 20,000 people turns on a 100-W lightbulb. Assume no other power in the city is being used. (a) If the utility company furnishes this total power at 120 V, calculate the current in the power lines from the utility to the city. (b) Calculate this current if the power company first steps up the voltage to 200,000 V. (c) How much heat is lost in each 1.00-m length of the power lines if the resistance of the lines is $5.00\times 10^{-4}\ \Omega/\text{m}$? Repeat this calculation for situation (a) and for situation (b). (d) If an individual line from the utility can handle only 100 A, how many lines are required to handle the current in each situation described?

Solution (a) The total power required by the city is

$$P = (2.00\times 10^4)(100\ \text{W}) = 2.00\times 10^6\ \text{W}$$

So, $$I = \frac{P}{V} = \frac{2.00\times 10^6\ \text{W}}{120\ \text{V}} = 1.67\times 10^4\ \text{A} \quad \Diamond$$

(b) $I = \frac{P}{V} = \frac{2.00\times 10^6\ \text{W}}{200000\ \text{V}} = 10.0\ \text{A}$ ◊

(c) If $I = 1.67 \times 10^4$ A then

$$P_{loss} = I^2R = (1.67 \times 10^4 \text{ A})^2(5.00 \times 10^{-4} \text{ }\Omega/\text{m}) = 1.39 \times 10^5 \text{ W/m} \quad \Diamond$$

If $I = 10.0$ A, then

$$P_{loss} = I^2R = (10.0 \text{ A})^2(5.00 \times 10^{-4} \text{ }\Omega/\text{m}) = 5.00 \times 10^{-2} \text{ W/m} \quad \Diamond$$

(d) Case a:

$$\text{number of lines needed} = \frac{I}{100 \text{ A/line}} = \frac{1.67 \times 10^4 \text{ A}}{100 \text{ A/line}} = 167 \text{ lines} \quad \Diamond$$

Case b:

$$\text{number lines needed} = \frac{I}{100 \text{ A/line}} = \frac{10.0 \text{ A}}{100 \text{ A/line}} = 0.10 \text{ line,}$$

or 1 line is more than sufficient. ◊

---

45. A standing wave interference pattern is set up by radio waves between two metal sheets positioned 2.00 m apart. This is the smallest distance between the plates that will produce a standing wave pattern. What is the fundamental frequency?

Solution In the fundamental mode there is a single loop in the standing wave between the plates. Thus, the distance between the plates is equal to half a wavelength, giving $\lambda = 2L = 2(2.00 \text{ m}) = 4.00 \text{ m}$.

Therefore, $f = \dfrac{c}{\lambda} = \dfrac{3.00 \times 10^8 \text{ m/s}}{4.00 \text{ m}} = 7.50 \times 10^7 \text{ Hz} = 75.0 \text{ MHz}$ ◊

---

49. Assume that the solar radiation incident on the Earth is 1340 W/m^2 (at the top of the Earth's atmosphere). Calculate the total power radiated by the Sun, taking the average separation between the Earth and the Sun to be 1.49×10^{11} m.

Solution $I = \frac{P}{A} = \frac{P}{4\pi r^2}$ so $P = (4\pi r^2)I = 4\pi(1.49 \times 10^{11}\text{ m})^2\left(1340\frac{\text{W}}{\text{m}^2}\right)$

or $P = 3.74 \times 10^{26}$ W ◊

---

53. A community plans to build a facility to convert solar radiation into electrical power. They require 1 MW of power, and the system to be installed has an efficiency of 30% (that is, 30% of the solar energy incident on the surface is converted to electrical energy). What must be the effective area of a perfectly absorbing surface used in such an installation, assuming a constant power per unit area of 1000 W/m^2 in the incident solar radiation?

Solution Power output = (power input)(efficiency)

Thus, $$\text{Power input} = \frac{\text{power out}}{\text{eff}} = \frac{10^6\text{ W}}{0.300} = 3.33 \times 10^6\text{ W}$$

and $$A = \frac{P}{I} = \frac{3.33 \times 10^6\text{ W}}{10^3\text{ W/m}^2} = 3.30 \times 10^3\text{ m}^2 \quad ◊$$

---

57. What are the wavelength ranges in the (a) AM radio band (540–1600 kHz) and (b) the FM radio band (88–108 MHz)?

Solution (a) For the AM band,

$$\lambda_{\text{max}} = \frac{c}{f_{\text{min}}} = \frac{3.00 \times 10^8\text{ m/s}}{540 \times 10^3\text{ Hz}} = 556\text{ m} \quad ◊$$

$$\lambda_{\text{min}} = \frac{c}{f_{\text{max}}} = \frac{3.00 \times 10^8\text{ m/s}}{1600 \times 10^3\text{ Hz}} = 190\text{ m} \quad ◊$$

(b) For the FM band,

$$\lambda_{\text{max}} = \frac{c}{f_{\text{min}}} = \frac{3.00 \times 10^8\text{ m/s}}{88.0 \times 10^6\text{ Hz}} = 3.40\text{ m} \quad ◊$$

$$\lambda_{min} = \frac{c}{f_{max}} = \frac{3.00 \times 10^8 \text{ m/s}}{108 \times 10^6 \text{ Hz}} = 2.80 \text{ m} \quad \lozenge$$

---

63. A 0.250-H inductor is connected to a capacitor and a 30.0-Ω resistor along with a 60.0-Hz, 30.0-V generator. (a) To what value would the capacitor have to be adjusted to produce resonance? (b) Find the voltage drop across the capacitor-and-inductor combination at resonance.

Solution (a) From $f_0 = \frac{1}{2\pi\sqrt{LC}}$, we find

$$C = \frac{1}{(2\pi f)^2 L} = \frac{1}{[2\pi(60.0 \text{ Hz})]^2(0.250 \text{ H})} = 2.81 \times 10^{-5} \text{ F} = 28.1\ \mu\text{F} \quad \lozenge$$

(b) At resonance, $X_L = X_C$ and they are 180° out of phase. Therefore, Z across the capacitor-inductor combination is zero, and the voltage drop across the combination of elements is also zero. ◊

---

67. A 0.700-H inductor is connected in series with a fluorescent lamp to limit the current drawn by the lamp. If the combination is connected to a 60.0-Hz, 120-V line, and if the voltage across the lamp is to be 40.0 V, what is the current in the circuit? (The lamp is a pure resistive load.)

Solution $X_L = 2\pi f L = 2\pi(60.0 \text{ Hz})(0.700 \text{ H}) = 264\ \Omega$

$$V = \sqrt{V_R{}^2 + V_L{}^2} \quad \text{or} \quad V_L = \sqrt{V^2 - V_R{}^2}$$

$$V_L = \sqrt{(120 \text{ V})^2 - (40.0 \text{ V})^2} = 113 \text{ V}$$

but $V_L = IX_L$ so $I = \frac{V_L}{X_L} = \frac{113 \text{ V}}{264\ \Omega} = 0.429 \text{ A}$ ◊

---

71. An inductor is to be made of a 5.00-m length of 1.00-mm-diameter copper wire wound on a coil of radius 3.00 cm. (The coil has an air core). (a) Estimate the length of the completed solenoid and number of turns on it. Find (b) the inductance of the completed coil and (c) the resistance of the completed coil. (d) Find the current that is drawn when this device is connected to a 60.0-Hz, 20.0-V rms source.

Solution (a) The circumference of the solenoid $= 2\pi r = 2\pi(0.030 \text{ m}) = 0.188 \text{ m}$

The number of turns that 5.00 m of wire can make is

$$N = \frac{5.00 \text{ m}}{0.188 \text{ m/turn}} = 26.5 \text{ turns}$$

The length of the solenoid is

$$\text{Length} = (\text{number of turns})(\text{wire diameter}) \quad \text{or}$$

$$\text{Length} = (26.5)(10^{-3} \text{ m}) = 2.65 \times 10^{-2} \text{ m} = 2.65 \text{ cm} \quad \Diamond$$

(b) From chapter 21, $L = \mu_0 N^2 A / \text{Length}$ or

$$L = \frac{(4\pi \times 10^{-7} \text{ N/A}^2)(26.5)^2 \pi (0.030 \text{ m})^2}{2.65 \times 10^{-2} \text{ m}} = 9.41 \times 10^{-5} \text{ H} \quad \Diamond$$

(c) $$R = \frac{\rho L}{A} = \frac{(1.70 \times 10^{-8} \ \Omega \cdot \text{m})(5.00 \text{ m})}{\pi (5.00 \times 10^{-4} \text{ m})^2} = 0.108 \ \Omega \quad \Diamond$$

(d) $$X_L = 2\pi f L = 2\pi (60.0 \text{ s}^{-1})(9.41 \times 10^{-5} \text{ H}) = 3.55 \times 10^{-2} \ \Omega$$

$$Z = \sqrt{R^2 + X_L{}^2} = \sqrt{(0.108 \ \Omega)^2 + (0.0355 \ \Omega)^2} = 0.114 \ \Omega$$

and then $$I = \frac{V}{Z} = \frac{20.0 \text{ V}}{0.114 \ \Omega} = 176 \text{ A} \quad \Diamond$$

CHAPTER SELF-QUIZ

1. When a series *RLC* circuit is in resonance, the maximum voltage across
   a. the resistor and inductance must be equal
   b. the resistor and capacitor must be equal
   c. the inductance and capacitor must be equal
   d. none of them must be equal

2. Find the resonant frequency for a series *RLC* circuit where $R = 10.0\ \Omega, C = 5.00\ \mu\text{F}$, and $L = 2.00$ mH.
   a. 998 Hz
   b. 1.592 kHz
   c. 2.45 kHz
   d. 11.3 kHz

3. A radio tuning circuit has a coil with an inductance of 5.00 mH. What must be the capacitance if the set is to be tuned to 980 kHz?
   a. 5.30×10^{-6} microfarads
   b. 4.70×10^{-3} microfarads
   c. 1.90×10^{-3} microfarads
   d. 5.10 microfarads

4. An ac voltage source is connected to a capacitor. The charge on the capacitor will be a maximum when
   a. the voltage and current in the circuit are both a maximum
   b. the voltage and current are both zero
   c. the voltage is a maximum and the current is zero
   d. the current is a maximum and the voltage is zero

5. The units of frequency times capacitance ($F \cdot s^{-1}$) is the same as the units for
   a. current (A)
   b. energy (J)
   c. resistance (R)
   d. conductivity (R^{-1})

6. When an ac voltage is connected to an inductance coil, the energy stored in the magnetic field of the coil will be a maximum when
   a. the voltage across the coil is a maximum
   b. the voltage across the coil is zero
   c. the product of voltage times current is a maximum
   d. there is never any energy stored in such a magnetic field

7. A resistor, inductor, and capacitor are connected in series, each with an effective (rms) voltage of 65.0 V, 140 V, and 80.0 V, respectively. What is the power factor in the circuit?
   a. 0.68
   b. 0.74
   c. 0.87
   d. 0.93

8. An *LC* circuit is set up to produce electromagnetic waves. If the inductor has a value of 2.00 mH and the capacitor a value of 1.26×10^{-12} F, what is the resonant frequency of the circuit?
   a. 10^6 Hz
   b. 10^7 Hz
   c. 100 MHz
   d. 10^9 Hz

9. At a distance of 10.0 km from a radio transmitter, the amplitude of the E-field is 0.20 volts/meter. What is the total power emitted by the radio transmitter?
   a. 10.0 kW
   b. 67.0 kW
   c. 140 kW
   d. 245 kW

10. What is the self-inductance of a coil which has an inductive reactance of 35.0 ohms at an angular frequency of 500 rad/s?
   a. 14.3 mH
   b. 11.0 mH
   c. 70.0 mH
   d. 71.5 mH

11. What is the capacitance of a capacitor which has a capacitive reactance of 85.0 ohms at an angular frequency of 500 rad/s?
   a. 1.85 μF
   b. 3.70 μF
   c. 9.00 μF
   d. 23.5 μF

12. A 100-kW radio station emits EM waves in all directions from an antenna on top of a mountain. What is the intensity in Watts/m^2 of the signal at a distance of 10.0 km?
   a. 8.00×10^{-5} W/m^2
   b. 8.00×10^{-6} W/m^2
   c. 3.00 m mW/m^2
   d. 0.80 W/m^2

22
Reflection and Refraction of Light

Chapter 22

REFLECTION AND REFRACTION OF LIGHT

The chief architect of the particle theory of light was Newton. With this theory he provided simple explanations of some known experimental facts concerning the nature of light, namely the laws of reflection and refraction.

In 1678 a Dutch physicist and astronomer, Christian Huygens (1629-1695), showed that a wave theory of light could also explain the laws of reflection and refraction. The wave theory did not receive immediate acceptance for several reasons. All the waves known at the time (sound, water, and so on) traveled through some sort of medium, but light from the Sun could travel to Earth through empty space. Furthermore, it was argued that if light were some form of wave, it would bend around obstacles; hence, we should be able to see around corners. It is now known that light does indeed bend around the edges of objects. This phenomenon, known as *diffraction,* is not easy to observe because light waves have such short wavelengths. For more than a century most scientists rejected the wave theory and adhered to Newton's particle theory. This was, for the most part, due to Newton's great reputation as a scientist.

NOTES FROM SELECTED CHAPTER SECTIONS

22.3 Huygens' Principle

Every point on a given wave front can be considered as a point source for a *secondary wavelet.* At some later time, the new position of the wave front is determined by the surface tangent to the set of secondary wavelets.

22.4 Reflection and Refraction

A line drawn perpendicular to a surface at the point where an incident ray strikes the surface is called the *normal line.* Angles of reflection and refraction are measured relative to the normal.

When an incident ray undergoes partial reflection and partial refraction, the incident, reflected and refracted rays are all *in the same plane.*

The *path of a light ray through a refracting surface is reversible.*

As light travels from one medium into another, the *frequency does not change.*

22.6 Dispersion and Prisms

Index of refraction is a function of wavelength.

22.9 Total Internal Reflection

Total internal reflection is possible only when light rays traveling in one medium are incident on an interface bounding a second medium of *lesser* index of refraction than the first. Total internal reflection of light occurs at angles of incidence $\theta \geq \theta_c$, where $n_1 > n_2$.

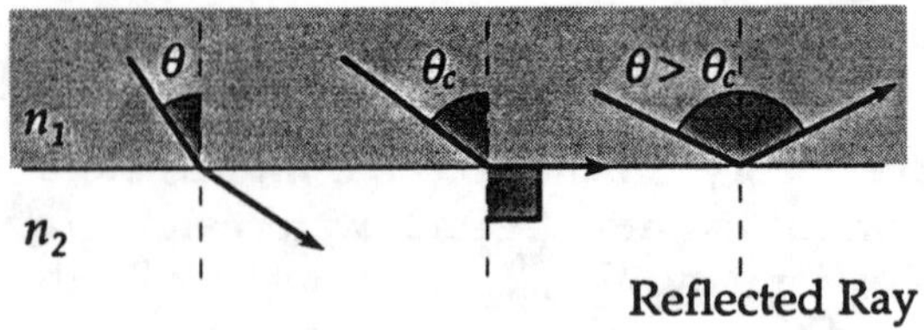

EQUATIONS AND CONCEPTS

The energy of a photon is proportional to the frequency of the associated electromagnetic wave.

$$E = hf \qquad (22.1)$$

Planck's constant.

$$h = 6.63 \times 10^{-34}\ \text{J}\cdot\text{s}$$

The law of reflection states that the angle of incidence (the angle measured between the incident ray and the normal line) equals the angle of reflection (the angle measured between the reflected ray and the normal line).

$$\theta_1' = \theta_1 \qquad (22.2)$$

This is one form of the statement of Snell's law. The angle of refraction (measured relative to the normal line) depends on the angle of incidence, and also on the ratio of the speeds of light in the two media on either side of the refracting surface.

$$\frac{\sin\theta_2}{\sin\theta_1} = \frac{v_2}{v_1} = \text{constant} \qquad (22.3)$$

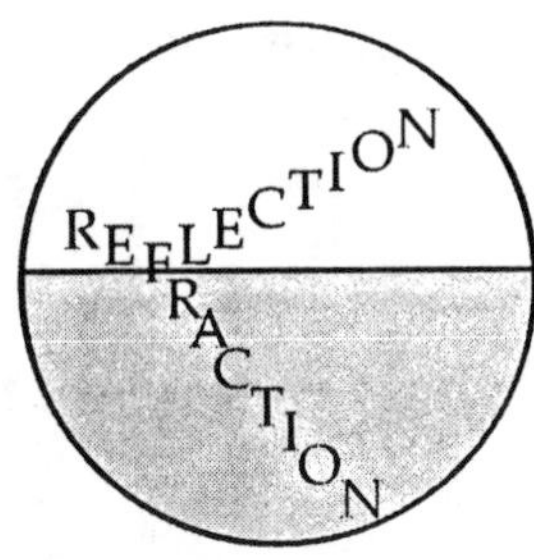

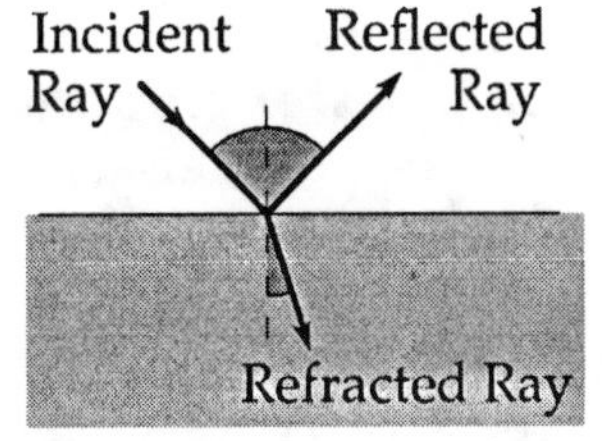

This is the most widely used and most practical form of Snell's law. This equation involves a parameter called the index of refraction, the value of which is characteristic of a particular medium. The index of refraction is defined in Equation 22.4 and Equation 22.7.

$$n_1 \sin\theta_1 = n_2 \sin\theta_2 \qquad (22.8)$$

Each transparent medium is characterized by a dimensionless number, the index of refraction which equals the ratio of the speed of light in the medium to the speed of light in vacuum.

$$n = \frac{c}{v} \qquad (22.4)$$

The frequency of a wave is characteristic of the source. Therefore, as light travels from one medium into another of different index of refraction, the frequency remains constant but the wavelength changes. The index of refraction of a given medium can be expressed as the ratio of the wavelength of light in vacuum to the wavelength in that medium.

$$n = \frac{\lambda_0}{\lambda_n} \qquad (22.7)$$

For angles of incidence equal to or greater than the critical angle, the incident ray will be totally internally reflected back into the first medium.

$$\sin\theta_c = \frac{n_2}{n_1} \qquad (22.9)$$

$$n_1 > n_2$$

Total internal reflection is possible only when a light ray is directed from a medium of high index of refraction into a medium of lower index of refraction.

Comment on internal reflection.

REVIEW CHECKLIST

- ▷ Describe the various experimental results which support the view of the dual nature of light (including Young's experiment and the photoelectric effect).
- ▷ Describe the methods used by Roemer and Fizeau for the measurement of c.
- ▷ Understand the conditions under which total internal reflection can occur in a medium and determine the critical angle for a given pair of adjacent media.
- ▷ Describe the process of dispersion of a beam of white light as it passes through a prism.
- ▷ Describe the conditions under which internal reflection of a light ray is possible; and calculate the critical angle for internal reflection at a boundary between two optical media of known indices of refraction. Describe the application of internal reflection to fiber optics techniques.

SOLUTIONS TO SELECTED END-OF-CHAPTER PROBLEMS

3. The Fizeau experiment is performed so that the round-trip distance for the light is 40.0 m. (a) Find the two lowest speeds of rotation that allow the light to pass through the notches. Assume that the wheel has 360 teeth and that the speed of light is 3.00×10^8 m/s. (b) Repeat for a round-trip distance of 4000 m.

Solution (a) The time for the light to travel 40.0 m is

$$t = \frac{40.0 \text{ m}}{3.00 \times 10^8 \text{ m/s}} = 1.33 \times 10^{-7} \text{ s}$$

At the lowest speed, the wheel will have turned through 1/360 rev in the time t.

Thus, $$\omega = \frac{\Delta\theta}{\Delta t} = \frac{1/360 \text{ rev}}{1.33\times10^{-7} \text{ s}} = 2.10\times10^{4} \text{ rev/s} \quad \Diamond$$

The next lowest speed occurs when the wheel turns through 2/360 rev in the time t.

$$\omega = \frac{\Delta\theta}{\Delta t} = \frac{2/360 \text{ rev}}{1.33\times10^{-7} \text{ s}} = 4.20\times10^{4} \text{ rev/s} \quad \Diamond$$

(b) The steps are identical to those used in part (a). The flight time of the light is 1.33×10^{-5} s, and the lowest speed is 2.10×10^{2} rev/s. The next lowest speed is 4.20×10^{2} rev/s. ◊

---

7. The angle between the two mirrors in Figure 22.30 is a right angle. The beam of light in the vertical plane P strikes mirror 1 as shown. (a) Determine the distance the reflected light beam travels before striking mirror 2. (b) In what direction does the light beam travel after being reflected from mirror 2?

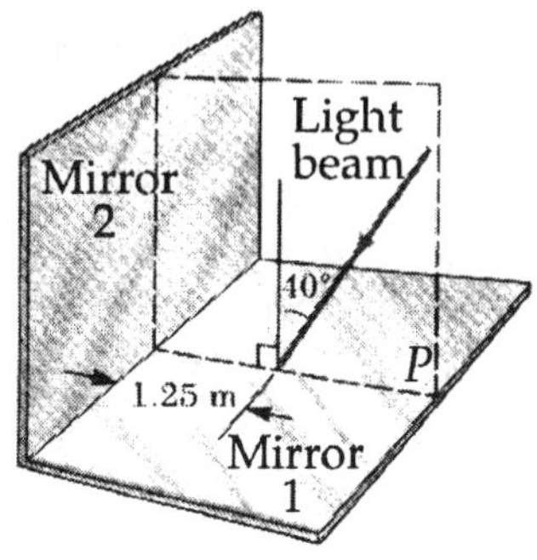

Figure 22.30

Solution

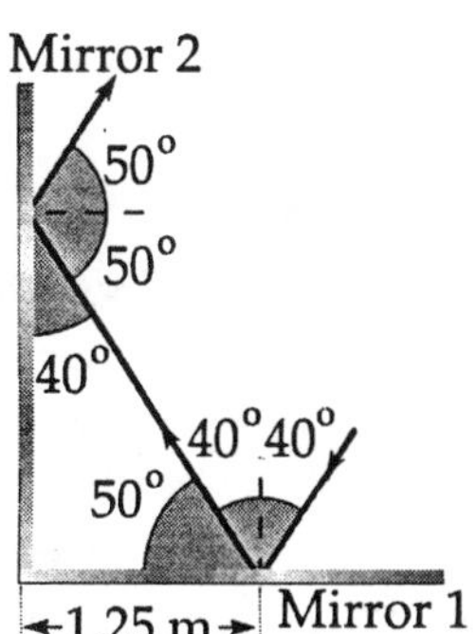

(a) Let d be the distance the light beam travels between the two mirrors. From geometry, $1.25 \text{ m} = d\sin 40.0°$

so $$d = \frac{1.25 \text{ m}}{\sin 40} = 1.94 \text{ m} \quad \Diamond$$

(b) At each reflection, the angle of reflection equals the angle of incidence. Therefore, after reflection from mirror 2, the reflected beam is at an angle of 50.0° above horizontal, or parallel to the incident beam. ◊

---

15. A narrow beam of sodium yellow light ($\lambda_0 = 589$ nm) is incident from air on a smooth surface of water at an angle of $\theta_1 = 35.0°$. Determine the angle of refraction, θ_2, and the wavelength of the light in water.

Solution $n_2 \sin\theta_2 = n_1 \sin\theta_1$ and $\sin\theta_2 = \left(\frac{n_1}{n_2}\right)\sin\theta_1$

$$\theta_2 = \sin^{-1}\left[\left(\frac{1}{4/3}\right)\sin 35.0°\right] = 25.5° \quad \Diamond$$

$$\lambda_2 = \left(\frac{n_1}{n_2}\right)\lambda_1 = \left(\frac{1}{4/3}\right)(589\text{ nm}) = 442\text{ nm} \quad \Diamond$$

---

17. A narrow beam of ultrasonic waves reflects off the liver tumor in Figure 22.34. If the wave speed is 10.0% less in the liver than in the surrounding medium, determine the depth of the tumor.

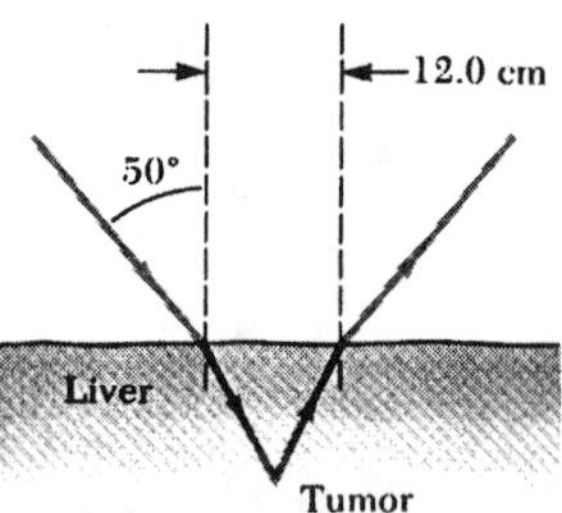

Figure 22.34

Solution Use Snell's law to calculate the angle of refraction, θ.

$$n_L \sin\theta = n_m \sin 50.0°$$

$$\theta = \sin^{-1}\left[\left(\frac{n_m}{n_L}\right)\sin 50.0°\right]$$

but $\frac{n_m}{n_L} = \frac{v_L}{v_m} = 0.900$

so $\theta = \sin^{-1}[(0.900)(0.766)] = 43.6°$

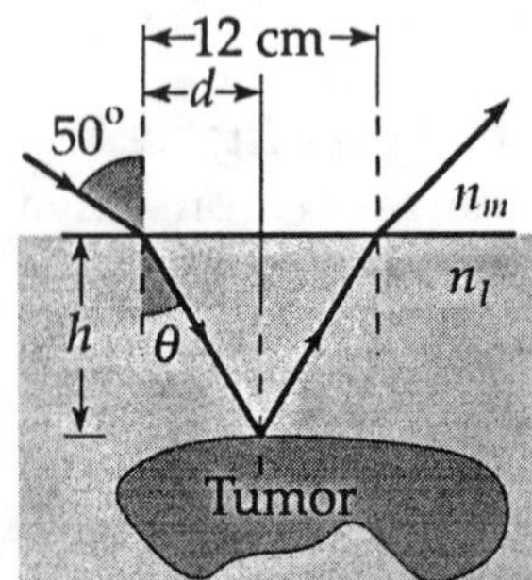

At the tumor, the beam reflects so that the angle of incidence equals the angle of reflection; and, therefore, by symmetry (as shown in the sketch) $d = 6.00$ cm.

$$h = \frac{d}{\tan\theta} = \frac{6.00 \text{ cm}}{\tan 43.6°} = 6.30 \text{ cm} \quad \Diamond$$

---

19. A ray of light strikes a flat 2.00-cm-thick block of glass (n = 1.50) at an angle of 30.0° with the normal (Fig. 22.35). Trace the light beam through the glass and find the angles of incidence and refraction at each surface.

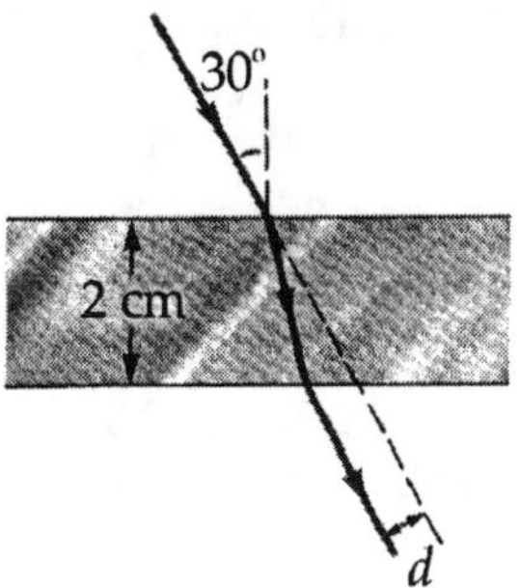

Figure 22.35

Solution At the first surface, call the air medium 1 and the glass medium 2; calculate the angle of refraction using Snell's law:

$$n_2 \sin\theta_2 = n_1 \sin\theta_1$$

$$\theta_2 = \sin^{-1}\left[\left(\frac{n_1}{n_2}\right)\sin\theta_1\right] = \sin^{-1}\left[\left(\frac{1}{1.50}\right)\sin 30.0°\right] = 19.5° \quad \Diamond$$

and $\sin\theta_2 = 0.333$ and $\theta_2 = 19.5°$

At the second surface, call medium 2 the glass and medium 3 the air into which the light exits. Again, using Snell's law, $n_3 \sin\theta_3 = n_2 \sin\theta_2$. Also, notice that the angle of incidence at the second (bottom) surface equals the angle of incidence at the first (top) surface. Therefore,

$$\theta_3 = \sin^{-1}\left[\left(\frac{n_2}{n_3}\right)\sin\theta_2\right] = \sin^{-1}\left[\left(\frac{1.50}{1}\right)\sin 19.5°\right] = 30.0° \quad \Diamond$$

As a result of the two refractions, the emerging ray is parallel to the incident ray.

---

23. A ray of light strikes the midpoint of one face of an equiangular glass prism (n = 1.50) at an angle of incidence of 30.0°. (a) Trace the path of the light ray through the glass, and find the angles of incidence and refraction at each surface. (b) If a small fraction of light is also reflected at each surface, find the angles of incidence and reflection at these surfaces.

Solution (a) At the first surface, the angle of incidence is 30.0° and the angle of refraction is found from Snell's law.

$$n_2 \sin\theta_2 = n_1 \sin\theta_1$$

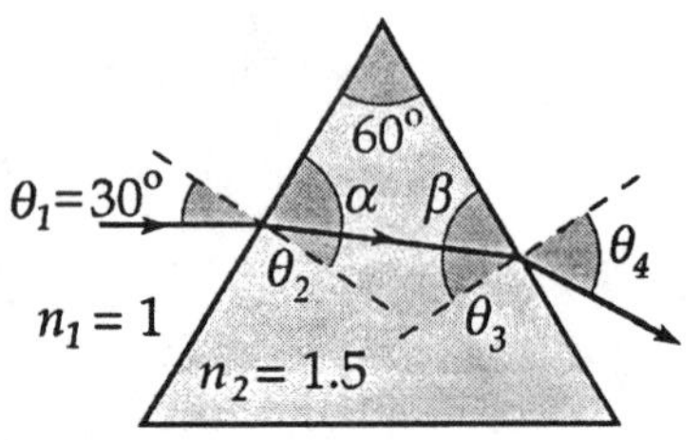

$1.50 \sin\theta_2 = 1.00 \sin 30.0°$ or $\theta_2 = 19.5°$ ◊

In order to find the value of θ_3, refer to the figure. At the first surface, we see

$\alpha + \theta_2 = 90.0°$ or $\alpha = 90.0° - 19.5° = 70.5°$

We also see that $\alpha + \beta + 60.0° = 180°$ Thus, $\beta = 49.5°$

Therefore, $\theta_3 = 90.0° - \beta = 90.0° - 49.5° = 40.5°$

θ_3 is the angle of incidence at the second surface. ◊

Snell's law gives us the angle of refraction at the second surface:

$n_4 \sin\theta_4 = n_3 \sin\theta_3$ or $1.00 \sin\theta_4 = 1.50$ in $40.5°$, so $\theta_4 = 77.1°$ ◊

(b) The angle of reflection at each surface is the same as the angle of incidence at that surface. ◊

---

25. Three sheets of plastic have unknown indices of refraction. Sheet 1 is placed on top of sheet 2, and a laser beam is directed onto the sheets from above so that it strikes the interface at an angle of 26.5° with the normal. The refracted beam in sheet 2 makes an angle of 31.7° with the normal. The experiment is repeated with sheet 3 on top of sheet 2, and with the same angle of incidence the refracted beam makes an angle of 36.7° with the normal. If the experiment is repeated again with sheet 1 on top of sheet 3, what is the expected angle of refraction in sheet 3? Assume the same angle of incidence.

Solution For the first placement, Snell's law becomes, with 1 referring to the upper sheet and 2 to the lower,

$n_1 \sin\theta_1 = n_2 \sin\theta_2$ or $n_2 = n_1 \dfrac{\sin 26.5°}{\sin 31.7°}$

or $n_2 = 0.849 n_1$ (1)

Given Conditions and Observed Results

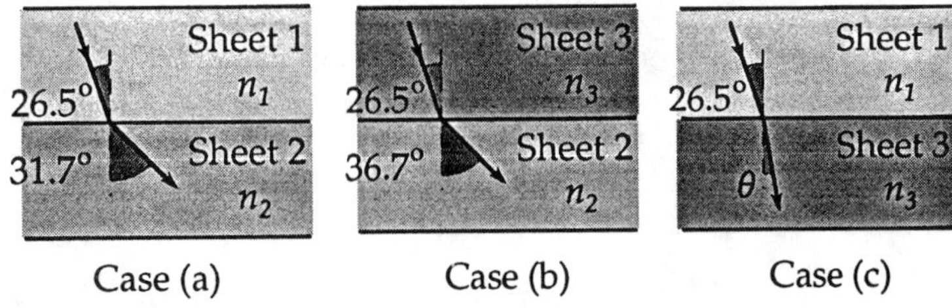

For the second placement, with 3 referring to the remaining sheet,

$$n_3 = n_2 \frac{\sin 36.7°}{\sin 26.5°} \quad \text{or} \quad n_3 = 1.339 n_2 \qquad (2)$$

Combining (1) and (2), we have

$$n_3 = 1.339 n_2 = (1.339)(0.849 n_1) = 1.137 n_1 \qquad (3)$$

Applying Snell's law to the final situation, we have

$$n_3 \sin\theta = n_1 \sin 26.5°$$

$$\sin\theta = \frac{n_1 \sin 26.5°}{n_3} = \frac{n_1 \sin 26.5°}{1.137 n_1} = 0.3924$$

and $$\theta = 23.1° \quad \lozenge$$

---

31. A cylindrical tank with an open top has a diameter of 3.00 m and is completely filled with water. When the setting Sun reaches an angle of 28.0° above the horizon, sunlight ceases to illuminate the bottom of the tank. How deep is the tank?

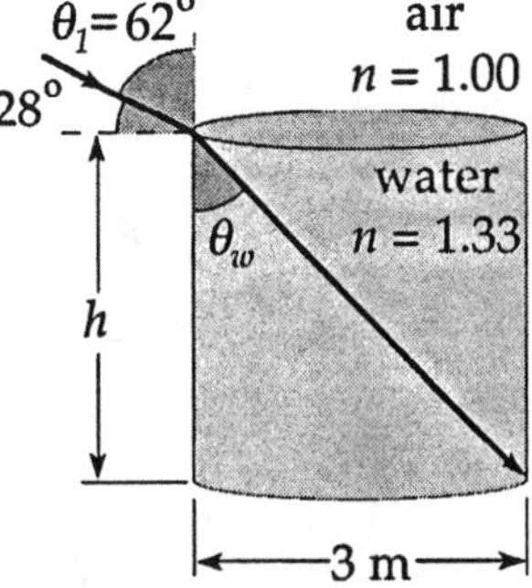

Solution When the Sun is 28.0° from horizon, incident rays are 62.0° from normal at air-water interface. (See the sketch at the right.)

Thus, $$n_w \sin\theta_w = n_{air} \sin\theta_1$$

$$\sin\theta_w = \frac{\sin 62.0°}{4/3} = 0.662 \quad \text{and} \quad \theta_w = 41.5°$$

$$\tan\theta_w = \frac{3.00 \text{ m}}{h} \quad \text{so} \quad h = \frac{3.00 \text{ m}}{\tan 41.5°} = 3.40 \text{ m} \quad \Diamond$$

---

33. Light of wavelength 400 nm is incident at an angle of 45.0° on acrylic and is refracted as it passes into the material. What wavelength of light incident on fused quartz at an angle of 45.0° would be refracted at exactly this same angle? (See Fig. 22.15.)

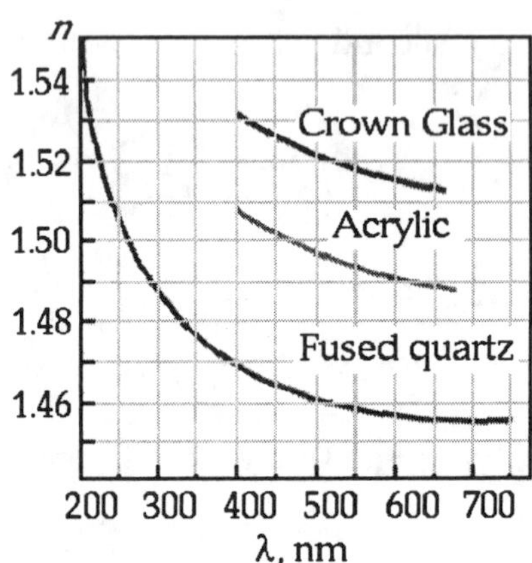

Figure 22.15

Solution For acrylic with λ = 400 nm, we find n = 1.507. (See Figure 22.15.) To be refracted at the same angle in the fused quartz, the wavelength used must have the same index of refraction as the 400 nm light had in the acrylic. From Figure 22.15, it is seen that the wavelength for which n = 1.507 in fused quartz is

$$\lambda = 245 \text{ nm} \quad \Diamond$$

---

39. A beam of light is incident from air on the surface of a liquid. If the angle of incidence is 30.0° and the angle of refraction is 22.0°, find the critical angle for the liquid when surrounded by air.

Solution

From Snell's law, we find the index of refraction of the fluid.

$$n_2 \sin\theta_2 = n_1 \sin\theta_1$$

or $$n_2 = \left(\frac{\sin\theta_1}{\sin\theta_2}\right) n_1 = \left(\frac{\sin 30.0°}{\sin 22.0°}\right)(1.00) = 1.335$$

Then, from $\sin\theta_c = \frac{n_2}{n_1}$, $\sin\theta_c = \frac{1}{1.335} = 0.749$, and $\theta_c = 48.5°$ ◊

---

43. A light ray is incident perpendicular to the long face (the hypotenuse) of a 45.0°–45.0°–90.0° prism surrounded by air, as shown in Figure 22.26b. Calculate the minimum index of refraction of the prism for which the ray will follow the path shown.

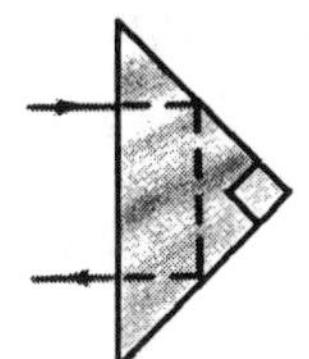

Figure 22.26b

Solution If the beam follows the path shown in Figure 22.26b, the angle of incidence when it strikes the short face of the prism is 45.0°. The index of refraction of the prism must be such that the angle 45.0° is greater than or equal to the critical angle. We find

$$\sin\theta_c = \frac{n_2}{n_1} = \frac{1}{n}$$

where n is the index of refraction of the prism material. We solve for n,

$$n = \frac{1}{\sin\theta_c}$$

However, if the critical angle is less than or equal to $\sin 45.0°$, then $\sin\theta_c$ is less than or equal to $\sin 45.0°$.

Therefore, $n \geq \frac{1}{\sin 45.0°} = \sqrt{2}$ ◊

---

47. A light pipe consists of a central strand of material surrounded by an outer coating. The interior portion of the pipe has an index of refraction of 1.60. If all rays striking the interior walls of the pipe with incident angles greater than 59.5° are subject to total internal reflection, what is the index of refraction of the coating?

Solution Given $\sin\theta_c = \frac{n_c}{n_p}$, so $n_c = n_p \sin\theta_c = 1.60 \sin 59.5° = 1.38$ ◊

---

49. Light is incident normally on a 1.00-cm layer of water that lies on top of a flat Lucite plate with a thickness of 0.500 cm. How much more time is required for light to pass through this double layer than is required to traverse the same distance in air? ($n_{\text{Lucite}} = 1.59$)

Solution Within each material, light travels with a speed, $v = \frac{c}{n}$. Let t_1 = time to traverse 1.00 cm of water.

$$t_1 = \frac{1.00 \times 10^{-2}\text{ m}}{\left(\frac{c}{n_{\text{water}}}\right)} = \frac{(1.00 \times 10^{-2}\text{ m})n_{\text{water}}}{c} = \frac{1}{c}\left(1.33 \times 10^{-2}\text{ m}\right)$$

Let t_2 = time to traverse 0.500 cm of Lucite.

$$t_2 = \frac{5.00 \times 10^{-3}\text{ m}}{\left(\frac{c}{n_{\text{Lucite}}}\right)} = \frac{(5.00 \times 10^{-3}\text{ m})(n_{\text{Lucite}})}{c} = \frac{1}{c}\left(7.95 \times 10^{-3}\text{ m}\right)$$

Let t_3 = time to traverse 1.50 cm of air.

$$t_3 = \frac{1.50 \times 10^{-2}\text{ m}}{\left(\frac{c}{n_{\text{air}}}\right)} = \frac{(1.50 \times 10^{-2}\text{ m})(n_{\text{air}})}{c} = \frac{1}{c}\left(1.50 \times 10^{-2}\text{ m}\right)$$

$$\Delta t = (t_1 + t_2) - t_3$$

$$\Delta t = \frac{1}{c}[1.33 + 0.795 - 1.50] \times 10^{-2}\text{ m} = 2.08 \times 10^{-11}\text{ s} \quad \lozenge$$

---

55. A thick plate of flint glass (n = 1.66) rests on top of a thick plate of transparent acrylic (n = 1.50). A beam of light is incident on the top surface of the flint glass at an angle of θ_i. The beam passes through the glass and the acrylic and emerges from the acrylic at an angle of 40.0° with respect to the normal. Calculate the value of θ_i. A sketch of the light path through the two plates of refracting material would be helpful.

Solution The sketch of the light path is shown. Using Snell's law, at the acrylic-air boundary we have

$$n_3 \sin\theta_3 = n_1 \sin\theta_4$$

$$(1.50)\sin\theta_3 = (1.00)\sin 40.0°$$

$$\sin\theta_3 = 0.4285; \qquad \theta_3 = 25.4°$$

At the glass-acrylic boundary, $n_2 \sin\theta_2 = n_3 \sin\theta_3$

$$(1.66)\sin\theta_2 = (1.50)\sin 25.4°$$

$$\sin\theta_2 = 0.3876; \qquad \theta_2 = 22.8°$$

At the air-glass boundary, $n_1 \sin\theta_1 = n_2 \sin\theta_2$

or $$(1.00)\sin\theta_i = (1.66)\sin 22.8°$$

$$\sin\theta_i = 0.6434; \qquad \theta_i = 40.0° \quad \lozenge$$

---

61. For this problem refer to Figure 22.16. For various angles of incidence, it can be shown that the angle δ is a minimum when the ray passes through the glass so that the ray is parallel to the base. A measurement of this minimum angle of deviation enables one to find the index of refraction of the prism material.

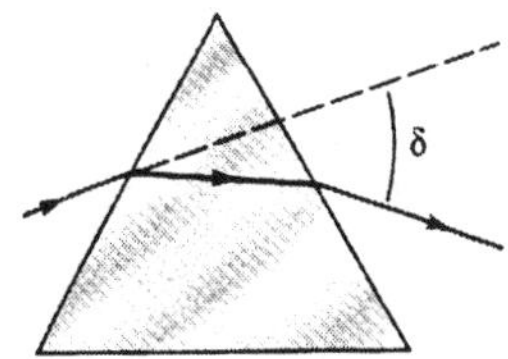

Figure 22.16

Show that n is given by the expression

$$n = \frac{\sin\left[\frac{1}{2}(A + \delta_{\text{min}})\right]}{\sin\left(\frac{A}{2}\right)}$$ where A is the apex angle of the prism.

Solution We have $\delta_{min} = \alpha + \varepsilon$

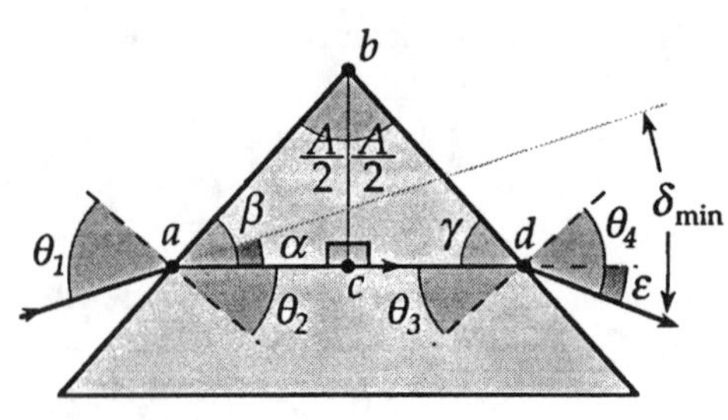

Also, $\beta + \theta_2 = 90.0°$ (1)

Use triangle *abc*.

$$\beta + \frac{A}{2} + 90.0° = 180°$$

or $\beta + \frac{A}{2} = 90.0°$ (2)

Comparing (1) and (2), we see that $\theta_2 = \frac{A}{2}$

Using triangle *bcd*, we have

$$\frac{A}{2} + 90.0° + \gamma = 180° \quad \text{or} \quad \gamma = 90.0° - \frac{A}{2} = 90.0° - \theta_2$$

Therefore, $\gamma + \theta_3 = 90.0°$ or $(90.0° - \theta_2) + \theta_3 = 90.0°$ which gives $\theta_3 = \theta_2$

We also have $1.00 \sin\theta_1 = n \sin\theta_2$ and $n \sin\theta_3 = 1.00 \sin\theta_4$

Since $\theta_3 = \theta_2$, these equations show that $\theta_4 = \theta_1$

Therefore, $\alpha = \theta_1 - \theta_2 = \theta_1 - \frac{A}{2}$ and $\varepsilon = \theta_4 - \theta_3 = \theta_1 - \theta_2 = \theta_1 - \frac{A}{2}$

Then, $\delta_{min} = \alpha + \varepsilon = \theta_1 - \frac{A}{2} + \theta_1 - \frac{A}{2} = 2\theta_1 - A$ or $2\theta_1 = A + \delta_{min}$

from which $\theta_1 = \frac{A + \delta_{min}}{2}$

Finally, from (1), $\sin\theta_1 = n \sin\theta_2$ we have $n = \frac{\sin\theta_1}{\sin\theta_2} = \frac{\sin\frac{(A + \delta_{min})}{2}}{\sin\left(\frac{A}{2}\right)}$ ◊

CHAPTER SELF-QUIZ

1. One phenomenon that demonstrates the particle nature of light is
   a. the photoelectric effect
   b. diffraction effects
   c. interference effects.
   d. the prediction by Maxwell's electromagnetic theory

2. Light in air enters a diamond (n = 2.42) at an angle of incidence of 30.0°. What is the angle of refraction inside the diamond?
   a. 11.9°
   b. 20.0°
   c. 23.8°
   d. 27.6°

3. An underwater scuba diver sees the Sun at an apparent angle of 45.0° from the vertical. How far is the Sun above the horizon? (n_{water} = 1.333)
   a. 9.90°
   b. 13.8°
   c. 19.5°
   d. 27.0°

4. A beam of light is incident upon a flat piece of glass (n = 1.50) at an angle of incidence of 30.0°. Part of the beam is transmitted and part is reflected. What is the angle between the reflected and transmitted rays?
   a. 79.5°
   b. 90.0°
   c. 130.5°
   d. 140°

5. Dispersion occurs when
   a. some materials bend light more than other materials
   b. a material slows down some wavelengths more than others
   c. a material changes some frequencies more than others
   d. light has different speeds in different materials

6. If the velocity of light through an unknown liquid is measured at 2.40×10^8 m/s, what is the index of refraction of this liquid? ($c = 3.00 \times 10^8$ m/s)
   a. 1.80
   b. 1.25
   c. 1.33
   d. 0.80

7. If the critical angle for internal reflection inside a certain transparent material is found to be 48.0°, what is the index of refraction of the material? (Air is outside the material).
   a. 1.35
   b. 1.48
   c. 1.49
   d. 0.743

8. Which of the following describes what will happen to a light ray incident on an air-to-glass boundary at less than the critical angle?
   a. total reflection
   b. total transmission
   c. partial reflection, partial transmission
   d. partial reflection, total transmission

9. A beam of light in air is incident at an angle of 35.0° to the surface of a rectangular block of clear plastic (n = 1.49). The light beam first passes through the block and re-emerges from the opposite side into air at what angle to the normal to that surface?
   a. 42.0°
   b. 23.0°
   c. 35.0°
   d. 59.0°

10. A beam of light in air is incident on the surface of a rectangular block of clear plastic (n = 1.49). If the velocity of the beam before it enters the plastic is 3.00×10^8 m/s, what is its velocity after emerging from the block?
   a. 3.00×10^8 m/s
   b. 1.93×10^8 m/s
   c. 2.01×10^8 m/s
   d. 1.35×10^8 m/s

11. A ray of light in air is incident on an air-to-glass boundary at an angle of 30.0° with the normal. If the index of refraction of the glass is 1.65, what is the angle of the refracted ray within the glass with respect to the normal?
   a. 56.0°
   b. 46.0°
   c. 30.0°
   d. 18.0°

12. A light ray in air is incident on an air-to-glass boundary at an angle of 45.0° and is refracted in the glass of 30.0° with the normal. What is the index of refraction of the glass?
   a. 2.13
   b. 1.74
   c. 1.23
   d. 1.41

23
Mirrors and Lenses

Chapter 23

MIRRORS AND LENSES

This chapter is concerned with the formation of images when plane and spherical waves fall on plane and spherical surfaces. Images can be formed by either reflection or refraction. Mirrors and lenses form images in both ways. In our study of mirrors and lenses, we continue to use the ray approximation and to assume that light travels in straight lines (in other words, we ignore diffraction). This chapter completes our study of geometric optics. In Chapters 25 and 26 we shall examine the construction and properties of some optical instruments that use mirrors and lenses.

NOTES FROM SELECTED CHAPTER SECTIONS

23.1 Plane Mirrors

The image formed by a plane mirror has the following properties:

1. The image is as far behind the mirror as the object is in front.

2. The image is unmagnified, virtual, and erect. (By erect, we mean that, if the object arrow points upward, so does the image arrow.)

3. The image has right-left reversal.

23.2 Images Formed by Spherical Mirrors

A spherical mirror is a reflecting surface which has the shape of a segment of a sphere.

Real images are formed at a point when *reflected light actually passes through the point*.

Virtual images are formed at a point when light rays *appear to diverge from the point*.

The point of intersection of any two of the following rays in a ray diagram for mirrors locates the image:

1. The first ray is drawn from the top of the object parallel to the optical axis and is reflected back through the focal point, F.

2. The second ray is drawn from the top of the object to the vertex of the mirror and is reflected with the angle of incidence equal to the angle of reflection.

3. The third ray is drawn from the top of the object through the center of curvature, C, which is reflected back on itself.

23.6 Thin Lenses

The following three rays form the ray diagram for a thin lens:

1. The first ray is drawn *parallel* to the optic axis. After being refracted by the lens, this ray passes *through* (or appears to come from) one of the focal points.

2. The second ray is drawn *through the center* of the lens. This ray continues in a *straight line.*

3. The third ray is drawn through the *focal point F,* and emerges from the lens *parallel to the optic axis.*

23.7 Lens Aberrations

Aberrations are responsible for the formation of imperfect images by lenses and mirrors. Spherical aberration is due to the variation in focal points for parallel incident rays that strike the lens at various distances from the optical axis. Chromatic aberration arises from the fact that light of different wavelengths focuses at different points when refracted by a lens.

EQUATIONS AND CONCEPTS

Lateral magnification is defined as the ratio of image height to the object height.

$$M = \frac{h'}{h} \quad (23.1)$$

Images formed by plane mirrors have the following properties: (i) the image is as far behind the mirror as the object is in front; (ii) the image is the same size as the object ($M = 1$), virtual, and erect; and (iii) the image has right-left reversal.

Comment on reflection by plane mirrors.

Chapter 23

For spherical mirrors the lateral magnification can also be expressed as the negative of the ratio of the image distance to the object distance.

$$M = \frac{h'}{h} = -\frac{q}{p} \qquad (23.2)$$

The mirror equation is used to determine the location of an image formed by reflection of paraxial rays from a spherical surface. The focal point of a spherical mirror is located midway between the center of curvature and the vertex of the mirror.

$$\frac{1}{p} + \frac{1}{q} = \frac{1}{f} \qquad (23.6)$$

$$f = \frac{R}{2} \qquad (23.5)$$

A magnified image of an object can be formed by a single spherical refracting surface of radius, R, which separates two media whose indices of refraction are n_1 and n_2.

$$\frac{n_1}{p} + \frac{n_2}{q} = \frac{n_2 - n_1}{R} \qquad (23.7)$$

$$M = \frac{h'}{h} = -\frac{n_1 q}{n_2 p} \qquad (23.8)$$

A special case is that of a virtual image formed by a plane refracting surface (when the radius, R, is infinite).

$$q = -\frac{n_2}{n_1} p \qquad (23.9)$$

The equations for magnification and image location for thin lenses are the same as the corresponding equations for spherical mirrors.

$$M = \frac{h'}{h} = -\frac{q}{p} \qquad (23.10)$$

$$\frac{1}{p} + \frac{1}{q} = \frac{1}{f} \qquad (23.11)$$

A major portion of this chapter is devoted to the development and presentation of equations which can be used to determine the location and nature of images formed by various optical components acting either singly or in combination. It is essential that these equations be used with the correct algebraic sign associated with each quantity involved. You must understand clearly the sign conventions for mirrors, refracting surfaces and lenses.

Comment on sign conventions.

The lens makers' equation can be used to calculate the focal length of a thin lens in terms of the radii of curvature of the front and back surfaces and the index of refraction of the lens material.

$$\frac{1}{f} = (n-1)\left(\frac{1}{R_1} - \frac{1}{R_2}\right) \qquad (23.12)$$

SUGGESTIONS, SKILLS, AND STRATEGIES

A major portion of this chapter is devoted to the development and presentation of equations which can be used to determine the location and nature of images formed by various optical components acting either singly or in combination. It is essential that these equations be used with the correct algebraic sign associated with each quantity involved. You must understand clearly the sign conventions for mirrors, refracting surfaces, and lenses. The following discussion represents a review of these sign conventions.

SIGN CONVENTIONS FOR MIRRORS

Equation: $\frac{1}{p} + \frac{1}{q} = \frac{1}{f}$

The front side of the mirror is the region on which light rays are incident and reflected.

p is + if the object is in front of the mirror (real object).
p is – if the object is in back of the mirror (virtual object).

q is + if the image is in front of the mirror (real image).
q is – if the image is in back of the mirror (virtual image).

Both f and R are + if the center of curvature is in front of the mirror (concave mirror).

Both f and R are – if the center of curvature is in back of the mirror (convex mirror).

If M is positive, the image is erect.
If M is negative, the image is inverted.

You should check the sign conventions as stated against the situations described in Figure 23.1.

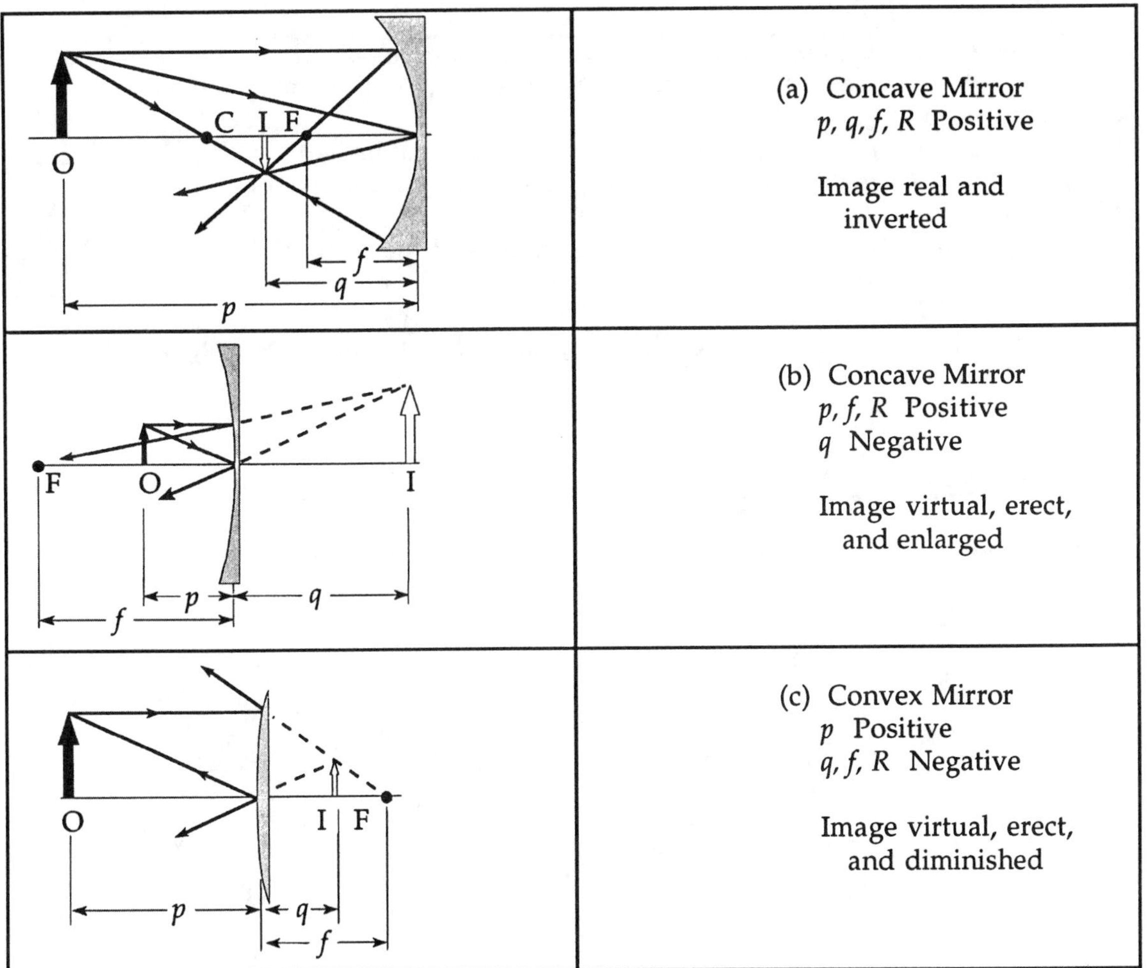

Figure 23.1 Figures describing sign conventions for mirrors.

SIGN CONVENTIONS FOR REFRACTING SURFACES

Equation: $$\frac{n_1}{p} + \frac{n_2}{q} = \frac{n_2 - n_1}{R}$$

In the following table, the *front* side of the surface is the side *from which the light is incident.*

p is + if the object is in front of the surface (real object). p is – if the object is in back of the surface (virtual object). q is + if the image is in back of the surface (real image). q is – if the image is in front of the surface (virtual image). R is + if the center of curvature is in back of the surface. R is – if the center of curvature is in front of the surface. n_1 refers to the index of the medium on the side of the interface from which the light comes. n_2 is the index of the medium into which the light is transmitted after refraction at the interface.

Review the above sign conventions for the situations shown in Figure 23.2.

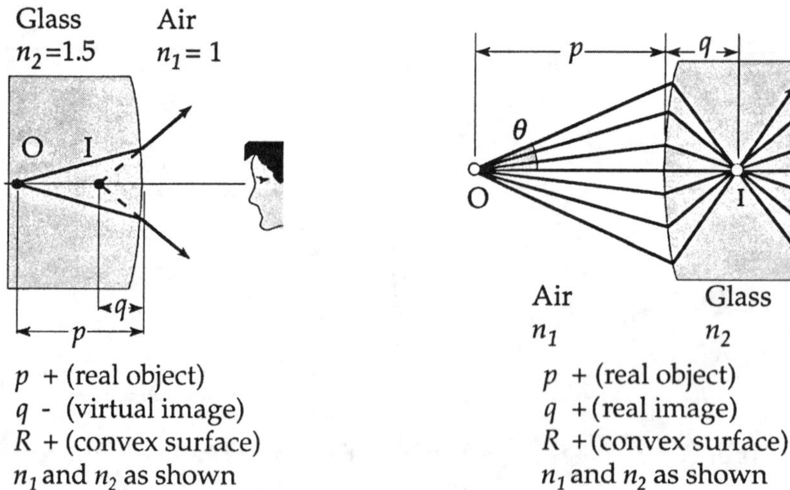

Figure 23.2 Figures describing sign conventions for refracting surfaces.

SIGN CONVENTIONS FOR THIN LENSES

Equations: $$\frac{1}{p}+\frac{1}{q}=\frac{1}{f}=(n-1)\left(\frac{1}{R_1}-\frac{1}{R_2}\right)$$

In the following table, the *front* of the lens is the *side from which the light is incident.*

p is + if the object is in front of the lens. p is – if the object is in back of the lens.
q is + if the image is in back of the lens. q is – if the image is in front of the lens.
R_1 and R_2 are + if the center of curvature is in back of the lens.
R_1 and R_2 are – if the center of curvature is in front of the lens.

The sign conventions for thin lenses are illustrated by the examples shown in Figure 23.3.

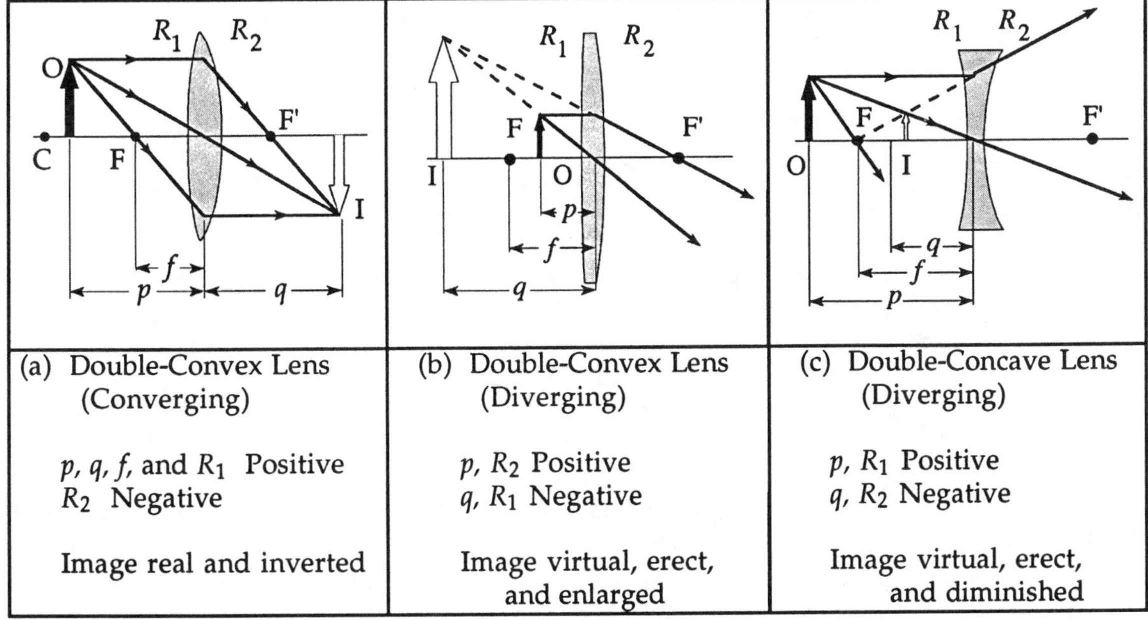

Figure 23.3 Figures describing the sign conventions for various thin lenses.

REVIEW CHECKLIST

- ▷ Identify the following properties which characterize an image formed by a lens or mirror system with respect to an object: position, magnification, orientation (i.e. inverted, erect or right-left reversal) and whether real or virtual.
- ▷ Understand the relationship of the algebraic signs associated with calculated quantities to the nature of the image and object: real or virtual, erect or inverted.
- ▷ Calculate the location of the image of a specified object as formed by a plane mirror, spherical mirror, plane refracting surface, spherical refracting surface, thin lens, or a combination of two or more of these devices. Determine the magnification and character of the image in each case.
- ▷ Construct ray diagrams to determine the location and nature of the image of a given object when the geometrical characteristics of the optical device (lens or mirror) are known.

SOLUTIONS TO SELECTED END-OF-CHAPTER PROBLEMS

3. A person walks into a room that has, on opposite walls, two plane mirrors producing multiple images. When the person is 5.00 ft from the mirror on the left wall and 10.0 ft from the mirror on the right wall, find the distances from the person to the first three images seen in the left-hand mirror.

Solution (1) The first image in the left mirror is 5.00 ft behind the mirror, or 10.0 ft from the position of the person. ◊

(2) The first image in the right mirror is located 10.0 ft behind the right mirror, but this location is 25.0 ft from the left mirror. Thus, the second image in the left mirror is 25.0 ft behind the mirror, or 30.0 ft from the person. ◊

(3) The first image in the left mirror forms an image in the right mirror. This first image is 20.0 ft from the right mirror, and thus, an image 20.0 ft behind the right mirror is formed. This image in the right mirror also forms an image in the left mirror. The distance from this image in the right mirror to the left mirror is 35.0 ft. The third image in the left mirror is, thus, 35.0 ft behind the mirror, or 40.0 ft from the person. ◊

---

9. A dentist uses a mirror to examine a tooth. The tooth is 1.00 cm in front of the mirror, and the image is formed 10.0 cm behind the mirror. Determine (a) the mirror's radius of curvature and (b) the magnification of the image.

Solution (a) $\frac{1}{p}+\frac{1}{q}=\frac{1}{f}$ so $\frac{1}{f}=\frac{p+q}{pq}$ or $q=\frac{fp}{p-f}$ according to the sign conventions, in this case p is positive and q is negative. Therefore,

$$f=\frac{(1.00\text{ cm})(-10.0\text{ cm})}{1.00\text{ cm}+(-10.0\text{ cm})}=1.11\text{ cm}$$

$R = 2f = 2.22$ cm ◊

(b) $M=-\frac{q}{p}=\frac{-10.0}{1.00}=10.0$ ◊

---

13. A spherical Christmas tree ornament is 6.00 cm in diameter. What is the magnification of an object placed 10.0 cm away from the ornament?

Solution From the thin-lens equation, $\frac{1}{p}+\frac{1}{q}=\frac{1}{f}$; $q=\frac{fp}{p-f}$

From the sign conventions for convex mirrors, we see that $R = -3.00$ cm and $f = -1.50$ cm. Also, $p = 10.0$ cm. Thus,

$$q=\frac{(-1.50\text{ cm})(10.0\text{ cm})}{10.0\text{ cm}-(-1.50\text{ cm})}=-1.30\text{ cm}$$

$$M=-\frac{q}{p}=-\frac{-1.30\text{ cm}}{10.0\text{ cm}}=0.130 \quad ◊$$

---

19. A convex spherical mirror with radius of curvature of 10.0 cm produces a virtual image one-third the size of the object. Where is the object located?

Solution A convex mirror has R and f negative.

Thus, $f = \frac{R}{2} = -5.00 \text{ cm}$

For a virtual image, q is negative; and for a convex mirror, M is positive. Therefore,

$$M = -\frac{q}{p} = \frac{1}{3} \quad \text{and} \quad q = -\frac{p}{3}$$

Therefore, use $\frac{1}{f} = \frac{1}{p} + \frac{1}{q}$ and substitute $q = -\frac{p}{3}$ to get

$$\frac{1}{p} + \left(\frac{-3}{p}\right) = \frac{1}{(-5.00 \text{ cm})} \quad \text{or} \quad \frac{2}{p} = \frac{1}{5} \quad \text{and} \quad p = 10.0 \text{ cm} \quad \Diamond$$

---

21. A concave mirror has a focal length of 40.0 cm. Determine the object position that results in an erect image four times the size of the object.

Solution For a concave mirror, R and f are positive. Also, for an erect image M is positive.

Therefore $M = -\frac{q}{p} = 4$ or $q = -4p$ and $f = \frac{R}{2} = 40.0 \text{ cm}$

$$\frac{1}{f} = \frac{1}{p} + \frac{1}{q} \quad \text{becomes} \quad \frac{1}{40.0 \text{ cm}} = \frac{1}{p} - \frac{1}{4p} = \frac{3}{4p} \quad \text{giving} \quad p = 30.0 \text{ cm} \quad \Diamond$$

---

23. What type of mirror is required to form, on a wall 2.00 m from the mirror, an image of an object placed 10.0 cm in front of the mirror? What is the magnification of the image?

Solution Since the image is projected on a screen, it is a real image. Therefore, q is positive. We have

$$\frac{1}{f} = \frac{1}{p} + \frac{1}{q} \quad \text{and} \quad f = \frac{pq}{p+q} = \frac{(10.0 \text{ cm})(200 \text{ cm})}{10.0 \text{ cm} + 200 \text{ cm}} = 9.52 \text{ cm}$$

Thus, because f is positive, we need a concave mirror. ◊

$$M = -\frac{q}{p} = -\frac{200\text{ cm}}{10.0\text{ cm}} = -20.0 \quad ◊$$

---

27. A colored marble is dropped into a large tank filled with benzene (n = 1.50). What is the depth of the tank if the apparent depth of the marble, viewed from directly above the tank, is 35.0 cm?

Solution We are given that q = –35.0 cm. The negative sign arises because we have a virtual image on the same side of the boundary as is the object.

$$p = -\frac{n_1}{n_2}q = -\frac{1.50}{1.00}(-35.0\text{ cm}) = 52.5\text{ cm}$$

The benzene is 52.5 cm deep. ◊

---

31. One end of a long glass rod (n = 1.50) is formed into the shape of a convex surface of radius 8.00 cm. An object is positioned in air along the axis of the rod. Find the image position that corresponds to each of the following object positions: (a) 20.0 cm, (b) 8.00 cm, (c) 4.00 cm, (d) 2.00 cm.

Solution Use $\frac{n_1}{p} + \frac{n_2}{q} = \frac{n_2 - n_1}{R}$ when $n_1 = 1.00$ and $n_2 = 1.50$

Solve this expression for q to get $q = \frac{Rn_2p}{p(n_2 - n_1) - Rn_1}$ and calculate the value of q for each given value of p.

(a) $$q = \frac{(8.00\text{ cm})(1.50)(20.0\text{ cm})}{(20.0\text{ cm})(1.50 - 1.00) - (8.00\text{ cm})(1.00)} = 120\text{ cm} \quad ◊$$

(b) $$q = \frac{(8.00\text{ cm})(1.50)(8.00\text{ cm})}{(8.00\text{ cm})(1.50 - 1.00) - (8.00\text{ cm})(1.00)} = -24.0\text{ cm} \quad ◊$$

(c) $$q = \frac{(8.00\text{ cm})(1.50)(4.00\text{ cm})}{(4.00\text{ cm})(1.50 - 1.00) - (8.00\text{ cm})(1.00)} = -8.00\text{ cm} \quad ◊$$

(d) $q = \dfrac{(8.00 \text{ cm})(1.50)(2.00 \text{ cm})}{(2.00 \text{ cm})(1.50 - 1.00) - (8.00 \text{ cm})(1.00)} = -3.43 \text{ cm}$ ◊

---

35. A contact lens is made of plastic with an index of refraction of 1.58. The lens has a focal length of +25.0 cm, and its inner surface has a radius of curvature of +18.0 mm. What is the radius of curvature of the outer surface?

Solution $\dfrac{1}{f} = (n-1)\left(\dfrac{1}{R_1} - \dfrac{1}{R_2}\right)$ becomes $\dfrac{1}{R_1} = \dfrac{1}{f(n-1)} + \dfrac{1}{R_2}$

$$R_1 = \frac{R_2 f(n-1)}{R_2 + f(n-1)} = \frac{(1.80 \text{ cm})(25.0 \text{ cm})(1.58 - 1)}{1.80 \text{ cm} + 25.0 \text{ cm}(1.58 - 1)} = 1.60 \text{ cm} \quad \lozenge$$

---

41. A diverging lens has a focal length of 20.0 cm. Locate the images for object distances of (a) 40.0 cm, (b) 20.0 cm, and (c) 10.0 cm. For each case, state whether the image is real or virtual, and erect or inverted; and find the magnification.

Solution Use the thin-lens equation, $\dfrac{1}{p} + \dfrac{1}{q} = \dfrac{1}{f}$, and solve for $q = \dfrac{f\,p}{p - f}$

to calculate the value of q for each of the stated object positions. Remember, that for a diverging lens, the focal length is *negative.*

(a) $q = \dfrac{(-20.0 \text{ cm})(40.0 \text{ cm})}{40.0 \text{ cm} - (-20.0 \text{ cm})} = -13.3 \text{ cm}$ ◊

$M = -\dfrac{q}{p} = -\dfrac{(-13.3 \text{ cm})}{40.0 \text{ cm}} = 0.333$ ◊

The image is virtual and erect. ◊

(b) $q = \dfrac{(-20.0 \text{ cm})(20.0 \text{ cm})}{20.0 \text{ cm} - (-20.0 \text{ cm})} = -10.0 \text{ cm}$ ◊

$$M = -\frac{q}{p} = -\frac{(-10.0 \text{ cm})}{20.0 \text{ cm}} = 0.500 \quad \Diamond$$

The image is virtual and erect. ◊

(c) $$q = \frac{(-20.0 \text{ cm})(10.0 \text{ cm})}{10.0 \text{ cm} - (-20.0 \text{ cm})} = -6.67 \text{ cm} \quad \Diamond$$

$$M = -\frac{q}{p} = -\frac{(-6.67 \text{ cm})}{10.0 \text{ cm}} = 0.667 \quad \Diamond$$

The image is virtual and erect. ◊

---

47. A diverging lens is used to form a virtual image of an object. The object is 80.0 cm to the left of the lens, and the image is 40.0 cm to the left of the lens. Determine the focal length of the lens.

Solution We are given that $p = 80.0$ cm and $q = -40.0$ cm.

Thus, from $\frac{1}{f} = \frac{1}{p} + \frac{1}{q}$,

$$f = \frac{pq}{p+q} = \frac{(80.0 \text{ cm})(-40.0 \text{ cm})}{(80.0 \text{ cm}) + (-40.0 \text{ cm})} = -80.0 \text{ cm} \quad \Diamond$$

---

49. An object's distance from a converging lines is 10 times the focal length. How far is the image from the focal point? Express the answer as a fraction of the focal length.

Solution Given $f > 0$ and $p = 10f$

Thus, $\frac{1}{q} = \frac{1}{f} - \frac{1}{p} = \frac{1}{f} - \frac{1}{10f}$ and $q = \frac{10}{9}f$

or, the image is $\frac{1}{9}f$ outside the focal point. ◊

---

53. Two converging lenses, each of focal length 15.0 cm, are placed 40.0 cm apart, and an object is placed 30.0 cm in front of the first. Where is the final image formed, and what is the magnification of the system?

Solution Consider the first lens (the one which is 30.0 cm from the object). Find the image position and magnification as

$$q_1 = \frac{f_1 p_1}{p_1 - f_1} = \frac{(15.0 \text{ cm})(30.0 \text{ cm})}{30.0 \text{ cm} - 15.0 \text{ cm}} = 30.0 \text{ cm} \quad \text{and} \quad M_1 = -\frac{q_1}{p_1} = -\frac{30.0 \text{ cm}}{30.0 \text{ cm}} = -1$$

Now consider the second lens. The image produced by the first lens becomes the object for this second lens. ◊

Thus, $p_2 = 40.0 \text{ cm} - q_1 = 40.0 \text{ cm} - 30.0 \text{ cm} = 10.0 \text{ cm}$.

$$q_2 = \frac{f_2 p_2}{p_1 - f_1} = \frac{(15.0 \text{ cm})(10.0 \text{ cm})}{10.0 \text{ cm} - 15.0 \text{ cm}} = -30.0 \text{ cm} \quad \text{and} \quad M_2 = -\frac{q_2}{p_2} = -\frac{(-30.0 \text{ cm})}{10.0 \text{ cm}} = 3$$

The overall magnification M is $M = (M_1)(M_2)(-1)(3) = -3$ ◊

---

59. Two converging lenses, having focal lengths of 10.0 cm and 20.0 cm are placed 50.0 cm apart as shown in Figure 23.38. The final image is to be located between the lenses at the position indicated. (a) How far to the left of the first lens should the object be positioned? (b) What is the overall magnification? (c) Is the final image erect or inverted?

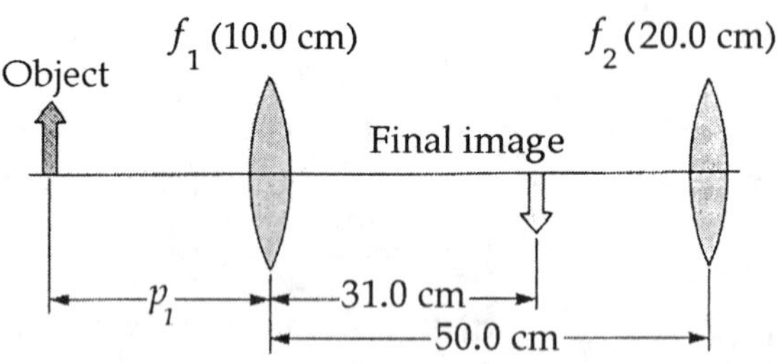

Figure 23.38

Solution (a) Start by calculating the object distance for the second lens.

$$f_2 = 20.0 \text{ cm}, \quad q_2 = -(50.0 \text{ cm} - 31.0 \text{ cm}) = -19.0 \text{ cm}$$

$$\frac{1}{p_2} + \frac{1}{q_2} = \frac{1}{f} \quad \text{becomes} \quad \frac{1}{p_2} - \frac{1}{19.0} = \frac{1}{20.0} \quad \text{from which} \quad p_2 = 9.74 \text{ cm}$$

$$q_1 = (50.0 \text{ cm} - 9.74 \text{ cm}) = 40.26 \text{ cm} \quad \text{and} \quad f_1 = 10.0 \text{ cm}$$

Now, treat the first lens:

$$\frac{1}{p_1}=\frac{1}{f_1}-\frac{1}{q_1}=\frac{1}{10.0\text{ cm}}-\frac{1}{40.26\text{ cm}} \quad \text{and} \quad p_1 = 13.3\text{ cm} \quad \Diamond$$

(b) $$M_1 = -\frac{q_1}{p_1} = -\frac{40.26\text{ cm}}{13.3\text{ cm}} = -3.03$$

$$M_2 = -\frac{q_2}{p_2} = -\frac{-19.0\text{ cm}}{9.74\text{ cm}} = 1.95$$

and $M = M_1 M_2 = -5.90$ ◊

(c) The overall magnification is less than 0, so the final image is inverted. ◊

---

61. The object in Figure 23.40 is midway between the lens and the mirror. The mirror's radius of curvature is 20.0 cm, and the lens has a focal length of –16.7 cm. Considering only the light that leaves the object and travels first toward the mirror, locate the final image formed by this system. Is this image real or virtual? Is it erect or inverted? What is the overall magnification?

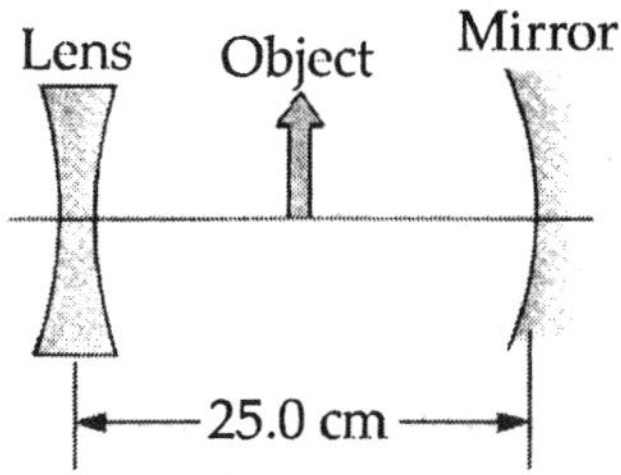

Figure 23.40

Solution First calculate the image distance for the mirror using the equation,

$$q_1 = \frac{f_1 p_1}{p_1 - f_1} = \frac{(10.0\text{ cm})(12.5\text{ cm})}{12.5\text{ cm} - 10.0\text{ cm}} = 50.0\text{ cm}$$

Since q_1 is *positive,* the image formed by the mirror is *to the left of the mirror.* This image serves as a *virtual object* for the lens; and the object distance for the lens is $p_2 = -25.0$ cm.

$$q_2 = \frac{f_2 p_2}{p_2 - f_2} = \frac{(-16.7\text{ cm})(-25.0\text{ cm})}{-25.0\text{ cm} - (-16.7\text{ cm})} = -50.3\text{ cm}$$

Therefore, the *final image* formed by the system is located 25.3 cm to the right of the mirror. ◊

$$M = M_1 M_2 = \left(-\frac{q_1}{p_1}\right)\left(-\frac{q_2}{p_2}\right)$$

$$M = \left(-\frac{50.0\text{ cm}}{12.5\text{ cm}}\right)\left(-\frac{-50.3\text{ cm}}{-25.0\text{ cm}}\right) = 8.05 \quad ◊$$

---

67. A converging lens of focal length 20.0 cm is separated by 50.0 cm from a converging lens of focal length 5.00 cm. (a) Find the final position of the image of an object placed 40.0 cm in front of the first lens. (b) If the height of the object is 2.00 cm, what is the height of the final image? Is it real or virtual? (c) Determine the image position of an object placed 5.00 cm in front of the two lenses in contact.

Solution (a) Find the image position produced by the first lens, L_1.

$$q_1 = \frac{f_1 p_1}{p_1 - f_1} = \frac{(20.0\text{ cm})(40.0\text{ cm})}{40.0\text{ cm} - 20.0\text{ cm}} = 40.0\text{ cm}$$

For the second lens, L_2,

$$p_2 = 10.0\text{ cm} \qquad \text{and} \qquad q_2 = \frac{f_2 p_2}{p_2 - f_2} = \frac{(5.00\text{ cm})(10.0\text{ cm})}{10.0\text{ cm} - 5.00\text{ cm}} = 10.0\text{ cm}$$

The final image position is 10.0 cm to the right of the second lens. ◊

(b) $$M = M_1 M_2 = \left(-\frac{q_1}{p_1}\right)\left(-\frac{q_2}{p_2}\right) = \left(-\frac{40.0\text{ cm}}{40.0\text{ cm}}\right)\left(-\frac{10.0\text{ cm}}{10.0\text{ cm}}\right) = 1$$

Therefore, the final image height is $h = 2.00$ cm. ◊

The final image is erect and real. ◊

(c) With the two lenses are in *contact*, the effective focal length is calculated using

$$\frac{1}{f_{eff}} = \frac{1}{f_1} + \frac{1}{f_2} \quad \text{or} \quad f_{eff} = \frac{f_1 f_2}{f_1 + f_2} = \frac{(20.0\text{ cm})(5.00\text{ cm})}{20.0\text{ cm} + 5.00\text{ cm}} = 4.00\text{ cm}$$

Then, $q = \frac{f_{eff}\,p}{p - f_{eff}} = \frac{(4.00\text{ cm})(5.00\text{ cm})}{5.00\text{ cm} - 4.00\text{ cm}} = 20.0\text{ cm}$ ◊

---

73. To work this problem, use the fact that the image formed by the first surface becomes the object for the second surface. Figure 23.44 shows a piece of glass whose index of refraction is 1.50. The ends are hemispheres with radii 2.00 cm and 4.00 cm, and the centers of the hemispherical ends are separated by a distance of 8.00 cm. A point object is located in air 1.00 cm from the left end of the glass. Find the location of the image of the object due to refraction at the two spherical surfaces.

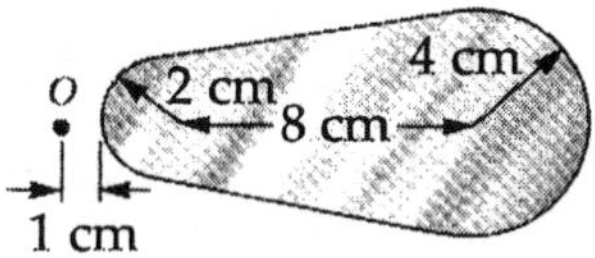

Figure 23.44

Solution We use $\frac{n_1}{p} + \frac{n_2}{q} = \frac{n_2 - n_1}{R}$ applied to each surface in turn.

At the first surface, we have $\frac{1}{1.00\text{ cm}} + \frac{1.50}{q_1} = \frac{1.50 - 1}{2.00\text{ cm}}$ or $q_1 = -2.00\text{ cm}$

Thus, the image is formed 2.00 cm in front of the surface, a distance of 16.0 cm from the second surface. This image becomes the object for the second surface (hence, $p_2 = +16.0\text{ cm}$).

We again use $\frac{n_1}{p} + \frac{n_2}{q} = \frac{n_2 - n_1}{R}$

$$\frac{1.50}{16.0\text{ cm}} + \frac{1}{q_2} = \frac{1.00 - 1.50}{-4.00\text{ cm}} \quad \text{or} \quad q_2 = 32.0\text{ cm}$$

The final image is a real image 32.0 cm past the second surface. ◊

CHAPTER SELF-QUIZ

1. Which of the following best describes the image of a concave mirror when the object is at a distance greater than twice the focal point distance from the mirror?
   a. virtual, erect; and magnification greater than one
   b. real, inverted; and magnification less than one
   c. virtual, erect; and magnification less than one
   d. real, inverted; and magnification greater than one

2. If a man's face is 30.0 cm in front of a concave shaving mirror creating an erect image 1.50 times as large as the object, what is the mirror's focal length?
   a. 12.0 cm
   b. 20.0 cm
   c. 70.0 cm
   d. 90.0 cm

3. An object is placed at a distance of 40.0 cm from a thin lens along the axis. If a virtual image forms at a distance of 50.0 cm from the lens, on the same side as the object, what is the focal length of the lens?
   a. 22.0 cm
   b. 45.0 cm
   c. 90.0 cm
   d. 200 cm

4. A glass block, for which $n = 1.48$, has a bubble blemish located 6.40 cm from one surface. At what distance from that surface does the image of the blemish appear to the outside observer?
   a. 3.20 cm
   b. 4.30 cm
   c. 7.60 cm
   d. 9.50 cm

5. Which of the following effects is the result of the fact that the index of refraction of glass will vary with wavelength?
   a. spherical aberration
   b. mirages
   c. chromatic aberration
   d. light scattering

6. Two thin lenses of focal lengths 15.0 and 20.0 cm, respectively, are placed in contact in an orientation so that their optic axes coincide. What is the focal length of the two in combination?
   a. 8.57 cm
   b. 17.5 cm
   c. 35.0 cm
   d. 60.0 cm

7. For a diverging lens with one flat surface, the radius of curvature for the curved surface is 10.0 cm. What must the index of refraction be so that the focal length is –10.0 cm?
   a. 1.50
   b. 2.00
   c. 3.00
   d. 0.50

8. If atmospheric refraction did not occur, how would the apparent time of sunrise and sunset be changed?
   a. both would be later
   b. both would be earlier
   c. sunrise would be later and sunset earlier
   d. sunrise would be earlier and sunset later

9. Parallel rays of light that hit a concave mirror will come together
   a. at the center of curvature
   b. at the focal point
   c. at a point half way to the focal point
   d. at infinity

10. A magnifying glass has a convex lens of focal length 15.0 cm. At what distance from a postage stamp should you hold this lens to get a magnification of +2?
    a. 10.0 cm
    b. 7.50 cm
    c. 5.00 cm
    d. 2.50 cm

11. When the reflection of an object is seen in a concave mirror, the image will
    a. always be real
    b. always be virtual
    c. may be either real or virtual
    d. will always be magnified

12. An object is 15.0 cm from the surface of a spherical Christmas tree ornament that is 5.00 cm in diameter. What is the magnification of the image?
    a. –0.055
    b. +0.033
    c. +0.077
    d. +0.154

24
Wave Optics

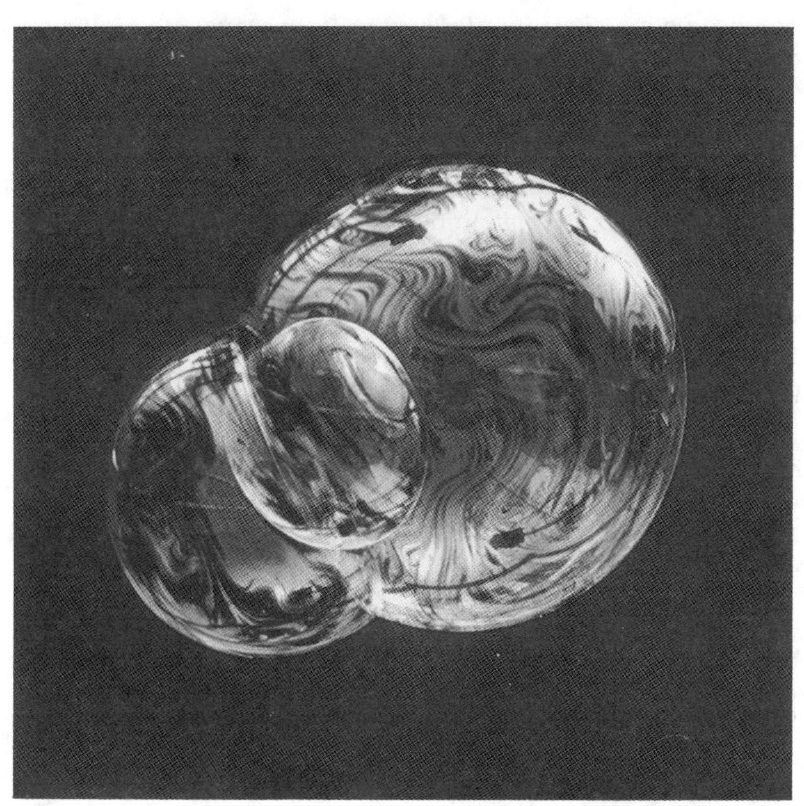

Chapter 24

WAVE OPTICS

Our discussion of light has thus far been concerned with what happens when light passes through a lens or reflects from a mirror. Because explanations of such phenomena rely on a geometric analysis of light rays, that part of optics is often called geometric optics. We now expand our study of light into an area called *wave optics.* The three primary topics we examine in this chapter are interference, diffraction, and polarization. These phenomena cannot be adequately explained with ray optics, but the wave theory leads us to satisfying descriptions.

NOTES FROM SELECTED CHAPTER SECTIONS

24.1 Conditions for Interference

In order to observe *sustained* interference in light waves, the following conditions must be met:

1. The sources must be coherent; they must maintain a *constant phase* with respect to each other.
2. The sources must be *monochromatic*—be of a *single wavelength.*
3. The *superposition principle* must apply.

24.2 Young's Double-Slit Experiment

A schematic diagram illustrating the geometry used in Young's double-slit experiment is shown in the figure below. The two slits S_1 and S_2 serve as coherent monochromatic sources. The *path difference* $\delta = r_2 - r_1 = d\sin\theta$.

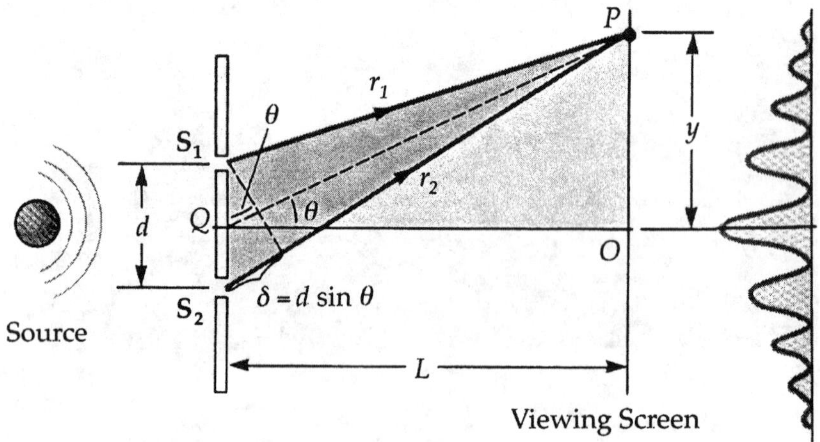

24.3 Change of Phase due to Reflection

An electromagnetic wave undergoes a *phase change of 180°* upon reflection from a medium that is *optically more dense* than the one in which it was traveling. There is also a 180° phase change upon reflection from a *conducting surface.*

24.4 Interference in Thin Films

Interference effects in thin films depend on the difference in length of path traveled by the interfering waves as well as any phase changes which may occur due to reflection. It can be shown by the Lloyd's mirror experiment that when light is reflected from a surface of index of refraction greater than the index of the medium in which it is initially traveling, the reflected ray undergoes a 180° (or π radians) phase change. In analyzing interference effects, this can be considered equivalent to the gain or loss of a half wavelength in path difference. Therefore, there are two different cases to consider: (a) a film surrounded by a common medium and (b) a thin film located between two different media. These cases are illustrated in the figures below.

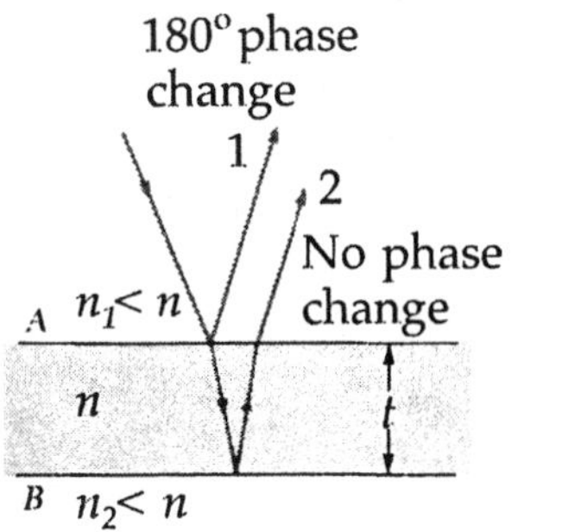

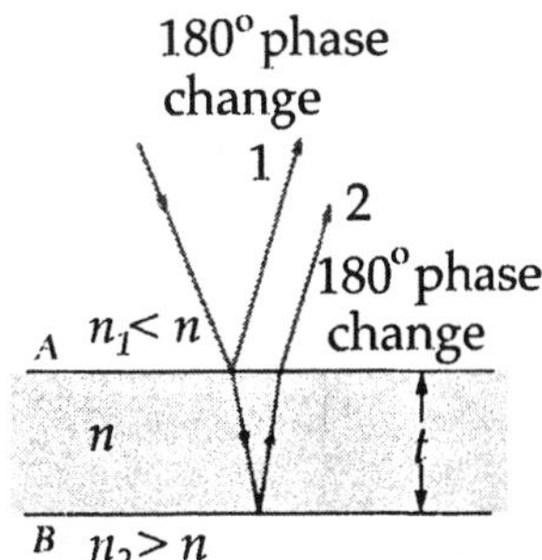

(a) Interference of light resulting from reflections at two surfaces of a thin film of thickness t, index of refraction n. (b) A thin film between two different media.

In case (a), where a phase change occurs at the top surface, the reflected rays will be in phase (constructive interference) if the thickness of the film is an odd number of quarter wavelengths—so that the path lengths will differ by an odd number of half wavelengths.

In case (b), the phase changes due to reflection at *both* the top and bottom surfaces are offsetting and, therefore, constructive interference for the reflected rays will occur when the film thickness is an integer number of half wavelengths—and, therefore, the path difference will be a whole number of wavelengths.

EQUATIONS AND CONCEPTS

In the arrangement used for the Young's double-slit experiment, two slits separated by a distance, d, serve as monochromatic coherent sources. The light intensity at any point on the screen is the resultant of light reaching the screen from both slits. Also, light from the two slits reaching any point on the screen (except the center) travel unequal path lengths. This difference in length of path is called the path difference.

$$\delta = r_2 - r_1 = d\sin\theta \qquad (24.1)$$

Bright fringes (constructive interference) will appear at points on the screen for which the path difference is equal to an integral multiple of the wavelength. The positions of bright fringes can also be located by calculating their vertical distance from the center of the screen (y). In each case, the number m is called the order number of the fringe. The central bright fringe ($\theta = 0, m = 0$) is called the zeroth-order maximum.

$$\delta = d\sin\theta = m\lambda \qquad (24.2)$$

$$m = 0,\ \pm 1,\ \pm 2, \ldots$$

$$y_{\text{bright}} = \left(\frac{\lambda L}{d}\right)m \qquad (24.5)$$

Dark fringes (destructive interference) will appear at points on the screen which correspond to path differences of an odd multiple of half wavelengths. For these points of destructive interference, waves which leave the two slits in phase arrive at the screen 180° out of phase.

$$\delta = d\sin\theta = \left(m + \frac{1}{2}\right)\lambda \qquad (24.3)$$

$$m = 0,\ \pm 1,\ \pm 2, \ldots$$

$$y_{\text{dark}} = \left(\frac{\lambda L}{d}\right)\left(m + \frac{1}{2}\right) \qquad (24.6)$$

The wavelength of light in a medium, λ_n, is less than the wavelength in free space, λ.

$$\lambda_n = \frac{\lambda}{n} \qquad (24.7)$$

When parallel light rays are incident on a slit of width, a, a diffraction pattern is formed on a screen in front of the slit. The condition for destructive interference (dark band) at a given point on the screen can be stated in terms of the angle θ (direction from the center of the slit to the point on the screen).

$$\sin\theta = m\frac{\lambda}{a} \qquad (24.11)$$

(Destructive interference)

$m = \pm 1, \ \pm 2, \ \pm 3, \ \ldots$

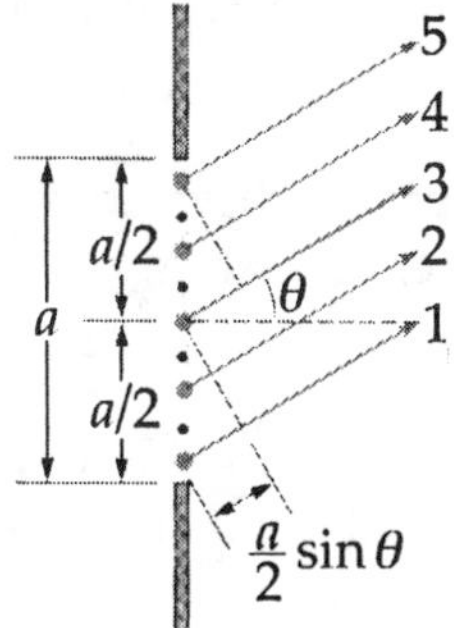

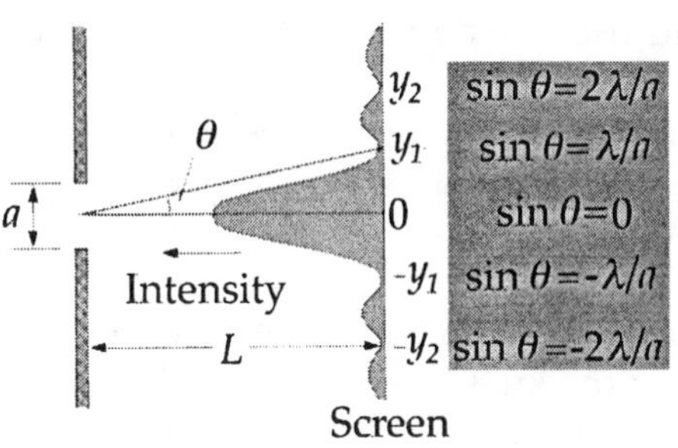

There is a broad central bright fringe at the center of the screen. There are secondary bright fringes (of progressively diminished intensity) located approximately midway between adjacent dark fringes.

Comment on single-slit diffraction.

Brewster's law gives the value of the polarizing angle for a surface of index of refraction, n. The polarizing angle is the angle of incidence for which the reflected beam is completely polarized. It is also true that when the angle of incidence is equal to the polarizing angle, the reflected ray is perpendicular to the transmitted ray.

$$n = \tan\theta_p \qquad (24.13)$$

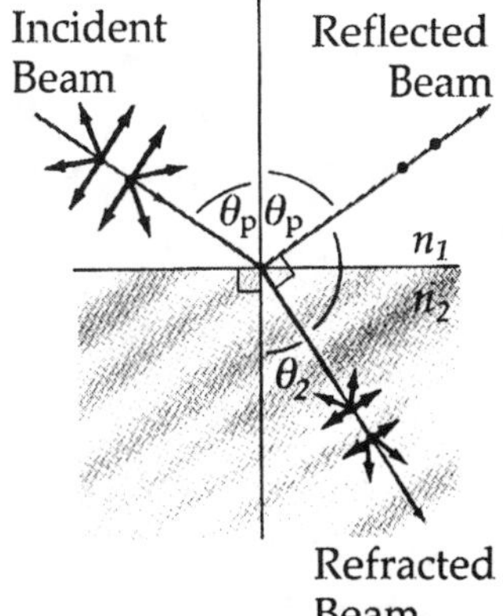

Chapter 24

SUGGESTIONS, SKILLS, AND STRATEGIES

The following features should be kept in mind while working thin-film interference problems:

1. Identify the thin film from which interference effects are being observed.

2. The type of interference that occurs in a specific problem is determined by the phase relationship between that portion of the wave reflected at the upper surface of the film and that portion reflected at the lower surface of the film.

3. Phase differences between the two portions of the wave occur because of differences in the distance traveled by the two portions and by phase changes occurring upon reflection.

   The wave reflected from the lower surface of the film has to travel a distance equal to twice the thickness of the film before it returns to the upper surface of the film where it interferes with that portion of the wave reflected at the upper surface. When *distances alone are considered,* if this extra distance is equal to an integral multiple of $\frac{1}{2}\lambda$, the interference will be constructive; if the extra distance equals $\frac{1}{4}\lambda$, $\frac{3}{4}\lambda$, and so forth, destructive interference will occur.

   Reflections may change the results above which are based solely on the distance traveled. When a wave traveling in a particular medium reflects off a surface having a higher index of refraction than the one it is in, a 180° phase shift occurs. This has the same effect as if the wave lost $\frac{1}{2}\lambda$. These losses must be considered in addition to those losses that occur because of the extra distance one wave travels over another.

4. When distance and phase changes upon reflection are both taken into account, the interference will be constructive if the waves are out of phase by an integral multiple of $\frac{1}{2}\lambda$. Destructive interference will occur when the phase difference is $\frac{1}{4}\lambda$, $\frac{3}{4}\lambda$, and so forth.

REVIEW CHECKLIST

▷ Describe Young's double-slit experiment to demonstrate the wave nature of light. Account for the phase difference between light waves from the two sources as they arrive at a given point on the screen. State the conditions for constructive and destructive interference in terms of each of the following: path difference, phase difference, distance from center of screen, and angle subtended by the observation point at the source midpoint.

▷ Account for the conditions of constructive and destructive interference in thin films considering both path difference and any expected phase changes due to reflection.

▷ Describe Fraunhofer diffraction produced by a single slit. Determine the positions of the maxima and minima in a single-slit diffraction pattern.

▷ Describe qualitatively the polarization of light by selective absorption, reflection, scattering, and double refraction. Also, make appropriate calculations using Brewster's law.

SOLUTIONS TO SELECTED END-OF-CHAPTER PROBLEMS

3. A pair of narrow, parallel slits separated by 0.250 mm are illuminated by the green component from a mercury vapor lamp (λ = 546.1 nm). The interference pattern is observed on a screen 1.20 m from the plane of the parallel slits. Calculate the distance (a) from the central maximum to the first bright region on either side of the central maximum and (b) between the first and second dark bands in the interference pattern.

Solution (a) The position of the first bright fringe, $m = 1$, is given by

$$y_{\text{bright}} = \frac{\lambda L}{d} = \frac{(546 \times 10^{-9}\ \text{m})(1.20\ \text{m})}{0.250 \times 10^{-3}\ \text{m}} = 2.62 \times 10^{-3}\ \text{m} = 2.62\ \text{mm} \quad \Diamond$$

(b) $y_{\text{dark}} = \left(m + \frac{1}{2}\right)\frac{\lambda L}{d}$ from which $\Delta y_{\text{dark}} = \left(1 + \frac{1}{2}\right)\frac{\lambda L}{d} - \left(0 + \frac{1}{2}\right)\frac{\lambda L}{d} = \frac{\lambda L}{d} = 2.62\ \text{mm} \quad \Diamond$

---

7. Light of wavelength 460 nm falls on two slits spaced 0.300 mm apart. What is the required distance from the slits to a screen if the spacing between the first and second dark fringes is to be 4.00 mm?

Solution For dark fringes, $y_{\text{dark}} = \left(m + \frac{1}{2}\right)\frac{\lambda L}{d}$ between consecutive fringes.

$$\Delta y = \frac{\lambda L}{d}$$ and therefore we have

$$L = \frac{(\Delta y)d}{\lambda} = \frac{(4.00\times10^{-3}\text{ m})(3.00\times10^{-4}\text{ m})}{460\times10^{-9}\text{ m}} = 2.61\text{ m} \quad \lozenge$$

---

11. Two radio antennas separated by 300 m, as shown in Figure 24.29 (following page), simultaneously transmit identical signals (assume waves) of the same wavelength. A radio in a car traveling due north receives the signals. (a) If the car is at the position of the second maximum, what is the wavelength of the signals? (b) How much farther must the car travel to encounter the next minimum in reception? (*Caution:* Avoid small angle approximations in this problem.)

Solution Note, with the conditions given, the small angle approximation *does not work well.* The approach to be used is outlined below.

(a) At $m = 2$ maximum,

$$\tan\theta = \frac{400\text{ m}}{1000\text{ m}} = 0.400$$

and $\theta = 21.8°$ so

$$\lambda = \frac{d\sin\theta}{m} = \frac{(300\text{ m})\sin 21.8°}{2}$$

$$\lambda = 55.7\text{ m} \quad \lozenge$$

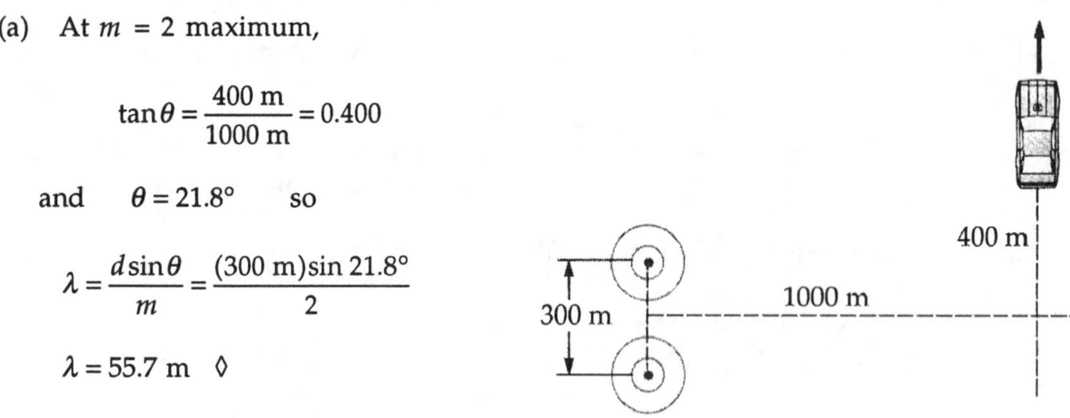

Figure 24.29

(b) The next minimum encountered is the $m = 2$ minimum, and at that point,

$$d\sin\theta = \left(m + \frac{1}{2}\right)\lambda \quad \text{which becomes} \quad d\sin\theta = \frac{5}{2}\lambda$$

or $$\sin\theta = \frac{m\lambda}{2d} = \frac{5(55.7\text{ m})}{2(300\text{ m})} = 0.464 \quad \text{and} \quad \theta = 27.7°$$

so $$y = (1000\text{ m})\tan 27.7° = 524\text{ m}$$

Therefore, the car must travel an additional 124 m. $\lozenge$

---

15. Radio waves from a star, of wavelength 250 m, reach a radio telescope by two separate paths. One is a direct path to the receiver, which is situated on the edge of a cliff by the ocean. The second is by reflection off the water. The first minimum of destructive interference occurs when the star is 25.0° above the horizon. Find the height of the cliff. (Assume no phase change on reflection.)

Solution (See sketch.)

We have $25.0° + 90.0° + \alpha + 25.0° = 180°$

Thus, $\alpha = 40.0°$

$$\delta = d_2 - d_1 = d_2 - d_2 \sin\alpha = d_2(1 - \sin 40.0°) = (0.357)d_2$$

For first order destructive interference, $\delta = \frac{\lambda}{2} = 125 \text{ m}$

Therefore, $(0.357)d_2 = 125 \text{ m}$ and $d_2 = 350 \text{ m}$

Then, $h = d_2 \sin 25.0° = (350 \text{ m})\sin 25.0° = 148 \text{ m}$ ◊

---

17. Suppose the film shown in Figure 24.7 has an index of refraction of 1.36 and is surrounded by air on both sides. Find the minimum thickness, other than zero, that will produce constructive interference in the reflected light when the film is illuminated by light of wavelength 500 nm.

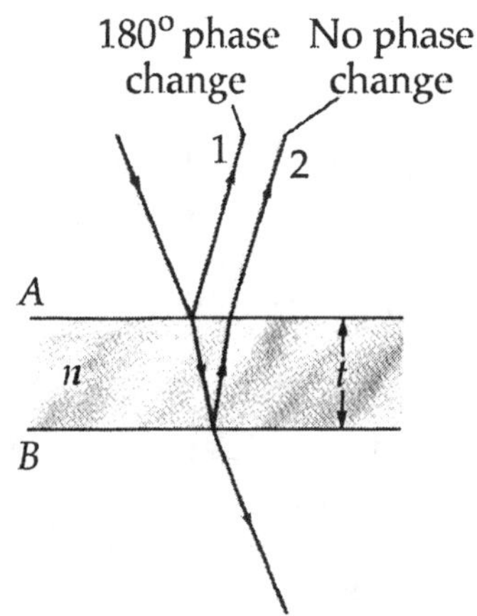

Figure 24.7

Solution The phase difference due to the difference in path lengths is $2n_f t$, where n_f is the index of refraction of the film. The phase difference due to reflection at the upper surface is $\frac{\lambda}{2}$.

For constructive interference, the total phase difference, $\delta = m\lambda$.

Thus, $2n_f t + \frac{\lambda}{2} = m\lambda$

For the minimum thickness, $m = 1$, and we have $t = \frac{\lambda}{4n_f} = \frac{500 \text{ nm}}{4(1.36)} = 91.9 \text{ nm}$ ◊

---

21. Two parallel glass plates are placed in contact and illuminated from above with light of wavelength 580 nm. As the plates are slowly moved apart and the reflected light is observed, darkness occurs at certain separations. What are the distances (other than, perhaps, zero) of the first three of these separations?

Solution The total phase difference is that due to reflection and that due to path difference. We have

$$2n_f t + \frac{\lambda}{2} = \left(m + \frac{1}{2}\right)\lambda \qquad \text{or} \qquad t = m\frac{\lambda}{2n_f} = m\frac{580 \text{ nm}}{2(1)} = (290 \text{ nm})\text{m}$$

When $m = 1$, $t = 290$ nm ◊

When $m = 2$, $t = 580$ nm ◊

When $m = 3$, $t = 870$ nm ◊

---

23. A beam of light of wavelength 580 nm passes through two closely spaced glass plates, as shown in Figure 24.30. For what minimum nonzero value of the plate separation, *d*, will the transmitted light be bright?

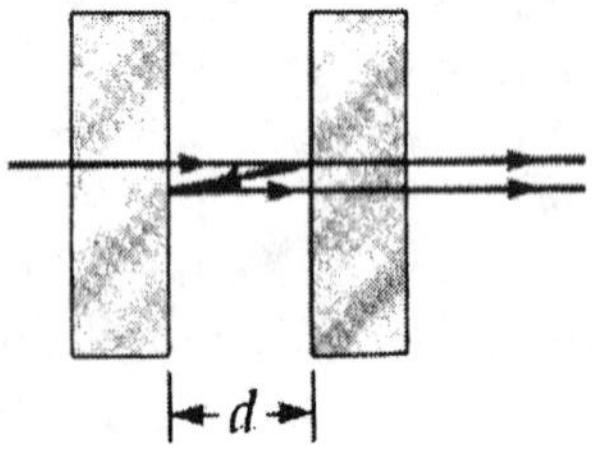

Figure 24.30

Solution Both reflections are of the same type. Thus, there is no phase difference due to the reflections. Therefore, for constructive interference, we have

$$\delta = 2n_f d = m\lambda$$

For the minimum value of d, $m = 1$; and

$$d = \frac{\lambda}{2} = \frac{580 \text{ nm}}{2} = 290 \text{ nm} \quad \Diamond$$

---

29. A thin 1.00×10^{-5}-cm-thick film of MgF_2 (n = 1.38) is used to coat a camera lens. Are any wavelengths in the visible spectrum intensified in the reflected light?

Solution There is a phase shift at both the upper and lower surfaces of the film.

For a bright fringe, path difference $= 2n_f t + 0 = m\lambda$

or $\lambda = \dfrac{2n_f t}{m} = \dfrac{2(1.38)(100 \text{ nm})}{m} = \dfrac{276 \text{ nm}}{m}$ which gives

$$\lambda = 276 \text{ nm}, \quad 138 \text{ nm}, \quad 92 \text{ nm}, \quad \text{etc.}$$

Thus, *no visible wavelengths* produce constructive interference. ◊

---

33. Two rectangular optically flat plates (n = 1.52) are in contact along one end and are separated along the other end by a sheet of paper that is 4.00×10^{-3} cm thick (Fig. 24.31). The top plate is illuminated by monochromatic light of wavelength 546.1 nm. Calculate the number of dark parallel bands crossing the top plate (including the dark band at zero thickness along the edge of contact between the two plates).

Figure 24.31

Solution For destructive interference,

$$\delta = 2n_f t + \frac{\lambda}{2} = \left(m + \frac{1}{2}\right)\lambda \quad \text{or} \quad 2n_f t = m\lambda \quad \text{and} \quad m = \frac{2n_f t}{\lambda}$$

Thus, $m = \dfrac{2(1.00)(4.00 \times 10^{-5} \text{ m})}{546.1 \times 10^{-9} \text{ m}} = 146.5$

Thus, the last dark fringe seen is the 146th order, and the total number of dark fringes seen (counting the zeroth order) is 146 + 1 = 147. ◊

---

41. A screen is placed 50.0 cm from a single slit, which is illuminated with light of wavelength 680 nm. If the distance between the first and third minima in the diffraction pattern is 3.00 mm, what is the width of the slit?

Solution We have for minimum in a single-slit diffraction pattern, $y = mL\left(\frac{\lambda}{a}\right)$

Thus, for the first dark band, $y_1 = \frac{\lambda}{a}(0.500 \text{ m})$

and the third dark band is at $y_3 = 3\frac{\lambda}{a}(0.500 \text{ m})$

So, $$\Delta y = y_3 - y_1 = 2\frac{\lambda}{a}(0.500 \text{ m}) = \frac{\lambda}{a}(1.00 \text{ m})$$

and $$a = \frac{\lambda}{\Delta y} = \frac{(1.00 \text{ m})(680 \times 10^{-9} \text{ m})}{3.00 \times 10^{-3} \text{ m}} = 0.227 \text{ mm} \quad \lozenge$$

---

43. Light of wavelength 587.5 nm illuminates a single 0.750-mm-wide slit. (a) At what distance from the slit should a screen be placed if the first minimum in the diffraction pattern is to be 0.850 mm from the central maximum? (b) Calculate the width of the central maximum.

Solution (a) At the first dark band,

$$\sin\theta = \frac{\lambda}{a} = \frac{5.875 \times 10^{-7} \text{ m}}{7.50 \times 10^{-4} \text{ m}} = 7.83 \times 10^{-4}$$

But also, $\sin\theta = \frac{y}{L}$ so $L = \frac{y}{\sin\theta}$

Thus, $$L = \frac{8.50 \times 10^{-4} \text{ m}}{7.83 \times 10^{-4}} = 1.09 \text{ m} \quad \lozenge$$

(b) The width of the central maximum = $2y$ = 2(0.850 mm) = 1.70 mm. ◊

---

47. A light beam is incident on heavy flint glass (n = 1.65) at the polarizing angle. Calculate the angle of refraction for the transmitted ray.

Solution $n_g = \tan\theta_p = 1.65$

$$\theta_p = \tan^{-1} n_g = \tan^{-1}(1.65) = 58.8°$$

and from Snell's law, $n_g \sin\theta_r = n_{\text{air}} \sin\theta_p$ which gives

$$\sin\theta_r = \frac{(1.00)(\sin 58.8°)}{1.65} = 0.5183 \quad \text{and} \quad \theta_r = 31.2° \quad ◊$$

---

53. The critical angle for total internal reflection for sapphire surrounded by air is 34.4°. Calculate the polarizing angle for sapphire if the light is incident from the air.

Solution The critical angle for total internal reflection is given by $\sin\theta_c = \frac{n_2}{n_1}$,

where n_1 is the index of refraction of the material in which the light travels, and n_2 is the index of the surrounding material. Thus,

$$n_1 = \frac{n_2}{\sin\theta_c} = \frac{1.00}{\sin 34.4°} = 1.77$$ This is the index of refraction of sapphire.

The polarizing angle is given by $\tan\theta_p = \frac{n_2}{n_1} = \frac{1.77}{1.00} = 1.77$, and $\theta_p = 60.5°$ ◊

---

55. When a monochromatic beam of light is incident from air at an angle of 37.0° with the normal on the surface of a glass block, it is observed that the refracted ray is directed at 22.0° with the normal. What angle of incidence from air would result in total polarization of the reflected beam?

Solution We first find the index of refraction of the glass by use of Snell's law.

$$n_1 \sin\theta_1 = n_g \sin\theta_2$$

$$n_g = \frac{n_1 \sin\theta_1}{\sin\theta_2} = \frac{(1.00)(\sin 37.0°)}{\sin 22.0°} = 1.606$$ from which $n_g = 1.606$

Then, $\tan\theta_p = \frac{n_2}{n_1} = \frac{1.606}{1.00} = 1.606$ and $\theta_p = 58.1°$ ◊

---

61. A beam containing light of wavelengths λ_1 and λ_2 is incident on a set of parallel slits. In the interference pattern, the fourth bright line of the λ_1 light occurs at the same position as the fifth bright line of the λ_2 light. If λ_1 is known to be 540 nm, what is the value of λ_2?

Solution Bright lines occur when $m\lambda = d\sin\theta$.

For the fifth order bright line of the λ_2 light, we have $5\lambda_2 = d\sin\theta$ (1)

For the fourth order bright line of λ_1, we have $4\lambda_1 = d\sin\theta$ (2)

From (1) and (2), we find $5\lambda_2 = 4\lambda_1$

Thus, $\lambda_2 = \frac{4}{5}\lambda_1 = 0.800(540\text{ nm}) = 432\text{ nm}$ ◊

---

63. Three polarizing plates whose planes are parallel are centered on a common axis. The directions of the transmission axes relative to the common vertical direction are shown in Figure 24.34. A plane-polarized beam of light with the plane of polarization parallel to the vertical reference direction is incident from the left on the first disk with intensity I_i = 10.0 units (arbitrary). Calculate the transmitted intensity, I_f, when $\theta_1 = 20.0°$, $\theta_2 = 40.0°$, and $\theta_3 = 60.0°$. (*Hint:* Refer to Problem 62 stated below.)

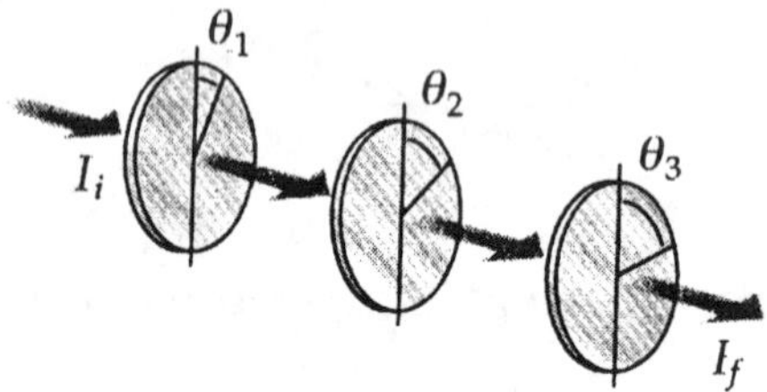

Figure 24.34

Problem 62. When a beam of light passes through a pair of polarizing plates (polarizer and analyzer), the intensity of the transmitted beam is a maximum when the transmission axes of the polarizer and the analyzer are parallel to each other ($\theta = 0°$ or $180°$) and zero when the axes are perpendicular ($\theta = 90.0°$). For any value of θ between 0° and 90.0°, the transmitted intensity is given by Malus' law: $I = I_i \cos^2\theta$

where I_i is the intensity of the light incident on the analyzer after the light has passed through the polarizer. (a) Polarized light is incident on the analyzer so that the direction of polarization makes an angle of 45.0° with respect to the transmission axes. What is the ratio I/I_i? (b) What should the angle between the transmission axes be to make $I/I_i = 1/3$?

Solution For $\theta_1 = 20.0°$, we have $I_2 = I_i \cos^2\theta = 10.0 \cos^2 20.0° = 8.83$ units

The light emerging from this plate is polarized at 20.0° with respect to the vertical. At the second plate, we have

$$I_3 = I_2 \cos^2(\theta_2 - \theta_1) = 8.83 \cos^2 20.0° = 7.80 \text{ units}$$

The light emerging from the second polarizer has its plane of polarization at 40.0° from the vertical. At the final plate,

$$I_f = I_3 \cos^2(\theta_3 - \theta_2) = 7.80 \cos^2 20.0° = 6.89 \text{ units} \quad \Diamond$$

---

67. A glass plate (n = 1.61) is covered with a thin, uniform layer of oil (n = 1.20). A light beam of variable wavelength from air is incident normally on the oil surface. Observation of the reflected beam shows destructive interference at 500 nm and constructive interference at 750 nm. From this information, calculate the thickness of the oil film.

Solution For destructive interference, $2t = \frac{\lambda_d}{n}\left(m + \frac{1}{2}\right)$ and for constructive interference, $2t = \frac{\lambda_b}{n}(m)$ where n is the index of refraction of the oil.

Eliminating m and solving for t, we find

$$t = \frac{\lambda_d}{4n\left(1 - \frac{\lambda_d}{\lambda_b}\right)} = \frac{500 \text{ nm}}{4(1.20)\left(1 - \frac{500}{750}\right)} = 313 \text{ nm}$$

CHAPTER SELF-QUIZ

1. A Young's double slit has a slit separation of 4.00×10^{-5} m on which a monochromatic light beam is directed. The resultant bright fringes on a screen 1.20 m from the double slit are separated by 2.15×10^{-2} m. What is the wavelength of this beam? (1 nanometer = 10^{-9} m)
   a. 573 nm
   b. 454 nm
   c. 717 nm
   d. 667 nm

2. What is the minimum thickness of a soap bubble film (n = 1.46) on which light of wavelength 500 nm shines, assuming one observes constructive interference of the reflected light?
   a. 63.0 nm
   b. 86.0 nm
   c. 125 nm
   d. 172 nm

3. A silicon monoxide thin film (n = 1.45) of thickness 97.4 nm is applied to a camera lens made of glass (n = 1.58). This will result in a destructive interference for reflected light of what wavelength?
   a. 720 nm
   b. 616 nm
   c. 565 nm
   d. 493 nm

4. A beam of unpolarized light in air strikes a flat piece of glass at an angle of incidence of 57.33°. If the reflected beam is completely polarized, what is the index of refraction of the glass?
   a. 1.60
   b. 1.56
   c. 1.52
   d. 2.48

5. In a Young's double-slit interference apparatus, by what factor is the distance between adjacent light and dark fringes changed when the wavelength of the source is tripled?
   a. 1/9
   b. 1/3
   c. 1.0
   d. 3.0

6. The dark spot observed in the center of a Newton's rings pattern is attributed to which of the following?
   a. polarization of light when reflected
   b. polarization of light when refracted
   c. phase shift of light when reflected
   d. phase shift of light when refracted

7. Light of wavelength 610 nm is incident on a slit of width 0.20 mm and an interference pattern is produced on a screen that is 1.50 m from the slit. What is the width of the central bright fringe? (1 nm = 10^{-9} m)
   a. 0.68 cm
   b. 0.92 cm
   c. 1.22 cm
   d. 1.35 cm

8. Polarization of light can be achieved using birefringent materials by which of the following processes?
   a. reflection
   b. double refraction
   c. selective absorption
   d. scattering

9. Waves from a radio station with a wavelength of 400 m arrive at a home receiver a distance 30.0 km away from the transmitter by two paths. One is a direct-line path and the second by reflection from a mountain directly behind the receiver. What is the minimum distance between the mountain and receiver so that destructive interference occurs at the location of the listener?
   a. 100 m
   b. 200 m
   c. 300 m
   d. 400 m

10. When light shines on a lens placed on a flat piece of glass, interference occurs which causes circular fringes called Newton's rings. The two beams that are interfering come from
    a. the top and bottom surface of the lens
    b. the top surface of the lens and the top surface of the piece of glass
    c. the bottom surface of the lens and the top surface of the piece of glass
    d. the top and bottom surface of the flat piece of glass

11. Polaroid material made from a dichroic substance polarizes light by
    a. selective absorption
    b. reflection
    c. double refraction
    d. scattering

12. A possible means for making an airplane radar-invisible is to coat the plane with an antireflective polymer. If radar waves have a wavelength of 3.00 cm and the index of refraction of the polymer is $n = 1.50$, how thick would the coating be?
    a. 1.00 mm
    b. 2.00 mm
    c. 5.00 mm
    d. 2.00 cm

25
Optical Instruments

Chapter 25

OPTICAL INSTRUMENTS

We use devices made from lenses, mirrors, or other optical components every time we put on a pair of eyeglasses, take a photograph, look at the sky through a telescope, and so on. In this chapter we examine how these and other optical instruments work. For the most part, our analyses will involve the laws of reflection and refraction and the procedures of geometric optics. However, to explain certain phenomena, we must use the wave nature of light.

NOTES FROM SELECTED CHAPTER SECTIONS

25.2 The Eye

The *near point* represents the closest distance for which the lens will produce a sharp image on the retina. This distance usually increases with age and has an average value of around 25.0 cm.

Hyperopia (farsightedness) is a defect of the eye that results either when the eyeball is too short or when the ciliary muscle is unable to change the shape of the lens enough to form a properly focused image. Myopia (nearsightedness) is caused either when the eye is longer than normal or when the maximum focal length of the lens is insufficient to produce a clearly focused image on the retina.

The power of a lens in diopters equals the inverse of the focal length in meters.

25.4 The Compound Microscope

The overall magnification of a compound microscope of length L is equal to the product of the magnification produced by the objective of focal length f_0 and the magnification produced by the eyepiece of focal length f_e.

25.5 The Telescope

There are two fundamentally different types of telescopes, both designed to aid in viewing distant objects, such as the planets in our solar system. The two classifications are (1) the *refracting telescope*, which uses a combination of lenses to form an image, and (2) the *reflecting telescope*, which uses a curved mirror and a lens to form an image.

25.6 Resolution of Single-Slit and Circular Apertures

When the central maximum of one image falls on the first minimum of another image, the images are said to be just resolved. This limiting condition of resolution is known as Rayleigh's criterion.

25.7 The Michelson Interferometer

The *Michelson interferometer* is an optical instrument that has great scientific importance. The interferometer is an ingenious device that splits a light beam into two parts and then recombines them to form an interference pattern. The device is used for obtaining accurate length measurements.

25.8 The Diffraction Grating

A *diffraction grating* consists of a large number of equally spaced, identical slits.

EQUATIONS AND CONCEPTS

The brightness of the image on a film depends on the ratio of the focal length to the diameter of the camera lens. This ratio, called the *f*-number, is a measure of the light concentrating power of the lens, and determines what is called the speed of the lens. A fast lens has a small *f*-number--usually a short focal length and a large diameter.

$$f\text{-number} \equiv \frac{f}{D} \qquad (25.1)$$

The power of a lens measured in diopters is equal to the inverse of the focal length measured in meters. The correct algebraic sign must be used with f. Optometrists and ophthalmologists usually prescribe lenses measured in diopters.

$$P = \frac{1}{f}$$

A simple magnifier increases the angular size of the object (the size of the angle subtended by the object at the eye). The angular magnification is the ratio of the

$$m \equiv \frac{\theta}{\theta_0} \qquad (25.2)$$

angle subtended by the image formed by a convex lens to the angle subtended by the object when it is placed at the near point of the eye (25.0 cm) with no lens.

This is an alternate form used to express the magnifying power of a simple lens in terms of the focal length of the lens.

$$m = 1 + \frac{25.0\text{ cm}}{f} \qquad (25.5)$$

This is the expression for the magnification of a simple lens when the object is placed at the focal point of the lens. In this case the image will be formed at infinity and will allow the eye to focus in a more relaxed manner.

$$m = \frac{25.0\text{ cm}}{f} \qquad (25.6)$$

A compound microscope contains an objective lens of short focal length f_o and an eyepiece of focal length f_e. The two lenses are separated by a distance L. When an object is located just beyond the focal point of the objective, the two lenses in combination form an enlarged, virtual, and inverted image of overall magnification, M. The negative sign indicates the inverted nature of the image.

$$M = M_1 m_e = -\frac{L}{f_o}\left(\frac{25.0\text{ cm}}{f_e}\right) \qquad (25.7)$$

The angular magnification of a telescope is equal to the ratio of the objective focal length to the eyepiece focal length. In this case, the two lenses are separated by a distance equal to the sum of their focal lengths. The negative sign indicates an inverted image.

$$m = -\frac{f_o}{f_e} \qquad (25.8)$$

Rayleigh's criterion states the condition for the resolution of two images of closely spaced sources. For a slit the angular separation between sources must be greater than the ratio of the wavelength to the slit width.

$$\theta_m \approx \frac{\lambda}{a} \qquad (25.9)$$

(slit)

In the case of a circular aperture, the minimum angular separation which can be resolved depends on the diameter of the aperture, D.

$$\theta_m = 1.22\frac{\lambda}{D} \qquad (25.10)$$

(circular aperture)

Rayleigh's criterion determines the limiting condition for resolution of adjacent sources. Under this criterion two sources are just resolved (seen as separate sources) when they are spaced so that the central maximum of the diffraction pattern of one source is located at the position of the first minimum of the diffraction pattern of the second source.

Comment on Rayleigh's criterion.

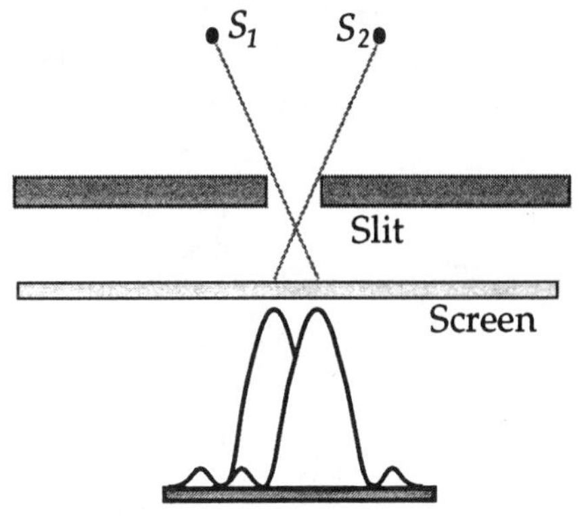

A diffraction grating of equally spaced parallel slits, separated by a distance, d, will produce an interference pattern in which there will be a series of maxima (bright lines) for each wavelength in the light source. The set of maxima due to wavelengths of all different values comprise a spectral order denoted by the number m.

$$d\sin\theta = m\lambda \qquad (25.11)$$

$m = 0, 1, 2, \ldots$

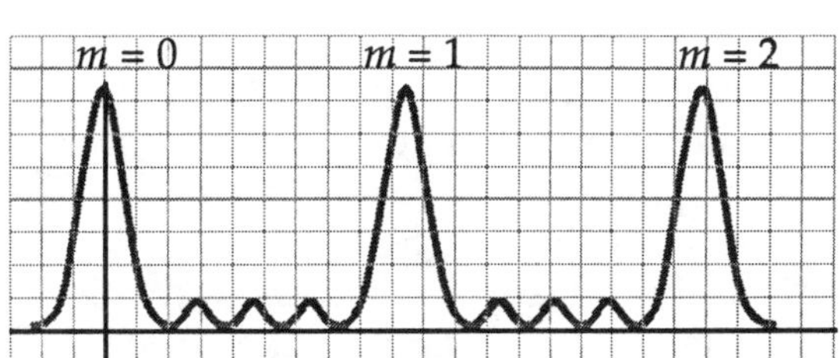

The resolving power of a diffraction grating increases as the number of lines illuminated increases; also, the resolving power is proportional to the order number in which the spectrum is observed.

$$R \equiv \frac{\lambda}{\lambda_2 - \lambda_1} = \frac{\lambda}{\Delta\lambda} \qquad (25.12)$$

$$R = Nm \qquad (25.13)$$

Comment on resolving power of a grating.

From Equation 25.12, it can be seen that a grating with a high resolving power can distinguish small differences between adjacent wavelengths. Equation 25.13 shows that in the zeroth order ($m = 0$), the resolution $R = 0$ (i.e. all wavelengths are indistinguishable for the zeroth order maximum). For example, if, in a given order (for which m is greater than zero), $R = 10{,}000$, the grating will produce a spectrum in which wavelengths differing in value by 1 part in 10,000 can be resolved.

REVIEW CHECKLIST

▷ Define the *f*-number of a camera lens and relate this criterion to shutter speed.

▷ Describe the geometry of the combination of lenses for each of several simple optical instruments: simple magnifier, compound microscope, reflecting telescope, and refracting telescope. Also, calculate the magnifying power for each instrument.

▷ Determine whether or not two sources under a given set of conditions are resolvable as defined by Rayleigh's criterion.

▷ Describe the technique employed in the Michelson interferometer for precise measurement of length based on known values for the wavelength of light.

▷ Determine the positions of the principal maxima in the interference pattern of a diffraction grating. Understand what is meant by the resolving power and the dispersion of a grating, and calculate the resolving power of a grating under specified conditions.

SOLUTIONS TO SELECTED END-OF-CHAPTER PROBLEMS

3. A photographic image of a building is 0.0920 m high. The image was made with a lens with a focal length of 52.0 mm. If the lens were 100 m from the building when the photograph was made, determine the height of the building.

Solution Magnification, $M = \frac{h'}{h} = \frac{q}{p}$ so $h = \left(\frac{p}{q}\right)h'$

But $\frac{1}{p} + \frac{1}{q} = \frac{1}{f}$ so $q = \frac{fp}{p-f}$ or $\frac{p}{q} = \frac{p-f}{f}$

Therefore, $$h = \left(\frac{p-f}{f}\right)h' = \left(\frac{100 \text{ m} - 0.052 \text{ m}}{0.052 \text{ m}}\right)0.0920 \text{ m}$$

height, $h = 177$ m ◊

---

9. The near point of an eye is 100 cm. (a) What focal length should the lens have so that the eye can clearly see an object 25.0 cm in front of it? (b) What is the power of the lens?

Solution (a) When an object is at 25.0 cm in front of the lens (p = 25.0 cm), the image must be virtual and 100 cm in front of the lens so that the eye can focus on it. (q = −100 cm)

Thus, $\frac{1}{p} + \frac{1}{q} = \frac{1}{f}$ becomes $f = \frac{pq}{q+p}$ from which $f = \frac{(25.0 \text{ cm})(100 \text{ cm})}{100 \text{ cm} + 25.0 \text{ cm}}$

or $f = 33.3$ cm ◊

(b) $P = \frac{1}{f} = \frac{1}{0.333 \text{ m}} = 3.00$ diopters ◊

---

13. An artificial lens is implanted in a person's eye to replace a diseased lens. The distance between the artificial lens and the retina is 2.80 cm. In the absence of the lens, the image of a distant object falls 2.53 cm behind the retina. The lens is designed to put the image of the distant object on the retina. What is the power of the implanted lens? (*Hint:* Consider the image formed by the eye as a virtual object.)

Solution Considering the image formed by the eye as a virtual object for the implanted lens, we have

$$p = -(2.53 \text{ cm} + 2.80 \text{ cm}) = -5.33 \text{ cm} \quad \text{and} \quad q = 2.80 \text{ cm}$$

Thus, $\frac{1}{p}+\frac{1}{q}=\frac{1}{f}$ becomes $f=\frac{pq}{q+p}=\frac{(-5.33\text{ cm})(2.80\text{ cm})}{(2.80\text{ cm}-5.33\text{ cm}}$ which gives

$$f=5.90\text{ cm} \quad \text{and} \quad P=\frac{1}{0.059\text{ m}}=17.0\text{ diopters} \quad \Diamond$$

---

17. An older person has far points of 1.00 m and near points of 0.667 m in both eyes. Bifocal lenses (Fig. 25.19) are prescribed. Determine the powers of the upper and lower portions of the bifocal lens.

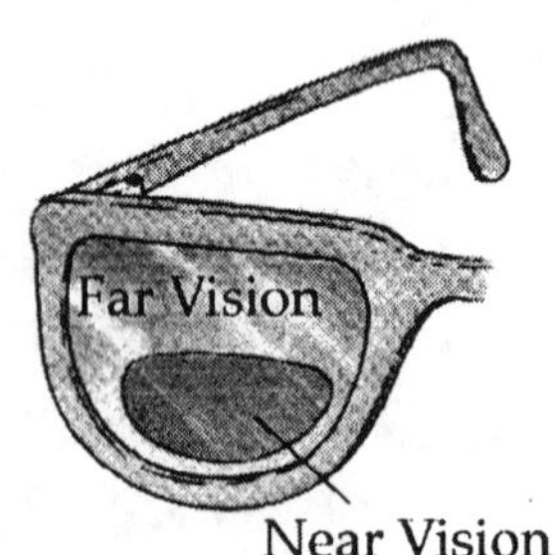

Figure 25.19

Solution The upper portion of the lens is to allow the wearer to see very distant objects. Thus, it must form virtual images of distant objects ($p=\infty$) at the far point of the eye (i.e., $q=-1.00$ m).

$$\frac{1}{p}+\frac{1}{q}=\frac{1}{f} \quad \text{becomes} \quad \frac{1}{\infty}+\frac{1}{-1.00\text{ m}}=\frac{1}{f}$$

so $$f=-1.00\text{ m} \quad \text{and} \quad P=\frac{1}{f}=\frac{1}{-1.00}=-1.00\text{ diopters} \quad \Diamond$$

The lower portion of the lens is to allow the wearer to view objects at the normal near point (25.0 cm) clearly. Thus, it must form a virtual image of 25.0 cm distant objects at the eyes' near point.

$$\frac{1}{p}+\frac{1}{q}=\frac{1}{f} \quad \text{becomes} \quad \frac{1}{0.250\text{ m}}+\frac{1}{-0.667\text{ m}}=\frac{1}{f}; \quad f=0.400\text{ m}$$

Thus, $$P=\frac{1}{f}=\frac{1}{0.400\text{ m}}=2.50\text{ diopters} \quad \Diamond$$

---

19. A leaf of length h is positioned 71.0 cm in front of a converging lens with a focal length of 39.0 cm. An observer views the image of the leaf from a position 1.26 m behind the lens, as shown in Figure 25.20 (on following page). (a) What magnification (ratio of image size to object size) is produced by the lens? (b) What angular magnification is achieved by viewing the image of the leaf rather than viewing the leaf directly?

Solution (a) First, locate the image being formed by the lens and the magnification produced by the lens.

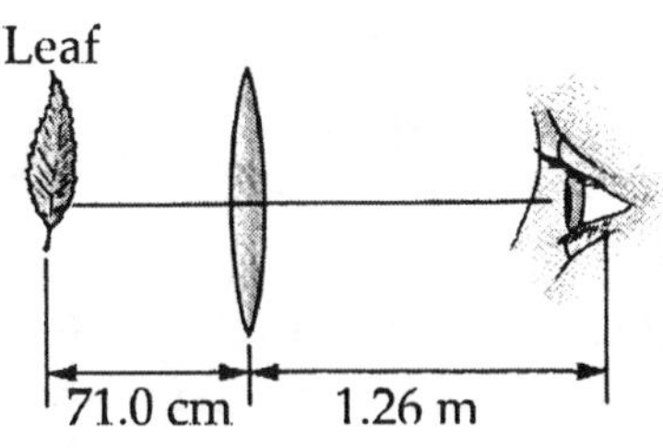

Figure 25.20

$$\frac{1}{p}+\frac{1}{q}=\frac{1}{f} \quad \text{becomes}$$

$$q=\frac{fp}{p-f}=\frac{(39.0 \text{ cm})(71.0 \text{ cm})}{71.0 \text{ cm}-39.0 \text{ cm}}$$

so $q = 86.5$ cm and $M=-\frac{q}{p}=-\frac{86.5}{71.0}=-1.22$ ◊

Thus, $h' = -1.22$ h ◊

(b) We have $\frac{\theta}{\theta_0}=\frac{\left(h'/d'\right)}{\left(h/d\right)}=\left(\frac{h'}{h}\right)\left(\frac{d}{d'}\right)=1.22\left(\frac{197 \text{ cm}}{39.5 \text{ cm}}\right)=6.08$ ◊

---

21. A biology student uses a simple magnifier to examine the structural features of the wing of an insect. The lens is mounted in a frame 3.50 cm above the work surface, and the image is formed 25.0 cm from the eye. What is the focal length of the lens?

Solution We are given that $p = 3.50$ cm and $q = -25.0$ cm

Thus, $\frac{1}{p}+\frac{1}{q}=\frac{1}{f}$ becomes $f=\frac{pq}{q+p}=\frac{(3.50 \text{ cm})(-25.0 \text{ cm})}{-25.0 \text{ cm} + 3.50 \text{ cm}}$ so $f = 4.07$ cm ◊

---

29. A microscope has an objective lens with a focal length of 16.22 mm and an eyepiece with a focal length of 9.50 mm. With the length of the barrel set at 29.0 cm, the diameter of a red blood cell's image subtends an angle of 1.43 mrad with the eye. If the final image distance is 29.0 cm from the eyepiece, what is the actual diameter of the red blood cell?

Solution First, find the size of the final image.

$$\text{angular size} = \frac{h_e}{\left|q_e'\right|} = \frac{h_e}{29.0 \text{ cm}} = 1.43\times10^{-3} \text{ rad} \quad \text{so} \quad h_e = 4.147\times10^{-2} \text{ cm}$$

Now apply thin-lens equation to each lens to find the magnification produced by each lens and hence the overall magnification.

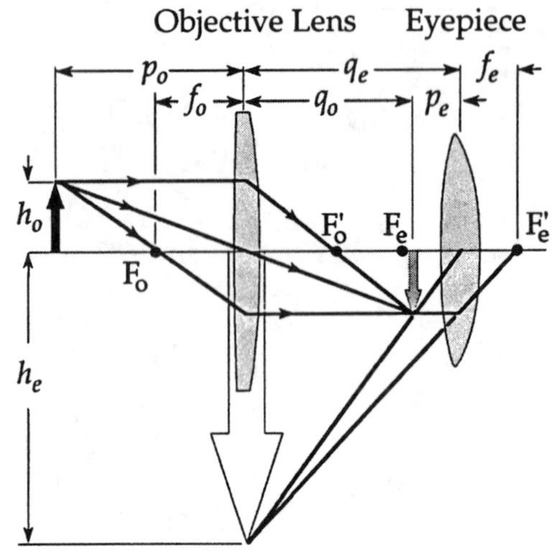

$$\frac{1}{p_e}+\frac{1}{q_e}=\frac{1}{f_e} \quad \text{becomes}$$

$$p_e = \frac{f_e q_e}{q_e - f_e} = \frac{(0.950 \text{ cm})(-29.0 \text{ cm})}{-29.0 \text{ cm} - 0.950 \text{ cm}}$$

$$p_e = 0.920 \text{ cm}$$

and

$$M_e = -\frac{q_e}{p_e} = -\frac{29.0 \text{ cm}}{0.92 \text{ cm}} = 31.5$$

For the objective lens, we have

$$\frac{1}{p_o}+\frac{1}{q_o}=\frac{1}{f_o} \quad \text{or} \quad p_o = \frac{f_o q_o}{q_o - f_o} = \frac{(1.622 \text{ cm})(-28.08 \text{ cm})}{-28.08 \text{ cm} - 1.622 \text{ cm}} \quad \text{so} \quad p_o = 1.721 \text{ cm}$$

Thus, the *magnitude* of the objective magnification is

$$M_o = \frac{q_o}{p_o} = \frac{28.08 \text{ cm}}{1.721 \text{ cm}} = 16.3$$

and the overall magnification is

$$M = M_e M_o = (31.5)(16.3) = 514$$

Therefore,

$$h_o = \frac{h_e}{M} = \frac{4.147\times10^{-2} \text{ cm}}{514} = 8.10\times10^{-2} \text{ cm} = 0.810\ \mu\text{m} \quad \Diamond$$

35. A certain telescope has an objective of focal length 1500 cm. If the Moon is used as an object, a 1.00-cm length of the image formed by the objective corresponds to what distance, in miles, on the Moon? Assume 3.80×10^8 m for the Earth-Moon distance.

Solution (See sketch.)

The length, x, of an object on the moon, is found as

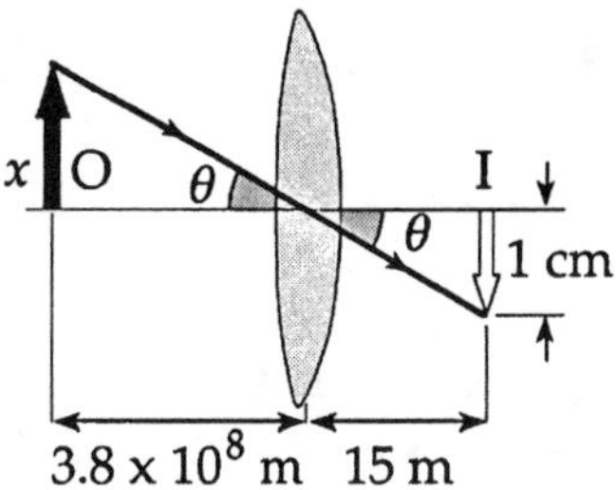

$$\frac{x}{3.80 \times 10^8 \text{ m}} = \frac{10^{-2} \text{ m}}{15.0 \text{ m}}$$

and $x = 2.53 \times 10^5 \text{ m} = 157 \text{ miles}$ ◊

---

37. If the distance from the Earth to the Moon is 3.80×10^8 m, what diameter would be required for a telescope objective to resolve a Moon crater 300 m in diameter? Assume a wavelength of 500 nm.

Solution The angular resolution needed is

$$\theta_m = \frac{s}{r} = \frac{300 \text{ m}}{3.80 \times 10^8 \text{ m}} = 7.90 \times 10^{-7} \text{ radians}$$

For a circular aperture,

$$\theta_m = 1.22\frac{\lambda}{D} \quad \text{so} \quad D = 1.22\frac{\lambda}{\theta_m}$$

and $$D = 1.22\frac{5.00 \times 10^{-7} \text{ m}}{7.90 \times 10^{-7} \text{ rad}} = 0.772 \text{ m} = 30.4 \text{ in}$$ ◊

---

41. A spy satellite circles the Earth at an altitude of 200 km and carries out surveillance with a special high-resolution telescopic camera having a lens diameter of 35.0 cm. If the angular resolution of this camera is limited by diffraction, estimate the separation of two small objects on the Earth's surface that are just resolved in yellow-green light (λ = 550 nm).

Solution If just resolved,

$$\theta = \theta_m = 1.22\frac{\lambda}{D} = 1.22\left(\frac{5500 \times 10^{-10}\text{ m}}{0.350\text{ m}}\right) = 1.92 \times 10^{-6}\text{ rad}$$

and $$d = \theta h = (1.92 \times 10^{-6}\text{ rad})(2.00 \times 10^{5}\text{ m}) = 0.383\text{ m} = 38.3\text{ cm} \quad \lozenge$$

---

45. (a) Calculate the limiting angle of resolution for the eye, assuming a pupil diameter of 2.00 mm, a wavelength of 500 nm in air, and an index of refraction for the eye of 1.33. (b) What is the maximum distance from the eye at which two points separated by 1.00 cm could be resolved?

Solution (a) $\lambda_{\text{medium}} = \dfrac{\lambda_{\text{vac}}}{n_{\text{medium}}} = \dfrac{500\text{ nm}}{1.33} = 375\text{ nm}$

Then, the limiting angle is $$\theta_m = 1.22\frac{\lambda}{D} = 1.22\left(\frac{3.75 \times 10^{-7}\text{ m}}{2.00 \times 10^{-3}\text{ m}}\right) = 2.29 \times 10^{-4}\text{ rad} \quad \lozenge$$

(b) $r = \dfrac{S}{\theta_m} = \dfrac{10^{-2}\text{ m}}{2.29 \times 10^{-4}\text{ rad}} = 43.7\text{ m} \quad \lozenge$

---

51. Light of wavelength 550 nm is used to calibrate a Michelson interferometer. By use of a micrometer screw, the platform on which one mirror is mounted is moved 0.180 mm. How many fringe shifts are counted?

Solution A fringe shift occurs when the mirror is moved a distance of $\frac{\lambda}{4}$. Thus, if the mirror is moved a distance $\Delta L = 0.180\text{ mm} = 1.80 \times 10^{-4}\text{ m}$ and the wavelength is $\lambda = 550\text{ nm} = 5.50 \times 10^{-7}\text{ m}$, the number of fringe shifts observed is

$$\text{Number of fringe shifts} = \frac{\Delta L}{\left(\frac{\lambda}{4}\right)} = \frac{4(\Delta L)}{\lambda} = \frac{2(1.80 \times 10^{-4}\text{ m})}{5.50 \times 10^{-7}\text{ m}} = 1.31 \times 10^{3} \quad \lozenge$$

---

53. A thin sheet of transparent material has an index of refraction of 1.40 and is 15.0 μm thick. When it is inserted along one arm of an interferometer in the path of 600-nm light, how many fringe shifts occur in the pattern?

Solution In a vacuum, the number of wavelengths in a distance t is $N = \dfrac{t}{\lambda_{\text{vac}}}$.

When this space is filled by a medium of refractive index n_m, the number of wavelengths in the same distance t is

$$N' = \frac{t}{\lambda_{\text{medium}}} = \frac{n_m t}{\lambda_{\text{vac}}}$$

Thus, the number of fringe shifts that will occur is

$$\text{Number of fringe shifts} = 4(\Delta N) = 4\left(\frac{n_m t}{\lambda_{\text{vac}}} - \frac{t}{\lambda_{\text{vac}}}\right) = \frac{4t}{\lambda_{\text{vac}}}(n_m - 1)$$

If $t = 15.0 \times 10^{-6}$ m, $\lambda = 600$ nm, and $n_m = 1.40$, we have

$$\text{Number of fringe shifts} = \frac{4(15.0 \times 10^{-6}\ \text{m})}{6.00 \times 10^{-7}\ \text{m}}(1.40 - 1) = 40.0 \quad \Diamond$$

---

57. Three discrete spectral lines occur at angles of 10.09°, 13.71°, and 14.77° in the first-order spectrum of a grating spectroscope. (a) If the grating has 3660 slits/cm, what are the wavelengths of the light? (b) At what angles are these lines found in the second-order spectra?

Solution (a) $d = \dfrac{1}{3660\ \text{lines/cm}} = 2.732 \times 10^{-4}\ \text{cm} = 2.732 \times 10^{-6}\ \text{m} = 2732\ \text{nm}$

The wavelength found at angle θ is $\lambda = \dfrac{d \sin\theta}{n}$

At $\theta = 10.09°$, $\lambda = 478.7$ nm $\Diamond$

At $\theta = 13.71°$, $\lambda = 647.6$ nm $\Diamond$

At $\theta = 14.77°$, $\lambda = 696.6$ nm $\Diamond$

(b) The grating spacing is $d = \dfrac{\lambda}{\sin\theta_1}$; and for the second order, $(2)\lambda = d\sin\theta_2$

Combining these equations gives $\sin\theta_2 = \dfrac{2\lambda}{d} = 2\lambda\left(\dfrac{\sin\theta_1}{\lambda}\right) = 2\sin\theta_1$

Therefore, if $\theta_1 = 10.09°$, then $\sin\theta_2 = 2\sin(10.09°)$ gives $\theta_2 = 20.51°$ ◊

Similarly, for $\theta_1 = 13.71°$, $\theta_2 = 28.30°$ ◊

and for $\theta_1 = 14.77°$, $\theta_2 = 30.66°$ ◊

---

61. Light containing two different wavelengths passes through a diffraction grating with 1200 slits/cm. On a screen 15.0 cm from the grating, the third-order maximum of the shorter wavelength falls on top of the first-order minimum of the longer wavelength. If the neighboring maxima of the longer wavelength are 8.44 mm apart on the screen, what are the wavelengths in the light? (*Hint:* Use the small angle approximation.)

Solution First, find the longer wavelength.

$$y = \left(\frac{n\lambda}{d}\right)L \quad \text{and} \quad y_2 - y_1 = \left(\frac{\lambda}{d}\right)L$$

$$\lambda = \frac{d(y_2 - y_1)}{L} \quad \text{and} \quad d = \frac{1}{1200\ \text{slits/cm}} = 8.333 \times 10^{-4}\ \text{cm} = 8333\ \text{nm}$$

$$\lambda_{\text{long}} = \frac{(8.333 \times 10^{-4}\ \text{cm})(0.844\ \text{cm})}{15.0\ \text{cm}}$$

$$\lambda_{\text{long}} = 4.689 \times 10^{-5}\ \text{cm} = 469\ \text{nm} \quad ◊$$

Now find the other wavelength. The third order of λ_{short} (at θ_3) coincides with the first order of λ_{long}. Thus,

$$3\lambda_{\text{short}} = d\sin\theta_3 = (1)\lambda_{\text{long}} \quad \text{and} \quad \lambda_{\text{short}} = \frac{1}{3}\lambda_{\text{long}} = \frac{469\ \text{nm}}{3} = 156\ \text{nm} \quad ◊$$

---

75. A fringe pattern is established in the field of view of a Michelson interferometer using light of wavelength 580 nm. A parallel-faced sheet of transparent material 2.50 μm thick is placed in front of one of the mirrors perpendicular to the incident and reflected light beams. An observer counts 12 fringe shifts. What is the index of refraction of the sheet?

Solution The number of wavelengths in a distance t in a vacuum is

$$N = \frac{t}{\lambda_{\text{vac}}}$$

If this space is then filled with material having a refractive index n, the number of wavelengths in that distance is now

$$N' = \frac{t}{\lambda} = \frac{\text{nt}}{\lambda_{\text{vac}}}$$

Since four fringe shifts occur when the path length increases by one wavelength, the number of fringe shifts observed as the material is introduced will be

$$\text{Number of fringe shifts} = 4(\Delta N) = 4\left(\frac{nt}{\lambda_{\text{vac}}} - \frac{t}{\lambda_{\text{vac}}}\right) = \frac{4t}{\lambda_{\text{vac}}}(n-1)$$

Solving for n gives

$$n = \left(1 + \frac{\lambda_{\text{vac}}}{4t}\right)(\text{number of fringe shifts})$$

Thus, if $t = 2.50\ \mu\text{m}$, $\lambda_{\text{vac}} = 580 \times 10^{-8}$ m, and 12 fringe shifts are observed,

$$n = 1 + \frac{580 \times 10^{-8}\ \text{m}(12)}{4(2.50 \times 10^{-6}\ \text{m})} = 1.696, \text{ the index of refraction} \quad \lozenge$$

CHAPTER SELF-QUIZ

1. A camera lens with a focal length of 2.00 cm and an aperture opening diameter of 0.40 cm has what *f*-number?
   a. 0.20
   b. 0.80
   c. 5.00
   d. 10.0

2. A converging lens will be prescribed by the eye doctor to correct which of the following?
   a. nearsightedness
   b. glaucoma
   c. farsightedness
   d. astigmatism

3. What is the magnification of a refracting telescope with objective and eyepiece lenses of focal lengths 90.0 cm and 2.00 cm, respectively?
   a. 30
   b. 45
   c. 60
   d. 180

4. The Michelson interferometer is a device that may be used to measure which of the following?
   a. magnifying power of lenses
   b. light wavelength
   c. atomic masses
   d. electron charge

5. Doubling the *f*-number of a camera lens will change the light intensity admitted to the film by what factor?
   a. 0.25
   b. 0.50
   c. 2.00
   d. 4.00

6. Which eye defect is corrected by a lens having different curvatures in two perpendicular directions?
   a. myopia
   b. presbyopia
   c. hyperopia
   d astigmatism

7. Two thin lenses in combination, placed in contact with each other along a common axis, have the respective powers of 45.0 and –15.0 diopters. What is their combined power?
   a. 15.0 diopters
   b. 30.0 diopters
   c. 60.0 diopters
   d. –22.5 diopters

8. A compound microscope has objective and eyepiece lenses of focal lengths 0.80 and 4.00 cm, respectively. If the microscope length is 15.0 cm, what is the maximum magnification?
   a. 3.20
   b. 6.30
   c. 48.0
   d. 117.0

9. What resolving power must a diffraction grating have in order to distinguish wavelengths of 630.1 and 631.7 nm?
   a. 315
   b. 394
   c. 630.9
   d. 788

10. A camera uses a
    a. converging lens to form a real image
    b. converging lens to form an imaginary image
    c. diverging lens to form a real image
    d. diverging lens to form an imaginary image

11. A magnifier uses a
    a. converging lens to form a real image
    b. converging lens to form a virtual image
    c. diverging lens to form a virtual image
    d. diverging lens to form a real image

12. A 1.70-m-tall woman stands 5.00 m in front of a camera with a 5.00-cm focal length lens. What is the size of the image formed on film?
    a. 3.40 cm
    b. 2.60 cm
    c. 1.70 cm
    d. 0.85 cm

26
Relativity

Chapter 26

RELATIVITY

Most of our everyday experiences and observations deal with objects that move at speeds much lower than the speed of light. Newtonian mechanics and the early ideas on space and time were formulated to describe the motion of such objects. As we saw in the chapters on mechanics, this formalism is very successful in describing a wide range of phenomena. Although Newtonian mechanics works very well at low speeds, it fails when applied to particles whose speeds approach that of light.

The theory of relativity represents one of the greatest intellectual achievements of the 20th century. With this theory, experimental observations over the range from $v = 0$ to velocities approaching the speed of light can be predicted. Newtonian mechanics, which was accepted for more than 200 years, is in fact a specialized case of Einstein's theory. This chapter introduces the special theory of relativity, with emphasis on some of the consequences of the theory. A discussion of general relativity and some of its consequences is presented in the essay that follows this chapter.

Special relativity covers such phenomena as the slowing down of clocks and the contraction of lengths in moving reference frames as measured by a stationary observer. In addition to these topics, we also discuss the relativistic forms of momentum and energy, terminating the chapter with the famous mass-energy equivalence formula, $E = mc^2$.

NOTES FROM SELECTED CHAPTER SECTIONS

26.1 Introduction

The special theory of relativity is based on two basic postulates:

1. The laws of physics are the same in all *inertial* reference systems.

2. The speed of light in vacuum is always measured to be 3.00×10^8 m/s, and the measured value is *independent* of the motion of the observer or of the motion of the source of light.

26.2 The Principle of Relativity

According to the principle of Newtonian relativity, the laws of mechanics are the same in all inertial frames of reference. Inertial frames of reference are those coordinate systems which are *at rest with respect to one another* or which move at constant velocity with respect to one another.

26.4 The Michelson-Morley Experiment

This experiment was designed to detect the velocity of the Earth with respect to the *hypothetical ether*. The outcome of the experiment was *negative*, contracting the ether hypothesis.

26.5 Einstein's Principle of Relativity

Einstein's special theory of relativity is based upon two postulates:

1. The laws of physics are the same in every inertial frame of reference.

2. The speed of light has the same value as measured by all observers, independent of the motion of the light source or observer.

The second postulate is consistent with the negative results of the Michelson-Morley experiment which failed to detect the presence of an ether and suggested that the speed of light is the same in all inertial frames.

26.6 Consequences of Special Relativity

Two events that are simultaneous in one reference frame are, in general, not simultaneous in another frame which is moving with respect to the first.

The *proper time* is always the time measured by a clock at rest in the frame of reference of the measurement. According to a stationary observer, a moving clock runs slower than an identical clock at rest. This effect is known as *time dilation.*

The *proper length* of an object is defined as the length of the object measured in the *reference frame in which the object is at rest.* The length of an object measured in a reference frame in which it is moving is always less than the proper length. This effect is known as *length contraction.* The contraction occurs only *along the direction of motion.*

26.7 Relativistic Momentum

To account for *relativistic effects,* it is necessary to modify the definition of momentum to satisfy the following conditions:

1. The relativistic momentum must be conserved in all collisions.

2. The relativistic momentum must approach the classical value, m_0v, as the quantity v/c approaches zero.

EQUATIONS AND CONCEPTS

As measured by a stationary observer, a moving clock runs slower than an identical stationary clock. This effect is known as time dilation. The time interval $\Delta t'$ is called the proper time and is always the time measured by an observer moving with the clock.

$$\Delta t = \frac{\Delta t'}{\sqrt{1 - v^2/c^2}} = \gamma \Delta t' \qquad (26.8)$$

$$\gamma = \frac{1}{\sqrt{1 - v^2/c^2}}$$

An observer moving with a speed v relative to an object will determine its length to be shorter than that measured by an observer at rest with respect to the object. The length measured in the reference frame in which the object is at rest is called the proper length. Note that length contraction is observed only along the direction of motion.

$$L = L'\sqrt{1 - v^2/c^2} \qquad (26.9)$$

This is the equation for relativistic addition of velocities in one-dimensional motion. In this equation, u is the velocity of an object relative to a frame of reference, S; u' is the velocity of the same object in a frame of reference S′ which is moving with velocity v relative to the first frame.

$$u = \frac{v + u'}{1 + \frac{vu'}{c^2}} \qquad (26.12)$$

This equation determines the momentum of a relativistic particle of mass m moving with a speed v. The equation satisfies the two essential requirements: (i) the relativistic momentum is conserved in all collisions, and (ii) the relativistic momentum approaches the classical value as the ratio v/c approaches zero.

$$p \equiv \frac{m_0 v}{\sqrt{1 - v^2/c^2}} = \gamma m_0 v \qquad (26.10)$$

The mass of an object increases as its speed increases. The mass measured by an observer at rest with respect to the object is its rest mass, m_0, and is less than the mass, m, measured by an observer moving with

$$m \equiv \frac{m_0}{\sqrt{1 - v^2/c^2}} \qquad (26.11)$$

a speed v relative to the object. It is possible to show from this equation that the speed of light is the ultimate speed of any object.

In these equations for relativistic energy and kinetic energy, m_0c^2 is the rest energy of the object and is independent of the object's speed. The term mc^2 is dependent on the object's speed (since mass increases with an increase in speed) and is the total energy of the object. These equations show that mass is a form of energy.

$$KE = mc^2 - m_0c^2 \qquad (26.13)$$

$$E = mc^2 = KE + m_0c^2 \qquad (26.14)$$

$$E = \frac{m_0c^2}{\sqrt{1 - v^2/c^2}} \qquad (26.15)$$

In cases where the speed of an object is not known, it is useful to have an expression which relates the relativistic energy, E, to the relativistic momentum, p. When an object is at rest, its total energy is equal to its rest energy.

$$E^2 = p^2c^2 + (m_0c^2)^2 \qquad (26.16)$$

This equation is an exact expression relating energy and momentum for photons which have a speed equal to c.

$$E = pc \qquad (26.17)$$

Einstein's theory of special relativity is based upon two postulates: (i) the laws of physics are the same in all inertial frames of reference (an inertial frame of reference is a non-accelerated frame); and (ii) the speed of light has the same value as measured by all observers, independent of the relative motion between light source and observer. The second postulate is consistent with the negative results of the Michelson-Morley experiment which failed to detect the presence of an ether.

Comment on basic postulates.

Chapter 26

REVIEW CHECKLIST

▷ State Einstein's two postulates of the special theory of relativity.

▷ Understand the Michelson-Morley experiment, its objectives, results, and the significance of its outcome.

▷ Understand the idea of simultaneity, and the fact that simultaneity is not an absolute concept. That is, two events which are simultaneous in one reference frame are not simultaneous when viewed from a second frame moving with respect to the first.

▷ Make calculations using the equations for time dilation, length contraction, and relativistic mass.

▷ State the correct relativistic expressions for the momentum, kinetic energy, and total energy of a particle. Make calculations using these equations.

SOLUTIONS TO SELECTED END-OF-CHAPTER PROBLEMS

1. Two airplanes fly paths I and II specified in Figure 26.5a. Both planes have air speeds of 100 m/s and fly a distance of 200 km. The wind blows at 20.0 m/s in the direction shown in the figure. Find (a) the time of flight to each city, (b) the time to return, and (c) the difference in total flight times.

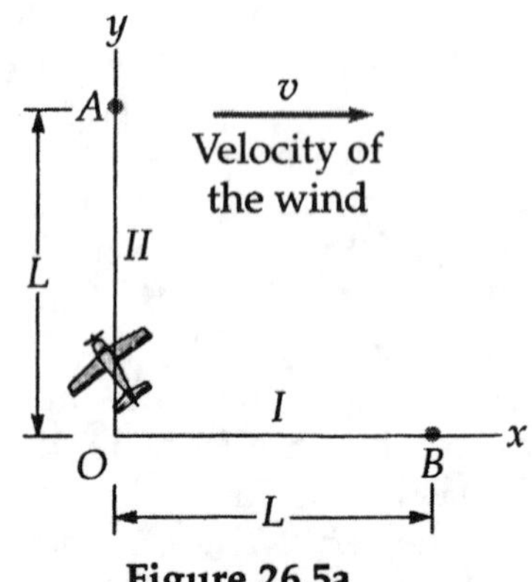

Figure 26.5a

Solution (a) Along path I, the velocity along the ground is

$$v_{\text{ground}} = 120 \text{ m/s}$$

Thus, $$t_{\text{I}} = \frac{200 \times 10^3 \text{ m}}{120 \text{ m/s}} = 1670 \text{ s} \quad \Diamond$$

Along path II (see sketch), we have

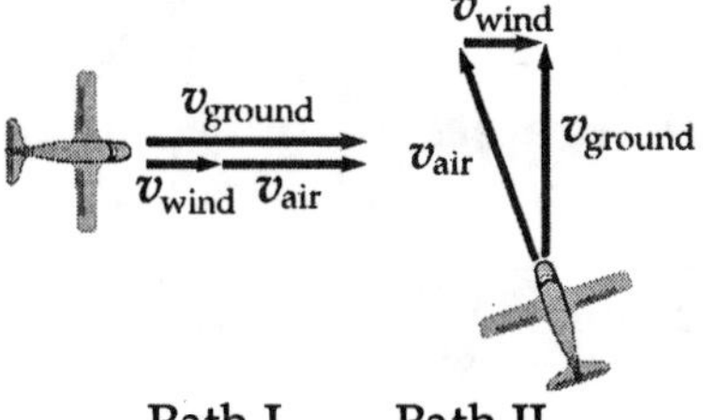

$$v_{air}{}^2 = v_{ground}{}^2 + v_{wind}{}^2$$

and

$$(100\text{ m/s})^2 = v_{ground}{}^2 + (20.0\text{ m/s})^2 \quad \text{gives}$$

$$v_{ground} = 98.0\text{ m/s}$$

Therefore, $$t_{II} = \frac{200\times10^3\text{ m}}{98.0\text{ m/s}} = 2040\text{ s} \quad \lozenge$$

(b) Along path I,

$$v_{ground} = -100\text{ m/s} + 20.0\text{ m/s} = -80.0\text{ m/s}$$

and $$t'_I = \frac{-200\times10^3\text{ m}}{-80.0\text{ m/s}} = 2500\text{ s} \quad \lozenge$$

Along path II,

$$v_{air}{}^2 = v_{ground}{}^2 + v_{wind}{}^2 = 98.0\text{ m/s}, \quad \text{as before.}$$

Thus, $$t'_{II} = \frac{200\times10^3\text{ m}}{98.0\text{ m/s}} = 2040\text{ s} \quad \lozenge$$

(c) $(t_I)_{total} = 1667\text{ s} + 2500\text{ s} = 4167\text{ s}$ and

$$(t_{II})_{total} = 2041\text{ s} + 2041\text{ s} = 4082\text{ s}$$

Thus, $$\Delta t = 4167\text{ s} - 4082\text{ s} = 85.0\text{ s} \quad \lozenge$$

7. If astronauts could travel at $v = 0.950c$, we on Earth would say it takes $(4.20/0.950) = 4.42$ years to reach Alpha Centauri, 4.20 lightyears away. The astronauts disagree. (a) How much time passes on the astronauts' clocks? (b) What is the distance to Alpha Centauri as measured by the astronauts?

Solution (a) The time required in the Earth's frame of reference is

$$T_1 = \frac{4.20 \text{ lightyear}}{0.950} = 4.42 \text{ yrs}$$

Thus, the time measured in the ship's reference frame

where $$\gamma = \frac{1}{\sqrt{1-(0.950)^2}} = 3.20$$

is $$T_2 = \frac{T_1}{\gamma} = \frac{4.42 \text{ yr}}{3.20} = 1.38 \text{ yrs} \quad \Diamond$$

(b) The length between Earth and the star as measured by astronauts is

$$L = \frac{L'}{\gamma} = \frac{4.20 \text{ ly}}{3.20} = 1.310 \text{ ly} \quad \Diamond$$

---

11. An unidentified flying object (UFO) flashes across the night sky. A UFO enthusiast on the top of Pike's Peak determines that the length of the UFO is 100 m along the direction of motion. If the UFO is moving with a speed of $0.900c$, how long is the UFO according to its pilot?

Solution $$\gamma = \frac{1}{\sqrt{1-\frac{v^2}{c^2}}} = \frac{1}{\sqrt{1-(0.900)^2}} = 2.294$$

The length measured by an observer in motion relative to the ship = 100 m = L.

The length measured by pilot (at rest relative to ship) = $L' = \gamma L = 2.294(100 \text{ m})$ or

$$L' = 229 \text{ m} \quad \Diamond$$

---

17. Observer A measures the length of two rods, one stationary, the other moving with a speed of 0.955*c*. She finds that the rods have the same length. A second observer B travels along with the moving rod. What is the ratio of the length of A's rod to the length of B's rod according to observer B?

Solution $\gamma = \dfrac{1}{\sqrt{1-\dfrac{v^2}{c^2}}} = \dfrac{1}{\sqrt{1-\dfrac{(0.955c)^2}{c^2}}} = 3.3\bar{7}15$

Let L_0 = the common length the observer measures for the two rods. Then, since rod A is at rest relative to this observer, the proper length of rod A is $L'_A = L_0$.

The proper length of B is given by $L'_B = \gamma L_0 = 3.3715\ L_0$.

The length the observer in the rest frame of B will measure for rod A is

$$L_{AB} = \frac{L'_A}{\gamma} = \frac{L_0}{3.3715} = 0.2966\ L_0$$

The ratio of lengths measured by the observer in B is

$$\frac{L_{AB}}{L_{BB}} = \frac{\text{measured length of A}}{\text{measured length of B}} = \frac{0.2966\ L_0}{3.3715\ L_0} = 0.0880 \quad \lozenge$$

---

21. Calculate the momentum of an electron moving with a speed of (a) 0.010*c*, (b) 0.500*c*, (c) 0.900*c*.

Solution $p = \gamma m_0 v$

(a) $v = 0.010c$, thus $\gamma = \dfrac{1}{\sqrt{1-\dfrac{v^2}{c^2}}} = \dfrac{1}{\sqrt{1-(0.010)^2}} = 1.00005 \approx 1$

Therefore, $p = (1)(9.11\times10^{-31}\ \text{kg})(0.010)(3.00\times10^{8}\ \text{m/s})$

or $p = 2.70\times10^{-24}\ \text{kg}\cdot\text{m/s} \quad \lozenge$

(b) Following the same steps as used in part (a), we find

$$\gamma = 1.16 \qquad \text{and} \qquad p = 1.60 \times 10^{-22}\ \text{kg}\cdot\text{m/s} \quad \Diamond$$

(c) $\gamma = 2.29$ and $p = 5.60 \times 10^{-22}\ \text{kg}\cdot\text{m/s}$ ◊

---

25. An unstable particle at rest breaks up into two fragments of *unequal mass.* The rest mass of the lighter fragment is 2.50×10^{-28} kg, and that of the heavier fragment is 1.67×10^{-27} kg. If the lighter fragment has a speed of 0.893*c* after the breakup, what is the speed of the heavier fragment?

Solution Relativistic momentum must be conserved. For total momentum to be zero after as it was before, we must have (with subscript 2 referring to the heavier fragment, and subscript 1 to the lighter fragment):

$$p_2 = p_1 \qquad \text{or} \qquad \gamma m_{02} v_2 = \gamma m_{01} v_1 = \frac{2.50 \times 10^{-28}\ \text{kg}}{\sqrt{1-(0.893)^2}}(0.893c)$$

This becomes

$$\frac{(1.67 \times 10^{-27}\ \text{kg})v_2}{\sqrt{1-\frac{v_2^{\,2}}{c^2}}} = (4.960 \times 10^{-28}\ \text{kg})c$$

Therefore,

$$v_2 = 0.285c \quad \Diamond$$

---

29. At low speeds, a charged object in a magnetic field moves in a circular path of radius $r = mv/qB$. If an electron moves in an orbit of radius 10.0 cm with a speed of 1.00×10^5 m/s, what will the radius be when its speed is 0.960c?

Solution The radius at high speeds is r_2, given by $r_2 = \frac{m_2 v_2}{qB}$.

At low speeds the radius is r_1, given by $r_1 = \frac{m_1 v_1}{qB}$.

The ratio of r_2 to r_1 is $\quad \frac{r_2}{r_1} = \frac{\gamma_2 m_0 v_2}{\gamma_1 m_0 v_1} = \frac{\gamma_2 v_2}{\gamma_1 v_1}$

But γ_1 is approximately equal to unity at low speeds. Thus,

$$r_2 = r_1\left(\frac{\gamma_2 v_2}{v_1}\right) = (0.100 \text{ m})\frac{\left(\frac{1}{\sqrt{1-(0.960)^2}}\right)(0.960)\left(3.00\times10^8 \frac{\text{m}}{\text{s}}\right)}{(1.00\times10^5 \text{ m/s})} = 1.00\times10^3 \text{ m} \quad \Diamond$$

33. A space vehicle is moving at a speed of 0.750*c* with respect to an external observer. An atomic particle is projected at 0.900*c* in the same direction as the ship's velocity with respect to an observer inside the vehicle. What is the speed of the projectile as seen by the external observer?

Solution We have $\quad u = \frac{u' + v}{1+\frac{u'v}{c^2}} = \frac{0.900c + 0.750c}{1+\frac{(0.900c)(0.750c)}{c^2}} = 0.985c \quad \Diamond$

39. A pulsar is a stellar object that emits light in short bursts. Suppose a pulsar with a speed of 0.950*c* approaches the Earth, and a rocket with a speed of 0.995*c* heads toward the pulsar (both speeds measured in the Earth's frame of reference). If the pulsar emits 10.0 pulses per second in its own frame of reference, at what rate are the pulses emitted in the rocket's frame of reference?

Solution First, find the velocity of the rocket relative to the pulsar.

$$u = \frac{v+u'}{1+\frac{vu'}{c^2}} = \frac{0.995c+0.950c}{1+\frac{(0.995c)(0.950c)}{c^2}} \quad \text{gives} \quad u = 0.99\bar{9}87c$$

$$\text{Proper time} = \text{period of pulsar in its own rest frame} = \frac{1}{10.0 \text{ Hz}} = 0.100 \text{ s}$$

Thus,

$$\Delta t_{\text{rocket}} = \frac{\Delta t_{\text{proper time}}}{\sqrt{1-\dfrac{v^2}{c^2}}} = \frac{0.100\text{ s}}{\sqrt{1-(0.99987)^2}} = 6.24\text{ s (period of pulsar as measured in rocket)}$$

or $\quad$ frequency as seen from rocket $= \dfrac{1}{6.24\text{ s}} = 0.160\text{ Hz}$ ◊

---

43. A proton moves with a speed of $0.950c$. Calculate its (a) rest energy, (b) total energy, and (c) kinetic energy.

Solution $\quad$ At $v = 0.950c$, $\quad \gamma = 3.20$

(a) $\quad E_0 = m_0c^2 = (1.67\times10^{-27}\text{ kg})(3.00\times10^8\text{ m/s})^2 = 1.50\times10^{-10}\text{ J} = 939\text{ MeV}$ ◊

(b) $\quad E = mc^2 = \gamma m_0c^2 = \gamma E_0 = (3.20)(939\text{ MeV}) = 3010\text{ MeV}$ ◊

(c) $\quad KE = E - E_0 = 3006\text{ MeV} - 939\text{ MeV} = 2070\text{ MeV}$ ◊

---

47. The Sun radiates approximately 4.00×10^{26} J of energy into space each second. (a) How much mass is converted into energy each second? (b) If the mass of the Sun is 2.00×10^{30} kg, how long can the Sun survive if the energy creation continues at the present rate?

Solution $\quad$ (a) From $E = mc^2$, we have

$$m = \frac{E}{c^2} = \frac{4.00\times10^{26}\text{ J}}{(3.00\times10^8\text{ m/s})^2} = 4.40\times10^9\text{ kg} \quad ◊$$

(b) $\quad t = \dfrac{\text{total mass}}{\text{rate of use}} = \dfrac{2.00\times10^{30}\text{ kg}}{4.44\times10^9\text{ kg/s}} = 4.50\times10^{20}\text{ s} = 1.40\times10^{13}\text{ years}$ ◊

---

51. A proton in a high-energy accelerator is given a kinetic energy of 50.0 GeV. Determine (a) the momentum and (b) the speed of the proton.

Solution $KE = E - E_0 = 50.0 \text{ GeV} = 5.00 \times 10^4 \text{ MeV}$

Therefore, $E = E_0 + KE = 939 \text{ MeV} + 5.00 \times 10^4 \text{ MeV} = 50{,}939 \text{ MeV}$

(a) $E^2 = (pc)^2 + E_0{}^2$ or $(pc)^2 = E^2 - E_0{}^2 = (50{,}939 \text{ MeV})^2 - (939 \text{ MeV})^2$

Thus, $pc = 5.09 \times 10^4 \text{ MeV}$ and $p = 5.09 \times 10^4 \text{ MeV}/c$ ◊

(b) From $E = \gamma E_0$, we have $\gamma = \frac{E}{E_0} = \frac{50939 \text{ MeV}}{939 \text{ MeV}} = 54.2$

From $\gamma = \frac{1}{\sqrt{1 - \frac{v^2}{c^2}}}$, $v = c\sqrt{1 - \frac{1}{\gamma^2}}$

$$v = c\sqrt{1 - \frac{1}{(54.2)^2}} = 0.999830c \quad ◊$$

---

53. Two futuristic rockets are on a collision course. The rockets are moving with speeds of $0.800c$ and $0.600c$ and are initially 2.52×10^{12} m apart as measured by Liz, an Earth observer, as in Figure 26.16. Both rockets are 50.0 m in length as measured by Liz. (a) What are their respective proper lengths? (b) What is the length of each rocket as measured by an observer in the other rocket? (c) According to Liz, how long before the rockets collide? (d) According to rocket I, how long before they collide? (e) According to rocket II, how long before they collide? (f) If both rocket crews are capable of total evacuation within 90 minutes (their own time), will there be any casualties?

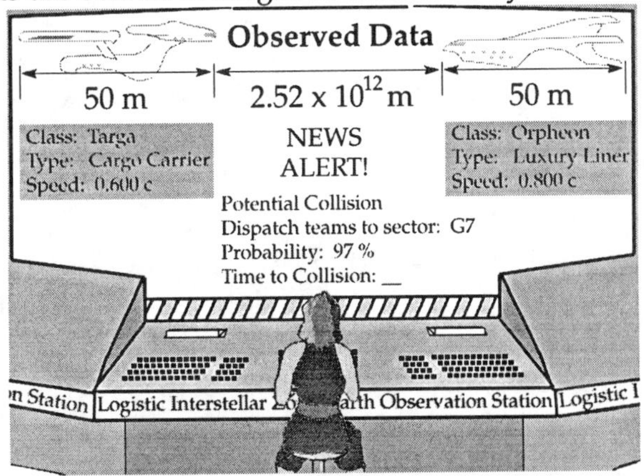

Solution (a) $L = \frac{L'}{\gamma}$ so $L' = \gamma L$ is the proper length where L is the measured length.

Relative to Earth, $v_1 = 0.800c$ gives $\gamma_1 = 1.67$, and $v_2 = -0.600c$ gives $\gamma_2 = 1.25$.

Since the measured length is L = 50.0 m for each rocket, their respective proper lengths are

$$L_1' = 1.67(50.0 \text{ m}) = 83.3 \text{ m} \quad \Diamond \qquad \text{and} \qquad L_2' = 1.25(50.0 \text{ m}) = 62.5 \text{ m} \quad \Diamond$$

(b) Compute the speed of one rocket relative to the other by considering an observer in rocket # 1. He sees the Earth moving with velocity $v = -0.800c$. The velocity addition equation then gives the velocity of rocket # 2 (relative to rocket #1) as

$$u_{12} = \frac{v + u'}{1 + \frac{vu'}{c^2}} = \frac{(-0.800c) + (-0.600c)}{1 + \frac{(-0.600c)(-0.800c)}{c^2}} = \frac{-1.400c}{1.48} = -0.946c \qquad (\#2 \text{ rel to } \#1)$$

Similarly, the velocity of #1 relative to #2 is $0.946c$. The γ factor for the relative motion between rockets is $\gamma_{12} = 3.08$.

The contracted length of rocket #2 as measured by an observer in #1 is

$$L_2 = \frac{L_2'}{\gamma_{12}} = \frac{62.5 \text{ m}}{3.08} = 20.3 \text{ m} \quad \Diamond$$

Similarly, the contracted length of #1 as measured by an observer in #2 is

$$L_1 = \frac{L_1'}{\gamma_{12}} = \frac{83.3 \text{ m}}{3.08} = 27.0 \text{ m} \quad \Diamond$$

(c) Liz sees the initial 2.52×10^{12} m-gap between the rockets decreasing at a rate of $1.400c$. Thus, according to her, the time before collision is

$$t = \frac{2.52 \times 10^{12} \text{ m}}{1.40(3.00 \times 10^{8} \text{ m/s})} = 6000 \text{ s} = 1.67 \text{ h} \quad \Diamond$$

(d) Liz, at rest in S, measures the proper length of the initial gap between the rockets. The contracted initial gap as measured by an observer in rocket #1 is

$$L_1 = \frac{L'}{\gamma_1} = \frac{2.52 \times 10^{12}\ \text{m}}{1.67} = 1.51 \times 10^{12}\ \text{m}$$

The time before collision according to this observer is

$$t_1 = \frac{L_1}{u_{12}} = \frac{1.51 \times 10^{12}\ \text{m}}{0.946(3.00 \times 10^8\ \text{m/s})} = 5328\ \text{s} = 1.48\ \text{h} \quad \Diamond$$

(e) The contracted initial gap between rockets as measured by an observer in rocket #2 is

$$L_2 = \frac{L}{\gamma_2} = \frac{2.52 \times 10^{12}\ \text{m}}{1.25} = 2.02 \times 10^{12}\ \text{m}$$

The time before collision according to this observer is

$$t_2 = \frac{L_2}{u_{12}} = \frac{2.02 \times 10^{12}\ \text{m}}{0.946(3.00 \times 10^8\ \text{m/s})} = 7104\ \text{s} = 1.97\ \text{h} \quad \Diamond$$

(f) Since it requires 90 min (1.50 h) on their own clock for each crew to evacuate, the crew of rocket #2 makes it, but the crew of rocket #1 does not. ◊

---

55. A radioactive nucleus moves with a speed of v relative to a laboratory observer. The nucleus emits an electron in the positive x direction with a speed of $0.700c$ relative to the decaying nucleus and a speed of $0.850c$ in the $+x$ direction relative to the laboratory observer. What is the value of v?

Solution $u = \dfrac{u' + v}{1 + \dfrac{u'v}{c^2}}$ becomes $v = c\left[\dfrac{c(u' - u)}{uu' - c^2}\right]$ from which

$$v = c\left[\frac{c(0.700c - 0.850c)}{(0.700c)(0.850c) - c^2}\right] = 0.370c \quad \Diamond$$

---

59. An alarm clock is set to sound in 10.0 h. At $t = 0$, the clock is placed in a spaceship moving with a speed of $0.750c$ (relative to the Earth). What distance, as determined by an Earth observer, does the spaceship travel before the alarm clock sounds?

Solution $$\gamma = \frac{1}{\sqrt{1-\frac{v^2}{c^2}}} = \frac{1}{\sqrt{1-(0.75)^2}} = 1.51$$

$$T = \gamma T' = 1.51(10.0 \text{ h}) = 15.1 \text{ h}$$

Therefore, $$d = vT = (0.750)(3.00\times 10^8 \text{ m/s})(15.1 \text{ h})(3600 \text{ s/h}) = 1.20\times 10^{13} \text{ m} \quad \Diamond$$

---

63. If a light source moves with a speed of v relative to an observer, there is a shift in the observed frequency analogous to the Doppler effect for sound waves. Show that the observed frequency, f_0, is related to the true frequency through the expression

$$f_0 = \sqrt{\frac{c \pm v_s}{c \mp v_s}}\, f$$

where the upper signs correspond to the source approaching the observer and the lower signs correspond to the source receding from the observer. (*Hint:* In the moving frame, the period is the proper time interval and is given by $T = 1/f$. Furthermore, the wavelength measured by the observer is given by $\lambda_0 = (c - v_s)T_0$, where T_0 is the period measured in the stationary frame.)

Solution $$f_0 = \frac{1}{T_0} = \frac{(c-v)}{\lambda_0} = \frac{(c-v)}{\frac{\lambda}{\gamma}} = \frac{(c-v)\gamma}{\frac{c}{f}} = \left(\frac{c-v}{c}\right)\left(\frac{1}{\sqrt{1-\frac{v^2}{c^2}}}\right)f$$

or $$f_0 = \left(\frac{c-v}{c}\right)\sqrt{\frac{c^2}{c^2-v^2}} = \sqrt{\frac{c-v}{c+v}}\,f \quad \Diamond$$

If source changes direction, v becomes $-v$ and the signs are reversed in numerator and denominator.

CHAPTER SELF-QUIZ

1. The observed relativistic length of a super rocket moving by the observer at $0.800c$ will be what factor times that of the measured rocket length if it were at rest?
   a. 0.45
   b. 0.60
   c. 0.80
   d. 1.33

2. According to a postulate of Einstein, which of the following describes the nature of the laws of physics as one observes processes taking place in various inertial frames of reference?
   a. laws are same only on inertial frames with zero velocity
   b. laws are same only on inertial frames moving at low velocities
   c. laws are same only on inertial frames moving at near speed of light
   d. laws are same on all inertial frames

3. A super fast freight train has a 20.0 m-long box car when measured at rest. What box car length does the ground observer measure when the train is going by at a speed of $0.650c$?
   a. 11.8 m
   b. 15.2 m
   c. 18.3 m
   d. 26.3 m

4. How fast would a rocket have to move past a ground observer if the latter were to observe a 1% length shrinkage in the rocket length? $(c = 3.00 \times 10^8 \text{ m/s})$
   a. 0.03×10^8 m/s
   b. 0.14×10^8 m/s
   c. 0.42×10^8 m/s
   d. 1.00×10^8 m/s

5. An unknown particle in an accelerator moving at a speed of 2.00×10^8 m/s has a measured relativistic mass of 2.00×10^{-26} kg. What must its rest mass be?
   a. 0.65×10^{-26} kg
   b. 0.81×10^{-26} kg
   c. 1.01×10^{-26} kg
   d. 1.49×10^{-26} kg

6. When one ton of TNT is exploded, approximately 4.50×10^9 J of energy is released. How much mass would this represent in a mass-to-energy conversion?
   ($c = 3.00 \times 10^8$ m/s)
   a. 1.50 kg
   b. 5.00×10^{-8} kg
   c. 5.30×10^4 kg
   d. 1.70×10^{-3} kg

7. The astronaut whose heart rate on Earth is 65 per minute increases his velocity to $v = 0.950c$. What is his heart rate now as measured by an Earth observer?
   a. 20 per minute
   b. 30 per minute
   c. 52 per minute
   d. 81 per minute

8. I am stationary in a reference system but, if my reference system is *not* an inertial reference system, then, *relative to me*, a system that is an inertial reference system must
   a. remain at rest
   b. move with constant velocity
   c. be accelerating
   d. none of the above

9. The short lifetime of muons created in the upper atmosphere of the Earth would not allow them to reach the surface of the Earth unless their lifetime increased by time dilation. From the reference system of the muons, the muons can reach the surface of the Earth because
   a. time dilation increases their velocity
   b. time dilation increases their energy
   c. length contraction decreases the distance to the Earth
   d. the relativistic speed of the Earth toward them is added to their velocity

10. A spaceship of mass 10^6 kg is to be accelerated to 0.600c. How much energy does this require?
   a. 1.13×10^{22} J
   b. 2.25×10^{22} J
   c. 3.38×10^{22} J
   d. 4.43×10^{22} J

11. At what speed would a clock have to be moving in order to run at a rate that is one-half the rate of a clock at rest?
   a. $0.670c$
   b. $0.750c$
   c. $0.870c$
   d. $0.950c$

12. In a typical color television tube, the electrons are accelerated through a potential difference of 25,000 volts. What speed do the electrons have when they strike the screen? ($q_e = 1.60 \times 10^{-19}$ C, $m_e = 9.10 \times 10^{-31}$ kg, and $c = 3.00 \times 10^8$ m/s)
   a. $v = 0.150c$
   b. $v = 0.300c$
   c. $v = 0.450c$
   d. $v = 0.600c$

27
Quantum Physics

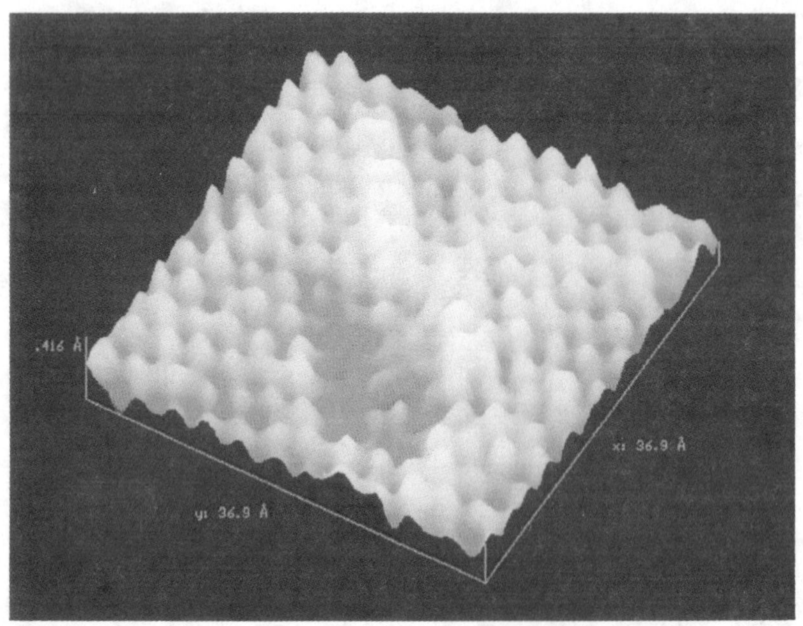

Chapter 27

QUANTUM PHYSICS

In the previous chapter, we discussed why Newtonian mechanics must be replaced by Einstein's special theory of relativity when dealing with particles whose speeds are comparable to the speed of light. Although many problems were indeed resolved by the theory of relativity in the early part of the 20th century, many experimental and theoretical problems remained unsolved. Attempts to explain the behavior of matter on the atomic level with the laws of classical physics were totally unsuccessful. Various phenomena, such as blackbody radiation, the photoelectric effect, and the emission of sharp spectra lines by atoms in a gas discharge tube, could not be understood within the framework of classical physics. We shall describe these phenomena because of their importance in subsequent developments.

Another revolution took place in physics between 1900 and 1930. This was the era of a new and more general formulation called *quantum mechanics.* This new approach was highly successful in explaining the behavior of atoms, molecules, and nuclei. As with relativity, the quantum theory requires a modification of our ideas concerning the physical world.

An extensive study of quantum theory is certainly beyond the scope of this book. This chapter is simply an introduction to the underlying ideas of quantum theory and the wave-particle nature of matter. We also discuss some simple applications of quantum theory, including the photoelectric effect, the Compton effect, and x-rays.

NOTES FROM SELECTED CHAPTER SECTIONS

27.1 Blackbody Radiation and Planck's Hypothesis

A *black body* is an ideal body that absorbs all radiation incident on it. Any body at some temperature T emits thermal radiation which is characterized by the properties of the body and its temperature. The spectral distribution of blackbody radiation at various temperatures is sketched in the figure. As the temperature increases, the intensity of the radiation (area under the curve) increases, while the peak of the distribution shifts to lower wavelengths.

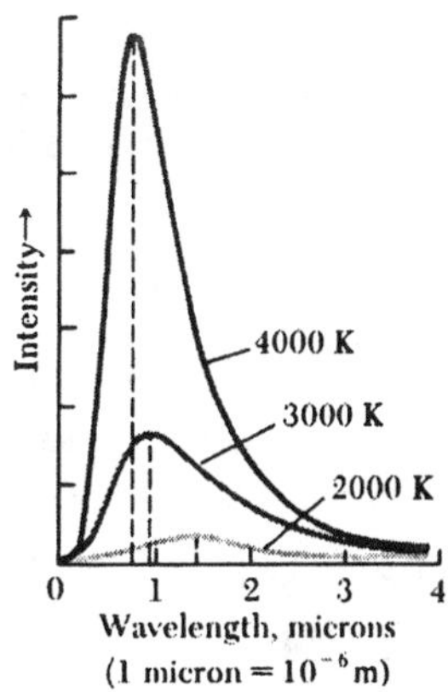

Classical theories failed to explain blackbody radiation. An empirical formula, proposed by Max Planck, is consistent with this distribution at all wavelengths. Planck made two basic assumptions in the development of this result:

(1) The oscillators emitting the radiation could only have *discrete energies* given by $E_n = nhf$, where n is a quantum number ($n = 1, 2, 3, \ldots$), f is the oscillator frequency, and h is Planck's constant.

(2) These oscillators can emit or absorb energy in discrete units called quanta (or photons), where the energy of a light quantum obeys the relation $E = hf$.

Subsequent developments showed that the quantum concept was necessary in order to explain several phenomena at the atomic level, including the photoelectric effect, the Compton effect, and atomic spectra.

27.2 The Photoelectric Effect

When light is incident on certain metallic surfaces, electrons can be emitted from the surfaces. This is called the photoelectric effect, discovered by Hertz. One cannot explain many features of the photoelectric effect using classical concepts. In 1905, Einstein provided a successful explanation of the photoelectric effect by extending Planck's quantum concept to include electromagnetic fields.

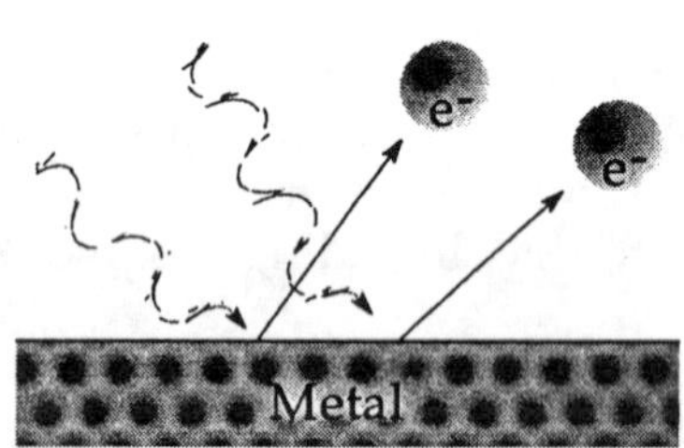

Several features of the photoelectric effect could not be explained with classical physics or with the wave theory of light. However, each of these features can be explained and understood on the basis of the photon theory of light. These observations and their explanations include:

1. No electrons are emitted if the incident light frequency falls below some cutoff frequency, f_c, which is characteristic of the material being illuminated. For example, in the case of sodium, $f_c = 5.50 \times 10^{14}$ Hz. This is inconsistent with the wave theory, which predicts that the photoelectric effect should occur at any frequency, provided the light intensity is high enough.

The fact that the photoelectric effect is not observed below a certain cutoff frequency follows from the fact that the energy of the photon must be greater than or equal to ϕ. If the energy of the incoming photon is not equal to or greater than ϕ, the electrons will never be ejected from the surface, regardless of the intensity of the light.

2. If the light frequency exceeds the cutoff frequency, a photoelectric effect is observed and the number of photoelectrons emitted is proportional to the light intensity. However, the maximum kinetic energy of the photoelectrons is independent of light intensity, a fact that cannot be explained by the concepts of classical physics.

   The fact that K_{max} is independent of the light intensity can be understood with the following argument. If the light intensity is doubled, the number of photons is doubled, which doubles the number of photoelectrons emitted. However, their kinetic energy, which equals $hf - \phi$, depends only on the light frequency and the work function, not on the light intensity.

3. The maximum kinetic energy of the photoelectrons increases with increasing light frequency. The fact that K_{max} increases with increasing frequency is easily understood with Equation 27.7.

4. Electrons are emitted from the surface almost instantaneously (less than 10^{-9} s after the surface is illuminated), even at low light intensities. Classically, one would expect that the electrons would require some time to absorb the incident radiation before they acquire enough kinetic energy to escape from the metal.

   Finally, the fact that the electrons are emitted almost instantaneously is consistent with the particle theory of light, in which the incident energy appears in small packets and there is a one-to-one interaction between photons and electrons. This is in contrast to having the energy of the photons distributed uniformly over a large area.

27.6 Compton Scattering

The Compton effect involves the scattering of an x-ray by an electron as shown in the following figure. The scattered x-ray undergoes a *change* in wavelength $\Delta\lambda$, called the Compton shift, which cannot be explained using classical concepts. By treating the x-ray as a photon (the quantum concept), the scattering process between the photon and electron predicts a shift in photon (x-ray) wavelength given by Equation 27.12, where θ is the angle between the incident and scattered x-ray and m is the mass of the electron. The formula is in excellent agreement with experimental results.

The figure below represents the data for Compton scattering at $\theta = 90.0°$ for scattering of x-rays from graphite. In this case, the Compton shift is $\Delta\lambda = 0.0236$ Å and λ_0 is the wavelength of the incident x-ray beam.

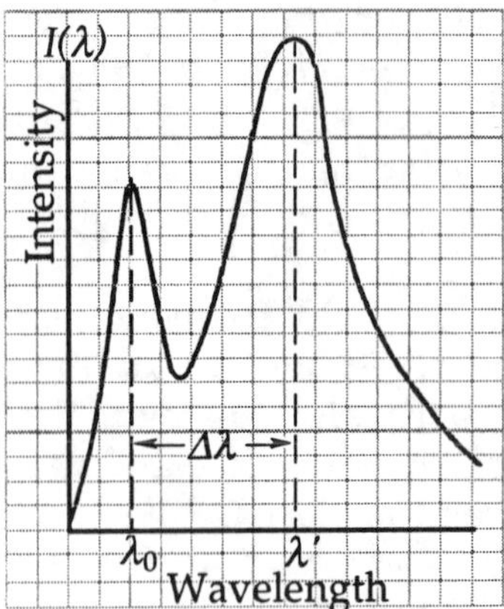

27.8 Photons and Electromagnetic Waves

The results of some experiments are better described on the basis of the photon model of light; other experimental outcomes are better described in terms of the wave model. The photon theory and the wave theory complement each other—*light exhibits both wave and photon characteristics.*

27.9 The Wave Properties of Particles

de Broglie postulated that a particle in motion has wave properties and a corresponding wavelength inversely proportional to the particle's momentum.

27.10 The Wave Function

The wave function is a complex valued quantity, the absolute square of which gives the probability of finding a particle at a given point at some instant; and the wave function contains all the information that can be known about the particle. The *Schrödinger equation* describes the manner in which matter waves change in time and space.

The probability of finding a certain value for a quantity (position, energy) is called the *expectation value* of the quantity.

27.11 The Uncertainty Principle

Quantum theory predicts that it is fundamentally impossible to make *simultaneous measurements* of a particle's position and velocity with infinite accuracy.

EQUATIONS AND CONCEPTS

The Wien displacement law properly describes the distribution of wavelengths in the energy spectrum emitted by a blackbody radiator.

$$\lambda_{max}T = 0.2898 \times 10^{-2}\ \text{m}\cdot\text{K} \quad (27.1)$$

A *black body* is an ideal body that absorbs all radiation incident on it. Any body at some temperature T emits thermal radiation which is characterized by the properties of the body and its temperature. As the temperature increases, the intensity of the radiation (area under the curve) increases, while the peak of the distribution shifts to shorter wavelengths.

Comment on blackbody radiation.

Vibrating molecules are characterized by discrete energy levels called quantum states. The positive integer n is called a quantum number.

$$E_n = nhf \quad (27.2)$$

$$h = 6.626 \times 10^{-34}\ \text{J}\cdot\text{s} \quad (27.3)$$

Molecules emit or absorb energy in discrete units of light energy called quanta. The energy of a quantum or photon corresponds to the energy difference between adjacent quantum states.

$$E = hf \quad (27.4)$$

When light is incident on certain metallic surfaces, electrons can be emitted from the surfaces. This is the photoelectric effect, and was discovered by Hertz. One cannot explain many of the features of the photoelectric effect using classical concepts. In 1905, Einstein provided a successful explanation of the photoelectric effect by extending Planck's quantum concept to include electromagnetic fields. In his model, Einstein assumed that light consists of a stream of particles called *photons* whose energy is given by $E = hf$, where h is Planck's constant and f is their frequency.

Comment on the photoelectric effect.

The maximum kinetic energy of the ejected photoelectron also depends on the work function of the metal, which is typically a few eV. This model is in excellent agreement with experimental results, including the prediction of a cutoff (or threshold) wavelength above which no photoelectric effect is observed.

$$KE_{max} = hf - \phi \quad (27.7)$$

$$\lambda_c = \frac{hc}{\phi} \quad (27.8)$$

In x-ray production, a series of "braking collisions" results in energy losses by a decelerating electron. Each increment of energy lost by the electron appears as a quantum of radiation in the x-ray spectrum of the target.

Comment on x-ray production.

If an electron is completely stopped in a single collision, the resulting photon will have the minimum wavelength possible for a given accelerating voltage.

$$\lambda_{min} = \frac{hc}{eV} \quad (27.10)$$

Bragg's law states the condition for constructive interference of x-rays diffracted by a crystal.

$$2d\sin\theta = m\lambda \quad (27.11)$$

$$(m = 1, 2, 3, \ldots)$$

The Compton effect involves the scattering of an x-ray by an electron. The scattered x-ray undergoes a change in wavelength called the Compton shift, which cannot be explained using classical concepts. By treating the x-ray as a photon (the quantum concept), the scattering process between the photon and electron predicts a shift in photon (x-ray) wavelength, where θ is the angle between the incident and scattered x-ray and m_0 is the mass of the electron. The formula is in excellent agreement with experimental results.

$$\Delta\lambda = \frac{h}{m_0 c}(1 - \cos\theta) \quad (27.12)$$

The quantity $h/(m_0 c)$ = 0.00243 nm is called the Compton wavelength.

The wavelength of a photon can be specified by its momentum.

$$p = \frac{h}{\lambda} \qquad (27.15)$$

In 1924, de Broglie proposed that any particle of momentum $p = mv$ has wavelike properties, and a wavelength given by $\lambda = h/(mv)$ called the de Broglie wavelength. The waves associated with material particles are called matter waves, and have a frequency which obeys the Einstein relation, $E = hf$.

$$\lambda = \frac{h}{mv} \qquad (27.16)$$

$$f = \frac{E}{h} \qquad (27.17)$$

The uncertainty principle, proposed by Heisenberg, deals with the limited precision to which one can make simultaneous measurements of the position and velocity of a particle. If Δx and Δp are the uncertainties in the position and momentum, respectively, the product $\Delta x \Delta p$ is always greater than or equal to a number of the order of Planck's constant. In other words, it is physically impossible to measure simultaneously the exact position and exact momentum of a particle.

$$\Delta x \Delta p \geq \frac{h}{4\pi} \qquad (27.18)$$

Another form of the uncertainty principle applies to the simultaneous measurement of energy and time.

$$\Delta E \Delta t \geq \frac{h}{4\pi} \qquad (27.19)$$

REVIEW CHECKLIST

▷ Describe the formula for blackbody radiation proposed by Planck, and the assumption made in deriving this formula.

▷ Describe the Einstein model for the photoelectric effect, and the predictions of the fundamental photoelectric effect equation for the maximum kinetic energy of photoelectrons. Recognize that Einstein's model of the photoelectric effect involves the photon concept ($E = hf$), and the fact that the basic features of the photoelectric effect are consistent with this model.

▷ Describe the Compton effect (the scattering of x-rays by electrons) and be able to use the formula for the Compton shift (Eq. 27.12). Recognize that the Compton effect can only be explained using the photon concept.

▷ Discuss the wave properties of particles, the de Broglie wavelength concept, and the dual nature of both matter and light.

▷ Discuss the manner in which the uncertainty principle makes possible a better understanding of the dual wave-particle nature of light and matter.

SOLUTIONS TO SELECTED END-OF-CHAPTER PROBLEMS

3. The threshold of dark-adapted (scotopic) vision is $4.00 \times 10^{-11}\ \text{W/m}^2$ at a central wavelength of 500 nm. If light with this intensity and wavelength enters the eye when the pupil is open to its maximum diameter of 8.50 mm, how many photons per second enter the eye?

Solution Energy of a single 500 nm photon:

$$E\gamma = hf = \frac{hc}{\lambda} = \frac{(6.63 \times 10^{-34}\ \text{J}\cdot\text{s})(3.00 \times 10^{8}\ \text{m/s})}{500 \times 10^{-9}\ \text{m}} = 3.98 \times 10^{-19}\ \text{J}$$

The energy entering the eye each second is

$$E = Pt = (IA)t = (4.00 \times 10^{-11}\ \text{W/m}^2)\frac{\pi}{4}(8.50 \times 10^{-3}\ \text{m})^2(1.00\ \text{s}) = 2.27 \times 10^{-15}\ \text{J}$$

The number of photons required to yield this energy is

$$n = \frac{E}{E\gamma} = \frac{2.27 \times 10^{-15}\ \text{J}}{3.98 \times 10^{-19}\ \text{J/photon}} = 5.70 \times 10^{3}\ \text{photons} \quad \Diamond$$

5. If the surface temperature of the Sun is 5800 K, find the wavelength that corresponds to the maximum rate of energy emission from the Sun.

Solution $\lambda_{max}T = 0.2898 \times 10^{-2}\ \text{m}\cdot\text{K}$ (Wien's Displacement Law)

Thus, $$\lambda_{max} = \frac{0.2898 \times 10^{-2}\ \text{m}\cdot\text{K}}{5800\ \text{K}} = 5.00 \times 10^{-7}\ \text{m} = 500\ \text{nm} \quad \Diamond$$

---

11. From the scattering of sunlight, Thomson calculated that the classical radius of the electron has a value of 2.82×10^{-15} m. If sunlight having an intensity of 500 W/m² falls on a disk with this radius, estimate the time required to accumulate 1.00 eV of energy. Assume that light is a classical wave and that the light striking the disk is completely absorbed. How does your estimate compare with the observation that photoelectrons are promptly (within 10^{-9} s) emitted?

Solution Energy needed = 1.00 eV = 1.60×10^{-19} J

The energy absorbed in time t is $E = Pt = (IA)t$, so

$$t = \frac{E}{IA} = \frac{1.60 \times 10^{-19}\ \text{J}}{\left[(500\ \text{J/s}\cdot\text{m}^2)(\pi)(2.82 \times 10^{-15}\ \text{m})^2\right]} = 1.28 \times 10^{7}\,\text{s} = 148\ \text{days} \quad \Diamond$$

---

17. Consider the metals lithium, iron, and mercury, which have work functions of 2.30 eV, 3.90 eV, and 4.50 eV, respectively. If light of wavelength 300 nm is incident on each of these metals, determine (a) which metals exhibit the photoelectric effect and (b) the maximum kinetic energy for the photoelectrons for those that exhibit the effect.

Solution The energy of a 300 nm wavelength photon is

$$E = h\frac{c}{\lambda} = 4.14\ \text{eV}$$

(a) In order for the photoelectric effect to occur, the energy of the photon must be greater than the work function. Thus, the effect will occur in lithium and iron. ◊

(b) For lithium, $KE_{max} = hf - \phi = 4.14 \text{ eV} - 2.30 \text{ eV} = 1.80 \text{ eV}$ ◊

For iron, $KE_{max} = hf - \phi = 4.14 \text{ eV} - 3.90 \text{ eV} = 0.240 \text{ eV}$ ◊

---

23. Calculate the minimum wavelength x-ray that can be produced when a target is struck by an electron that has been accelerated through a potential difference of (a) 15.0 kV; (b) 100 kV.

Solution We have $\lambda_{min} = \frac{hc}{eV} = \frac{1243 \text{ nm} \cdot \text{eV}}{eV} = \frac{1243 \text{ nm} \cdot \text{V}}{V}$

(a) $\lambda_{min} = \frac{1243 \text{ nm} \cdot \text{V}}{15000 \text{ V}} = 0.083 \text{ nm}$ ◊

(b) $\lambda_{min} = \frac{1243 \text{ nm} \cdot \text{V}}{1.00 \times 10^5 \text{ V}} = 0.012 \text{ nm}$ ◊

---

29. A monochromatic x-ray beam is incident on a NaCl crystal surface where $d = 0.353$ nm. The second-order maximum in the reflected beam is found when the angle between the incident beam and the surface is 20.5°. Determine the wavelength of the x-rays.

Solution We have $\lambda = \frac{2d\sin\theta}{n} = \frac{2(0.353 \text{ nm})\sin 20.5°}{(2)} = 0.124 \text{ nm}$ ◊

---

35. A beam of 0.68-nm photons undergoes Compton scattering from free electrons. What are the energy and momentum of the photons that emerge at a 45.0° angle with respect to the incident beam?

Solution $\Delta\lambda = \lambda_c(1 - \cos\theta)$ becomes

$$\lambda = \lambda_0 + \lambda_c(1 - \cos\theta) = \lambda_0 + 0.00243(1 - \cos\theta)$$

If $\lambda_0 = 0.680$ nm and $\theta = 45.0°$, the wavelength of the scattered photons is

$$\lambda = 0.680 \text{ nm} + (2.43 \times 10^{-3} \text{ nm})(1.00 - \cos 45.0°) = 0.6807 \text{ nm}$$

The energy of these photons is

$$E = \frac{hc}{\lambda} = \frac{(6.63 \times 10^{-34} \text{ J} \cdot \text{s})(3.00 \times 10^8 \text{ m/s})}{(0.6807 \times 10^{-9} \text{ m})} = 2.92 \times 10^{-16} \text{ J} \quad \lozenge$$

or E = 1.80 keV, and the momentum is

$$p = \frac{E}{c} = 9.74 \times 10^{-25} \text{ kg} \cdot \text{m/s} = 1.80 \text{ keV}/c \quad \lozenge$$

---

37. After a 0.80 nm x-ray photon scatters from a free electron, the electron recoils with a speed equal to 1.40×10^6 m/s. (a) What was the Compton shift in the photon's wavelength? (b) Through what angle was the photon scattered?

Solution $KE = \frac{1}{2}mv^2 = \frac{1}{2}(9.11 \times 10^{-31} \text{ kg})(1.40 \times 10^6 \text{ m/s})^2 = 8.93 \times 10^{-19} \text{ J}$ but

$$KE = E_\gamma - E'_\gamma = \frac{hc}{\lambda} - \frac{hc}{\lambda'} = hc\left(\frac{\lambda' - \lambda}{\lambda'\lambda}\right) \approx hc\left(\frac{\Delta\lambda}{\lambda^2}\right)$$

(a) Thus, $\Delta\lambda = \dfrac{\lambda^2(KE)}{hc} = \dfrac{(8.00 \times 10^{-10} \text{ m})^2(8.93 \times 10^{-19} \text{ J})}{(6.63 \times 10^{-34} \text{ J} \cdot \text{s})(3.00 \times 10^8 \text{ m/s})}$ or

$$\Delta\lambda = 2.87 \times 10^{-12} \text{ m} = 2.87 \times 10^{-3} \text{ nm} \quad \lozenge$$

(b) From the Compton equation,

$$\Delta\lambda = \lambda_C(1 - \cos\theta) \quad \text{when} \quad \lambda_C = 2.43 \times 10^{-3} \text{ nm}$$

$$\cos\theta = \left(1 - \frac{\Delta\lambda}{\lambda_C}\right) = 1 - \frac{2.87 \times 10^{-3} \text{ nm}}{2.43 \times 10^{-3} \text{ nm}} = -0.1811$$

and $\theta = 100°$ $\lozenge$

---

41. How much total kinetic energy will an electron-positron pair have if produced by a photon of energy 3.00 MeV?

Solution The kinetic energy of the pair will be the energy of the photon used to produce them, less the combined rest energy of the two. Thus,

$$KE = E - 2E_0 = 3.00 \text{ MeV} - 2(0.511 \text{ MeV}) = 1.98 \text{ MeV} \quad \lozenge$$

---

49. Find the de Broglie wavelength of a ball whose mass is 0.200 kg just before it strikes the Earth after being dropped from a building 50.0 m tall.

Solution From conservation of energy, $\frac{1}{2}mv^2 = mgh$ or

$$v = \sqrt{2gh} = \sqrt{2(9.80 \text{ m/s}^2)(50.0 \text{ m})} = 31.3 \text{ m/s}$$

Thus, $p = mv = (0.200 \text{ kg})(31.3 \text{ m/s}) = 6.26 \text{ kg}\cdot\text{m/s}$ and

$$\lambda = \frac{h}{p} = \frac{6.63\times10^{-34} \text{ J}\cdot\text{s}}{6.26 \text{ kg}\cdot\text{m/s}} = 1.06\times10^{-34} \text{ m} \quad \lozenge$$

---

53. A monoenergetic beam of electrons is incident on a single slit of width 0.500 nm. A diffraction pattern is formed on a screen 20.0 cm from the slit. If the distance between successive minima of the diffraction pattern is 2.10 cm, what is the energy of the incident electrons?

Solution $\frac{y}{L} = m\left(\frac{\lambda}{a}\right)$ gives $y_2 - y_1 = 2\left(\frac{L\lambda}{a}\right) - \frac{L\lambda}{a} = \frac{L\lambda}{a}$, or

$$\lambda = \frac{(\Delta y)a}{L} = \frac{(2.10 \text{ cm})(0.500 \text{ nm})}{20.0 \text{ cm}} = 0.0525 \text{ nm}$$

Thus, $$p = \frac{h}{\lambda} = \frac{6.63\times10^{-34} \text{ J}\cdot\text{s}}{(5.25\times10^{-11} \text{ kg})} = 1.26\times10^{-23} \text{ kg}\cdot\text{m/s}$$

Classically, $KE = \frac{p^2}{2m_0} = \frac{(1.26\times10^{-23}\text{ kg}\cdot\text{m/s})^2}{2(9.11\times10^{-31}\text{ kg})} = 8.75\times10^{-17}\text{ J} = 546\text{ eV}$

Relativistically, $KE = \sqrt{p^2c^2 + E_0{}^2} - E_0 = 8.748\times10^{-17}\text{ J} = 546\text{ eV}$ ◊

---

55. A 50.0-g ball moves at 30.0 m/s. If its speed is measured to an accuracy of 0.10%, what is the minimum uncertainty in its position?

Solution $(\Delta x)(\Delta p) \ge \frac{h}{4\pi}$

We have $v = 30.0\text{ m/s}$ and $\Delta v = 0.10\%(v) = 3.00\times10^{-2}\text{ m/s}$

Thus, $\Delta p = (50.0\times10^{-3}\text{ kg})(3.00\times10^{-2}\text{ m/s}) = 1.50\times10^{-3}\text{ kg}\cdot\text{m/s}$ and

$$\Delta x \ge \frac{h}{4\pi(\Delta p)} = \frac{6.63\times10^{-34}\text{ J}\cdot\text{s}}{4\pi(1.50\times10^{-3}\text{ kg}\cdot\text{m/s})} = 3.50\times10^{-32}\text{ m} \quad ◊$$

---

61. (a) Show that the kinetic energy of a nonrelativistic particle can be written in terms of its momentum as $KE = p^2/2m$. (b) Use the results of (a) to find the smallest kinetic energy of a proton confined to a 10^{-15}-m nucleus.

Solution (a) $KE = \frac{1}{2}mv^2 = \frac{(mv)^2}{2m} = \frac{p^2}{2m}$ ◊

(b) $\Delta p \ge \frac{h}{4\pi(\Delta x)} = \frac{6.63\times10^{-34}\text{ J}\cdot\text{s}}{4\pi(10^{-15}\text{ m})} = 5.25\times10^{-20}\text{ kg}\cdot\text{m/s}$

The smallest momentum is also of the order of Δp.

Thus, $p = 5.28\times10^{-20}\text{ kg}\cdot\text{m/s}$ and

$$KE_{\text{min}} = \frac{p_{\text{min}}^2}{2m} = \frac{(5.28\times10^{-20}\text{ kg}\cdot\text{m/s})^2}{2(1.67\times10^{-27}\text{ kg})} = 8.34\times10^{-13}\text{ J} = 5.20\text{ MeV} \quad ◊$$

---

65. How many photons are emitted per second by a 100-W sodium lamp if the wavelength of sodium light is 589.3 nm?

Solution The energy of one photon is $E = hf = \dfrac{hc}{\lambda} = 3.37 \times 10^{-19}\ \text{J}$

The energy emitted per second $= (\text{power})t = (100\ \text{J/s})(1.00\ \text{s}) = 100\ \text{J}$

Thus, the number of photons emitted is

$$N = \frac{\text{energy from lamp}}{\text{energy per photon}} = \frac{100\ \text{J}}{3.37 \times 10^{-19}\ \text{J}} = 2.96 \times 10^{20} \quad \Diamond$$

---

69. Photons of wavelength 450 nm are incident on a metal. The most energetic electrons ejected from the metal are bent into a circular arc of radius 20.0 cm by a magnetic field whose strength is 2.00×10^{-5} T. What is the work function of the metal?

Solution Recall that the orbit of a charged particle in a magnetic field is given by

$$r = \frac{mv}{qB} \quad \text{or} \quad mv = qrB$$

Now, $$p_{max} = mv_{max} = (1.60 \times 10^{-19}\ \text{C})(0.200\ \text{m})(2.00 \times 10^{-5}\ \text{T}) = 6.40 \times 10^{-25}\ \text{kg} \cdot \text{m/s}.$$

and $$KE_{max} = \frac{p_{max}^2}{2m} = \frac{(6.40 \times 10^{-25}\ \text{kg} \cdot \text{m/s})^2}{2(9.11 \times 10^{-31}\ \text{kg})} = 2.25 \times 10^{-19}\ \text{J} = 1.41\ \text{eV}$$

The photon energy is $E = hf = \dfrac{hc}{\lambda} = 2.76\ \text{eV}$

So, $$\phi = E - KE_{max} = 2.76\ \text{eV} - 1.41\ \text{eV} = 1.35\ \text{eV} \quad \Diamond$$

CHAPTER SELF-QUIZ

1. According to the de Broglie hypothesis, which of the following statements is applicable to the wavelength of a moving particle?
   a. directly proportional to its energy
   b. directly proportional to its momentum
   c. inversely proportional to its energy
   d. inversely proportional to its momentum

2. As the temperature of a radiation-emitting black body becomes higher, what happens to the peak wavelength of the radiation?
   a. increases
   b. decreases
   c. remains constant
   d. is directly proportional to temperature

3. According to Einstein, what is true of the stopping potential for a photoelectric current as the wavelength of incident light becomes shorter?
   a. increases
   b. decreases
   c. remains constant
   d. stopping potential is directly proportional to wavelength

4. What is the wavelength of a monochromatic light beam where the photon energy is 2.00 eV? ($h = 6.63 \times 10^{-34}$ J·s, $c = 3.00 \times 10^{8}$ m/s, 1.00 nm $= 10^{-9}$ m, and 1.00 eV $= 1.60 \times 10^{-19}$ J)
   a. 414 nm
   b. 621 nm
   c. 746 nm
   d. 829 nm

5. What is the de Broglie wavelength for a proton ($m = 1.67 \times 10^{-27}$ kg) moving at a speed of 5.00×10^5 m/s? ($h = 6.63 \times 10^{-34}$ J·s)
   a. 1.10×10^{-12} m
   b. 0.42×10^{-12} m
   c. 1.80×10^{-12} m
   d. 0.79×10^{-12} m

6. The Compton experiment demonstrated which of the following when an x-ray photon collides with an electron?
   a. momentum is conserved
   b. energy is conserved
   c. momentum and energy are both conserved
   d. wavelength of scattered photon equals that of incident photon

7. Light of wavelength 480 nm is incident on a metallic surface with a resultant photoelectric stopping potential of 0.55 V. What is the maximum kinetic energy of the emitted electrons? ($h = 6.63 \times 10^{-34}$ J·s, $c = 3.00 \times 10^8$ m/s, 1.00 nm $= 10^{-9}$ m, and 1.00 eV = 1.60×10^{-19} J)
   a. 3.19 eV
   b. 0.55 eV
   c. 2.04 eV
   d. 2.59 eV

8. The Sun's surface temperature is 5800 K and the peak wavelength in its radiation is 500 nm. What is the surface temperature of a distant star where the peak wavelength is 475 nm?
   a. 5510 K
   b. 5626 K
   c. 6105 K
   d. 6350 K

9. According to Einstein, increasing the brightness of a beam of light without changing its color will increase the
   a. number of photons
   b. energy of each photon
   c. speed of the photons
   d. frequency of the photons

10. According to the principle of complementarity, everything acts in a given experiment as if it is either a wave or a particle. In which experiment is the wave aspect exhibited?
   a. the Davisson and Germer experiment
   b. the photoelectric effect
   c. pair production
   d. Compton scattering

11. How much energy (in eV) does a photon of red light (λ = 640 nm) have? ($h = 6.626 \times 10^{-34}$ J·s and 1.00 eV = 1.60×10^{-19} J)
   a. 3.20 eV
   b. 2.50 eV
   c. 1.90 eV
   d. 1.30 eV

12. An electron microscope operates with electrons of kinetic energy 40.0 keV. What is the wavelength of these electrons? ($h = 6.626 \times 10^{-34}$ J·s, 1.00 eV = 1.60×10^{-19} J, and $m_e = 9.10 \times 10^{-31}$ kg)
   a. 0.50×10^{-10} m
   b. 7.17×10^{-11} m
   c. 6.14×10^{-12} m
   d. 3.07×10^{-13} m

28
Atomic Physics

Chapter 28

ATOMIC PHYSICS

A large portion of this chapter is concerned with the study of the hydrogen atom. Although the hydrogen atom is the simplest atomic system, it is especially important.

We first discuss the Bohr model of hydrogen, which helps us understand many features of hydrogen but fails to explain many finer details of atomic structure. Next we examine the hydrogen atom from the viewpoint of quantum mechanics and the quantum numbers used to characterize various atomic states. Additionally, we examine the physical significance of the quantum numbers and the effect of a magnetic field on certain quantum states. The Pauli exclusion principle is also presented. This physical principle is extremely important in understanding the properties of complex atoms and the arrangement of elements in the periodic table. Finally, we apply our knowledge of atomic structure to describe the mechanisms involved in the production of x-rays and the operation of a laser.

NOTES FROM SELECTED CHAPTER SECTIONS

28.1 Early Models of the Atom

One of the first indications that there was a need for modification of the Bohr theory became apparent when improved spectroscopic techniques were used to examine the spectral lines of hydrogen. It was found that many of the lines in the Balmer and other series were not single lines at all. Instead, each line was actually a group of lines spaced very close together. An additional difficulty arose when it was observed that, in some situations, certain single spectral lines were split into three closely spaced lines when the atoms were placed in a strong magnetic field.

28.3 The Bohr Theory of Hydrogen

The basic postulates of the Bohr model of the hydrogen atom are as follows:

1. The electron moves in circular orbits about the nucleus (the planetary model of the atom) under the influence of the Coulomb force of attraction between the electron and the positively charged nucleus.

2. The electron can exist only in very specific orbits; hence, the states are *quantized* (Planck's quantum hypothesis). The allowed orbits are those for which the angular momentum of the electron about the nucleus is an integral multiple of $\hbar = h/2\pi$, where h is Planck's constant.

3. When the electron is in one of its allowed orbits, it does not radiate energy; hence, the atom is stable. Such stable orbits are called stationary states.

4. The atom radiates energy only when the electron "jumps" from one allowed stationary orbit to another. This postulate states that the energy given off by an atom is carried away by a photon of energy *hf*.

It is important to understand the behavior of the hydrogen atom as an atomic system for the following reasons:

1. Much of what is learned about the hydrogen atom with its single electron can be extended to such single-electron ions as He^+ and Li^{2+}, which are hydrogen-like in their atomic structure.

2. The hydrogen atom is an ideal system for performing precise tests of theory against experiment and for improving our overall understanding of atomic structure.

3. The quantum numbers used to characterize the allowed states of hydrogen can be used to describe the allowed states of more complex atoms. This enables us to understand the periodic table of the elements, which is one of the greatest triumphs of quantum mechanics.

4. The basic ideas about atomic structure must be well understood before we attempt to deal with the complexities of molecular structures and the electronic structure of solids.

In addition to providing a theoretical derivation of the line spectrum, Bohr also explained:
(a) the limited number of lines seen in the absorption spectrum of hydrogen compared to the emission spectrum,
(b) the emission of x-rays from atoms,
(c) the chemical properties of atoms in terms of the electron shell model,
(d) how atoms associate to form molecules.

28.4 Modification of the Bohr Theory

In the three-dimensional problem of the hydrogen atom, three quantum numbers are required for each stationary state, corresponding to the three independent degrees of freedom for the electron.

The three quantum numbers which emerge from the theory are represented by the symbols n, l, and m_l. The quantum number n is called the *principal quantum number,* l is called the *orbital quantum number,* and m_l is called the *orbital magnetic quantum number.*

There are certain important relationships between these quantum numbers, as well as certain restrictions on their values. These restrictions are:

The values of n can range from 1 to ∞.
The values of l can range from 0 to $n - 1$.
The values of m_l can range from $-l$ to l.

For historical reasons, *all states with the same principal quantum number are said to form a shell.* These shells are identified by the letters K, L, M, . . . , which designate the states for which $n = 1, 2, 3, \ldots$. Likewise, *the states having the same values of n and l are said to form a subshell.* The letters *s, p, d, f, g, h,* . . . are used to designate the states for which $l = 0, 1, 2, 3, \ldots$.

28.7 The Spin Magnetic Quantum Number

In order to completely describe a quantum state of the hydrogen atom, it is necessary to include a fourth quantum number, m_s, called the *spin magnetic quantum number.* The *spin magnetic quantum number* m_s accounts for the two closely spaced energy states corresponding to the two possible orientations of electron spin. This quantum number can have only two values, $\pm\frac{1}{2}$. In effect, this doubles the number of allowed states specified by the quantum numbers n, l, and m_l.

28.9 The Exclusion Principle and the Periodic Table

The *exclusion principle* states that no two electrons can exist in identical quantum states. This means that no two electrons in a given atom can be characterized by the same set of quantum numbers at the same time.

Hund's rule states that when an atom has orbitals of equal energy, the order in which they are filled by electrons is such that a maximum number of electrons will have unpaired spins.

X-rays are emitted by atoms when an electron undergoes a transition from an outer shell into an electron vacancy in one of the inner shells. Transitions into a vacant state in the K shell give rise to the K series of spectral lines, transitions into a vacant state in the L shell create the L series of lines, and so on. The x-ray spectrum of a metal target consists of a set of sharp characteristic lines superimposed on a broad, continuous spectrum.

28.11 Atomic Transitions

Stimulated absorption occurs when light with photons of an energy which matches the energy separation between two atomic energy levels is absorbed by an atom.

An atom in an excited state has a certain probability of returning to its original energy state. This process is called *spontaneous emission*.

When a photon with an energy equal to the excitation energy of an excited atom is incident on the atom, it can increase the probability of de-excitation. This is called *stimulated emission* and results in a second photon of energy equal to that of the incident photon.

28.12 Lasers and Holography

The following three conditions must be satisfied in order to achieve laser action:

1. The system must be in a state of *population inversion* (that is, more atoms in an excited state than in the ground state).

2. The excited state of the system must be a *metastable state*, which means its lifetime must be long compared with the usually short lifetimes of excited states. When such is the case, stimulated emission will occur before spontaneous emission.

3. The emitted photons must be confined in the system long enough to allow them to stimulate further emission from other excited atoms. This is achieved by the use of reflecting mirrors at the ends of the system. One end is made totally reflecting, and the other is slightly transparent to allow the laser beam to escape.

EQUATIONS AND CONCEPTS

When the electron undergoes a transition from one allowed orbit to another, the frequency of the emitted photon is proportional to the difference in energies of the initial and final states.

$$E_i - E_f = hf \qquad (28.3)$$

The Bohr model of the atom assumed that the electron orbited the nucleus in a circular path under the influence of the coulomb force of attraction. Also, the angular momentum of the electron about the nucleus must be quantized in units of $nh/(2\pi)$.

$$mvr = n\hbar \quad (28.4)$$

$$n = 1, 2, 3, \ldots$$

While in one of the allowed orbits or stationary states (determined by quantization of the orbital angular momentum), the electron does not radiate energy.

The total energy of the hydrogen atom $(KE + PE)$ depends on the radius of the allowed orbit of the electron.

$$E = -\frac{ke^2}{2r} \quad (28.8)$$

The electron can exist only in certain allowed orbits, the value of which can be expressed in terms of the Bohr radius, a_0.

$$r_n = \frac{n^2\hbar^2}{mke^2} \quad (28.9)$$

$$n = 1, 2, 3, \ldots$$

This is the *Bohr radius* which corresponds to $n = 1$.

$$a_0 = \frac{\hbar^2}{mke^2} = 0.0529 \text{ nm} \quad (28.10)$$

This is a general expression for the radius of any orbit in the hydrogen atom.

$$r_n = n^2 a_0 = n^2(0.0529 \text{ nm}) \quad (28.11)$$

When numerical values of the constants are used, the energy level values can be expressed in units of electron volts (eV).

$$E_n = -\frac{13.6}{n^2} \text{ eV} \quad (28.13)$$

The lowest energy state or ground state corresponds to the principle quantum number $n = 1$. The energy level approaches $E = 0$ as r approaches infinity. This is the ionization energy for the atom.

Comment on energy levels.

A photon of frequency, f, (and wavelength λ) is emitted when an electron undergoes a transition from an initial energy level to a final lower level.

$$f = \frac{E_i - E_f}{h} \qquad (28.14)$$

$$f = \frac{mk^2e^4}{4\pi\hbar^3}\left(\frac{1}{n_f^2} - \frac{1}{n_i^2}\right)$$

$$\frac{1}{\lambda} = R\left(\frac{1}{n_f^{\,2}} - \frac{1}{n_i^{\,2}}\right) \quad \text{where}$$

R is the theoretical value of the Rydberg constant.

$$R = \frac{mk^2e^4}{4\pi c\hbar^3} \qquad (28.16)$$

The lines observed in the hydrogen spectrum can be arranged into series corresponding to assigned values of principal quantum numbers of the initial and final states.

Comment on the line spectra of hydrogen.

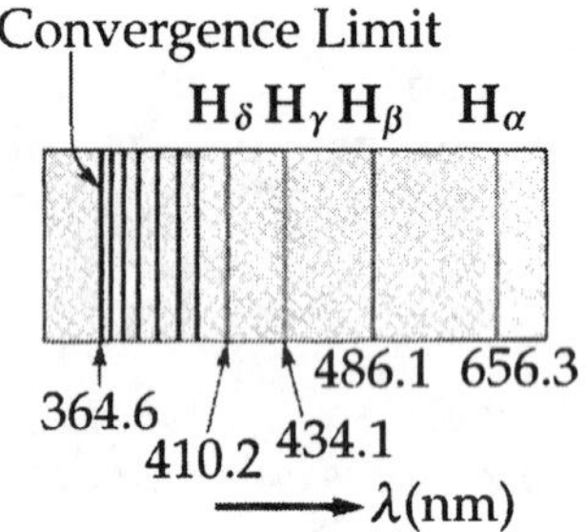

For the Lyman series,
$n_f = 1$ and $n_i = 2, 3, 4, \ldots$

For the Balmer series,
$n_f = 2$ and $n_i = 3, 4, 5, \ldots$

For the Paschen series,
$n_f = 3$ and $n_i = 4, 5, 6, \ldots$

For the Brackett series,
$n_f = 4$ and $n_i = 5, 6, 7, \ldots$

In the case of very large values of the principal quantum number, the energy differences between adjacent levels approach zero and essentially a continuous range (as opposed to a quantized set) of energy values of the emitted photon is

Comment on the correspondence principle.

possible. In this limit of very large quantum numbers, the classical model of the atom is reasonably accurate.

In addition to the principal quantum, n, other quantum numbers are necessary to completely and accurately specify the possible energy levels in the hydrogen atom and also in more complex atoms.

Comment on quantum numbers.

All energy states with the same principal quantum number, n, form a shell. These shells are identified by the spectroscopic notation K, L, M, . . . corresponding to $n = 1, 2, 3, \ldots$.

n can range from 1 to ∞.

The orbital quantum number, l, determines the allowed value of orbital angular momentum. All energy states having the same values of n and l form a subshell. The letter designations $s, p, d, f, \ldots$ correspond to values of $l = 1, 2, 3, 4, \ldots$.

l can range from 0 to $(n - 1)$.

The magnetic orbital quantum number determines the possible orientations of the electron's orbital angular momentum vector in the presence of an external magnetic field.

m_l can range from $-l$ to l.

The spin magnetic quantum number can have only two values which in turn correspond to the two possible directions of the electron's intrinsic spin (either parallel or antiparallel to the direction of the orbital angular momentum vector).

$$m_s = \pm\frac{1}{2}$$

No two electrons in an atom can have the same set of quantum numbers n, l, m_l, and m_s.

Comment on the exclusion principle.

REVIEW CHECKLIST

▷ State the basic postulates of the Bohr model of the hydrogen atom.

▷ Sketch an energy level diagram for hydrogen (include assignment of values of the principle quantum number, n), show transitions corresponding to spectral lines in the several known series, and make calculations of wavelength values.

▷ For each of the quantum numbers, n, l (the orbital quantum number), m_l (the orbital magnetic quantum number), and m_s (the spin magnetic quantum number): (i) qualitatively describe what each implies concerning atomic structure, (ii) state the allowed values which may be assigned to each, and the number of allowed states which may exist in a particular atom corresponding to each quantum number.

▷ State the Pauli exclusion principle and describe its relevance to the periodic table of the elements. Show how the exclusion principle leads to the known electronic ground state configuration of the light elements.

SOLUTIONS TO SELECTED END-OF-CHAPTER PROBLEMS

3. The "size" of the nucleus in Rutherford's model of the atom is about 10^{-15} m. (a) Determine the repulsive electrostatic force between two protons separated by this distance. (b) Determine (in MeV) the electrostatic potential energy of the pair of protons.

Solution (a) Using $F = k\frac{q_1 q_2}{r^2}$ gives

$$F = (9.00\times 10^9 \text{N}\cdot\text{m}^2/\text{C}^2)\frac{(1.60\times 10^{-19}\text{ C})^2}{(10^{-15}\text{ m})^2} = 230 \text{ N} \quad \Diamond$$

(b) Using $V = k\frac{q_1 q_2}{r}$, we get

$$V = 2.30\times 10^{-13}\text{ J} = 1.44\times 10^6\text{ eV} = 1.40\text{ MeV} \quad \Diamond$$

5. Show that the Balmer series formula (Equation 28.1) can be written as

$$\lambda = \frac{364.5n^2}{n^2 - 4}\ \text{nm} \qquad \text{where } n = 3, 4, 5, \ldots.$$

Solution Start with Balmer's equation, $\frac{1}{\lambda} = R_H\left(\frac{1}{2^2} - \frac{1}{n^2}\right)$ or

$$\lambda = \frac{(4n^2/R_H)}{(n^2 - 4)}$$

Substituting $R_H = 1.0973732 \times 10^7\ \text{m}^{-1}$, we obtain

$$\lambda = \frac{(3.645 \times 10^{-7}\ \text{m})n^2}{n^2 - 4} = \frac{364.5n^2}{n^2 - 4}\ \text{nm}, \quad \text{where } n = 3, 4, 5, \ldots \ \lozenge$$

---

13. A hydrogen atom initially in its ground state (n = 1) absorbs a photon and ends up in the state for which n = 3. (a) What is the energy of the absorbed photon? (b) If the atom returns to the ground state, what photon energies could the atom emit?

Solution (a) From the energy level sketch shown, we find the energy of the absorbed photon, $E = E_3 - E_1$, is

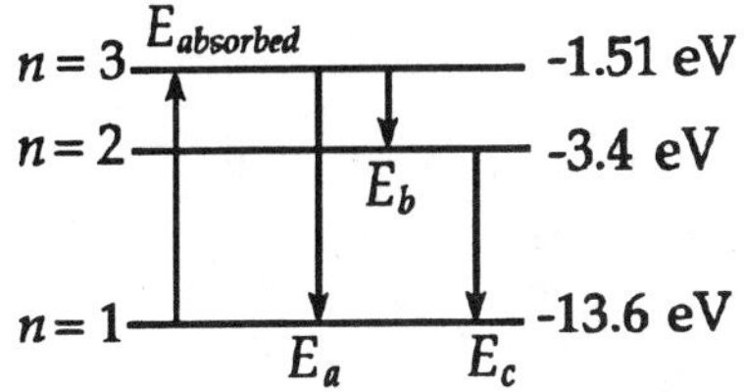

$$E = -1.51\ \text{eV} - (-13.6\ \text{eV}) \quad \text{or}$$

$$E = 12.09\ \text{eV} \ \lozenge$$

(b) The three possible transitions are shown in the energy level diagram. The energies are:

$$E_a = E_i - E_f = -1.51\ \text{eV} - (-13.6\ \text{eV}) = 12.09\ \text{eV} \ \lozenge$$

$$E_b = E_i - E_f = -1.51\ \text{eV} - (-3.40\ \text{eV}) = 1.89\ \text{eV} \ \lozenge$$

$$E_c = E_i - E_f = -3.40\ \text{eV} - (-13.6\ \text{eV}) = 10.2\ \text{eV} \ \lozenge$$

---

17. Show that the speed of the electron in the nth Bohr orbit in hydrogen is given by

$$v_n = \frac{ke^2}{n\hbar}$$

Solution Starting with $\frac{1}{2}mv^2 = \frac{ke^2}{2r}$, we have $v^2 = \frac{ke^2}{mr}$

Using $r_n = \frac{n^2\hbar^2}{mke^2}$, gives

$$v_n^2 = \frac{ke^2}{m\left(\frac{n^2\hbar^2}{mke^2}\right)} \quad \text{or} \quad v_n = \frac{ke^2}{n\hbar} \quad \Diamond$$

---

23. Four possible transitions for a hydrogen atom are listed below.

(A) $n_i = 2;\ n_f = 5$ (B) $n_i = 5;\ n_f = 3$

(C) $n_i = 7;\ n_f = 4$ (D) $n_i = 4;\ n_f = 7$

(a) Which transition will emit the shortest wavelength photon? (b) For which transition will the atom gain the most energy? (c) For which transition(s) does the atom lose energy?

Solution $\Delta E = 13.6\left(\frac{1}{n_i^2} - \frac{1}{n_f^2}\right)$ (See problem 28.22.)

For $\Delta E > 0$, we have absorption, and for $\Delta E < 0$, we have emission.

(A) for $n_i = 2$ and $n_f = 5$ $\Delta E = 2.86$ eV (absorption)

(B) for $n_i = 5$ and $n_f = 3$ $\Delta E = -0.97$ eV (emission)

(C) for $n_i = 7$ and $n_f = 4$ $\Delta E = -0.57$ eV (emission)

(D) for $n_i = 4$ and $n_f = 7$ $\Delta E = 0.57$ eV (absorption)

(a) $E = \frac{hc}{\lambda}$, so the shortest wavelength is emitted in transition B. ◊

(b) Atom gains most energy in transition A. ◊

(c) Atom loses energy in transitions B and C. ◊

---

25. An electron is in the first Bohr orbit of hydrogen. Find (a) the speed of the electron, (b) the time required for the electron to circle the nucleus, and (c) the current in amps corresponding to the motion of the electron.

Solution (a) In the first Bohr orbit, we have $n = 1$ and $r = 0.529 \times 10^{-10}$ m.

From the quantization of angular momentum, we have $v_n = \frac{n\hbar}{mr_n}$ or

$$v_1 = \frac{n\hbar}{mr_0} = \frac{1.055 \times 10^{-34} \text{ J} \cdot \text{s}}{(9.11 \times 10^{-31} \text{ kg})(0.529 \times 10^{-10})} = 2.19 \times 10^6 \text{ m/s}$$ ◊

(b) $t = \frac{2\pi r}{v} = \frac{2\pi(0.529 \times 10^{-10} \text{ m})}{2.19 \times 10^6 \text{ m/s}} = 1.52 \times 10^{-16} \text{ s}$ ◊

(c) $I = \frac{\Delta Q}{\Delta t} = \frac{e}{t} = \frac{1.60 \times 10^{-19} \text{ C}}{1.52 \times 10^{-16} \text{ s}} = 1.05 \times 10^{-3} \text{ A}$ ◊

---

29. (a) Find the energy of the electron in the ground state of doubly ionized lithium, which has an atomic number Z = 3. (b) Find the radius of this ground state orbit.

Solution (a) From $E_n = -\frac{Z^2(13.6 \text{ eV})}{n^2}$, we have (with Z = 3 and $n = 1$),

$$E_1 = -9(13.6 \text{ eV}) = -122.4 \text{ eV}$$ ◊

(b) From $r_n = \frac{n^2 r_0}{Z}$, with $n = 1$ and Z = 3),

$$r_1 = \frac{0.529 \times 10^{-10}\ \text{m}}{3} = 1.76 \times 10^{-11}\ \text{m} \quad \lozenge$$

---

33. (a) Substitute numerical values into Equation 28.19 to find a value for the Rydberg constant for singly ionized helium, He^+. (b) Use the result of Part (a) to find the wavelength associated with a transition from the $n = 2$ state to the $n = 1$ state of He^+. (c) Identify the region of the electromagnetic spectrum associated with this transition.

Solution

(a) We have $$R = \frac{Z^2 mk_e^2 e^4}{r\pi c\hbar^2} = Z^2 R_{\text{hydrogen}}$$

For singly ionized helium, Z = 2; so

$$R = 4R_{\text{hydrogen}} = 4(1.0974 \times 10^7\ \text{m}^{-1}) = 4.39 \times 10^7\ \text{m}^{-1} \quad \lozenge$$

(b) $$\frac{1}{\lambda} = R\left(\frac{1}{n_f^2} - \frac{1}{n_i^2}\right) = 4.39 \times 10^7\ \text{m}^{-1}\left(\frac{1}{1} - \frac{1}{4}\right) = 3.29 \times 10^7\ \text{m}^{-1} \quad \text{or}$$

$$\lambda = 3.03 \times 10^{-8}\ \text{m} = 30.3\ \text{nm} \quad \lozenge$$

(c) This wavelength is in the deep ultraviolet region. ◊

---

35. The *p*-meson has a charge of $-e$, a spin quantum number of 1, and a mass 1507 times that of the electron. If the electrons in atoms were replaced by *p*-mesons, list the possible sets of quantum numbers for *p*-mesons in the 3*d* subshell.

Solution The 3*d* subshell has $l = 2$, and $n = 3$. Also, we have $s = 1$. Therefore, we can have $n = 3$, $l = 2$, $m_l = -2, -1, 0, 1, 2$, $s = 1$, and $m_s = -1, 0, 1$, leading to the following table:

n	l	m_l	s	m_s
3	2	-2	1	-1
3	2	-2	1	0
3	2	-2	1	+1
3	2	-1	1	-1
3	2	-1	1	0
3	2	-1	1	+1
3	2	0	1	-1
3	2	0	1	0
3	2	0	1	+1
3	2	+1	1	-1
3	2	+1	1	0
3	2	+1	1	+1
3	2	+2	1	-1
3	2	+2	1	0
3	2	+2	1	+1

---

37. When the principal quantum number is $n = 4$, how many different values of (a) l and (b) m_e are possible?

Solution We list the quantum numbers as:

$$\begin{array}{lll} n = 4 & l = 3 & m_l = \pm 3, \pm 2, \pm 1, \ 0 \\ & l = 2 & m_l = \pm 2, \pm 1, \ 0 \\ & l = 1 & m_l = \pm 1, \ 0 \\ & l = 0 & m_l = 0 \end{array}$$

(a) There are 4 different values for l. ◊

(b) There are 16 different combinations, but only 7 different (or distinct) values for m_l. ◊

---

41. Suppose two electrons in the same system each have $l = 0$ and $n = 3$. (a) How many states would be possible if the exclusion principle were inoperative? (b) List the possible states, taking the exclusion principle into account.

Solution $n = 3, l = 0, m_l = 0$, and $m_s = \pm\frac{1}{2}$

(a) If the exclusion principle were inoperative, so electrons could be in identical states, there would be 4 possible states for the system as shown:

state		electron 1 (n,l,m_l,m_s)		electron 2 (n,l,m_l,m_s)
1		$3,0,0,+\frac{1}{2}$		$3,0,0,+\frac{1}{2}$
2		$3,0,0,+\frac{1}{2}$		$3,0,0,-\frac{1}{2}$
3		$3,0,0,-\frac{1}{2}$		$3,0,0,+\frac{1}{2}$
4		$3,0,0,-\frac{1}{2}$		$3,0,0,-\frac{1}{2}$

(b) Taking the exclusion principle into account, it is seen that states 1 and 4 above are not allowed, leaving only the remaining two states. ◊

---

45. The K series of the discrete spectrum of tungsten contains wavelengths of 0.0185 nm, 0.0209 nm, and 0.0215 nm. The K-shell ionization energy is 69.5 keV. Determine the ionization energies of the L, M, and N shells. Sketch the transitions.

Solution $E = \frac{hc}{\lambda} = \frac{1241.6 \text{ eV} \cdot \text{nm}}{\lambda} = \frac{1.2416 \text{ keV} \cdot \text{nm}}{\lambda}$

For $\lambda_1 = 0.0185$ nm, $E = 67.11$ keV

$\lambda_2 = 0.0209$ nm, $E = 59.4$ keV

$\lambda_3 = 0.0215$ nm, $E = 57.7$ keV

The ionization energy for K shell = 69.5 keV; so, the ionization energies for the other shells are

L shell = 11.8 keV ◊; M shell = 10.1 keV ◊; N shell = 2.39 keV ◊

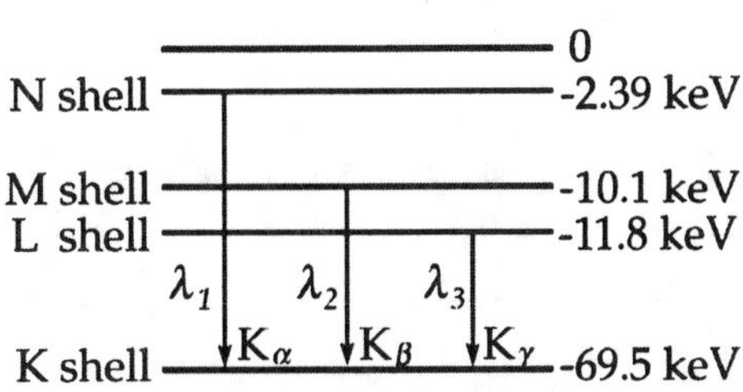

49. An electron is in the nth Bohr orbit of the hydrogen atom. (a) Show that the time it takes the electron to circle the nucleus once can be expressed as $T = \tau n^3$, and determine the numerical value of τ_0. (b) On the average, an electron in the n = 2 orbit will exist in that orbit for about 10^{-8} s before it jumps down to the n = 1 ground-state orbit. How many revolutions about the nucleus are made by the electron before it jumps to the ground state? (c) If one revolution of the electron about the nucleus is defined as an "electron year" (analogous to an Earth year being one revolution of the Earth around the Sun), does the electron in the n = 2 orbit "live" very long? Explain. (d) How does the above calculation support the "electron cloud" concept?

Solution (a) The time for one complete orbit is $T = \dfrac{2\pi r}{v}$

From Bohr's quantization postulate, $L = mvr = n\,\hbar$, we see that $v = \dfrac{n\hbar}{mr}$

Thus, the orbital period becomes $$T = \frac{2\pi m r^2}{n\hbar} = \frac{2\pi m (a_0 n^2)^2}{n\hbar} = \frac{2\pi m a_0^2}{\hbar} n^3$$

or $T = \tau_0 n^3$ where $$\tau_0 = \frac{2\pi m a_0^2}{\hbar} = \frac{2\pi(9.11\times10^{-31}\text{ kg})(0.0529\times10^{-9}\text{ m})^2}{(1.05\times10^{-34}\text{ J}\cdot\text{s})}$$

$$\tau_0 = 1.50\times10^{-16}\text{ s} \quad \Diamond$$

(b) With n = 2, we have $T = 4\tau_0 = 8(1.525\times10^{-16}\text{ s}) = 1.22\times10^{-15}\text{ s}$

Thus, if the electrons stay in the n = 2 state for 10^{-8} s, it will make

$$\frac{10^{-8}\text{ s}}{1.22\times10^{-15}\text{ s/rev}} = 8.20\times10^{6}\text{ revolutions of the nucleus} \quad \Diamond$$

(c) Yes, for 8.20×10^6 "electron years" ◊

(d) With that many revolutions, it appears like the electron (or at least some part of it) is present everywhere in the $n = 2$ orbit. ◊

---

53. Mercury's ionization energy is 10.39 eV. The three longest wavelengths of the absorption spectrum of mercury are 253.7 nm, 185.0 nm, and 158.5 nm. (a) Construct an energy-level diagram for mercury. (b) Indicate all possible emission lines, starting from the highest energy level on the diagram. (c) Disregarding recoil, determine the minimum speed an electron must have in order to make an inelastic collision with a mercury atom.

Solution (a & b)

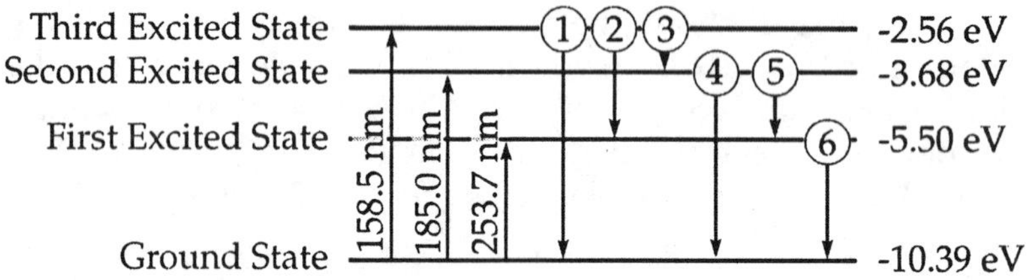

Third Excited State — -2.56 eV
Second Excited State — -3.68 eV
First Excited State — -5.50 eV
Ground State — -10.39 eV
158.5 nm
185.0 nm
253.7 nm
1 2 3 4 5 6

$$E = \frac{hc}{\lambda} = \frac{1241.6 \text{ eV} \cdot \text{nm}}{\lambda} \quad \text{so}$$

for $\lambda = 253.7$ nm, $E = 4.894$ eV

$\lambda = 185.0$ nm, $E = 6.711$ eV

$\lambda = 158.5$ nm, $E = 7.833$ eV

The wavelengths of the emission lines shown are $\lambda_1 = 158.5$ nm, $\lambda_2 = 422$ nm,

$\lambda_3 = 1109$ nm, $\lambda_4 = 185$ nm, $\lambda_5 = 682$ nm, and $\lambda_6 = 253.7$ nm ◊

(c) To have an inelastic collision, we must excite the atom from ground state to the first excited state; and electron must have a minimum kinetic energy of 10.39 eV – 5.50 eV = 4.89 eV, so

$$v = \sqrt{\frac{2(4.89 \text{ eV})(1.60 \times 10^{-19} \text{ J/eV})}{9.11 \times 10^{-31} \text{kg}}} = 1.31 \times 10^6 \text{m/s} \quad ◊$$

---

55. An electron has a de Broglie wavelength equal to the diameter of the hydrogen atom. (a) What is the kinetic energy of the electron? (b) How does this energy compare with the ground-state energy of the hydrogen atom?

Solution Since $\lambda = d = 2r_0 = 2(0.0529 \text{ nm}) = 0.1058 \text{ nm}$, the electron is nonrelativistic and we can use

$$p = \frac{h}{\lambda} \quad \text{and} \quad KE = \frac{p^2}{2\,\text{m}}$$

$$KE = \frac{h^2}{2m\lambda^2} = \frac{(6.626 \times 10^{-34} \text{ J}\cdot\text{s})^2}{2(9.10 \times 10^{-31} \text{ kg})(0.1058 \times 10^{-9})^2}$$

$$KE = 2.155 \times 10^{-17} \text{ J} = 135 \text{ eV}$$

This is about 10 times as large as the ground-state energy of hydrogen, which is 13.6 eV. ◊

---

57. Use Bohr's model of the hydrogen atom to show that when the atom makes a transition from the state n to the state $n - 1$, the frequency of the emitted light is given by

$$f = \frac{2\pi^2 mk^2e^4}{h^3}\left(\frac{2n-1}{(n-1)^2 n^2}\right)$$

Solution In the Bohr model,

$$hf = \Delta E = \frac{4\pi^2 mk^2e^4}{2h^2}\left(\frac{1}{(n-1)^2} - \frac{1}{n^2}\right)$$

which reduces to

$$f = \frac{2\pi^2 mk^2e^4}{h^3}\left(\frac{2n-1}{(n-1)^2 n^2}\right) \quad ◊$$

CHAPTER SELF-QUIZ

1. The restriction that no more than one electron may occupy a given quantum state in an atom was first stated by which of the following scientists?
   a. Bohr
   b. de Broglie
   c. Heisenberg
   d. Pauli

2. The ionization energy for the hydrogen atom is 13.6 eV. What is the energy of a photon that is emitted, as a hydrogen atom makes a transition between the $n = 5$ and $n = 3$ states?
   a. 0.544 eV
   b. 0.970 eV
   c. 1.51 eV
   d. 10.2 eV

3. When a cool gas is placed between a glowing wire filament source and a diffraction grating, the resultant spectrum from the grating is which one of the following?
   a. line emission
   b. line absorption
   c. continuous
   d. monochromatic

4. The Balmer series of hydrogen is comprised of transitions from higher levels to the $n = 2$ level. If the first line in that series has a wavelength 653 nm, what wavelength corresponds to the transition from $n = 5$ to $n = 2$?
   a. 390 nm
   b. 432 nm
   c. 503 nm
   d. 630 nm

5. If the radius of the electron orbit in the $n = 1$ level of the hydrogen atoms is 0.053 nm, what is its radius for the $n = 4$ level? (Assume the Bohr model is valid.)
   a. 0.11 nm
   b. 0.21 nm
   c. 0.48 nm
   d. 0.85 nm

6. The quantum mechanical model of the hydrogen atom requires that if the orbital quantum number = 4, how many possible substates are permitted?
   a. 4
   b. 8
   c. 16
   d. 18

7. Atoms which absorb ultraviolet photons and in turn emit photons in the visible range are used in which type of device?
   a. incandescent lamp
   b. radar
   c. fluorescent lamp
   d. electron microscope

8. The four visible colors emitted by hydrogen atoms are produced by electrons that
   a. start in the ground state
   b. end up in the ground state
   c. start in the level with $n = 2$
   d. end up in the level with $n = 2$

9. In the Bohr model of the atom, the orbits where electrons move fastest have
   a. the least energy
   b. the most energy
   c. the biggest radius
   d. the greatest angular momentum

10. The familiar yellow light from a sodium vapor street lamp results from the $3p \rightarrow 3s$ transition in Na. Determine the wavelength of the light given off if the energy difference $E_{3p} - E_{3p}s = 2.10$ eV.
    a. 560 nm
    b. 575 nm
    c. 590 nm
    d. 600 nm

11. A hydrogen atom in the ground state absorbs a 12.1-eV photon. To what level is the electron promoted? (The ionization energy of hydrogen is 13.6 eV.)
    a. $n = 2$
    b. $n = 3$
    c. $n = 4$
    d. $n = 5$

12. An energy of 13.6 eV is needed to ionize an electron. From the ground state of a hydrogen atom, what wavelength is needed if a photon accomplishes this task?
    a. 60 nm
    b. 70 nm
    c. 80 nm
    d. 90 nm

29

Nuclear Physics

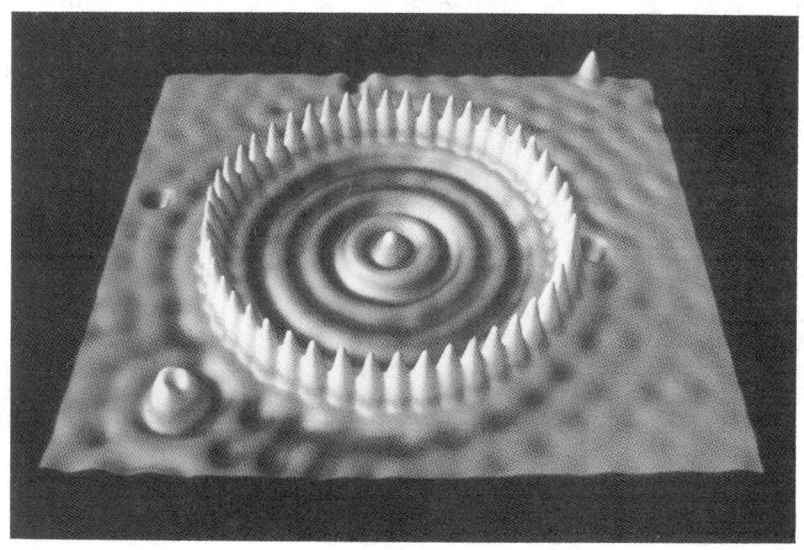

Chapter 29

NUCLEAR PHYSICS

In 1896, the year that marks the birth of nuclear physics, Henri Becquerel (1852-1908) discovered radioactivity in uranium compounds. Pioneering work by Rutherford showed that the radiation was of three types, which he called alpha, beta, and gamma rays. These types are classified according to the nature of their electric charge and according to their ability to penetrate matter.

In 1911 Rutherford and his students, Geiger and Marsden, performed a number of important scattering experiments which established that the nucleus of an atom can be regarded as essentially a point mass and point charge and that most of the atomic mass is contained in the nucleus. Furthermore, such studies demonstrated a wholly new type of force, the *nuclear force,* which is predominant at distances of less than about 10^{-14} m and zero at great distances.

In this chapter we discuss the properties and structure of the atomic nucleus. We start by describing the basic properties of nuclei and follow with a discussion of the phenomenon of radioactivity. Finally, we explore nuclear reactions and the various processes by which nuclei decay.

NOTES FROM SELECTED CHAPTER SECTIONS

29.1 Some Properties of Nuclei

Important quantities in the description of nuclear properties are:

1. The *atomic number,* Z, which equals the number of protons in the nucleus.
2. The *neutron number,* N, which equals the number of neutrons in the nucleus.
3. The *mass number,* A, which equals the number of nucleons (neutrons plus protons) in the nucleus.

The nuclei of all atoms of a particular element contain the same number of protons but often contain different numbers of neutrons. Nuclei that are related in this way are called *isotopes.* The isotopes of an element have the same Z value but different N and A values.

The *atomic mass unit,* u, is defined such that the mass of the isotope ^{12}C is exactly 12 u.

Experiments have shown that most nuclei are approximately spherical and all *have nearly the same density.* The stability of nuclei is due to the *nuclear force.* This is a *short range, attractive* force which acts between all nuclear particles.

Nuclei have *intrinsic angular momentum* which is quantized by the *nuclear spin quantum number* which may be integer or half integer.

The *magnetic moment of the nucleus* is measured in terms of the *nuclear magneton*. When placed in an external magnetic field, nuclear magnetic moments precess with a frequency called the *Larmor precessional frequency.*

29.2 Binding Energy

The total mass of a nucleus is always less than the sum of the masses of its individual nucleons. The *binding energy* of the nucleus is mass difference multiplied by c^2.

29.3 Radioactivity

There are three processes by which a radioactive substance can undergo decay: alpha (α) decay, where the emitted particles are ^{4}He nuclei; beta (β) decay, in which the emitted particles are either electrons or positrons; and gamma (γ) decay, in which the emitted "rays" are high-energy photons. A positron is a particle similar to the electron in all respects except that it has a charge of $+e$ (the antimatter twin of the electron). The symbol β^- is used to designate an electron, and β^+ designates a positron.

The three types of radiation have quite different penetrating powers. Alpha particles barely penetrate a sheet of paper, beta particles can penetrate a few millimeters of aluminum, and gamma rays can penetrate several centimeters of lead.

29.4 The Decay Processes

Alpha decay can occur because, according to quantum mechanics, some nuclei have barriers that can be penetrated by the alpha particles (the tunneling process). This process is energetically more favorable for those nuclei having a large excess of neutrons. A nucleus can undergo beta decay in two ways. It can emit either an electron (β^-) and an antineutrino ($\overline{\nu}$) or a positron (β^+) and a neutrino (ν). In the electron-capture process, the nucleus of an atom absorbs one of its own electrons (usually from the K shell) and emits a neutrino.

The neutrino has the following properties:
1. It has zero electric charge.

2. It has a rest mass smaller than that of the electron; and, in fact, its mass may be zero (although recent experiments suggest that this may not be true).
3. It has a spin of $\frac{1}{2}$, which satisfies the law of conservation of angular momentum.
4. It interacts very weakly with matter and is therefore very difficult to detect.

In gamma decay, a nucleus in an excited state decays to its ground state and emits a gamma ray. The Q value (disintegration energy) is the energy released as a result of the decay process.

29.5 Natural Radioactivity

Radioactive nuclei are generally classified into two groups: (1) unstable nuclei found in nature, which give rise to what is called *natural radioactivity*, and (2) nuclei produced in the laboratory through nuclear reactions, which exhibit *artificial radioactivity*.

29.6 Nuclear Reactions

Nuclear reactions are events in which collisions change the identity or properties of nuclei. The total energy released as a result of a nuclear reaction is called the *reaction energy*, Q.

An *endothermic reaction* is one in which Q is negative and the minimum energy for which the reaction will occur is called the *threshold energy*.

EQUATIONS AND CONCEPTS

Most nuclei are approximately spherical in shape and have an average radius which is proportional to the cube root of the mass number or total number of nucleons. This means that the volume is proportional to A and that all nuclei have nearly the same density.

$$r = r_0 A^{1/3} \quad (29.1)$$

The number of radioactive nuclei in a given sample which undergo decay during a time interval Δt depends on the number

$$\Delta N = -\lambda N \Delta t \quad (29.2)$$

of nuclei present. The number of decays depends also on the decay constant which is characteristic of a particular isotope.

The decay rate or activity of a sample of radioactive nuclei is defined as the number of decays per second.

$$R = \lambda N \tag{29.3}$$

Activity can be expressed in units of becquerels or curies.

$$1\ \text{Ci} \equiv 3.70 \times 10^{10}\ \text{decays/s} \tag{29.6}$$

$$1\ \text{Bq} = 1\ \text{decay/s} \tag{29.7}$$

The number of nuclei in a radioactive sample decreases exponentially with time. The plot of number of nuclei, N, versus elapsed time, t, is called a decay curve.

$$N = N_0 e^{-\lambda t} \tag{29.4}$$

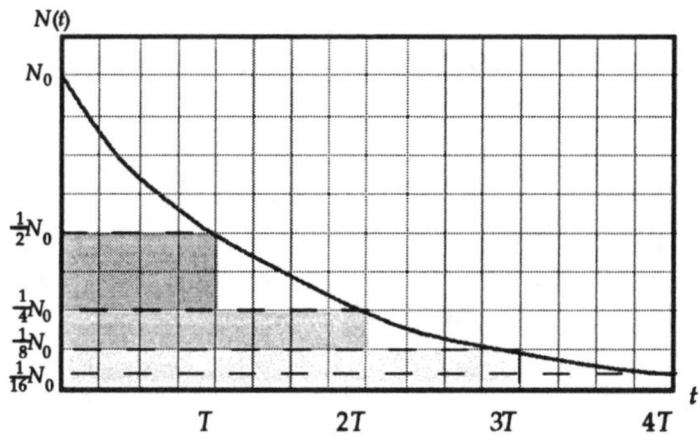

The half-life is the time required for half of a given number of radioactive nuclei to decay.

$$T_{1/2} = \frac{\ln 2}{\lambda} = \frac{0.693}{\lambda} \tag{29.5}$$

When a nucleus decays by alpha emission, the parent nucleus loses two neutrons and two protons.

$${}^{238}_{92}\text{U} \rightarrow {}^{234}_{90}\text{Th} + {}^{4}_{2}\text{He} \tag{29.8}$$

In order for alpha emission to occur, the mass of the parent nucleus must be greater than the combined mass of the daughter nucleus and the emitted alpha particle.

Comment on alpha decay.

The mass difference is converted into energy and appears as kinetic energy shared (unequally) by the alpha particle and the daughter nucleus.

When a radioactive nucleus undergoes beta decay, the daughter nucleus has the same mass number as the parent nucleus but the charge number (or atomic number) increases by one. The electron that is emitted is created within the parent nucleus by a process which can be represented by a neutron transformed into a proton and an electron.

$${}^{14}_{6}C \rightarrow {}^{14}_{7}N + {}^{0}_{-1}e \qquad (29.9)$$

$${}^{1}_{0}n \rightarrow {}^{1}_{1}p + {}^{0}_{-1}e \qquad (29.10)$$

The total energy released in beta decay is greater than the combined kinetic energies of the electron and the daughter nucleus. This difference in energy is associated with a third particle called a neutrino.

Comment on beta decay.

Nuclei which undergo alpha or beta decay are often left in an excited energy state. The nucleus returns to the ground state by emission of one or more photons. Gamma ray emission results in no change in mass number or atomic number.

$${}^{12}_{5}B \rightarrow {}^{12}_{6}C^{*} + {}^{0}_{-1}e \qquad (29.13)$$

$${}^{12}_{6}C^{*} \rightarrow {}^{12}_{6}C + \gamma \qquad (29.14)$$

Artificial transmutations can be induced by bombarding stable target nuclei with energetic particles.

$${}^{4}_{2}He + {}^{14}_{7}N \rightarrow {}^{17}_{8}O + {}^{1}_{1}H \qquad (29.16)$$

The quantity of energy required to balance the equation representing a nuclear reaction (e.g. Eq. 29.16) is called the Q value of the reaction. The Q value can be calculated in terms of the total mass of the reactants minus the total mass of the products or as the kinetic energy of the products minus the kinetic energy of the reactants. Q is positive in the case of exothermic reactions and negative for endothermic reactions.

Comment on energy conservation in nuclear reactions: Q values.

Endothermic reactions are characterized by a threshold energy (or minimum kinetic energy of the incoming particle) which is required in order for energy and momentum to be conserved. In this equation, m is the mass of the incident particle and M is the mass of the target particle.

$$KE_{\min} = \left(1 + \frac{m}{M}\right)|Q| \qquad (29.19)$$

SUGGESTIONS, SKILLS, AND STRATEGIES

The rest energy of a particle is given by $E = mc^2$. It is therefore often convenient to express the unified mass unit in terms of its energy equivalent, $1\ \text{u} = 1.660\,559 \times 10^{-27}$ kg or $1\ \text{u} = 931.50\ \text{MeV}/c^2$. When masses are expressed in units of u, energy values are then $E = m(931.50\ \text{MeV/u})$.

Equation 29.4 can be solved for the particular time t after which the number of remaining nuclei will be some specified fraction of the original number N_0. This can be done by taking the ln of each side of Equation 29.4 to find $t = \left(\frac{1}{\lambda}\right)\ln\left(\frac{N_0}{N}\right)$.

REVIEW CHECKLIST

- ▷ Account for nuclear binding energy in terms of the Einstein mass-energy relationship. Describe the basis for energy release by fission and fusion in terms of the shape of the curve of binding energy per nucleon vs. mass number.
- ▷ Identify each of the components of radiation that are emitted by the nucleus through natural radioactive decay and describe the basic properties of each. Write out typical equations to illustrate the processes of transmutation by alpha and beta decay.
- ▷ State, and apply to the solution of related problems, the formula which expresses decay rate as a function of decay constant and number of radioactive nuclei. Also apply the exponential formula which expresses the number of remaining radioactive nuclei left as a function of elapsed time, decay constant, or half-life, and the initial number of nuclei.
- ▷ Calculate the Q value of given nuclear reactions and determine the threshold energy of endothermic reactions.

SOLUTIONS TO SELECTED END-OF-CHAPTER PROBLEMS

1. Find the radius of a nucleus of (a) ${}^{4}_{2}\text{He}$ and (b) ${}^{238}_{92}\text{U}$.

Solution (a) $r = r_0 A^{1/3} = (1.20 \times 10^{-15} \text{ m})(4)^{1/3} = 1.90 \times 10^{-15} \text{ m}$ ◊

(b) $r = r_0 A^{1/3} = (1.20 \times 10^{-15} \text{ m})(238)^{1/3} = 7.44 \times 10^{-15} \text{ m}$ ◊

---

7. Consider the hydrogen atom to be a sphere of radius equal to the Bohr radius, 0.53×10^{-10} m, and calculate the approximate value of the ratio of the nuclear density to the atomic density.

Solution The mass of the hydrogen atom is approximately equal to the mass of the proton which is 1.67×10^{-27} kg. If the radius of the atom is 0.530×10^{-10} m, the volume is

$$V = \frac{4}{3}\pi r^3 = 6.24 \times 10^{-31} \text{ m}^3$$

Therefore, the density of the atom is

$$\rho_a = \frac{m}{V} = \frac{1.67 \times 10^{-27} \text{ kg}}{6.24 \times 10^{-31} \text{ kg/m}^3} = 2.68 \times 10^3 \text{ kg/m}^3$$

and the ratio of the nuclear density to the atomic density is

$$\frac{\rho_n}{\rho_a} = \frac{2.30 \times 10^{17} \text{ kg/m}^3}{2.68 \times 10^3 \text{ kg/m}^3} = 8.60 \times 10^{13} \quad ◊$$

---

9. Use energy methods to calculate the distance of closest approach for a head-on collision between an alpha particle with an initial energy of 0.50 MeV and a gold nucleus $\left({}^{197}\text{Au}\right)$ at rest. Assume the gold nucleus remains at rest during the collision.

Solution We have $PE)_f = KE)_i$ or $\dfrac{kQ_1Q_2}{r_{min}} = 0.500 \text{ MeV}$

For an alpha particle, $Q = 2e,$ and for gold, $Q = 79e.$

So, $$\frac{k(2e)(79e)}{r_{min}} = (0.50)(1.60\times10^{-13}\text{ J}) \quad \text{or} \quad r_{min} = \frac{k(2e)(79e)}{(0.500)(1.60\times10^{-13}\text{ J})}$$

$$r_{min} = \frac{(9.00\times10^{9}\text{ N}\cdot\text{m}^2/\text{C}^2)(158)(1.60\times10^{-19}\text{ C})^2}{(0.50)(1.60\times10^{-13}\text{ J})} = 4.60\times10^{-13}\text{ m} \quad \Diamond$$

---

13. An α particle (Z = 2, mass 6.64×10^{-27} kg) approaches to within 1.00×10^{-14} m of a carbon nucleus (Z = 6). What is the (a) maximum Coulomb force on the α particle, (b) acceleration of the α particle at this point, and (c) potential energy of the α particle at this point?

Solution (a) $$F = k\frac{Q_1Q_2}{r^2} = (9.00\times10^{9}\text{ N}\cdot\text{m}^2/\text{C}^2)\frac{(2)(6)(1.60\times10^{-19}\text{ C})^2}{(1.00\times10^{-14}\text{ m})^2} = 27.6\text{ N} \quad \Diamond$$

(b) $$a = \frac{F}{m} = \frac{27.6\text{ N}}{6.64\times10^{-27}\text{ kg}} = 4.16\times10^{27}\text{ m/s}^2 \quad \Diamond$$

(c) $$PE = k\frac{Q_1Q_2}{r} = (9.00\times10^{9}\text{ N}\cdot\text{m}^2/\text{C}^2)\frac{(2)(6)(1.60\times10^{-19}\text{ C})^2}{(1.00\times10^{-14}\text{ m})} \quad \text{or}$$

$$PE = 2.76\times10^{-13}\text{ J} = 1.73\text{ MeV} \quad \Diamond$$

---

21. Two isotopes having the same mass number are known as isobars. Calculate the difference in binding energy per nucleon for the isobars ${}^{23}_{11}\text{Na}$ and ${}^{23}_{12}\text{Mg}$. How do you account for this difference?

Solution For ${}^{23}_{11}\text{Na}$, $Z = 11$ and $A - Z = 12$

Thus, $$E_b = (11m_H + 12m_n - 22.989\,770\,\mu)(931.5\text{ MeV}/\mu)$$

$$E_b = (0.200\,285\,\mu)(931.5\text{ MeV}/\mu) = 186.6\text{ MeV} \quad \text{and}$$

$$\frac{E_b}{A} = 8.112 \text{ MeV/nucleon}$$

For ${}^{23}_{12}\text{Mg}$, $Z = 12$ and $A - Z = 11$

Thus, $$E_b = (12m_H + 11m_n - 22.994\,127\,\mu)(931.5 \text{ MeV}/\mu)$$

$$E_b = (0.195\,088\,\mu)(931.5 \text{ MeV}/\mu) = 181.7 \text{ MeV}$$ and

$$\frac{E_b}{A} = 7.901 \text{ MeV/nucleon}$$

Thus, $$\Delta\left(\frac{E_b}{A}\right) = 0.211 \text{ MeV/nucleon} \ \Diamond$$

The difference is mostly due to increased Coulomb repulsion, due to the extra proton in ${}^{23}_{12}\text{Mg}$.

---

27. Suppose that you start with 1.00×10^{-3} g of a pure radioactive substance and 2.0 h later determine that only 0.25×10^{-3} g of the substance remains. What is the half-life of this substance?

Solution Using $M = M_0 e^{-\lambda t}$, we have $e^{-\lambda t} = \dfrac{M}{M_0}$

or (taking logarithm of each side of equation), $-\lambda t = \ln\left(\dfrac{M}{M_0}\right)$ and $\lambda = -\dfrac{1}{t}\ln\left(\dfrac{M}{M_0}\right)$

so $\lambda = -\dfrac{1}{2h}\ln\left(\dfrac{0.250 \times 10^{-3} \text{ g}}{10^{-3} \text{ g}}\right) = 0.693 \text{ h}^{-1}$ and

$$T_{1/2} = \frac{\ln 2}{\lambda} = \frac{0.693}{0.693 \text{ h}^{-1}} = 1.00 \text{ h} \ \Diamond$$

---

33. Tritium has a half-life of 12.33 years. What percentage of the nuclei in a tritium sample will decay in five years?

Solution $\lambda = \frac{0.693}{T_{1/2}} = \frac{0.693}{12.33\text{ y}} = 5.62 \times 10^{-2}\text{ y}^{-1}$

At $t = 5.00$ y, $\frac{N}{N_0} = e^{-\lambda t} = e^{-(5.62 \times 10^{-2}\text{ y}^{-1})(5.00\text{ y})} = 0.755$

$\frac{N}{N_0}$ is the fraction of the original which will *remain* after 5.00 y.

Thus, the fraction which has *decayed* = $1 - 0.755 = 0.245$, or the % decayed = 24.5 %. ◊

---

35. A building has become accidentally contaminated with radioactivity. The longest-lived material in the building is strontium-90 ($^{90}_{38}$Sr, atomic mass 89.9077). If the building initially contained 5.00 kg of this substance and the safe level is less than 10.0 counts/min, how long will the building be unsafe?

Solution $N_0 = \frac{\text{mass present}}{\text{mass of nucleus}} = \frac{5.00\text{ kg}}{(89.9077\ \mu)(1.66 \times 10^{-27}\text{ kg}/\mu)}$

$N_0 = 3.35 \times 10^{25}$ nuclei

$\lambda = \frac{0.693}{T_{1/2}} = \frac{0.693}{(28.8\text{ y})} = 2.4063 \times 10^{-2}\text{ y}^{-1} = 4.575 \times 10^{-8}\text{ min}^{-1}$

(Half-life value is taken from Appendix B.)

$R_0 = \lambda N_0 = (4.575 \times 10^{-8}\text{ min}^{-1})(3.35 \times 10^{25}) = 1.533 \times 10^{18}$ counts/min

So, $\frac{R}{R_0} = e^{-\lambda t} = \frac{10.0}{1.533 \times 10^{18}\text{ counts/min}} = 6.525 \times 10^{-18}$ and

$\lambda t = \ln(6.525 \times 10^{-18}) = 39.57$ giving $t = 1.65 \times 10^{3}$ y ◊

---

39. Find the energy released in the alpha decay of ${}^{238}_{92}\text{U}$. The following mass values will be useful:

$$M\left({}^{238}_{92}\text{U}\right) = 238.050\,786 \text{ u}$$

$$M\left({}^{234}_{90}\text{Th}\right) = 234.043\,583 \text{ u}$$

$$M\left({}^{4}_{2}\text{He}\right) = 4.002\,603 \text{ u}$$

Solution ${}^{238}_{92}\text{U} \to {}^{4}_{2}\text{He} + {}^{234}_{90}\text{Th}$

$E = (\Delta m)c^2 = \left[(M_{238\text{U}}) - (M_{4\text{He}} + M_{234\text{Th}})\right](931.5 \text{ MeV}/\mu)$ or

$E = [238.050\,786\ \mu - (4.002\,602\ \mu + 234.043\,583\ \mu)](931.5 \text{ MeV}/\mu)$ giving

$E = (0.0046\ \mu)(931.5 \text{ MeV}/\mu) = 4.28 \text{ MeV}$ ◊

---

41. Determine which of the following suggested decays can occur spontaneously:

(a) ${}^{40}_{20}\text{Ca} \to {}^{0}_{1}\text{e} + {}^{40}_{19}\text{K}$ and (b) ${}^{144}_{60}\text{Nd} \to {}^{4}_{2}\text{He} + {}^{140}_{58}\text{Ce}$

Solution (a) ${}^{40}_{20}\text{Ca} \to {}^{0}_{+1}\text{e} + {}^{40}_{19}\text{K}$

For positron decay,

$$\Delta m = M_p - M_d - 2m_e = 39.962\,591\ \mu - 39.964\,00\ \mu - 2(0.000\,549\ \mu)$$

$$\Delta m < 0$$

Hence, the decay cannot occur spontaneously. ◊

(b) ${}^{144}_{60}\text{Nd} \to {}^{4}_{2}\text{He} + {}^{140}_{58}\text{Ce}$

For alpha decay,

$$\Delta m = M_p - (M_d + m_\alpha) = 143.910\,096\ \mu - (139.905\,44\ \mu + 4.002\,603\ \mu)$$ or

$$\Delta m = 2.053 \times 10^{-3}\ \mu.$$

The mass difference is greater than zero, so the decay can occur spontaneously. ◊

---

45. A piece of charcoal used for cooking is found at the remains of an ancient campsite. A 1.00-kg sample of carbon from the wood has an activity of 2.00×10^{-3} decays per minute. Find the age of the charcoal.

Solution The original activity of the sample is R_0 and the present activity is R.

$$R = R_0 e^{-\lambda t}$$

Taking logarithm of each side of equation,

$$t = -\frac{1}{\lambda}\ln\left(\frac{R}{R_0}\right) \qquad (1)$$

$$R_0 = (1000\ \text{g})\left(15\frac{\text{decays}}{\text{min}\cdot\text{g}}\right) = 15000\ \frac{\text{decays}}{\text{min}}$$

$$R = 2000\frac{\text{decays}}{\text{min}}$$

$$\lambda = \frac{\ln 2}{T_{1/2}} = \frac{0.693}{3.01\times 10^{9}\ \text{min}} = 2.30\times 10^{-10}\ \text{min}^{-1}$$

Therefore, the above values in Equation (1),

$$t = \left(-\frac{1}{2.30\times 10^{-10}\ \text{min}^{-1}}\right)\ln\left(\frac{2000}{15000}\right) = 8.76\times 10^{9}\ \text{min}$$

or $\quad t = 1.66\times 10^{4}\ \text{y}$ ◊

---

49. (a) Determine the product of the reaction ${}^{7}_{3}\text{Li} + {}^{4}_{2}\text{He} \rightarrow \,?\, + \text{n}$.
(b) What is the Q value of the reaction?

Solution (a) ${}^{7}_{3}\text{Li} + {}^{4}_{2}\text{He} \rightarrow {}^{10}_{5}\text{B} + {}^{1}_{0}\text{n}$ ◊

(b) $\Delta m = (M_{\text{Li}} + m_{\alpha}) - (M_{B} + m_{n})$ or

$\Delta m = (7.016\,005\ \mu + 4.002\,603\ \mu) - (10.012\,938\ \mu + 1.008\,665\ \mu)$ giving

$\Delta m = -0.002\,995\ \mu = -4.97 \times 10^{-30}\ \text{kg}$ and

$Q = \Delta mc^2 = -4.475 \times 10^{-13}\ \text{J} = -2.80\ \text{MeV}$ ◊

---

55. A beam of 6.61 MeV protons is incident on a target of ${}^{27}_{13}\text{Al}$. Those that collide produce the reaction

$$\text{p} + {}^{27}_{13}\text{Al} \rightarrow {}^{27}_{14}\text{Si} + \text{n}$$

(${}^{27}_{14}\text{Si}$ has a mass of 26.986 721 u.) Neglect any recoil of the product nucleus and determine the kinetic energy of the emerging neutrons.

Solution Neglect recoil of product nucleus. (i.e., Do not require momentum conservation). Then energy balance gives

$$KE_{\text{emerging}} = KE_{\text{incident}} + Q$$

To find Q, $\Delta m = (m_p + M_{\text{Al}}) - (M_{\text{Si}} + m_n)$ gives

$\Delta m = (1.007\,825\ \mu + 26.981\,541\ \mu) - (26.986\,721\ \mu + 1.008\,665\ \mu)$ or

$\Delta m = -6.02 \times 10^{-3}\ \mu = -9.993 \times 10^{-30}\ \text{kg}$ and

$Q = \Delta mc^2 = -8.994 \times 10^{-13}\ \text{J} = -5.621\ \text{MeV}$

Thus, $KE_{\text{emerging}} = 6.61\ \text{MeV} - 5.621\ \text{MeV} = 0.989\ \text{MeV}$ ◊

---

59. Find the threshold energy that the incident neutron must have to product the reaction

$$^1_0\text{n} + ^4_2\text{He} \rightarrow ^2_1\text{H} + ^3_1\text{H}$$

Solution $^1_0\text{n} + ^4_2\text{H} \rightarrow ^2_1\text{H} + ^3_1\text{H}$

The Q value is

$$Q = \left[(M_n + M_{\text{He}}) - (M_{2\text{H}} + M_{3\text{H}})\right](931.5 \text{ Mev}/\mu) \quad \text{or}$$

$$Q = \left[(1.008\,665\,\mu + 4.002\,603\,\mu) - (2.014\,102\,\mu + 3.016\,049\,\mu)\right](931.5 \text{ MeV}/\mu)$$

$$\text{Q} = (-0.018\,883\,\mu)(931.5 \text{ MeV}/\mu) = -17.6 \text{ MeV}$$

Thus, $$KE_{\text{min}} = \left(1 + \frac{m_{\text{incident projectile}}}{m_{\text{target nucleus}}}\right)|Q| = \left(1 + \frac{1.008\,665}{4.002\,603}\right)(17.6 \text{ MeV})$$

or $$KE_{\text{min}} = 22.0 \text{ MeV} \quad \lozenge$$

---

65. A medical laboratory stock solution is prepared with an initial activity due to ^{24}Na of 2.50 mCi/ml, and 10.0 ml of the stock solution is diluted at $t_0 = 0$ to a working solution whose total volume is 250 ml. After 48 h, a 5.00-ml sample of the working solution is monitored with a counter. What is the measured activity?
(*Note* that 1 ml = 1 milliliter.)

Solution Let R_0 equal the total activity withdrawn from the stock solution.

$$R_0 = (2.50 \text{ mCi/ml})(10 \text{ ml}) = 25.0 \text{ mCi}$$

Let R_0' equal the initial activity per unit volume of the working solution.

$$R_0' = \frac{25.0 \text{ mCi}}{250 \text{ ml}} = 0.100 \text{ mCi/ml}$$

After 48 hours the activity per unit volume of the working solution will be

$$R' = R_0' e^{-\lambda t} = (0.100 \text{ mCi/ml}) e^{-(0.693/15 \text{ h})(48.0 \text{ h})} = 0.011 \text{ mCi/ml}$$

and the activity in the sample will be

$$R = (0.011 \text{ mCi/ml})(5.00 \text{ ml}) = 0.055 \text{ mCi} \quad \lozenge$$

---

67. A fission reactor is hit by a nuclear weapon, causing 5.00×10^6 Ci of ^{90}Sr ($T_{1/2} = 28.7$ years) to evaporate into the air. The ^{90}Sr falls out over an area of 10^4 km^2. How long will it take the activity of the ^{90}Sr to reach the agriculturally "safe" level of 2.00 μCi/m^2?

Solution $R = R_0 e^{-\lambda t}$ so (take logarithm of each side of equation), $t = -\frac{1}{\lambda}\ln\left(\frac{R}{R_0}\right)$

The original activity per unit area is

$$R_0 = \frac{5.00 \times 10^6 \text{ Ci}}{10^{10} \text{ m}^2} = 5.00 \times 10^{-4} \text{ Ci/m}^2$$

and the final desired activity is

$$R = 2.00 \times 10^{-6} \text{ Ci/m}^2$$

The decay constant is

$$\lambda = \frac{0.693}{T_{1/2}} = \frac{0.693}{28.7 \text{ y}} = 2.414 \times 10^{-2} \text{ y}^{-1}$$

Therefore,

$$t = \left(-\frac{1}{2.414 \times 10^{-2} \text{ y}^{-1}}\right)\ln\left(\frac{2.00 \times 10^{-6}}{5.00 \times 10^{-4}}\right) \quad \text{or} \quad t = 299 \text{ y} \quad \lozenge$$

---

71. Many radioisotopes have important industrial, medical, and research applications. One of these is ^{60}Co, which has a half-life of 5.2 years and decays by the emission of a beta particle (energy 0.31 MeV) and two gamma photons (energies 1.17 MeV and 1.33 MeV). A scientist wishes to prepare a ^{60}Co sealed source that will have an activity of at least 10 Ci after 30 months of use. What is the minimum initial mass of ^{60}Co required?

Solution

$R = 10.0$ Ci, so

$$R = 10.0\ \text{Ci}(3.70 \times 10^{10}\ \text{decays/s/Ci}) = 3.70 \times 10^{11}\ \text{decays/s}$$

$$\lambda = \frac{\ln 2}{T_{1/2}} = \frac{0.693}{5.27\ \text{y}} = 0.1315\ \text{y}^{-1} = 4.16 \times 10^{-9}\ \text{s}^{-1}$$

Use $R = R_0 e^{-\lambda t}$

Therefore, $R_0 = Re^{+\lambda t}$ and when $t = 2.50$ y,

$$R_0 = \left(3.70 \times 10^{11}\ \frac{\text{decays}}{\text{s}}\right) e^{(0.1315\ \text{y}^{-1})(2.50\ \text{y})} = 5.14 \times 10^{11}\ \frac{\text{decays}}{\text{s}}$$

The initial number of nuclei, $N_0 = \dfrac{R_0}{\lambda}$

$$N_0 = \frac{5.14 \times 10^{11}\ \text{decays/s}}{4.16 \times 10^{-9}\ \text{s}^{-1}} = 1.236 \times 10^{20}\ \text{decays (or nuclei)}$$

Initial mass of ^{60}Co, M = (mass per nucleus)(number of nuclei)

$$M = mN_0 = (60\ \mu/\text{nucleus})(1.66 \times 10^{-27}\ \text{kg}/\mu)(1.236 \times 10^{20}\ \text{nuclei})$$

$$M = 1.23 \times 10^{-5}\ \text{kg} = 12.3\ \text{mg} \quad \Diamond$$

CHAPTER SELF-QUIZ

1. A radioactive material initially is observed to have an activity of 800 counts/sec. If four hours later it is observed to have an activity of 200 counts/sec, what is its half-life?
   a. 1 hour
   b. 2 hours
   c. 4 hours
   d. 8 hours

2. Radium-226 decays to Radon-222 by emitting which of the following?
   a. beta
   b. alpha
   c. gamma
   d. positron

3. An endothermic nuclear reaction occurs as a result of the collision of two reactant nuclei. If the Q value of this reaction is –2.17 MeV, which of the following describes the minimum kinetic energy needed in the reactant nuclei if the reaction is to occur?
   a. equal to 2.17 MeV
   b. greater than 2.17 MeV
   c. less than 2.17 MeV
   d. exactly half of 2.17 MeV

4. The ratio of the numbers of neutrons to protons in the nucleus of naturally occurring isotopes tends to vary with atomic number in what manner?
   a. increases with greater atomic number
   b. decreases with greater atomic number
   c. is maximum for atomic number = 60
   d. remains constant for entire range of atomic numbers

5. The beta emission process results in the daughter nucleus differing in what manner from the parent?
   a. atomic mass increases by one
   b. atomic number decreases by two
   c. atomic number increases by one
   d. atomic mass decreases by two

6. What energy must be added or given off in a reaction where one hydrogen atom and two neutrons are combined to form a tritium atom? (atomic masses for each: hydrogen, 1.007 825; neutron, 1.009 665; tritium, 3.016 049; also, $1\ \text{u} \times c^2 = 931.5\ \text{MeV}$)
   a. 8.50 MeV added
   b. 8.50 MeV given off
   c. 10.3 MeV given off
   d. 10.3 MeV added

7. Tritium has a half-life of 12.3 years. How many years will elapse when the radioactivity of a tritium sample diminishes to 10% of its original value?
   a. 31 years
   b. 41 years
   c. 84 years
   d. 123 years

8. If there are 128 neutrons in Pb-210, how many neutrons are found in the nucleus of Pb-206?
   a. 122
   b. 124
   c. 126
   d. 130

9. What particle is emitted when P-32 decays to S-32? (Atomic numbers of P and S are, respectively, 15 and 16.)
   a. alpha
   b. beta
   c. positron
   d. gamma quantum

10. The binding energy of a nucleus is equal to
    a. the energy needed to remove one of the nucleons
    b. the average energy with which any nucleon is bound in the nucleus
    c. the energy needed to separate all the nucleons from each other
    d. the mass of the nucleus times c^2.

11. The half-life of radioactive iodine-137 is 8 days. Find the number of ^{131}I nuclei necessary to produce a sample of activity 1.0 μCi.
    (1 Curie = 3.70×10^{10} decays/second)
    a. 4.60×10^9
    b. 3.70×10^{10}
    c. 7.60×10^{12}
    d. 8.10×10^{13}

12. What is the Q-value for the reaction $^9\text{Be} + \alpha \rightarrow {}^{12}\text{C} + \text{n}$?
    ($m_\alpha = 4.0026$ u, $m_{\text{Be}} = 9.01218$ u, $m_{\text{C}} = 12.0000$ u, $m_n = 1.008\,665$ u, and $1\text{ u} = 931.5\text{ MeV}/c^2$)
    a. 8.40 MeV
    b. 7.30 MeV
    c. 6.20 MeV
    d. 5.70 MeV

30
Nuclear Energy and Elementary Particles

NUCLEAR ENERGY AND ELEMENTARY PARTICLES

In this concluding chapter we discuss the two means by which energy can be derived from nuclear reactions. These two techniques are fission, in which a nucleus of large mass number splits, or fissions, into two smaller nuclei; and fusion, in which two light nuclei fuse to form a heavier nucleus. In either case, there is a release of large amounts of energy.

We end our study of physics by examining the known subatomic particles and the fundamental interactions that govern their behavior. We also discuss the current theory of elementary particles, which states that all matter in nature is constructed from only two families of particles: quarks and leptons. Finally, we describe how clarifications of such models might help scientists understand the evolution of the Universe.

NOTES FROM SELECTED CHAPTER SECTIONS

30.1 Nuclear Fission

Nuclear fission occurs when a heavy nucleus, such as ^{235}U, splits, or fissions, into two smaller nuclei. In such a reaction, *the total rest mass of the products is less than the original rest mass.*

The sequence of events in the fission process is

1. The ^{235}U nucleus captures a thermal (slow-moving) neutron.

2. This capture results in the formation of $^{236}U^*$, and the excess energy of this nucleus causes it to undergo violent oscillations.

3. The $^{236}U^*$ nucleus becomes highly distorted, and the force of repulsion between protons in the two halves of the dumbbell shape tends to increase the distortion.

4. The nucleus splits into two fragments, emitting several neutrons in the process.

30.2 Nuclear Reactors

A nuclear reactor is a system designed to maintain a *self-sustained chain reaction.*

The *reproduction constant* K is defined as the average number of neutrons released from each fission event that will cause another event. In a power reactor, it is necessary to maintain a value of K close to 1. Under this condition, the reactor is said to be *critical.*

Important factors and processes relative to the design and operation of a nuclear reactor include: neutron leakage, regulating neutron energies, neutron capture by nonfissioning nuclei, control of the power level, and provisions to ensure reactor safety.

30.3 Nuclear Fusion

Nuclear fusion is a process in which two light nuclei combine to form a heavier nucleus. A great deal of energy is released in such a process. The major obstacle in obtaining useful energy from fusion is the large Coulomb repulsive force between the charged nuclei at close separations. Sufficient energy must be supplied to the particles to overcome this Coulomb barrier and thereby enable the nuclear attractive force to take over.

The temperature at which the power generation exceeds the loss rate is called the *critical ignition temperature.* The confinement time is the time the interacting ions are maintained at a temperature equal to or greater than the ignition temperature.

30.5 The Fundamental Forces in Nature

There are four fundamental forces in nature: *strong* (hadronic), *electromagnetic, weak,* and *gravitational.* The strong force is the force between nucleons that keeps the nucleus together. The weak force is responsible for beta decay. The electromagnetic and weak forces are now considered to be manifestations of a single force called the *electroweak* force.

The fundamental forces are described in terms of particle or quanta exchanges which *mediate* the forces. The electromagnetic force is mediated by photons, which are the quanta of the electromagnetic field. Likewise, the strong force is mediated by field particles called *gluons,* the weak force is mediated by particles called the W and Z *bosons,* and the gravitational force is mediated by quanta of the gravitational field called *gravitons.*

30.6 Positrons and Other Antiparticles

An antiparticle and a particle have the same mass, but opposite charge. Furthermore, other properties may have opposite values such as lepton number and baryon number. It is possible to produce particle-antiparticle pairs in nuclear reactions if the available energy is greater than $2mc^2$, where m is the rest mass of the particle (or antiparticle).

Pair production is a process in which a gamma ray with an energy of at least 1.02 MeV interacts with a nucleus and an electron-positron pair is created.

Pair annihilation is an event in which an electron and a positron can annihilate to produce two gamma rays, each with an energy of at least 0.511 MeV.

30.7 Mesons and the Beginning of Particle Physics

The interaction between two particles can be represented in a diagram called a *Feynman diagram.*

30.8 Classification of Particles

All particles (other than photons) can be classified into two categories: *hadrons* and *leptons.*

There are two classes of hadrons: *mesons* and *baryons,* which are grouped according to their masses and spins. It is believed that hadrons are composed of units called *quarks* which are more fundamental in nature.

Leptons have no structure or size and are therefore considered to be truly elementary particles.

30.9 Conservation Laws

In all reactions and decays, quantities such as energy, linear momentum, angular momentum, electric charge, baryon number, and lepton number are strictly conserved. Certain particles have properties called *strangeness* and *charm.* These unusual properties are conserved only in those reactions and decays that occur via the strong force.

Whenever a nuclear reaction or decay occurs, the sum of the baryon numbers before the process must equal the sum of the baryon numbers after the process.

The sum of the electron-lepton numbers before a reaction or decay must equal the sum of the electron-lepton numbers after the reaction or decay.

30.10 Strange Particles and Strangeness

Whenever a nuclear reaction or decay occurs, the sum of the strangeness numbers before the process must equal the sum of the strangeness numbers after the process.

30.12 Quarks—Finally

Recent theories in elementary particle physics have postulated that all hadrons are composed of smaller units known as *quarks*. Quarks have a fractional electric charge and a baryon number of 1/3. There are six flavors of quarks, up (u), down (d), strange (s), charmed (c), top (t), and bottom (b). All baryons contain three quarks, while all mesons contain one quark and one antiquark.

According to the theory of *quantum chromadynamics*, quarks have a property called *color*, and the strong force between quarks is referred to as the *color force.*

EQUATIONS AND CONCEPTS

The fission of an uranium nucleus by bombardment with a low energy neutron results in the production of fission fragments and typically two or three neutrons. The energy released in the fission event appears in the form of kinetic energy of the fission fragments and the neutrons.

$$^{1}_{0}\text{n} + ^{235}_{92}\text{U} \rightarrow ^{236}_{92}\text{U}^{*} \rightarrow \text{X} + \text{Y} + \text{neutrons} \tag{30.1}$$

These fusion reactions seem to be most likely to be used as the basis of the design and operation of a fusion power reactor.

$$^{2}_{1}\text{H} + ^{2}_{1}\text{H} \rightarrow ^{3}_{2}\text{He} + ^{1}_{0}\text{n} \tag{30.4}$$

$$(Q = 3.27\text{ MeV})$$

$$^{2}_{1}\text{H} + ^{2}_{1}\text{H} \rightarrow ^{3}_{1}\text{H} + ^{1}_{1}\text{H}$$

$$(Q = 4.03\text{ MeV})$$

$$^{2}_{1}\text{H} + ^{3}_{1}\text{H} \rightarrow ^{4}_{2}\text{He} + ^{1}_{0}\text{n}$$

$$(Q = 17.59\text{ MeV})$$

Lawson's criterion states the conditions under which a net power output of a fusion reactor is possible. In these expressions, n is the plasma density (number of ions per cubic cm) and t is the plasma confinement time (the time during which the interacting ions are maintained at a temperature equal to or greater than that required for the reaction to proceed).

$$n\tau \geq 10^{14}\ \mathrm{s/cm^3} \qquad (30.5)$$

(D – T interaction)

$$n\tau \geq 10^{16}\ \mathrm{s/cm^3}$$

(D - D interaction)

Pions and muons are very unstable particles. Shown is the sequence of decays for π^-.

$$\pi^- \rightarrow \mu^- + \overline{\nu} \qquad (30.6)$$

$$\mu^- \rightarrow \mathrm{e} + \nu + \overline{\nu}$$

REVIEW CHECKLIST

▷ Write an equation which represents a typical fission event and describe the sequence of events which occurs during the fission process. Use data obtained from the binding energy curve to estimate the disintegration energy of a typical fission event.

▷ Describe the basis of energy release in fusion and write out several nuclear reactions which might be used in a fusion-powered reactor.

▷ Outline the classification of elementary particles, and mention several characteristics of each group.

▷ Know the broad classification of particles and the characteristic properties of the several classes (relative mass value, spin, decay mode).

▷ Determine whether or not a suggested decay can occur based on the conservation of baryon number and the conservation of lepton number. Determine whether or not a predicted reaction/decay will occur based on the conservation of strangeness for the strong and electromagnetic interactions.

SOLUTIONS TO SELECTED END-OF-CHAPTER PROBLEMS

1. Find the energy released in the following fission reaction:

$${}^{1}_{0}\mathrm{n} + {}^{235}_{92}\mathrm{U} \rightarrow {}^{144}_{56}\mathrm{Ba} + {}^{89}_{36}\mathrm{Kr} + 3\,{}^{1}_{0}\mathrm{n}$$

Solution $\,^1_0\text{n} + \,^{235}_{92}\text{U} \rightarrow \,^{144}_{56}\text{Ba} + \,^{89}_{36}\text{Kr} + 3\,^1_0\text{n}$

$$Q = (\Delta m)c^2 = [(M_n + M_U) - (M_{\text{Ba}} + M_{\text{Kr}} + 3m_n)](931.5\text{ MeV}/\mu)$$

$$Q = [235.043\,925\,\mu - 143.922\,673\,\mu - 88.917\,563\,\mu - 2(1.008\,665\,\mu)](931.5\text{ MeV}/\mu)$$

$$Q = (0.186\,359\,\mu)(931.5\text{ MeV}/\mu) = 174\text{ MeV} \quad \Diamond$$

---

7. How many grams of ^{235}U must undergo fission to operate a 1000-MW power plant for one day if the conversion efficiency is 30.0%? (Assume 208 MeV released per fission event.)

Solution The total energy output of the plant is

$$E = (\text{power})t = (1000\times10^6\text{ J/s})(86,400\text{ s}) = 8.64\times10^{13}\text{ J}$$

and the total energy which must be released is

$$E_r = \frac{E}{0.300} = \frac{8.64\times10^{13}\text{ J}}{0.300} = 2.88\times10^{14}\text{ J}$$

The number of events required is

$$N = \frac{E_r}{Q} = \frac{2.88\times10^{14}\text{ J}}{(208\text{ MeV/event})(1.60\times10^{-13}\text{ J/MeV})} = 8.65\times10^{24}\text{ events}$$

The number of moles of uranium required is

$$8.65\times10^{24}\text{ nuclei}\frac{1\text{ mol}}{6.02\times10^{23}\text{ nuclei}} = 14.4\text{ mol}$$

and the mass required is

$$m = (14.4\text{ mol})\frac{235\text{ g}}{\text{mol}} = 3.38\times10^3\text{ g} = 3.38\text{ kg} \quad \Diamond$$

---

11. Suppose that the water exerts an average frictional drag of 1.00×10^5 N on a nuclear-powered ship. How far can the ship travel per kilogram of fuel if the fuel consists of enriched uranium containing 1.70% of the fissionable isotope ^{235}U, and the ship's engine has an efficiency of 20.0%? (Assume 208 MeV released per fission event.)

Solution The mass of U-235 present in 1.00 kg of fuel is

$$M_{U235} = 0.017(1.00 \text{ kg}) = 1.70 \times 10^{-2} \text{ kg} = 17.0 \text{ g}$$

The number of U-235 atoms (and nuclei) present is

$$N = 17.0 \text{ g}\left(\frac{6.02 \times 10^{23} \text{ atoms/mol}}{235 \text{ g/mol}}\right) = 4.355 \times 10^{22} \text{ nuclei}$$

Thus, in using 1.00 kg of fuel, 4.355×10^{22} fissions occur with the release of 208 MeV per fission. The energy released is

$$E = 4.355 \times 10^{22} \text{ fissions}(208 \text{ MeV/fission})(1.60 \times 10^{-13} \text{ J/MeV}) = 1.45 \times 10^{12} \text{ J}$$

Energy used effectively = (energy released)(efficiency)

$$E_{\text{used}} = (1.45 \times 10^{12} \text{ J})(0.200) = 2.90 \times 10^{11} \text{ J}$$

Work done equals energy used, and $W = FS$.

Therefore, $$S = \frac{W}{F} = \frac{2.90 \times 10^{11} \text{ J}}{10^5 \text{ N}} = 2.90 \times 10^6 \text{ m}$$

$$S = 2.90 \times 10^3 \text{ km} = 1.80 \times 10^3 \text{ mi} \quad \Diamond$$

---

17. Of all the hydrogen nuclei in the ocean, 0.0156% are deuterium. (a) How much deuterium could be obtained from 1.00 gal of ordinary tap water? (b) If all of this deuterium could be converted to energy through reactions such as the first reaction in Equation 30.4, how much energy could be obtained from the 1.00 gal of water? (c) The energy released through the burning of 1.00 gal of gasoline is approximately 2.00×10^8 J. How many gallons of gasoline would have to be burned in order to produce the same amount of energy as was obtained from the 1.00 gal of water?

Solution (a) 1 gallon = 3786 cm^3 = 3786 g of water

The number of moles of water present is

$$n = \frac{3786 \text{ g}}{18.0 \text{ g/mol}} = 210.3 \text{ mol}$$

The number of water molecules present is

$$N = 210.3 \text{ mol}(6.02 \times 10^{23} \text{ molecules/mol}) = 1.27 \times 10^{26} \text{ molecules}$$

The number of hydrogen atoms is (diatomic molecule)

$$N_{\text{atoms}} = 2(1.27 \times 10^{26}) = 2.53 \times 10^{26} \text{ hydrogen atoms}$$

The number of deuterium nuclei is

$$N_{\text{deut}} = 1.56 \times 10^{-4}(2.53 \times 10^{26} \text{ hydrogen atoms}) \quad \text{or}$$

$$N_{\text{deut}} = 4.00 \times 10^{22} \text{ deuterium nuclei} \quad \Diamond$$

(b) The reaction, ${}^2_1\text{H} + {}^2_1\text{H} \rightarrow {}^3_2\text{He} + {}^1_0\text{n}$, has a Q value of 3.27 MeV

The number of reactions which can occur equals

$$\frac{1}{2} \text{ (number of deuterium nuclei)} = 1.98 \times 10^{22} \text{ reactions}$$

Thus, the total energy released is

$$E = (3.27 \text{ MeV/reaction})(1.98 \times 10^{22} \text{ reactions})(1.60 \times 10^{-13} \text{ J/MeV})$$

or $$E = 1.04 \times 10^{10} \text{ J} \quad \Diamond$$

(c) Number of gallons of gasoline required is

$$N_{\text{gal}} = \frac{1.03 \times 10^{10} \text{ J}}{2.00 \times 10^{8} \text{ J/gal}} = 52.0 \text{ gallons} \quad \Diamond$$

25. If a π^0 at rest decays into two γ's, what is the energy of each of the γ's?

Solution Total momentum before = 0 = total momentum after. Thus, the photons must go in opposite directions and have equal momenta, and hence equal energies since $E_\gamma = p_\gamma c$. Thus, from mass energy conservation,

$$E_0 = 2E_\gamma \text{ (where } E_0 \text{ is the rest energy of the } \pi^0\text{)} \quad \text{or}$$

$$E_\gamma = \frac{1}{2}E_0 = \frac{1}{2}(135 \text{ MeV}) = 67.5 \text{ MeV for each gamma ray} \quad \Diamond$$

---

27. The neutral ρ meson decays by the strong interaction into two pions according to $\rho^0 \to \pi^+ + \pi^-$, with a half-life of about 10^{-23} s. The neutral K meson also decays into two pions according to $K^0 \to \pi^+ + \pi^-$, but with a much longer half-life of about 10^{-10} s. How do you explain these observations?

Solution The $\rho^0 \to \pi^+ + \pi^-$ decay must occur via the strong interaction. ◊

The $K^0 \to \pi^+ + \pi^-$ decay must occur via the weak interaction. ◊

---

29. For the following two reactions, the first may occur but the second cannot. Explain.

(1) $K^0 \to \pi^+ + \pi^-$ (can occur)

(2) $\Lambda^0 \to \pi^+ + \pi^-$ (cannot occur)

Solution

(1) $K^0 \to \pi^+ + \pi^-$

Strangeness, $+1 \to 0 + 0$ Baryon number, $0 \to 0 + 0$

Lepton number, $0 \to 0$ Charge, $0 \to +1 - 1$

Does not violate any absolute conservation law, and violates strangeness by only 1 unit. Thus, it can occur via the weak interaction. ◊

(2) $\Lambda^0 \rightarrow \pi^+ + \pi^-$ Baryon number, $-1 \rightarrow 0 + 0$

Since all interactions conserve Baryon number, this process <u>cannot occur</u>. ◊

---

33. Fill in the missing particle. Assume that (a) occurs via the strong interaction while (b) and (c) involve the weak interaction.

(a) $K^+ + p \rightarrow \_\_ + p$ (b) $\Omega^- \rightarrow \_\_ + \pi^-$ (c) $K^+ \rightarrow \_\_ + \mu^+ + \nu_\mu$

Solution (a) $K^+ + p \rightarrow \underline{\ ?\ } + p$

The strong interaction conserves everything.

Baryon number, $0 + 1 \rightarrow B + 1$ so $B = 0$

Charge, $+1 + 1 \rightarrow Q + 1$ so $Q = +1$

Lepton numbers, $0 + 0 \rightarrow L + 0$ so $L_e = L_\mu = L_\tau = 0$

Strangeness, $+1 + 0 \rightarrow S + 0$ so $S = 1$

The conclusion is that the particle must be positively charged, a non-baryon, with strangeness of +1. Of particles in Table 30.2, it can only be the K^+. Thus, this is an elastic scattering process.

The weak interaction conserves all but strangeness, and $\Delta S = \pm 1$. ◊

(b) $\Omega^- \rightarrow \underline{\ ?\ } + \pi^-$

Baryon number, $+1 \rightarrow B + 0$ so $B = 1$

Charge, $-1 \rightarrow Q - 1$ so $Q = 0$

Lepton numbers, $0 \rightarrow L + 0$ so $L_e = L_\mu = L_\tau = 0$

Strangeness, $-3 \rightarrow S + 0$ so $\Delta S = 1$ and $S = -2$

The particle must be a neutral baryon with strangeness of –2. Thus, it is the Ξ^0. ◊

(c) $K^+ \rightarrow \underline{?} + \mu^+ + \nu_\mu$

Baryon number, $0 \rightarrow B + 0 + 0$ so $B = 0$

Charge, $+1 \rightarrow Q + 1 + 0$ so $Q = 0$ L_e, $0 \rightarrow L_e + 0 + 0$ so $L_e = 0$

L_μ, $0 \rightarrow L_\mu - 1 + 1$ so $L_\mu = 0$ L_τ, $0 \rightarrow L_\tau + 0 + 0$ so $L_\tau = 0$

Strangeness, $1 \rightarrow S + 0 + 0$ so $\Delta S = \pm 1$ (for weak interaction), so $S = 0$

The particle must be a neutral meson with strangeness $= 0 \Rightarrow \pi^0$. ◊

---

37. Determine the type of neutrino or antineutrino involved in each of the following processes:

(a) $\pi^+ \rightarrow \pi^0 + e^+ + ?$ (b) $? + p \rightarrow \mu^- + p + \pi^+$

(c) $\Lambda^0 \rightarrow p + \mu^- + ?$ (d) $\tau^+ \rightarrow \mu^+ + ? + ?$

Solution The relevant conservation laws are: $\Delta L_e = 0$, $\Delta L_\mu = 0$, $\Rightarrow \Delta L_\tau = 0$

(a) $\pi^+ \rightarrow \pi^0 + e^+ + ?$ L_e: $0 \rightarrow 0 - 1 + L_e$ so $L_e = 1$, $\Rightarrow$ we have a ν_e ◊

(b) $? + p \rightarrow \mu^- + p + \pi^+$ L_μ: $L_\mu + 0 \rightarrow +1 + 0 + 0$ so $L_\mu = 1$, $\Rightarrow$ we have a ν_μ ◊

(c) $\Lambda^0 \rightarrow p + \mu^- + ?$ L_μ: $0 \rightarrow 0 + 1 + L_\mu$ so $L_\mu = -1$, $\Rightarrow$ we have a $\bar{\nu}_\mu$ ◊

(d) $\tau^+ \rightarrow \mu^+ + ? + ?$ L_μ: $0 \rightarrow -1 + L_\mu$ so $L_\mu = 1$, $\Rightarrow$ we have a ν_μ ◊

L_τ: $+1 \rightarrow 0 + L_\tau$ so $L_\tau = 1$, $\Rightarrow$ we have a $\bar{\nu}_\tau$ ◊

Conclusion for (d): $L_\mu = 1$ for one particle, and $L_\tau = 1$ for the other particle.

We have ν_μ and $\bar{\nu}_\tau$ ◊

---

39. Two protons approach each other with equal and opposite velocities. What is the minimum kinetic energy of each of the protons if they are to produce a π^+ meson at rest in the reaction,

$$p + p \rightarrow p + n + \pi^+$$

Solution Total momentum before = 0 and they must have zero momentum after. Also, they have equal original energies. Since the π^+ is at rest, the p and n must go in opposite directions with equal magnitude momenta. For minimum energy of incident protons, the p and n will be at rest afterward. Mass-energy conservation gives

$$2E_p = E_{0p} + E_{0n} + E_{0\pi} = 938.3 \text{ MeV} + 939.6 \text{ MeV} + 139.6 \text{ MeV} \quad \text{or}$$

$$E_p = 1008.75 \text{ MeV} \quad \text{but} \quad E_p = KE + E_0, \quad \text{and}$$

$$KE = E_p - E_0 = 1008.75 - 938.3 \text{ MeV} = 70.5 \text{ MeV} \ \lozenge$$

---

41. The quark compositions of the K^0 and Λ^0 particles are $d\bar{s}$ and uds, respectively. Show that the charge, baryon number, and strangeness of these particles equal the sums of these numbers for the quark constituents.

Solution (a)

	K^0	d	$\bar{s}$	total
strangeness	1	0	1	1
baryon number	0	$\frac{1}{3}$	$-\frac{1}{3}$	0
charge	0	$-\frac{1}{3}e$	$\frac{1}{3}e$	0

(b)

	Λ^0	u	d	s	total
strangeness	−1	0	0	−1	−1
baryon number	1	$\frac{1}{3}$	$\frac{1}{3}$	$\frac{1}{3}$	1
charge	0	$\frac{2}{3}e$	$-\frac{1}{3}e$	$-\frac{1}{3}e$	0

43. What is the electrical charge of the baryons with the quark compositions (a) $\overline{uud}$ and (b) $\overline{udd}$? What are these baryons called?

Solution (a) $\overline{uud}$: $\text{charge} = \left(-\frac{2}{3}e\right) + \left(-\frac{2}{3}e\right) + \left(\frac{1}{3}e\right) = -e$

This is the antiproton. ◊

(b) $\overline{udd}$: $\text{charge} = \left(-\frac{2}{3}e\right) + \left(\frac{1}{3}e\right) + \left(\frac{1}{3}e\right) = 0$

This is the antineutron. ◊

---

53. Calculate the mass of ^{235}U required to provide the total energy requirements of a nuclear submarine during a 100-day patrol, assuming a constant power demand of 100000 kW and a conversion efficiency of 30%. (Assume that the average energy released per fission is 208 MeV.)

Solution The energy output required = (power)*t*, or

$$E_{\text{output}} = (10^8 \text{ J/s})(100 \text{ days})(86{,}400 \text{ s/day}) = 8.64 \times 10^{14} \text{ J}$$

The total energy input required is $\dfrac{8.64 \times 10^{14} \text{ J}}{0.300} = 2.88 \times 10^{15} \text{ J}$

Assuming an average of 208 MeV per fission of U-235, the number of fission events, or nuclei required, is

$$N = \frac{2.88 \times 10^{15} \text{ J}}{(208 \text{ MeV/fission})(1.60 \times 10^{-13} \text{ J/MeV})} = 8.65 \times 10^{25} \text{ nuclei}$$

The number of moles used = $\dfrac{8.65 \times 10^{25} \text{ atoms}}{6.02 \times 10^{23} \text{ atoms/mol}} = 143.8 \text{ mol}$

and the mass of this much uranium is $M = (143.8 \text{ mol})(0.235 \text{ kg/mol}) = 33.8 \text{ kg}$ ◊

CHAPTER SELF-QUIZ

1. If the controlled fusion process were to be proven feasible, which of the following would be the main source of fuel?
   a. air
   b. water
   c. coal
   d. petroleum

2. When several small pieces of fissionable material are combined into one large spherical body, what happens to the overall surface area?
   a. increases
   b. decreases
   c. remains the same
   d. becomes larger or smaller depending on the element

3. Nuclear fusion involves combining nuclei of elements that fall into what category?
   a. metals
   b. non-metals
   c. high atomic numbers
   d. low atomic numbers

4. The moderator material in a nuclear fission reactor is intended to fulfill which of the following purposes?
   a. absorb neutrons
   b. create new neutrons
   c. accelerate neutrons
   d. decelerate neutrons

5. Assume that (i) the energy released per fission event of U-235 is 208 MeV and (ii) 30% of the nuclear energy released in a power plant is ultimately converted to usable electrical energy. Approximately how many fission events will occur in one second in order to provide the 2.00 kW electrical power needs of a typical home? ($1\text{ eV} = 1.60\times10^{-19}$ J)
   a. 3.00×10^{10} fissions per sec
   b. 2.00×10^{14} fissions per sec
   c. 5.00×10^{18} fissions per sec
   d. 3.00×10^{21} fissions per sec

6. Cadmium control rods used in a nuclear fission reactor are intended to serve what purpose?
   a. absorb neutrons
   b. create neutrons
   c. accelerate neutrons
   d. decelerate neutrons

7. A proton and a neutron combine in a fusion process to form a stable deuterium nucleus. Which of the following statements best applies in describing the mass of the deuterium nucleus?
   a. less than sum of proton and neutron masses
   b. equal to sum of proton and neutron masses
   c. greater than sum of proton and neutron masses
   d. exactly equal to twice proton mass

8. A fast neutron will lose more kinetic energy if it collides with
   a. a lead atom
   b. a carbon atom
   c. a hydrogen atom
   d. an electron

9. Which type of particle is most massive?
   a. leptons
   b. baryons
   c. mesons
   d. photons

10. The spin of all quarks is
   a. 0
   b. 1/2
   c. 1
   d. 1/3 or 2/3

11. Which of the following particle reactions cannot occur?
   a. $\overline{n} \rightarrow \overline{p} + e^{+} + \nu$
   b. $p + p \rightarrow p + p + p + \overline{p}$
   c. $\pi^{0} \rightarrow \gamma + \gamma$
   d. $p + p \rightarrow p + \pi^{+}$

12. Which of the following reactions violates conservation of strangeness?
   a. $\Omega^{-} \rightarrow 3\gamma$
   b. $\overline{p} + p \rightarrow \overline{\Lambda^{0}} + \Lambda^{0}$
   c. $\overline{n} \rightarrow e^{+} + \overline{p} + \nu_{e}$
   d. $e^{+} + e^{-} \rightarrow 2\gamma$

ANSWERS TO CHAPTER SELF-QUIZ

Chapter 1	1. **b**	2. **d**	3. **d**	4. **c**	5. **a**	6. **c**
	7. **b**	8. **c**	9. **a**	10. **d**	11. **c**	12. **b**
Chapter 2	1. **a**	2. **b**	3. **a**	4. **b**	5. **d**	6. **d**
	7. **d**	8. **b**	9. **a**	10. **a**	11. **a**	12. **c**
Chapter 3	1. **b**	2. **c**	3. **d**	4. **d**	5. **b**	6. **d**
	7. **d**	8. **a**	9. **d**	10. **c**	11. **a**	12. **d**
Chapter 4	1. **b**	2. **a**	3. **b**	4. **b**	5. **a**	6. **d**
	7. **a**	8. **c**	9. **b**	10. **a**	11. **c**	12. **d**
Chapter 5	1. **c**	2. **d**	3. **d**	4. **a**	5. **a**	6. **c**
	7. **c**	8. **d**	9. **c**	10. **a**	11. **b**	12. **a**
Chapter 6	1. **b**	2. **d**	3. **d**	4. **d**	5. **a**	6. **b**
	7. **a**	8. **a**	9. **b**	10. **b**	11. **d**	12. **b**
Chapter 7	1. **b**	2. **a**	3. **c**	4. **d**	5. **d**	6. **d**
	7. **b**	8. **b**	9. **a**	10. **a**	11. **b**	12. **a**
Chapter 8	1. **c**	2. **b**	3. **c**	4. **b**	5. **b**	6. **b**
	7. **b**	8. **b**	9. **d**	10. **d**	11. **c**	12. **a**
Chapter 9	1. **a.**	2. **a**	3. **d**	4. **c**	5. **d**	6. **c**
	7. **a**	8. **b**	9. **d**	10. **d**	11. **a**	12. **a**
Chapter 10	1. **c**	2. **b**	3. **d**	4. **a**	5. **a**	6. **d**
	7. **a**	8. **c**	9. **c**	10. **b**	11. **b**	12. **b**
Chapter 11	1. **a**	2. **d**	3. **c**	4. **d**	5. **d**	6. **b**
	7. **a**	8. **d**	9. **d**	10. **b**	11. **b**	12. **b**
Chapter 12	1. **d**	2. **a**	3. **c**	4. **a**	5. **d**	6. **b**
	7. **b**	8. **c**	9. **c**	10. **d**	11. **b**	12. **a**
Chapter 13	1. **c**	2. **b**	3. **c**	4. **a**	5. **b**	6. **d**
	7. **d**	8. **d**	9. **b**	10. **b**	11. **c**	12. **c**
Chapter 14	1. **d**	2. **a**	3. **b**	4. **c**	5. **a**	6. **c**
	7. **d**	8. **d**	9. **c**	10. **a**	11. **a**	12. **d**
Chapter 15	1. **d**	2. **c**	3. **d**	4. **a**	5. **b**	6. **b**
	7. **c**	8. **c**	9. **b**	10. **a**	11. **a**	12. **a**

Chapter						
Chapter 16	1. d	2. b	3. b	4. a	5. d	6. d
	7. a	8. a	9. a	10. c	11. c	12. d
Chapter 17	1. c	2. c	3. b	4. d	5. c	6. a
	7. b	8. d	9. b	10. c	11. b	12. c
Chapter 18	1. d	2. a	3. a	4. a	5. b	6. a
	7. b	8. d	9. a	10. b	11. b	12. c
Chapter 19	1. b	2. b	3. c	4. b	5. b	6. b
	7. a	8. a	9. c	10. a	11. c	12. c
Chapter 20	1. a	2. d	3. b	4. d	5. d	6. a
	7. d	8. b	9. b	10. b	11. b	12. a
Chapter 21	1. c	2. b	3. a	4. c	5. d	6. b
	7. b	8. c	9. b	10. c	11. d	12. a
Chapter 22	1. a	2. a	3. c	4. c	5. b	6. b
	7. a	8. c	9. c	10. a	11. d	12. d
Chapter 23	1. b	2. d	3. d	4. b	5. c	6. a
	7. b	8. c	9. b	10. b	11. c	12. c
Chapter 24	1. c	2. b	3. c	4. b	5. d	6. c
	7. b	8. b	9. b	10. c	11. a	12. c
Chapter 25	1. c	2. c	3. b	4. b	5. a	6. d
	7. b	8. d	9. b	10. a	11. b	12. c
Chapter 26	1. b	2. d	3. b	4. c	5. d	6. b
	7. a	8. c	9. c	10. b	11. c	12. b
Chapter 27	1. d	2. b	3. a	4. b	5. d	6. c
	7. b	8. c	9. a	10. a	11. c	12. c
Chapter 28	1. d	2. a	3. b	4. b	5. d	6. d
	7. c	8. d	9. a	10. c	11. b	12. d
Chapter 29	1. b	2. b	3. b	4. a	5. c	6. c
	7. b	8. b	9. b	10. c	11. b	12. d
Chapter 30	1. b	2. b	3. d	4. d	5. b	6. a
	7. a	8. c	9. b	10. b	11. d	12. a

▷ The torque associated with a force has a magnitude equal to the force times the moment arm; and the value of the torque depends on the origin about which it is evaluated. Also, the net torque on a rigid body about some axis is proportional to the angular acceleration; that is, $\tau = I\alpha$, where I is the moment of inertia about the axis about which the net torque is evaluated.

▷ The work-energy theorem can be applied to a rotating rigid body. That is, the net work done on a rigid body rotating about a fixed axis equals the change in its rotational kinetic energy; and the *law of conservation of mechanical energy* can be used in the solution of problems involving rotating rigid bodies.

▷ The time rate of change of the angular momentum of a rigid body rotating about an axis is proportional to the net torque acting about the axis of rotation. This is the rotational analog of Newton's second law.

SOLUTIONS TO SELECTED END-OF-CHAPTER PROBLEMS

1. If the torque required to loosen a nut that is holding a flat tire in place on a car has a magnitude of 40.0 N·m, what *minimum* force must be exerted by the mechanic at the end of a 30-cm lug wrench to accomplish the task?

Solution In order to exert the minimum force, the force must be applied perpendicular to the wrench. With the pivot at the nut, we have

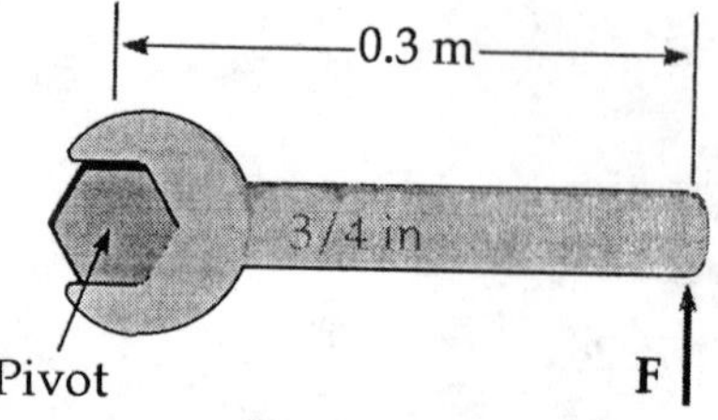

$$\tau = Fd \qquad \text{or} \qquad 40.0\ \text{N}\cdot\text{m} = F(0.300\ \text{m})$$

So, $F = 133$ N ◊

---

5. A steel band exerts a horizontal force of 80.0 N on a tooth at the point B in Figure 8.32. What is the torque on the root of the tooth about point A?

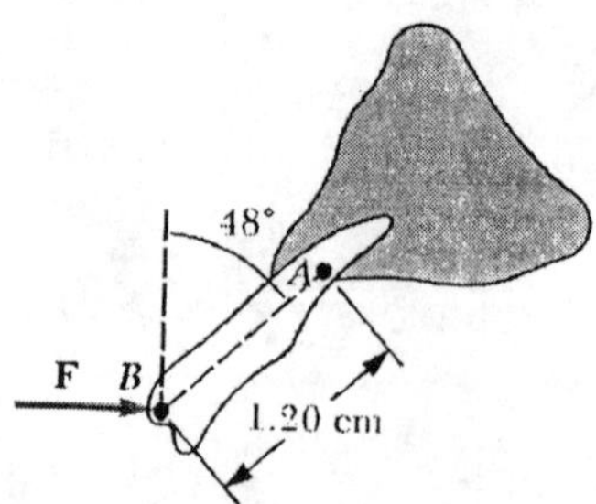

Figure 8.32

Solution The lever arm is

$$d = (1.20\times10^{-2}\ \text{m})\cos48° = 8.03\times10^{-3}\ \text{m}$$

and the torque is

$$\tau = Fd = (80.0\ \text{N})(8.03\times10^{-3}\ \text{m}) = 0.642\ \text{N}\cdot\text{m (CCW)} \quad ◊$$
